IMPERIAL UNITS

ac	acre
bbl	barrel
cu ft	cubic foot
cu in.	cubic inch
cu yd	cubic yard
cwt	hundred weight
fbm	foot board measure
ft	foot or feet
gal	gallon(s)
in.	inch(es)
lb	pound
lf	linear foot (feet)
mi	mile(s)
mph	miles per hour
psi	pounds per square inch
sq ft	square foot (feet)
sq in.	square inch(es)
sq yd	square yard(s)
mf bm	thousand foot board measure
m gal	thousand gallons
yd	yard(s)

METRIC UNITS

C	Celsius
cm	centimeter
ha	hectare
kg	kilogram(s)
km	kilometer(s)
kN	kilonewton(s)
kPa	kilopascal(s)
L	liter(s)
m	meter(s)
m^2	square meter
m^3	cubic meter
mm	millimeter(s)
t	tonne

DATE DUE

SEP 1 2 2022	

SURVEYING
with Construction Applications

Sixth Edition

Barry F. Kavanagh
Seneca College, Emeritus

Upper Saddle River, New Jersey
Columbus, Ohio

Library of Congress Cataloging-in-Publication Data

Kavanagh, Barry F.
  Surveying with construction applications/Barry F. Kavanagh.-6th ed.
    p. cm.
  Includes bibliographical references and index.
  ISBN 0-13-170932-1
    1. Surveying. 2. Building sites. I. Title.

TA625.K38 2007
526.9—dc22

2005053455

Acquisitions Editor: Tim Peyton
Editorial Assistant: Nancy Kesterson
Production Editor: Holly Shufeldt
Design Coordinator: Diane Ernsberger
Cover Designer: Bryan Huber
Cover photo: Getty One
Insert Designer: Candace Rowley
Production Manager: Deidra Schwartz
Marketing Manager: Mark Marsden
Senior Marketing Coordinator: Liz Farrell
Marketing Assistant: Les Roberts

This book was set in Times Roman by Carlisle Publishing Services. It was printed and bound by
R.R. Donnelley & Sons Company. The cover was printed by The Lehigh Press, Inc.

Pearson Education Ltd.
Pearson Education Singapore Pte. Ltd.
Pearson Education Canada, Ltd.
Pearson Education—Japan

Pearson Education Australia Pty. Limited
Pearson Education North Asia Ltd.
Pearson Educación de Mexico, S.A. de C.V.
Pearson Education Malaysia Pte. Ltd.

10 9 8 7 6 5 4 3
ISBN 0-13-170932-1

Preface

Many technological advances have occurred in surveying since *Surveying with Construction Applications* was first published. This sixth edition is up to date, with the latest advances in instrumentation technology, field-data capture, and data-processing techniques. Although surveying is becoming much more efficient and automated, the need for a clear understanding of the principles underlying all forms of survey measurement remains unchanged.

General surveying principles and techniques, used in all branches of surveying, are presented in Chapters 1 to 9, while contemporary applications for the construction of most civil projects are covered in Chapters 10 to 17. With this organization, the text is useful not only for the student, but it can also be used as a handy reference for the graduate who may choose a career in civil/survey design or construction. The glossary has been expanded to include new terminology. Every effort has been made to remain on the leading edge of new developments in techniques and instrumentation, while maintaining complete coverage of traditional techniques and instrumentation.

Chapter 2 has been refreshed in several areas, and the section on EDM without reflecting prisms has been rewritten. The section on digital levels in Chapter 3 has been rewritten, and the adjustment section in Chapter 4 has been revised to include tests and adjustments for theodolites and total stations. Chapter 5, "Total Stations," has been revised in the areas of instrument capabilities, errors, field techniques, handheld instrumentation, and integrated total station/GPS instruments. Chapter 7 has been expanded to include the topic of geomatics and to discuss in more depth modern techniques of survey computations and plan/drawing preparation. The national map is introduced in the GIS section. Chapters 8 ("Satellite Positioning") and 9 ("Horizontal Control Surveys") have been extensively revised to reflect current technological advances and the increasing dependence on GPS for horizontal control.

Chapters 10 to 17 in Part II continue to provide extensive coverage of highway curves, and the construction surveying techniques used in highway construction, municipal street construction, pipeline and tunnel construction, culvert and bridge construction, building construction, and quantity and final surveys. In addition, a new chapter (Chapter 10) has been added to explain the use of machine guidance and control in the provision of line and grade for civil construction projects. Chapter 10 also discusses the concepts underlying three-dimensional data files—the foundation of modern machine guidance and control practices.

The Instructor's Manual includes solutions for end-of-chapter problems; a typical evaluation scheme; subject outlines (two terms or two-semester programs); term assignments, sample instruction class handouts for instrument use, etc.; and midterm and final tests. Also included is a PowerPoint (CD) presentation that can be used as an aid in presenting text material and as a source for overhead transparencies.

To access supplementary materials online, instructors need to request an instructor access code. Go to **www.prenhall.com,** click the **Instructor Resource Center** link, and then click **Register Today** for an instructor access code. Within 48 hours after registering you will receive a confirming e-mail including an instructor access code. Once you have received your code, go to the site and log on for full instructions on downloading the materials you wish to use.

Technology continues to expand; improvements to field equipment, data-processing techniques, and construction practices in general will inevitably continue. Surveyors must keep up with these dynamic events. I hope that students, by using this text, will be completely up to date in this subject area and will be readily able to cope with the technological changes that continue to occur. Comments and suggestions about the text are welcomed and can be e-mailed to me at *barry_kavanagh@sympatico.ca*

Barry F. Kavanagh

Acknowledgments

The author is grateful for the comments and suggestions offered by professors who have adopted the text for class use. Particular thanks are due to Dianne H. Kay, Southern Illinois University, Edwardsville, who reviewed the manuscript for this edition. Thanks are also due the faculty and staff of the School of Civil and Resources Engineering Technology, Seneca College, for their generous assistance and support.

The following surveying, engineering, and equipment manufacturing companies provided generous assistance:

Amberg Measuring Technique, Switzerland
American Association of State Highway and Transportation Officials (AASHTO),
    Washington, D.C.
American Augers Inc., Loudan, Ohio
American Concrete Pipe Association, Vienna, Virginia
American Congress on Surveying and Mapping, Bethesda, Maryland
American Society for Photogrammetry and Remote Sensing, Bethesda, Maryland
American Society of Civil Engineers, New York, New York
Applanix, Richmond Hill, Ontario
Ausran, Alexandra Hills, Australia
Caterpillar Inc., Peoria, Illinois

Corrugated Steel Pipe Association, Cambridge, Ontario
CST/ Berger Corporation, Watseka, Illinois
Department of Energy, Mines and Resources, Ottawa, Canada
Euclid-Hitachi Heavy Equipment Ltd., Cleveland, Ohio
Geomatics Canada, Ottawa, Ontario
J.I. Case Co., Chicago, Illinois
John Deere, Moline, Illinois
Laser Atlanta, Norcross, Georgia
Laser Technology Inc., Englewood, Colorado
Leica Geosystems, Norcross, Georgia
Marathon Letourneau Co., Longview, Texas
Marshall, Macklin, Monoghan, Surveyors and Engineers, Markham, Ontario
MicroSurvey Software Inc., Westbank, British Columbia
Ministry of Transportation, Toronto, Ontario
National Geodetic Survey (NGS), Silver Spring, Maryland
National Swedish Institute for Building Research, Gavle, Sweden
Nikon, Inc., Melville, New York
Optech, Toronto, Ontario
Pacific Crest Corporation, Santa Clara, California
Pentax Corporation, Englewood, Colorado
Position Inc., Calgary, Alberta
The Robbins Company, Solon, Ohio
Sokkia Corporation, Olathe, Kansas
Topcon Instrument Corporation, Pleasanton, California
Trimble, Sunnyvale, California
Tripod Data Systems (TDS), Corvallis, Oregon
U.S. Census Bureau
U.S. Coast Guard
U.S. Department of the Interior, Geological Survey, Reston, Virginia
USGS National Center for Earth Resources Observation and Science (EROS),
    Sioux Falls, South Dakota
Wild Heerbruug, Heerbruug, Switzerland
XYZ Works, Masonville, Colorado

Contents

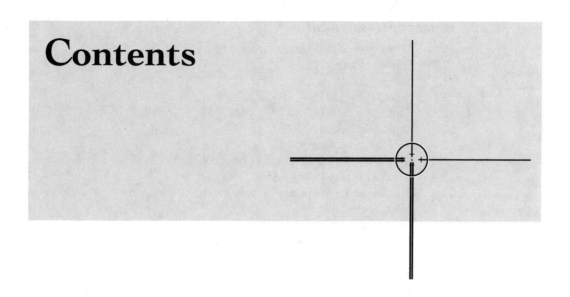

Preface   iii

Acknowledgments   iv

PART I   SURVEYING PRINCIPLES   1

    1   *Surveying Fundamentals   2*

        1.1   Surveying Defined   2
        1.2   Surveying: General Background   4
        1.3   Control Surveys   4
        1.4   Preliminary Surveys   4
        1.5   Surveying Instruments   5
        1.6   Construction Surveys   6
        1.7   Distance Measurement   6
        1.8   Angle Measurement   9
        1.9   Position Measurement   9
        1.10   Units of Measurement   10
        1.11   Stationing   12
        1.12   Types of Construction Projects   12
        1.13   Random and Systematic Errors   13
        1.14   Accuracy and Precision   14
        1.15   Mistakes   15
        1.16   Field Notes   15
        Review Questions   16

2 *Distance Measurement* 18

2.1 Methods of Linear Measurement 18
2.2 Gunter's Chain 19
2.3 Tapes 20
2.4 Steel Tapes 20
2.5 Taping Accessories and Their Use 23
2.6 Taping Techniques 27
2.7 Taping Corrections 31
2.8 Systematic Taping Errors and Corrections 31
2.9 Random Taping Errors 35
2.10 Techniques for "Ordinary" Taping Precision 36
2.11 Mistakes in Taping 37
2.12 Field Notes for Taping 39
2.13 Electronic Distance Measurement (EDM) 40
2.14 Electronic Angle Measurement 40
2.15 Principles of Electronic Distance Measurement (EDM) 42
2.16 EDM Characteristics 44
2.17 Prisms 45
2.18 EDM Instrument Accuracies 46
2.19 EDM Operation 47
2.20 Geometry of Electronic Distance Measurements 50
2.21 EDM Without Reflecting Prisms 53
     Problems 55

3 *Leveling* 57

3.1 General Background 57
3.2 Theory of Differential Leveling 58
3.3 Curvature and Refraction 60
3.4 Types of Surveying Levels 62
3.5 Leveling Rods 67
3.6 Definitions for Differential Leveling 71
3.7 Techniques of Leveling 72
3.8 Benchmark Leveling (Vertical Control Surveys) 76
3.9 Profile and Cross-Section Leveling 78
3.10 Reciprocal Leveling 83
3.11 Peg Test 84
3.12 Three-Wire Leveling 86
3.13 Trigonometric Leveling 87
3.14 Level Loop Adjustments 88
3.15 Suggestions for Rod Work 89
3.16 Suggestions for Instrument Work 90
3.17 Mistakes in Leveling 91
     Problems 92

4   *Angles and Theodolites*   97

4.1   General Background   97
4.2   Reference Directions for Vertical Angles   97
4.3   Meridians   97
4.4   Horizontal Angles   98
4.5   Theodolites   99
4.6   Electronic Theodolites   102
4.7   Theodolite/Total Station Setup   106
4.8   Repeating Optical Theodolites   107
4.9   Angle Measurement with an Optical Theodolite   109
4.10   Direction Optical Theodolites   109
4.11   Angles Measured with a Direction Theodolite   110
4.12   Geometry of the Theodolite and Total Station   111
4.13   Adjustment of the Theodolite and Total Station   112
4.14   Laying Off Angles   117
4.15   Prolonging a Straight Line (Double Centering)   118
4.16   Bucking-In (Interlining)   120
4.17   Intersection of Two Straight Lines   120
4.18   Prolonging a Measured Line by Triangulation over an Obstacle   121
4.19   Prolonging a Line Past an Obstacle   123
        Review Questions   124

5   *Total Stations*   125

5.1   General Background   125
5.2   Total Station Capabilities   125
5.3   Total Station Field Techniques   131
5.4   Field Procedures for Total Stations in Topographic Surveys   139
5.5   Field-Generated Graphics   146
5.6   Construction Layout Using Total Stations   149
5.7   Motorized Total Stations   153
5.8   Summary of Modern Total Station Characteristics and Capabilities   158
5.9   Instruments Combining Total Station Capabilities and GPS Receiver Capabilities   159
5.10   Portable/Handheld Total Stations   161
        Review Questions   162

6   *Traverse Surveys and Computations*   164

6.1   General Background   164
6.2   Balancing Field Angles   167
6.3   Meridians   168
6.4   Bearings   169
6.5   Azimuths   172
6.6   Latitudes and Departures   176
6.7   Traverse Precision and Accuracy   182

Contents

6.8   Compass Rule Adjustment   183

6.9   Effects of Traverse Adjustments on Measured Angles and Distances   185

6.10   Omitted Measurement Computations   186

6.11   Rectangular Coordinates of Traverse Stations   187

6.12   Area of a Closed Traverse by the Coordinate Method   191

    Problems   *193*

7   *An Introduction to Geomatics*   197

7.1   Geomatics Defined   197

7.2   Branches of Geomatics   197

7.3   Data-Collection Branch   199

7.4   Design and Plotting   205

7.5   Contours   215

7.6   Aerial Photography   222

7.7   Airborne and Satellite Imagery   233

7.8   Remote-Sensing Satellites   245

7.9   Geographic Information System (GIS)   246

7.10   Database Management   252

7.11   Metadata   253

7.12   Spatial Entities or Features   254

7.13   Typical Data Representation   254

7.14   Spatial Data Models   256

7.15   GIS Data Structures   258

7.16   Topology   261

7.17   Remote Sensing Internet Web Sites and Further Reading   262

    Review Questions   *263*

    Problems   *264*

8   *Satellite Positioning*   268

8.1   General Background   268

8.2   Global Positioning System   270

8.3   Receivers   272

8.4   Satellite Constellations   274

8.5   GPS Satellite Signals   278

8.6   Position Measurements   279

8.7   Errors   286

8.8   Continuously Operating Reference Station (CORS)   288

8.9   Canadian Active Control System   290

8.10   Survey Planning   290

8.11   GPS Field Procedures   296

8.12   GPS Applications   302

8.13   Vertical Positioning   308

8.14   Conclusion   313

8.15   GPS Glossary   313

8.16   Recommended Readings   315
Review Questions   316

9   *Horizontal Control Surveys*   317

9.1   General Background   317
9.2   Plane Coordinate Grids   325
9.3   Lambert Projection   332
9.4   Transverse Mercator Grid System   332
9.5   Universal Transverse Mercator (UTM) Grid System   333
9.6   Use of Grid Coordinates   341
9.7   Illustrative Examples   350
9.8   Horizontal Control Techniques   357
9.9   Project Control   358
Review Questions   368
Problems   368

PART II   CONSTRUCTION APPLICATIONS: GENERAL BACKGROUND   369

10   *Modern Construction Surveying Practices*   370

10.1   General Background   370
10.2   Grade   371
10.3   Machine Guidance and Control   371
10.4   Three-Dimensional Data Files   378
10.5   Summary of the 3D Design Process   380
10.6   Web Site References for Data Collection, DTM, and Civil Design   382
Review Questions   382

11   *Highway Curves*   383

11.1   Route Surveys   383
11.2   Circular Curves: General Background   384
11.3   Circular Curve Geometry   385
11.4   Circular Curve Deflections   391
11.5   Chord Calculations   393
11.6   Metric Considerations   394
11.7   Field Procedure   394
11.8   Moving up on the Curve   395
11.9   Offset Curves   396
11.10   Compound Circular Curves   403
11.11   Reverse Curves   405
11.12   Vertical Curves: General Background   406
11.13   Geometric Properties of the Parabola   407
11.14   Computation of the High or the Low Point on a Vertical Curve   408

11.15   Computing a Vertical Curve   409
11.16   Design Considerations   412
11.17   Spiral Curves: General Background   414
11.18   Spiral Curve Computations   416
11.19   Spiral Layout Procedure Summary   423
11.20   Approximate Solution for Spiral Problems   430
11.21   Superelevation: General Background   432
11.22   Superelevation Design   432
        Review Questions   441
        Problems   442

12 *Highway Construction Surveys   444*

12.1   Preliminary (Preengineering) Surveys   444
12.2   Highway Design   448
12.3   Highway Construction Layout   448
12.4   Clearing, Grubbing, and Stripping Topsoil   454
12.5   Placement of Slope Stakes   455
12.6   Layout for Line and Grade   459
12.7   Grade Transfer   460
12.8   Ditch Construction   464
        Review Questions   464

13 *Municipal Street Construction Surveys   466*

13.1   General Background   466
13.2   Classification of Roads and Streets   467
13.3   Road Allowances   468
13.4   Road Cross Sections   468
13.5   Plan and Profile   468
13.6   Establishing Centerline (℄)   469
13.7   Establishing Offset Lines and Construction Control   473
13.8   Construction Grades for a Curbed Street   475
13.9   Street Intersections   480
13.10   Sidewalk Construction   482
13.11   Site Grading   483
        Problems   485

14 *Pipeline and Tunnel Construction Surveys   491*

14.1   Pipeline Construction   491
14.2   Sewer Construction   493
14.3   Layout for Line and Grade   495
14.4   Catch-Basin Construction Layout   504
14.5   Tunnel Construction   505
        Problems   511

15 *Culvert and Bridge Construction Surveys* *516*

15.1   Culvert Construction   516
15.2   Culvert Reconstruction   516
15.3   Bridge Construction: General Background   519
15.4   Contract Drawings   519
15.5   Layout Computations   528
15.6   Offset Distance Computations   528
15.7   Dimension Verification   529
15.8   Vertical Control   531
15.9   Cross Sections for Footing Excavations   533
     Review Questions   *533*

16 *Building Construction Surveys* *534*

16.1   Building Construction: General Background   534
16.2   Single-Story Construction   534
16.3   Multistory Construction   545
     Review Questions   *552*

17 *Quantity and Final Surveys* *553*

17.1   Construction Quantity Measurements: General Background   553
17.2   Area Computations   554
17.3   Area by Graphical Analysis   561
17.4   Construction Volumes   567
17.5   Cross Sections, End Areas, and Volumes   569
17.6   Prismoidal Formula   573
17.7   Volume Computations by Geometric Formulas   575
17.8   Final (As-Built) Surveys   575
     Problems   *577*

Appendix A Trigonometry and Coordinate Geometry Review *581*

A.1   Trigonometric Definitions and Identities   581
A.2   Coordinate Geometry   585

Appendix B Surveying and Mapping Web Sites *595*

Appendix C Glossary *597*

Appendix D Typical Field Projects *609*

D.1   Field Notes   609
D.2   Project 1: Building Measurements   610
D.3   Project 2: Experiment to Determine Normal Tension   612
D.4   Project 3: Field Traverse Measurements with a Steel Tape   613
D.5   Project 4: Differential Leveling   614

D.6   Project 5: Traverse Angle Measurements and Closure Computations   616
D.7   Project 6: Topographic Survey   617
D.8   Project 7: Building Layout   624
D.9   Project 8: Horizontal Curve   625
D.10   Project 9: Pipeline Layout   626

Appendix E Answers to Selected Problems   629

Appendix F Steel Tape Corrections   631

F.1   Erroneous Tape-Length Corrections   631
F.2   Temperature Corrections   632
F.3   Tension and Sag Corrections   634
       Problems   636

Appendix G Early Surveying   639

G.1   Evolution of Surveying   639
G.2   Dumpy Level   644
G.3   The Engineer's Vernier Transit   647
G.4   Stadia   655
       Problems   667

Appendix H Illustrations of Machine Control and of Various Data-Capture Techniques   671

Index   673

FIELD NOTE INDEX

Page	Figure Number	Figure Title
38	2.21	Taping field notes for a closed traverse
39	2.22	Taping field notes for building dimensions
75	3.15	Leveling field notes and arithmetic check (data from Figure 3.14)
80	3.19	Profile field notes
81	3.21	Cross-section notes (municipal format)
82	3.22	Cross-section notes (highway format)
87	3.28	Survey notes for three-wire leveling
105	4.8	Field notes for angles by repetition (closed traverse)
112	4.13	Field notes for angles by direction
147	5.17	Field notes for total station graphics descriptors—MicroSurvey Software Inc. codes
148	5.18	Field notes for total station graphics descriptors—Sokkia codes
166	6.3	Field notes for open traverse
167	6.4	Field notes for closed traverse
203	7.3	Topographic field notes. (a) Single baseline. (b) Split baseline.
204	7.4	Original topographic field notes, 1907 (distances shown are in chains).
295	8.19	Station visibility diagram
297	8.20	GPS field log
362	9.25	Field notes for control point directions and distances
363	9.26	Prepared polar coordinate layout notes
473	13.5	Property markers used to establish centerline
557	17.1	Example of the method for recording sodding payment measurements
558	17.2	Field notes for fencing payment measurements
559	17.3	Example of field-book entries regarding removal of sewer pipe, etc.
560	17.4	Example of field notes for pile driving
610	D.1	Typical field book layout
611	D.2	Sample taping field notes for Project 1 (building dimensions)
612	D.3	Sample field notes for Project 3 (traverse distances)
615	D.4	Sample field notes for Project 4 (differential leveling)
617	D.5	Sample field notes for Project 5 (traverse angles)
618	D.6	Sample field notes for Project 6 (topography tie-ins)
619	D.7	Sample field notes for Project 6 (topography cross sections)
621	D.9	Sample field notes for Project 6 (topography by theodolite/EDM)
622	D.10	Sample field notes for Project 6 (topography by total station)
623	D.11	Sample field notes for Project 7 (building layout)
665	G.19	Stadia field notes

Contents

PART I
Surveying
Principles

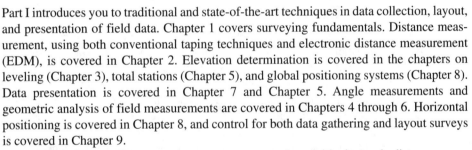

Part I introduces you to traditional and state-of-the-art techniques in data collection, layout, and presentation of field data. Chapter 1 covers surveying fundamentals. Distance measurement, using both conventional taping techniques and electronic distance measurement (EDM), is covered in Chapter 2. Elevation determination is covered in the chapters on leveling (Chapter 3), total stations (Chapter 5), and global positioning systems (Chapter 8). Data presentation is covered in Chapter 7 and Chapter 5. Angle measurements and geometric analysis of field measurements are covered in Chapters 4 through 6. Horizontal positioning is covered in Chapter 8, and control for both data gathering and layout surveys is covered in Chapter 9.

Although most distance measurements are now done with electronic distance measurement (EDM) techniques, many applications still exist for steel taping on short-distance measurements, which are often found in construction layouts. Taping correction techniques can be found in Appendix F.

Chapter 1

Surveying Fundamentals

1.1 Surveying Defined

Surveying is the art and science of taking field measurements on or near the surface of the earth. Survey field measurements include horizontal and slope distances, vertical distances, and horizontal and vertical angles. In addition to measuring distances and angles, surveyors can measure position as given by the northing, easting, and elevation of a survey station by using global positioning systems (GPSs) and remote sensing techniques. In addition to taking measurements in the field, the surveyor can derive related distances and directions through geometric and trigonometric analysis.

Once a survey station has been located by angle and distance techniques, or by positioning techniques, the surveyor then attaches to that survey station (in handwritten or electronic field notes) a suitable identifier or attribute that describes the nature of the survey station. In Chapter 7, you will see that attribute data for a survey station can be expanded from a simple descriptive label to include a wide variety of related information that can be tagged specifically to that survey station.

Since the 1980s, the term **geomatics** has come into popular usage to describe the computerization and digitization of data collection, data processing, data analysis, and data output. Geomatics includes traditional surveying as its cornerstone, but it also reflects the now-broadened scope of measurement science and information technology. Figure 7.1 shows a computerized surveying data model. The illustration in that figure gives you a sense of the diversity of the integrated scientific activities now covered by the term *geomatics*.

The vast majority of engineering and construction projects are so limited in geographic size that the surface of the earth is considered to be a plane for all X (east) and Y (north) dimensions. Z dimensions (height) are referred to a datum, usually mean sea level. Surveys that ignore the curvature of the earth for horizontal dimensions are called **plane surveys**. Surveys that cover a large geographic area—for example, state or provincial boundary surveys—must have corrections made to the field measurements so that these measurements reflect the curved (ellipsoidal) shape of the earth. These surveys are called **geodetic surveys**. The Z dimensions (**orthometric heights**) in geodetic surveys are also referenced to a datum—usually mean sea level.

In the past, geodetic surveys were very precise surveys of great magnitude, for example, national boundaries and control networks. Modern surveys (data-gathering, control, and layout) utilizing global positioning systems (GPSs) are geodetic surveys based on the ellipsoidal shape of the earth and referenced to the geodetic reference system (GRS80) ellipsoid. Such survey measurements must be translated mathematically from ellipsoidal coordinates and ellipsoidal heights to plane grid coordinates and to orthometric heights (referenced to mean sea level) before being used in leveling and other local surveying projects.

Engineering or construction surveys that span long distances (e.g., highways, railroads) are treated as plane surveys, with corrections for curvature being applied at regular intervals (e.g., at 1-mile intervals or at township boundaries). **Engineering surveying** is defined as those activities involved in the planning and execution of surveys for the location, design, construction, maintenance, and operation of civil and other engineered projects.* Such activities include:

1. Preparation of surveying and related mapping specifications.
2. Execution of photogrammetric and field surveys for the collection of required data, including topographic and hydrographic data.
3. Calculation, reduction, and plotting (manual and computer-aided) of survey data for use in engineering design.
4. Design and provision of horizontal and vertical control survey networks.
5. Provision of line and grade and other layout work for construction and mining activities.
6. Execution and certification of quality control measurements during construction.
7. Monitoring of ground and structural stability, including alignment observations, settlement levels, and related reports and certifications.
8. Measurement of material and other quantities for inventory, economic assessment, and cost accounting purposes.
9. Execution of as-built surveys and preparation of related maps, plans, and profiles upon completion of the project.
10. Analysis of errors and tolerances associated with the measurement, field layout, and mapping or other plots of survey measurements required in support of engineered projects.

Engineering surveying does not include surveys for the retracement of existing land ownership boundaries or the creation of new boundaries. These activities are reserved for licensed property surveyors—also known as land surveyors or cadastral surveyors.

---

*Definition adapted from the definition of *engineering surveying* as given by the American Society of Civil Engineers (ASCE) in their *Journal of Surveying Engineering* in 1987.

1.2 Surveying: General Background

Surveys are usually performed for one of two reasons. First, surveys are made to collect data, which can then be drawn to scale on a plan or map (**preliminary surveys** or **preengineering surveys**); second, field surveys are made to lay out dimensions taken from a design plan and thus define precisely the location of the proposed construction facility. The layouts of proposed property lines and corners as required in land division are called **layout surveys**; the layouts of proposed construction features are called **construction surveys**. Preliminary and construction surveys for the same area must have this one characteristic in common: measurements for both surveys must be referenced to a common base for X, Y, and Z dimensions. The establishment of a base for horizontal and vertical measurements is known as a **control survey**.

1.3 Control Surveys

Control surveys establish reference points and reference lines for preliminary and construction surveys. Vertical reference points, called benchmarks, are established using leveling surveys (Chapter 3) or global positioning system (GPS) surveys (Chapter 8). Horizontal control surveys (Chapter 9) can be tied into (1) state or provincial coordinate grid monuments, (2) property lines, (3) roadway centerlines, and (4) arbitrarily placed baselines or grids. These topics are discussed in some detail in Chapters 7 and 11.

1.4 Preliminary Surveys

Preliminary surveys (also known as preengineering surveys, location surveys, or data-gathering surveys) are used to collect measurements that locate the position of natural features, such as trees, rivers, hills, valleys, and the like, and the position of built features, such as roads, structures, pipelines, and so forth. Measured tie-ins can be taken by any of the following techniques.

1.4.1 Rectangular Tie-Ins

The rectangular tie-in (also known as the right-angle offset tie) was one of the most widely used field location techniques for preelectronic surveying techniques. This technique, when used to locate point P in Figure 1.1(a) to baseline AB, requires distance AC (or BC), where C is on AB at 90 degrees to point P, and it also requires measurement CP.

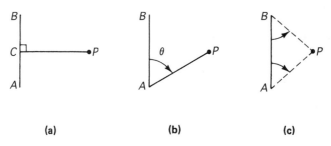

(a)   (b)   (c)

FIGURE 1.1 Location ties.

1.4.2 Polar Tie-Ins

Polar tie-ins (also known as the angle/distance technique) are now the most widely used location technique (see Chapter 5). Here, point P is located from point A on baseline AB by measuring angle θ and distance AP. See Figure 1.1(b).

1.4.3 Intersection Tie-Ins

This technique is useful in specialized location surveys. Point P in Figure 1.1(c) is located to baseline AB either by measuring angles from A and B to P or by swinging out arc lengths AP and BP until they intersect. The angle intersection technique is useful for near-shore marine survey locations using theodolites or total stations set on shore control points. The distance arc intersection technique is an effective method for replacing "lost" survey points from preestablished reference ties.

1.4.4 Positioning Tie-Ins

Topographic features can also be tied in using remote sensing techniques (see Chapter 7) and GPS survey techniques (see Chapter 8).

1.5 Surveying Instruments

The instruments most commonly used in field surveying are (1) steel or fiberglass tape, (2) theodolite, (3) total station, (4) level and rod, and (5) GPS receiver. Steel tapes are relatively precise measuring instruments and are used mostly for short measurements in both preliminary and layout surveys (see Figure 1.3). Steel tapes and their use are discussed in detail in Chapter 2.

Theodolites (sometimes called transits—short for transiting theodolites) are instruments designed for use in measuring horizontal and vertical angles and for establishing linear and curved alignments in the field. The theodolite has evolved through four distinct phases:

1. An open-faced, vernier-equipped (for angle determination) theodolite was commonly called a transit. A plumb bob was used to center the transit over the station mark. See Figures G.8 and G.9. Vernier transits are discussed in detail in Section G.3.

2. In the 1950s, the vernier transit gave way to the optical theodolite. This instrument came equipped with optical scales, permitting direct digital readouts or micrometer-assisted readouts. An optical plummet was used to center the instrument over the station mark. See Figures 4.4 to 4.7.

3. Electronic theodolites first appeared in the 1960s. These instruments used photoelectric sensors capable of sensing vertical and horizontal angles and displaying horizontal and vertical angles in degrees, minutes, and seconds. Optical plummets (and later, laser plummets) are used to center the instrument over the station mark. See Figure 1.7. Optical and electronic theodolites are discussed in detail in Chapter 4.

4. The total station appeared in the 1980s. This instrument combines electronic distance measurement (EDM), which was developed in the 1950s, with an electronic theodolite. In addition to electronic distance and angle measuring capabilities, this instrument is equipped with a central processor, which enables the computation of horizontal and vertical positions. The central processor also monitors instrument status and performs a wide variety of surveying applications. Total stations measure horizontal and vertical angles as well as horizontal and vertical distances. All data can be captured into electronic field books or as on-board storage as the data are received. See Figure 1.6. Total stations are described in detail in Chapter 5.

The level and rod are used to determine differences in elevation and elevations in a wide variety of surveying, mapping, and engineering applications. See Figures 1.4 and 1.5. Levels and rods are discussed in Chapter 3.

Global positioning system (GPS) receivers (see Figures 8.2 to 8.6) capture signals transmitted by four or more NAVSTAR satellites to determine position coordinates (e.g., northing, easting, and elevation) of a survey station. GPS positioning is discussed in Chapter 8.

Positions of ground points and surfaces can also be collected using various remote-sensing techniques (e.g., panchromatic, multispectral, lidar, and radar) utilizing both satellite and airborne platforms (see Chapter 7).

1.6 Construction Surveys

Construction surveys provide **line and grade** for a wide variety of construction projects, for example, highways, streets, pipelines, bridges, buildings, and site grading. Construction layout marks the horizontal location (line) as well as the vertical location or elevation (grade) for the proposed work. The builder can measure from the surveyor's markers to the exact location of each component of the facility to be constructed. Layout markers can be wood stakes, steel bars, nails with washers, spikes, chiseled marks in concrete, and so forth. Modern layout techniques also permit the contractor to position construction equipment for line and grade using machine guidance techniques involving lasers, total stations, and GPS receivers (see Chapter 10, Sections 10.3 to 10.6). When commencing a construction survey, it is important that the surveyor use the same control survey points as those used for the preliminary survey on which the construction design was based.

1.7 Distance Measurement

Distances between two points can be **horizontal, slope**, or **vertical** and are recorded in feet or meters. See Figures 1.2 and 1.3. Horizontal and slope distances can be measured with a fiberglass or steel tape or with an electronic distance-measuring device. When surveying, the horizontal distance is always required for plan-plotting purposes. A distance measured on slope can be trigonometrically converted to its horizontal equivalent by using either the slope angle or the difference in elevation (vertical distance) between the two points.

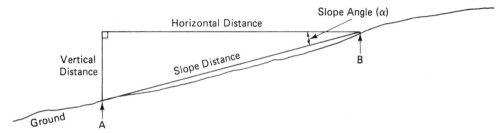

FIGURE 1.2 Distance measurement.

FIGURE 1.3 Preparing to measure to a stake tack, using a plumb bob and steel tape.

Vertical distances can be measured with a tape, as in construction work. They are more usually measured with a surveyors' level and rod (Figures 1.4 and 1.5) or with a total station (Figure 1.6).

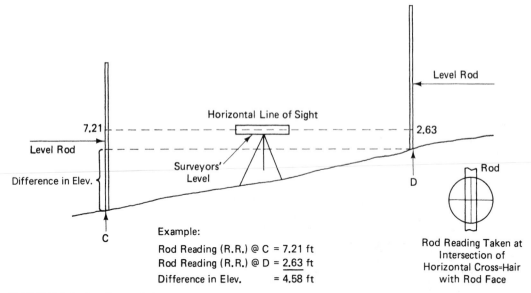

FIGURE 1.4 Leveling technique.

Example:

Rod Reading (R.R.) @ C = 7.21 ft
Rod Reading (R.R.) @ D = 2.63 ft
Difference in Elev. = 4.58 ft

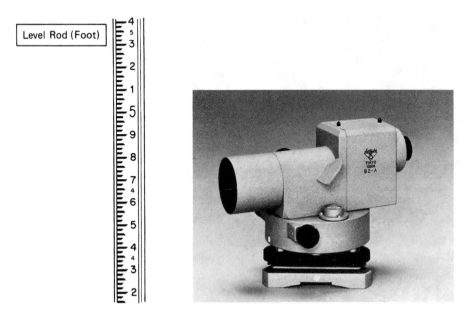

FIGURE 1.5 Level and rod. (Courtesy of SOKKIA Corp.)

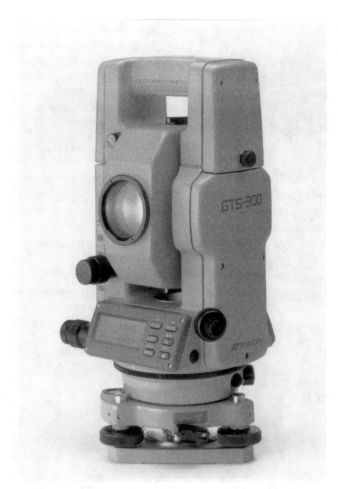

FIGURE 1.6 Topcon GTS 300 total station. The 300 series instruments have angle accuracies from 1 second to 10 seconds and single prism distances from 1,200 m (3,900 ft) to 2,400 m (7,900 ft) at 2 mm + 2 ppm accuracy. (Courtesy of Topcon Positioning Systems, Inc.)

1.8   Angle Measurement

Horizontal and vertical angles can be measured with a theodolite or total station. Theodolites are manufactured to read angles to the closest minute, 20 seconds, 10 seconds, 6 seconds, or 1 second. Figure 1.7 shows a 20-second electronic theodolite. Slope angles can also be measured with a clinometer (see Chapter 2). The precision of this instrument is typically 10 minutes.

1.9   Position Measurement

The position of a natural or built entity can be determined by using a global positioning system (GPS) receiver, which is simultaneously tracking four or five positioning satellites.

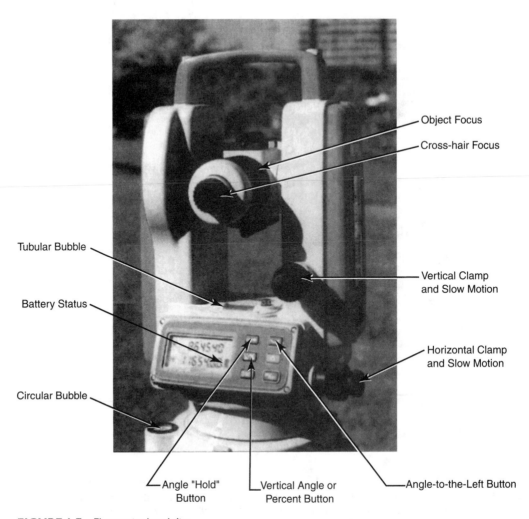

FIGURE 1.7 Electronic theodolite.

The position can be expressed in geographic or grid coordinates, along with ellipsoidal or orthometric elevations (in feet or meters).

Position can also be recorded using airborne and satellite imagery. Such imagery includes aerial photography, lidar imaging, radar imaging, and spectral scanning (see Chapter 7).

1.10 Units of Measurement

Although the foot system of measurement has been in use in the United States from colonial days until the present, the metric system has been making steady inroads. The Metric Conversion Act of 1975 made conversion to the metric system largely voluntary, but subsequent amendments and government actions have now made use of the metric system mandatory for all federal agencies as of September 1992. By January 1994, the metric system was required in the design of many federal facilities. Many states' departments of transportation have also commenced the switch to the metric system for field work and highway design. Although the enthusiasm for metric use in the United States seems to have waned in recent years, both metric units and English units are used in this text because both units are now in wide use.

The complete changeover to the metric system will take many years, perhaps several generations. The impact of all this on the American surveyor is that, from now on, most surveyors will have to be proficient in both the foot and the metric systems. Additional equipment costs in this dual system are limited mostly to measuring tapes and leveling rods.

System International (SI) units are a modernization (1960) of the long-used metric units. This modernization included a redefinition of the meter (international spelling: metre) and the addition of some new units (e.g., Newton; see Table 2.1).

Table 1.1 describes and contrasts metric and foot units. Degrees, minutes, and seconds are used almost exclusively in both metric and foot systems; however, in some

Table 1.1 MEASUREMENT DEFINITIONS AND EQUIVALENCIES

Linear Measurement	Foot Units
1 mile = 5,280 feet	1 foot = 12 inches
= 1,760 yards	1 yard = 3 feet
= 320 rods	1 rod = 16½ feet
= 80 chains	1 chain = 66 feet
1 acre = 43,560 ft^2 = 10 square chains	1 chain = 100 links

Linear Measurement		Metric (SI) Units
1 kilometer	=	1,000 meters
1 meter	=	100 centimeters
1 centimeter	=	10 millimeters
1 decimeter	=	10 centimeters
1 hectare (ha)	=	10,000 m^2
1 square kilometer	=	1,000,000 m^2
	=	100 hectares

Foot to Metric Conversion	
1 ft = 0.3048 m (exactly)	1 inch = 25.4 mm (exactly)*
1 km = 0.62137 mile	
1 hectare (ha) = 2.471 acres	
1 km^2 = 247.1 acres	

Angular Measurement	
1 revolution = 360°	1 revolution = 400.0000 gon†
1 degree = 60′ (minutes)	
1 minute = 60″ (seconds)	

*Prior to 1959, the United States used the relationship 1 m = 39.37 in., which resulted in a U.S. survey foot of 0.3048006 m.
†Used in some European countries.

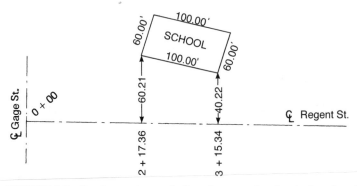

FIGURE 1.8 Baseline stations and offset distances, showing the location of the school on Regent St.

European countries, the circle has also been graduated into 400 gon (also called grad). In that system, angles are expressed to four decimals (e.g., a right angle = 100.0000 gon).

1.11 Stationing

In surveying, measurements are often taken along a baseline and at right angles to that base-line. Distances along a baseline are referred to as **stations** or **chainages,** and distances at right angles to the baseline (offset distances) are simple dimensions. The beginning of the survey baseline—the zero end—is denoted as 0 + 00; a point 100 ft (m) from the zero end is denoted as 1 + 00; a point 156.73 ft (m) from the zero end is 1 + 56.73; and so on.

In the preceding discussion, the full stations are at 100-ft (m) intervals, and the half stations are at even 50-ft (m) intervals. Twenty-meter intervals are often used as the key partial station in the metric system for preliminary and construction surveys. With the ongoing changeover to metric units, most municipalities have kept the 100-unit station (i.e., 1 + 00 = 100 meters), whereas highway agencies have adopted the 1,000-unit station (i.e., 1 + 000 = 1,000 meters).

Figure 1.8 shows a school building tied in to the centerline (℄) of Regent St. The fig-ure also shows the ℄ (baseline) distances as stations, and the offset distances as simple dimensions.

1.12 Types of Construction Projects

The first part of this text covers the surveying techniques common to most surveying endeavors. The second part of the text is devoted to construction surveying applications—an area that accounts for much surveying activity. Listed below are the types of construc-tion projects that depend a great deal on the construction surveyor or engineering surveyor for the successful completion of the project:

1. Streets and highways
2. Drainage ditches
3. Intersections and interchanges
4. Sidewalks

5. High- and low-rise buildings
6. Bridges and culverts
7. Dams and weirs
8. River channelization
9. Sanitary landfills
10. Mining—tunnels, shafts
11. Gravel pits, quarries
12. Storm and sanitary sewers
13. Water and fuel pipelines
14. Piers and docks
15. Canals
16. Railroads
17. Airports
18. Reservoirs
19. Site grading, landscaping
20. Parks, formal walkways
21. Heavy equipment locations (millwright)
22. Electricity transmission lines

1.13 Random and Systematic Errors

An **error** is the difference between a measured, or observed, value and the "true" value. No measurement can be performed perfectly (except for counting), so every measurement must contain some error. Errors can be minimized to an acceptable level by the use of skilled techniques and appropriately precise equipment. For the purposes of calculating errors, the "true" value of a dimension is determined statistically after repeated measurements have been taken.

Systematic errors are defined as those errors for which the magnitude and the algebraic sign can be determined. The fact that these errors can be determined allows the surveyor to eliminate them from the measurements and thus further improve the accuracy. An example of a systematic error is the effect of temperature on a steel tape. If the temperature is quite warm, the steel expands, and thus the tape is longer than normal. For example, at 83°F, a 100-ft steel tape can expand to 100.01 ft, a systematic error of 0.01 ft. Knowing this error, the surveyor can simply subtract 0.01 ft each time the tape is used at that temperature.

Random errors are associated with the skill and vigilance of the surveyor. Random errors (also known as accidental errors) are introduced into each measurement mainly because no human can perform perfectly. Random errors can be illustrated by the following example. Let's say that point B is to be located a distance of 109.55 ft from point A. The tape is only 100.00 ft long, so an intermediate point must first be set at 100.00 ft, and then 9.55 ft must be measured from the intermediate point. Random errors occur as the surveyor is marking out 100.00 ft. The actual mark may be off a bit; that is, the mark may actually be made at 99.99 or 99.98, and so on. When the final 9.55 ft are measured out, two more opportunities for error exist: the lead surveyor will have the same opportunity for error as existed at the 100.00 mark, and the rear surveyor may introduce a random error by inadvertently holding something other than 0.00 ft (for example, 0.01) on the intermediate mark.

This example illustrates two important characteristics of random errors. First, the magnitude of the random error is unknown. Second, because the surveyor is estimating too high (or too far right) on one occasion and probably too low (or too far left) on the next occasion, random errors tend to cancel out over the long run.

A word of caution: large random errors, possibly due to sloppy work, also tend to cancel out. Thus, they give the appearance of accurate work to work that could be highly inaccurate.

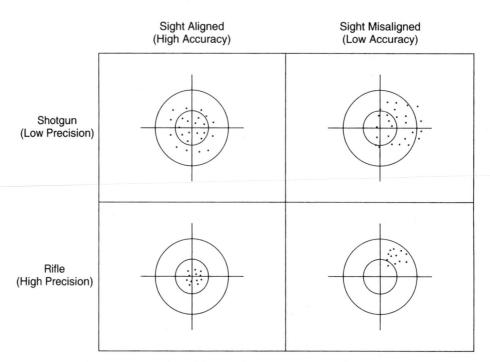

FIGURE 1.9 An illustration of the difference between accuracy and precision.

1.14 Accuracy and Precision

Accuracy is the relationship between the value of a measurement and the "true" value of the dimension being measured; the greater the accuracy, the smaller the error. **Precision** describes the degree of refinement with which the measurement is made. For example, a distance measured four times with a steel tape by skilled personnel will be more precise than the same distance measured twice by unskilled personnel using a fiberglass tape. Figure 1.9 illustrates the difference between accuracy and precision by showing the results of target shooting using both a high-precision rifle and a low-precision shotgun.

The **accuracy ratio** of a measurement or a series of measurements is the ratio of the error of closure to the distance measured. The error of closure is the difference between the measured location and its theoretically correct location. Because relevant systematic errors and mistakes can and should be eliminated from all survey measurements, the error of closure will normally be composed of random errors.

To illustrate, a distance is measured and found to be 196.33 ft. The distance was previously known to be 196.28 ft. The error is 0.05 ft in a distance of 196.28 ft:

$$\text{Accuracy ratio} = \frac{0.05}{196.28} = \frac{1}{3,926} \approx \frac{1}{3,900}$$

The accuracy ratio is expressed as a fraction whose numerator is 1 and whose denominator is rounded to the closest 100 units. Many engineering surveys are specified at

1/3,000 and 1/5,000 levels of accuracy; property surveys are often specified at 1/5,000 and 1/7,500 levels of accuracy. With polar layouts now being used more often, the coordinated control stations needed for this type of layout must be established using techniques giving higher orders of accuracy (e.g., 1/10,000, 1/15,000, etc.). Sometimes the accuracy ratio, or error ratio, is expressed in parts per million (ppm). One ppm is simply the ratio of 1/1,000,000; 50 ppm is 50/1,000,000, or 1/20,000. See Table 3.1 and Tables 9.2 through 9.5 for various survey specifications and standards.

1.15 Mistakes

Mistakes are blunders made by survey personnel. Examples of mistakes are transposing figures (recording a value of 86 as 68), miscounting the number of full tape lengths in a long measurement, and measuring to or from the wrong point. You should be aware that mistakes will occur. Mistakes must be discovered and eliminated, preferably by the people who made them. *All survey measurements are suspect until they have been verified.* Verification may be as simple as repeating the measurement, or verification may result from geometric or trigonometric analysis of related measurements. As a rule, all measurements are immediately checked or repeated. This immediate repetition enables the surveyor to eliminate most mistakes and at the same time to improve the precision of the measurement.

1.16 Field Notes

One of the most important aspects of surveying is the taking of neat, legible, and complete field notes. The notes will be used to plot scale drawings of the area surveyed and also to provide a permanent record of the survey proceedings. Modern surveys, employing electronic *data collectors*, automatically store point-positioning angles, distances, and attributes, which will later be transferred to the computer. Surveyors have discovered that some handwritten field notes are also valuable for these modern surveys. (See also Section 5.10.)

An experienced surveyor's notes should be complete, without redundancies; be arranged to aid comprehension; and be neat and legible to ensure that the correct information is conveyed. Sketches are used to illustrate the survey and thus help remove possible ambiguities.

Handwritten field notes are placed in bound field books or in loose-leaf binders. Loose-leaf notes are preferred for small projects because they can be filed alphabetically by project name or in order by number. Bound books are advantageous on large projects, such as highway construction or other heavy construction operations, where the data can readily fill one or more field books.

1.16.1 Requirements for Bound Books

Bound field books should include the following information:

1. Name, address, and phone number should be in ink on the outside cover.
2. Pages are numbered throughout.
3. Space is reserved at the front of the field book for a title, an index, and a diary.
4. Each project must show the date, title, surveyors' names, and instrument numbers.

1.16.2   Requirements for Loose-Leaf Books

Loose-leaf field books should include the following information:

1. Name, address, and phone number should be in ink on the binder.
2. Each page must be titled and dated, and must be identified by project number, surveyors' names, and instrument numbers.

1.16.3   Requirements for All Field Notes

All field notes, whether bound into books or organized into loose-leaf binders, should follow this checklist:

1. Entries should be in pencil, written with 2H to 4H lead (lead softer than 2H will cause unsightly smears on the notes).
2. All entries are neatly printed. Uppercase letters can be used throughout, or they can be reserved for emphasis.
3. All arithmetic computations must be checked and signed.
4. Although sketches are not scale drawings, they are drawn roughly to scale to help order the inclusion of details.
5. North arrows are placed so that they are pointing upward on the page.
6. Sketches are not freehand; straightedges and curve templates are used for all line work.
7. Do not crowd information on the page. Crowded information is one of the chief causes of poor field notes.
8. Mistakes in the entry of measured data are to be carefully lined out, not erased.
9. Mistakes in entries other than measured data (e.g., descriptions, sums or products of measured data) may be erased and reentered neatly.
10. If notes are copied, they must be clearly labeled as such so that they are not thought to be field notes.
11. Lettering on sketches is to be read from the bottom of the page or from the right side; any other position is upside down.
12. Note keepers verify all given data by repeating the data aloud as they enter the data in their notes; the surveyor who originally gave the data to the note keeper listens and responds to the verification callout.
13. If the data on an entire page are to be voided, the word VOID, together with a diagonal line, is placed on the page. A reference page number is shown for the new location of the relevant data.

Review Questions

1.1 Describe the four different procedures used to locate a physical feature in the field so that it can be plotted later in its correct position on a scaled plan.

1.2 Describe how a very precise measurement can be inaccurate.

1.3 How do plane surveys and geodetic surveys differ?

1.4 How is a total station different from an electronic theodolite?

1.5 How can you ensure that a survey measurement is free of mistakes?

1.6 Why do surveyors usually convert measured slope distances to their horizontal equivalents?

1.7 Describe the term *error*. How does this term differ from *mistake*?

1.8 What is the difference between a layout survey and a preliminary survey?

Chapter 2

Distance Measurement

2.1 Methods of Linear Measurement

Survey measurements can be completed using both direct and indirect techniques, or through calculations. An example of a direct technique is the application of a graduated tape against the marks to be measured. An example of an indirect technique is the measurement of the phase differences between the transmitted and reflected light waves used by the electronic distance measurement (EDM) instruments. An example of a calculated measurement occurs when the desired measurement (perhaps over water) is one side of a triangle whose other sides and angles measured have been measured.

2.1.1 Pacing

Pacing is a useful method of approximate measure. Surveyors can determine the length of pace that, for them, can be comfortably repeated (for convenience, some surveyors use a 3-ft stride). Pacing is particularly useful when looking for survey markers in the field. The plan distance from a found marker to another marker can be paced off to aid in locating that marker. Another important use for pacing is for a rough check of all key points in construction layouts.

2.1.2 Odometer

Automobile odometer readings can be used to measure from one fence line to another when they intersect the road right-of-way. These readings are precise enough to differentiate rural fence lines and can thus assist in identifying platted property lines. This information is useful when collecting information to begin a survey. Odometers are also used on measuring wheels that are simply rolled along the ground on the desired route; this approximate technique is employed where low-order precision is acceptable. For example, surveyors from the assessor's office often check property frontages this way, and police officers sometimes use this technique when preparing sketches of automobile accident scenes.

2.1.3 Electronic Distance Measurement (EDM)

Most EDM instruments function by sending a light wave along the path to be measured, and then the instrument measures the phase difference between the transmitted light wave and the light wave as it is reflected back to its source. Pulse laser EDMs operate by measuring the time for a laser pulse to be transmitted to a reflector and then returned to the EDM; with the velocity of light programmed into the EDM, the distance to the reflector and back is quickly determined.

2.1.4 Subtense Bar

A subtense bar is a tripod-mounted bar with targets precisely 2.000 m apart. The targets are kept precisely 2.000 m apart by the use of invar wires under slight tension. The subtense bar is positioned over the point and then positioned perpendicular to the survey line. A theodolite (1-second capability) is used to measure the angle between the targets. The longer the distance, the smaller the angle. This technique is accurate (1/5,000) at short distances, that is, less than 500 ft. This instrument was used to obtain distances over difficult terrain, for example, across freeways, water, or steep slopes. See Figure 2.1. Field use of this instrument has declined with the widespread use of EDM instruments. Subtense bars have recently been used calibrating baselines in electronic coordinate determination: a technique of precise positioning using two or more electronic theodolites interfaced to a computer. This technique has been used, for example, to position robotic welding machines on automobile assembly lines.

2.2 Gunter's Chain

The measuring device in popular use during settlement of North America (eighteenth and nineteenth centuries) was the Gunter's chain; it was 66 ft long and subdivided into 100 links. It has more than historical interest to surveyors because many property descriptions still on file include dimensions in chains and links. This chain, named after its inventor (Edmund Gunter, 1581–1626), was uniquely suited for work in English units:

$$
\begin{aligned}
1 \text{ chain} &= 100 \text{ links} \\
1 \text{ rod} &= 25 \text{ links} \\
4 \text{ rods} &= 1 \text{ chain} \\
80 \text{ chains} &= 1 \text{ mile} \\
10 \text{ sq chains} &= 1 \text{ acre} \ (10 \times 66^2 = 43{,}560 \text{ ft}^2)
\end{aligned}
$$

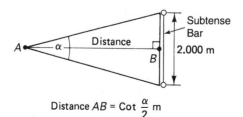

FIGURE 2.1 Subtense bar.

Because Gunter's chains were used in many of the original surveys of North America, most of the continent's early legal plans and records contain dimensions in chains and links. Present-day surveyors occasionally must use these old plans and must make conversions to feet or meters.

■ EXAMPLE 2.1

An old plan shows a dimension of 5 chains, 32 links. Convert this value to (a) feet and (b) meters.

Solution

(a) $5.32 \times 66 = 351.12$ ft

(b) $5.32 \times 66 \times 0.3048 = 107.02$ m

2.3   Tapes

Woven tapes made of linen, Dacron, and the like, can have fine copper strands interwoven to provide strength and to limit deformation due to long use and moisture. See Figure 2.2. Measurements taken near electric stations should be made with dry nonmetallic or fiberglass tapes. Fiberglass tapes have now come into widespread use.

All tapes come in various lengths, the 100-ft and 30-m tapes being the most popular, and are used for many types of measurements where high precision is not required. All woven tapes should be checked periodically (e.g., against a steel tape) to verify their precision.

Many tapes are now manufactured with foot units on one side and metric units on the reverse side. Foot-unit tapes are graduated in feet, 0.10 ft, and 0.05 ft or in feet, inches, and ¼-inches. Metric tapes are graduated in meters, centimeters (0.01 m), and ½-centimeters (0.005 m).

2.4   Steel Tapes

2.4.1   General Background

Not so many years ago, most precise measurements were made using steel tapes. Although electronic distance measurement (EDM) is now favored because of its high precision and the quickness of repeated measurements, even over rough ground, there are some drawbacks when EDM is used for single distances, particularly in the short distance situations that regularly crop up in engineering applications. The problems with EDM in a short distance situation are twofold: (1) the time involved in setting up the EDM and the prism, and (2) the unreliable accuracies in some short distance situations. For example, if a single check distance is required in a construction layout, and if the distance is short (especially distances less than one tape-length), it is much quicker to obtain the distance through taping. Also, because most EDMs now have stated accuracies in the range of ± [5 mm + 5 ppm] to ± [2 mm + 2 ppm], the 5-mm to 2-mm errors occur regardless of the length of distance measured. These errors have little impact on long distances but can severely impair the measurement of short distances. For example, at a distance of 10.000 m, an error of 0.005

(a)

(b)

Graduated in 10ths and Metric

7   8   9   1 F   1

0·2   ▲   0·3

Printed on two sides—one side in 10ths and 100ths
of a foot; the second side in metric with increments
in meters, cm, and 2 mm.

Graduated in 8ths and Metric

8   9   10   11   1 F   1

0·2   ▲   0·3

Printed on two sides—one side in feet, inches, and
8ths; the second side in metric with increments in
meters, cm, and 2 mm.

(c)

FIGURE 2.2   Fiberglass tapes. (a) Closed case. (b) Open reel. (c) Tape graduations. (Courtesy of
CST/Berger, Illinois)

m limits accuracy to 1:2,000. The opportunity for additional errors can occur when centering the EDM instrument over the point (e.g., an additional 0.001 to 0.002 m for well-
adjusted laser plummets), and when centering the prism over the target point, either by
using tribrach-mounted prisms or by using prism-pole assemblies. The errors here can
range from 2 mm for well-adjusted optical/laser plummets to several millimeters for well-
adjusted prism-pole circular bubble levels. When laser or optical plummets are poorly
adjusted and/or when prism-pole levels are poorly adjusted, serious errors can occur in distance measurements—errors that do diminish in severity as the distance measured
increases. For these reasons, the steel tape remains a valuable tool for the surveyor.

    Steel tapes (see Figure 2.3) are manufactured in both foot and metric units and come
in a variety of lengths, graduations, and unit weights. Commonly used foot-unit tapes are
100-, 200-, and 300-ft lengths, the 100-ft length being the most widely used. Commonly
used metric-unit tapes are 20-, 30-, 50-, and 100-m lengths. The 30-m length is the most
widely used because it closely resembles the 100-ft length tape in field characteristics.

    Generally, lightweight tapes are graduated throughout and are used on the reel; heavier
tapes are designed for use off the reel (drag tapes) and do not have continuous small-interval

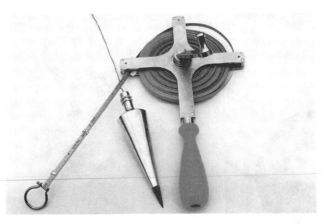

FIGURE 2.3 Steel tape and plumb bob.

markings. Drag tapes are popular in route surveys (highways, railways, etc.), whereas light-weight tapes are more popular in building and municipal works.

Invar tapes are composed of 35 percent nickel and 65 percent steel. This alloy has a very low coefficient of thermal expansion, which made this tape useful for pre-EDM distance measurement. Steel tapes are occasionally referred to as chains, a throwback to early surveying practice.

2.4.2 Types of Readouts

Steel tapes are normally graduated in one of three ways; consider a distance of 38.82 ft (m):

1. The tape is graduated throughout in feet and hundredths (0.01) of a foot, or in meters and millimeters (see Figures 2.3 and 2.4). The distance (38.82 ft) is read directly from the steel tape.
2. The cut tape is marked throughout in feet, with the first and last foot graduated in tenths and hundredths of a foot [see Figure 2.4(a)]. The metric cut tape is marked throughout in meters and decimeters, with the first and last decimeters graduated in millimeters. A measurement is made with the cut tape by one surveyor holding the even-foot mark (39 ft in this example). This arrangement allows the other surveyor to read a distance on the first foot (decimeter), which is graduated in hundredths of a foot (millimeters). For example, the distance from *A* to *B* in Figure 2.4(a) is determined by holding 39 ft at *B* and reading 0.18 ft at *A*. Distance *AB* = 38.82 ft (i.e., 39 ft cut 0.18). Because each measurement involves this type of mental subtraction, care and attention are required from the surveyor to avoid unwelcome blunders.
3. The add tape is also marked throughout in feet, with the last foot graduated in hundredths of a foot. An additional foot, graduated in hundredths, is included prior to the zero mark. For metric tapes, the last decimeter and the extra before-the-zero decimeter

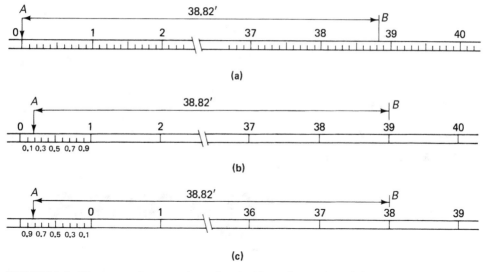

FIGURE 2.4 Various steel tape markings (hundredth marks not shown). (a) Fully graduated tape. (b) Cut tape. (c) Add tape.

are graduated in millimeters. The distance *AB* in Figure 2.4(b) is determined by holding 38 ft at *B* and reading 0.82 ft at *A*. Distance *AB* is 38.82 ft (i.e., 38 ft add 0.82).

As noted, cut tapes have the disadvantage of creating opportunities for subtraction mistakes; the add tapes have the disadvantage of forcing the surveyor to adopt awkward measuring stances when measuring from the zero mark. The full meter add tape is the most difficult to use correctly because the surveyor must fully extend his or her left (right) arm (which is holding the end of the tape) to position the zero mark on the tape over the ground point. The problems associated with both add and cut tapes can be eliminated if, instead, the surveyor uses tapes graduated throughout. These tapes are available in both drag- and reel-type tapes.

2.5 Taping Accessories and Their Use

2.5.1 Plumb Bob

Plumb bobs are normally made of brass and weigh from 8 to 18 oz, with 10 and 12 oz being the most common. Plumb bobs are used in taping to transfer from tape to ground (or vice versa) when the tape is being held off the ground to maintain its horizontal alignment. Plumb bobs are also used routinely to provide theodolite sightings. See Figures 2.5 and 2.9.

2.5.2 Hand Level

The hand level (see Figure 2.6) can be used to keep the steel tape horizontal when measuring. The surveyor at the lower elevation holds the hand level, and a sight is taken back at the higher-elevation surveyor. For example, if the surveyor with the hand level is sighting

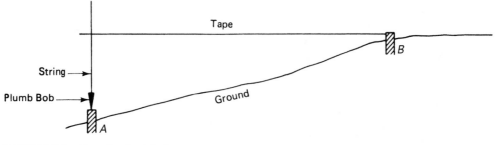

FIGURE 2.5 Use of a plumb bob.

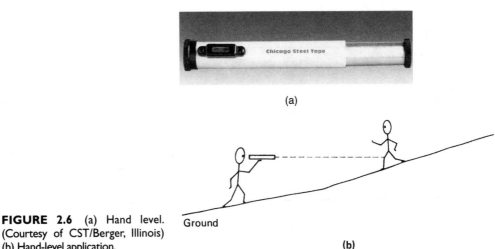

(a)

(b)

FIGURE 2.6 (a) Hand level. (Courtesy of CST/Berger, Illinois) (b) Hand-level application.

with the instrument cross hair horizontally on his or her partner's waist, and if both are roughly the same height, then the surveyor with the hand level is lower by the distance from eye to waist. The low end of the tape is held that distance off the ground (using a plumb bob), the high end of the tape being held on the mark.

Figure 2.7(a) shows an Abney hand level (clinometer). In addition to the level bubble and cross hair found on the standard hand level, the clinometer can take vertical angles (to the closest 10 minutes). The clinometer is used mainly to record vertical angles for slope distance reduction to horizontal or to determine the heights of objects. These two applications are illustrated in Examples 2.2 and 2.3.

■ EXAMPLE 2.2

For height determination, a value of 45° is placed on the scale. The surveyor moves back and forth until the point to be measured is sighted [e.g., the top of the building shown in Figure 2.7(b)]. The distance from the observer to the surveyor at the base of the building is measured with a fiberglass tape. The hand level can be set to zero to determine where the surveyor's eye height intersects the building wall. The partial height of the building (h_1) is equal to this measured distance (i.e., h_1/measured distance = tan 45° = 1). If the distance h_2 [eye height mark above the ground; see Figure 2.7(b)] is now added to the measured distance, the height of the building (in this example) is found. In addition to determining the heights of buildings, this technique is particularly useful in determining the heights of electric power lines for highway construction clearances. In Figure 2.7(b), if the measured distance to the second surveyor at the building is 63.75 ft, and if the horizontal line of sight hits the building (the clinometer scale is set to zero for horizontal sights) at 5.55 ft (h_2), the height of the building is 63.75 + 5.55 = 69.30 ft.

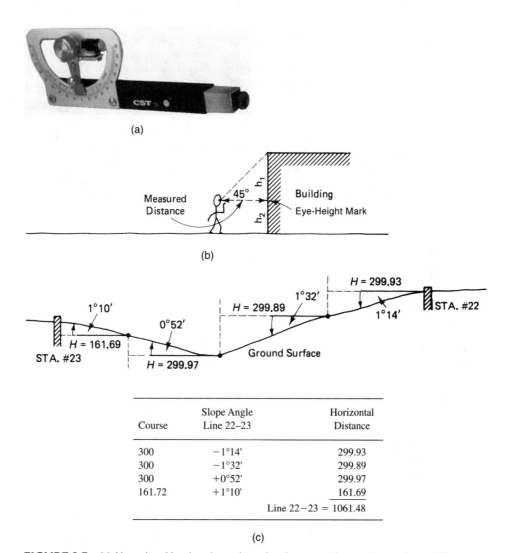

Course	Slope Angle Line 22–23	Horizontal Distance
300	−1°14'	299.93
300	−1°32'	299.89
300	+0°52'	299.97
161.72	+1°10'	161.69
	Line 22−23 =	1061.48

(c)

FIGURE 2.7 (a) Abney hand level; scale graduated in degrees with a vernier reading to 10 minutes. (Courtesy of CST/Berger, Illinois) (b) Abney hand-level application in height determination. (c) Abney hand-level typical application in taping.

■ **EXAMPLE 2.3**

The clinometer is useful when working on route surveys for which extended-length tapes (300 ft or 100 m) are being used. The long tape can be used in slope position under proper tension. (Because the tape will be touching the ground in many places, it will be mostly supported, and the tension required will not be too high.) The clinometer can be used to measure the slope angle for each tape measurement. These slope angles and related slope distances can be used later to compute the appropriate horizontal distances. The angles, measured distances, and computed distances for a field survey are summarized in Figure 2.7(c).

2.5.3 Additional Taping Accessories

Range poles are 6-ft wooden or aluminum poles with steel points. The poles are usually painted red and white in alternate 1-ft sections. Range poles are used in taping and theodolite work to provide alignment sights. See Figure 2.8(a). The clamp handle [see Figure 2.8(b)] helps grip the tape at any intermediate point without bending or distorting the tape. Tension handles [see Figure 2.8(c)] are used in precise work to ensure that the appropriate tension is being applied. They are graduated to 30 lb, in ½-lb graduations (50 N = 11.24 lb).

Chaining pins (marking arrows) come in sets of eleven. They are painted alternately red and white and are 14 to 18 in. long. Chaining pins are used to mark intermediate points on the ground. In route surveying, the whole set of pins is used to mark out the centerline.

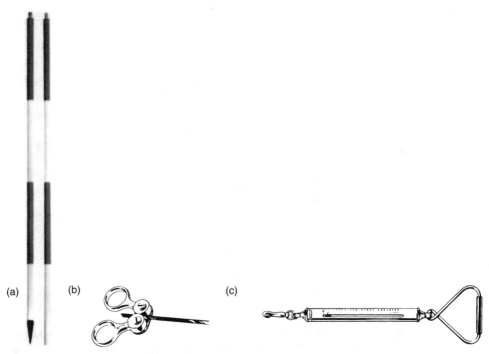

(a)   (b)   (c)

FIGURE 2.8   Taping accessories. (a) Two-section range pole. (b) Tape clamp handle. (c) Tension handle.

FIGURE 2.9 Plumb bob cord target used to provide a transit sighting.

The rear surveyor is responsible for checking the number of whole tape lengths by keeping an accurate count of the pins collected. Eleven pins are used to measure out 1,000 ft.

Tape repair kits are available so that broken tapes can be put back into service quickly. The repair kits come in three main varieties: (1) punch pliers and repair eyelets, (2) steel punch block and rivets, and (3) tape repair sleeves. The second technique (punch block) is the only method that gives lasting repair; although the technique is simple, great care must be exercised to ensure that the integrity of the tape is maintained.

Plumb bob targets are also used to provide alignment sights (see Figure 2.9). The plumb bob string is threaded through the upper and lower notches so that the target center-line is superimposed on the plumb bob string. The target, which can be adjusted up and down the string for optimal sightings, is preferred to the range pole because of its portability—it fits in the surveyor's pocket.

2.6   Taping Techniques

Taping is normally performed with the tape held horizontally. If the distance to be measured is across smooth, level land, the tape can simply be laid on the ground and the end mark lined up against the initial survey marker; the tape is properly aligned and tensioned, and then the zero mark on the tape can be marked on the ground. If the distance between two marked points is to be measured, the tape is read as already described in Section 2.4.2.

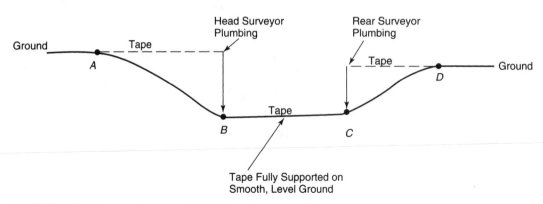

FIGURE 2.10 Horizontal taping; plumb bob used at one end.

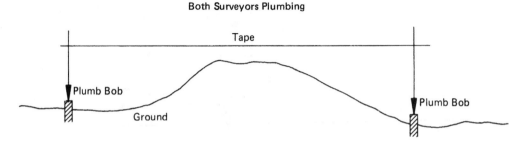

FIGURE 2.11 Horizontal taping; plumb bob used at both ends.

If the distance to be measured is across sloping or uneven land, then at least one end of the tape must be raised off the ground to keep the tape horizontal. The raised end of the tape is referenced back to the ground mark with the aid of a plumb bob (see Figure 2.10). Normally the only time that both ends of the tape are plumbed is when the ground—or other obstruction—rises between the marks being measured (see Figure 2.11).

2.6.1 Measuring Procedures

The measurement begins with the head surveyor carrying the zero end of the tape forward toward the final point. He or she continues walking until the tape has been unwound and the rear surveyor calls, "Tape," thus alerting the head surveyor to stop walking and to prepare for measuring. If a drag tape is being used, the tape is removed from the reel and a leather thong is attached to the reel end of the tape (the zero end is already equipped with a leather thong). If the tape is not designed to come off the reel, the winding handle is folded to the lock position so that the reel can be used to help hold the tape. The rear surveyor can keep the head surveyor on line by sighting a range pole or other target that has been erected at the final mark. In precise work, these intermediate marks can be aligned by theodolite.

The rear surveyor holds the appropriate graduation (e.g., 100.00 ft, or 30.000 m) against the mark from which the measurement is being taken. The head surveyor, after ensuring that the tape is straight, slowly increases tension to the proper amount and then marks the ground with a chaining pin or other marker. Once the mark has been made, both surveyors repeat the measuring procedure to check the measurement. If necessary, corrections are made and the check procedure is repeated.

If the ground is not horizontal (determined by estimation or by use of a hand level), one or both surveyors must use a plumb bob. When plumbing, the tape is often held at waist height, although any height between the shoulders and the ground is common. Holding the tape above the shoulders creates more chance for error because the surveyor must move his or her eyes up and down to include both the ground mark and the tape graduation in his or her field of view. When the surveyor's eyes are on the tape, the plumb bob may move off the mark, and when his or her eyes are on the ground mark, the plumb bob string may move off the correct tape graduation.

The plumb bob string is usually held on the tape with the left thumb (for right-handed people); take care not to cover the graduation mark completely because, as the tension is increased, it is not unusual for the surveyor to take up some of the tension with the left thumb, causing the thumb to slide along the tape. If the graduations have been covered completely with the left thumb, the surveyor is not aware that the thumb (and thus the string) has moved, resulting in an erroneous measurement. When plumbing, hold the tape close to the body to provide good leverage for applying or holding tension, and to transfer accurately from tape to ground, and vice versa. If the rear surveyor is using a plumb bob, he or she shouts out "Mark," or some other word indicating that at that instant in time, the plumb bob is right over the mark. If the head surveyor is also using a plumb bob, he or she must wait until both plumb bobs are simultaneously over their respective marks.

You will discover that plumbing and marking are difficult aspects of taping. You may find it difficult to hold the plumb bob steady over the point and at the same time apply the appropriate tension. To help steady the plumb bob, hold it only a short distance above the mark and continually touch down to the point. This momentary touching down dampens the plumb bob oscillations and generally steadies the plumb bob. Do not allow the plumb bob point to rest on the ground or other surface because you will obtain an erroneous measurement.

2.6.2 Breaking Tape

A slope is sometimes too steep to permit an entire tape length to be held horizontal. When this occurs, shorter measurements are taken, each with the tape held horizontal; these shorter measurements are then totaled to provide the overall dimension. This technique, called breaking tape, must be done with greater care because the extra marking and measuring operations provide that many more opportunities for the occurrence of random errors—errors associated with marking and plumbing. There are two common methods of breaking tape. First, the head surveyor takes the zero end forward one tape length and then walks back to a point where he or she can hold the tape horizontal with the rear surveyor; if working downhill, the tape can be held at shoulder height. With a plumb bob, the ground can be marked at an even foot or meter graduation (say, 80 ft or 25 m). The rear surveyor then comes forward and holds that same graduation on the ground mark while the head surveyor moves forward to the next pre-selected tape graduation, which is also to be held at

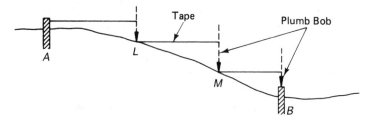

FIGURE 2.12 Breaking tape.

shoulder height (say, 30 ft or 10 m). This procedure is repeated until the head surveyor can hold the zero graduation and thus complete one full tape length. In the second method of breaking tape, the head surveyor can proceed forward only until his or her shoulders are horizontal with the rear surveyor's knees or feet and can then mark the zero-end on the ground. The rear surveyor, who is probably holding an even foot or meter right on the mark, calls out that value, which is then recorded. This process is repeated until the whole distance has been measured and all the intermediate measurements are totaled for the final answer. See Figure 2.12, which shows the distance *AB*, comprising the increments *AL*, *LM*, and *MB*.

2.6.3 Taping Summary

The rear surveyor follows these steps when taping:

1. Aligns the head surveyor by sighting to a range pole or other target, which is placed at the forward station.
2. Holds the tape on the mark, either directly or with the aid of a plumb bob. He or she calls out "Mark," or some other word to signal that the tape graduation—or the plumb bob point marking the tape graduation—is momentarily on the mark.
3. Calls out the station and tape reading for each measurement and listens for verification from the head surveyor.
4. Keeps a count of all full tape lengths included in each overall measurement.
5. Maintains the equipment (e.g., wipes the tape clean at the conclusion of the day's work or as conditions warrant).

The head surveyor follows these steps during taping:

1. Carries the tape forward, ensuring that the tape is free of loops, which can lead to kinks and tape breakage.
2. Prepares the ground surface for the mark (e.g., clears away grass, leaves, etc.).
3. Applies proper tension after first ensuring that the tape is straight.
4. Places marks (e.g., chaining pins, wood stakes, iron bars, nails, rivets, cut crosses).
5. Takes and records measurements of distances and other factors (e.g., temperature).
6. Supervises the taping work.

2.7 Taping Corrections

2.7.1 General Background

As noted in Section 1.12, no measurements can be perfectly performed; thus, all measurements (except for counting) must contain some errors. Surveyors must use measuring techniques that minimize random errors to acceptable levels and they must make corrections to systematic errors that can affect the accuracy of the survey. Typical taping errors are summarized below:

Systematic Taping Errors*	Random Taping Errors†
1. Slope	1. Slope
2. Erroneous length	2. Temperature
3. Temperature	3. Tension and sag
4. Tension and sag	4. Alignment
	5. Marking and plumbing
*See Section 2.8.	†See Section 2.9.

2.7.2 Standard Conditions for the Use of Steel Tapes

Tape manufacturers, noting that steel tapes behave differently in various temperature, tension, and support situations, specify the accuracy of their tapes under the following standard conditions:

Foot System	Metric System
1. Temperature = 68°F	1. Temperature = 20°C
2. Tape fully supported throughout	2. Tape fully supported throughout
3. Tape under a tension of 10 lb	3. Tape under a tension of 50 N (newtons)
	(a 1 lb force = 4.448 N, so 50 N = 11.24 lbs)

Field conditions usually dictate that some or all of the above standard conditions cannot be met. The temperature is seldom exactly 68°F (20°C), and because many measurements are taken on a slope, the condition of full support is also not regularly fulfilled when one end of the tape is plumbed.

2.8 Systematic Taping Errors and Corrections

The previous section outlined the standard conditions for a steel tape to give precise results. The standard conditions referred to a specific temperature and tension and to a condition of full support. In addition, the surveyor must be concerned with horizontal versus slope distances and with ensuring that the actual taping techniques are sufficiently precise to provide the desired accuracy. Systematic errors in taping are: slope, erroneous length, temperature, tension, and sag. The effects of slope and tension/sag on field measurements are discussed in the next section; the techniques for computing corrections for errors in erroneous length, temperature, tension, and sag are discussed in Appendix F.

2.8.1 Slope Corrections

Survey distances can be measured either horizontally or on a slope. Survey measurements are usually shown on a plan; if they are taken on a slope, they must then be converted to horizontal distances (plan distances) before they can be plotted. To convert slope distances to their horizontal equivalents, the slope angle (θ), the zenith angle ($90 - \theta$), or the vertical distance (V) must also be known:

$$\frac{H\,(\text{horizontal})}{S\,(\text{slope})} = \cos\theta \tag{2.1}$$

Equation 2.1 can also be written as follows:

$$\frac{H}{S} = \sin(90 - \theta) \quad \text{or} \quad H = S\sin(90 - \theta)$$

where θ is the angle of inclination and $(90 - \theta)$ is the zenith angle.

$$H = \sqrt{S^2 - V^2} \tag{2.2}$$

where V is the difference in elevation. See Example 2.4, part (c).

Slope can also be defined as gradient, or rate of grade. The gradient is expressed as a ratio of the vertical distance over the horizontal distance; this ratio, when multiplied by 100, gives a percentage gradient. For example, if the ground rises 2 ft (m) in 100 ft (m), it is said to have a 2 percent gradient (i.e., $2/100 \times 100 = 2$). If the ground rises 2 ft (m) in 115 ft (m), it is said to have a 1.74 percent gradient (i.e., $2/115 \times 100 = 1.739$).

If the elevation of a point on a gradient is known (see Figure 2.13), the elevation of any other point on that gradient can be calculated as follows:

$$\text{Difference in elevation} = 150 \times \left(\frac{2.5}{100}\right) = -3.75$$
$$\text{Elevation at } 1 + 50 = 564.22 - 3.75 = 560.47 \text{ ft}$$

If the elevations of two points, as well as the distance between them (see Figure 2.14), are known, the gradient between can be calculated as follows:

$$\text{Elevation difference} = 5.40$$
$$\text{Distance} = 337.25$$
$$\text{Gradient} = \frac{5.40}{337.25} \times 100 = +1.60\,\%$$

Station	Elevation
0 + 00	564.22 ft
(Gradient = −2.5%)	
1 + 50	Required

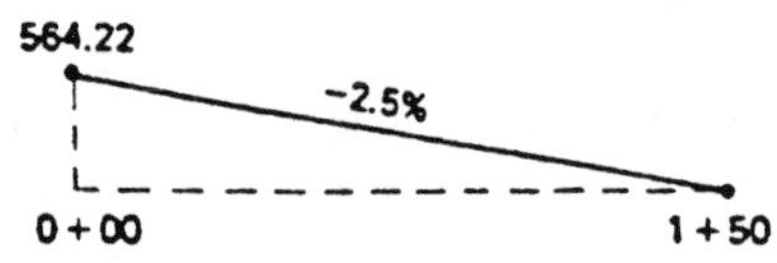

FIGURE 2.13 Computation of the elevation of station 1 + 50.

Station	Elevation
1 + 00	471.37
4 + 37.25	476.77

FIGURE 2.14 Gradient computation.

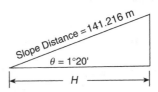

FIGURE 2.15 Horizontal distance computation.

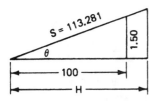

FIGURE 2.16 Slope angle determination.

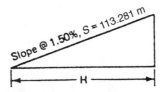

FIGURE 2.17 Horizontal distance computation.

■ **EXAMPLE 2.4** *Slope Corrections*

(a) Given the slope distance (S) and slope angle θ, use Equation 2.1 and Figure 2.15 to find the horizontal distance (H):

$$\frac{H}{S} = \cos\theta$$

$$H = S\cos\theta$$

$$= 141.216 \cos 1°20'$$

$$= 141.178 \text{ m}$$

(b) Given the slope distance (S) and the gradient (slope), find the horizontal distance. See Figures 2.16 and 2.17. First, find the vertical angle (θ).

$$\frac{1.50}{100} = \tan\theta$$

$$\theta = 0.85937°$$

Second, use Equation 2.1 to determine the horizontal distance (H).

$$\frac{H}{113.281} = \cos 0.85937°$$

$$H = 113.268 \text{ m}$$

Sec. 2.8   Systematic Taping Errors and Corrections

33

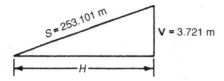

FIGURE 2.18 Horizontal distance computation (metric).

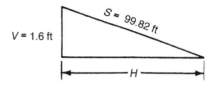

FIGURE 2.19 Horizontal distance computation (foot units).

(c) Metric units: Given the slope distance (*S*) and difference in elevation (*V*), use Equation 2.2 and Figure 2.18 to find the horizontal distance (*H*).

$$H = \sqrt{S^2 - V^2}$$
$$= \sqrt{(253.101^2 - 3.721^2)}$$
$$= 253.074 \text{ m}$$

Foot units: Given the slope distance (*S*) and the difference in elevation (*V*), use Equation 2.2 and Figure 2.19 to find the horizontal distance (*H*).

$$H = \sqrt{(99.82^2 - 1.6^2)}$$
$$= 99.807 \text{ ft}$$
$$= 99.8 \text{ ft}$$

2.8.2 Tension/Sag

The error in measurement due to sag can sometimes be eliminated by increasing the applied tension. Tension that eliminates sag errors is known as **normal tension**. Normal tension ranges from about 19 lb (light 100-ft tapes) to 31 lb (heavy 100-ft tapes).

$$P_n = \frac{0.204 \, W\sqrt{AE}}{\sqrt{P_n - P_s}} \qquad (2.3)$$

This formula gives a value for P_n that eliminates the error caused by sag. The formula is solved by making successive approximations for P_n until the equation is satisfied. This formula is not used often because of the difficulties in determining the individual tape characteristics. See Table F.1 for the units employed in the tension and sag correction formulas found in Appendix F.

■ **EXAMPLE 2.5** *Experiment to Determine Normal Tension*

Normal tension can be determined experimentally for individual tapes, as shown in the following procedure:

1. Lay the tape flat on a horizontal surface; an indoor corridor is ideal.
2. Select (or mark) a well-defined point on the surface at which the 100-ft mark is held.
3. Attach a tension handle at the zero end of the tape. Apply standard tension— say, 10 lb—and mark the surface at 0.00 ft.
4. Repeat the process, switching personnel duties and ensuring that the two marks are, in fact, exactly 100.00 ft apart.
5. Raise the tape off the surface to a comfortable height (waist). While the surveyor at the 100-ft end holds a plumb bob over the point, the surveyor at the zero end slowly increases tension until his or her plumb bob is also over the mark. The tension read from the tension handle will be normal tension for that tape. The readings are repeated several times, and the average results are used for subsequent field work.

The most popular steel tapes (100 ft) now in use require a normal tension of about 24 lb. For most 30-m steel tapes now in use (lightweight), a normal tension of 90 N (20 lb) is appropriate. For structural and bridge surveys, very lightweight 200-ft tapes are available and can be used with a comfortable normal tension—in some cases, about 28 lb.

2.9 Random Taping Errors

Random errors occur because surveyors cannot measure perfectly. A factor of estimation is always present in all measuring activity (except counting). In the previous section, various systematic errors were discussed; in each of the areas discussed, there was also the opportunity for random errors to occur. For example, the temperature problems all require the determination of temperature. If the temperature used is the estimated air temperature, a random error could be associated with the estimation. The temperature of the tape can also be significantly different from that of the air. In the case of sag and tension, random errors can exist when one is estimating applied tension or even estimating between graduations on a spring balance. The higher the precision requirements, the greater must be the care taken in determining these parameters.

In addition to the above-mentioned random errors are perhaps more significant random errors associated with alignment, plumbing, marking, and estimating horizontal positions. Alignment errors occur when the tape is inadvertently aligned off the true path (see Figure 2.20). Usually a rear surveyor can keep the head surveyor on line by sighting a range

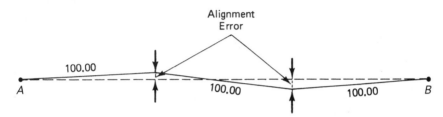

FIGURE 2.20 Alignment errors.

pole marking the terminal point. It would take an alignment error of about 1½ ft to produce an error of 0.01 ft in 100 ft. Because it is not difficult to keep the tape aligned by eye to within a few tenths of a foot (0.2 to 0.3 ft), alignment is not normally a major concern. Note that, although most random errors are compensating, alignment errors are cumulative (misalignment can randomly occur on the left or on the right, but in both cases, the result of the misalignment is to make the measurement too long). Alignment errors can be nearly eliminated on precise surveys by using a theodolite to align all intermediate points.

Marking and plumbing errors are often the most significant of all random taping errors. Even experienced surveyors must exercise great care to place a plumbed mark accurately within 0.02 ft of true value—in a distance of 100 ft. Horizontal measurements, taken with the tape fully supported on the ground, can be determined more accurately than measurements that are taken on a slope and require the use of plumb bobs. Also rugged terrain conditions that require many breaks in the taping process will cause marking and plumbing errors to multiply significantly.

Errors are also introduced when surveyors estimate a horizontal tape position for a plumbed measurement. The effect of this error is identical to that of the alignment error previously discussed, although the magnitude of these errors is often larger than alignment errors. Skilled surveyors can usually estimate a horizontal position to within 1 ft (0.3 m) over a distance of 100 ft (30 m). Even experienced surveyors can be seriously in error, however, when measuring across side-hills, where one's perspective with respect to the horizon can be distorted. Using a hand level can largely eliminate these errors.

2.10 Techniques for "Ordinary" Taping Precision

"Ordinary" taping is referred to as taping at the level of 1/5,000 accuracy. The techniques used for "ordinary" taping, once mastered, can easily be maintained. It is possible to achieve an accuracy level of 1/5,000 with little more effort than is required to attain the 1/3,000 level. Because the bulk of all taping measurements is at either the 1/3,000 or the 1/5,000 level, experienced surveyors often use 1/5,000 techniques even for 1/3,000 level work. This practice permits good measuring work habits to be reinforced continually without appreciably increasing surveying costs. Because of the wide variety of field conditions

Table 2.1 SPECIFICATIONS FOR 1/5,000 ACCURACY

Source of Error	Maximum Effect on One Tape Length	
	100 ft	30 m
Temperature estimated to closest 7°F (4°C)	±0.005 ft	±0.0014 m
Care is taken to apply at least normal tension (lightweight tapes), and tension is known to within 5 lb (20 N)	±0.006 ft	±0.0018 m
Slope errors are no larger than 1 ft/100 ft (0.30 m/30 m)	±0.005 ft	±0.0015 m
Alignment errors are no larger than 0.5 ft/100 ft (0.15 m/30 m)	±0.001 ft	±0.0004 m
Plumbing and marking errors are at a maximum of 0.015 ft/100 ft (0.0046 m/30 m)	±0.015 ft	±0.0046 m
Length of tape is known to within ±0.005 ft (0.0015 m)	±0.005 ft	±0.0015 m

that can be encountered, absolute specifications cannot be prescribed. The specifications in Table 2.1 can be considered typical for "ordinary" 1/5,000 taping.

To determine the total random error in one tape length, take the square root of the sum of the squares of the individual maximum anticipated errors, as shown in the following example:

Feet	Meters
0.005^2	0.0014^2
0.006^2	0.0018^2
0.005^2	0.0015^2
0.001^2	0.0004^2
0.015^2	0.0046^2
0.005^2	0.0015^2
0.000337	0.000031

$$\text{Error} = \sqrt{(0.000337)} = 0.018 \text{ ft} \quad \text{or} \quad \text{Error} = \sqrt{(0.000031)} = 0.0056 \text{ m}$$
$$\text{Accuracy} = 0.018/100 = 1/5,400 \quad \text{or} \quad \text{Accuracy} = 0.0056/30 = 1/5,400$$

2.11 Mistakes in Taping

If errors can be associated with inexactness, mistakes must be thought of as being blunders. Whereas errors can be analyzed and to some degree predicted, mistakes are unpredictable. Just one undetected mistake can nullify the results of an entire survey; thus, it is essential to perform the work so that you minimize the opportunity for mistakes to occur and also allow for verification of the results.

Setting up and then rigorously following a standard method of performing the measurement minimizes the opportunities for the occurrence of mistakes. The more standardized and routine the measurement manipulations, the more likely it is that the surveyor will spot a mistake. The immediate double-checking of all measurements reduces the opportunities for mistakes to go undetected and at the same time increases the precision of the measurement. In addition to checking all measurements immediately, the surveyor is constantly looking for independent methods of verifying the survey results.

Gross mistakes can often be detected by comparing the survey results with distances scaled (or read) from existing plans. The simple check technique of pacing can be a valuable tool for rough verification of measured distances—especially construction layout distances. The possibilities for verification are limited only by the surveyor's diligence and imagination.

Common mistakes encountered in taping are the following:

1. Measuring to or from the wrong marker.
2. Reading the tape incorrectly or transposing figures (e.g., reading 56 instead of 65).
3. Losing proper count of the number of full tape lengths involved in a measurement.

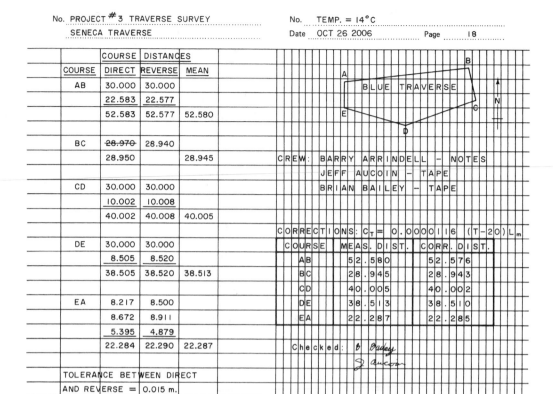

No. PROJECT #3 TRAVERSE SURVEY
SENECA TRAVERSE

No. TEMP. = 14°C
Date OCT 26 2006 Page 18

COURSE	COURSE DISTANCES		
	DIRECT	REVERSE	MEAN
AB	30.000	30.000	
	22.583	22.577	
	52.583	52.577	52.580
BC	~~28.970~~	28.940	
	28.950		28.945
CD	30.000	30.000	
	10.002	10.008	
	40.002	40.008	40.005
DE	30.000	30.000	
	8.505	8.520	
	38.505	38.520	38.513
EA	8.217	8.500	
	8.672	8.911	
	5.395	4.879	
	22.284	22.290	22.287
TOLERANCE BETWEEN DIRECT AND REVERSE =	0.015 m.		

BLUE TRAVERSE

CREW: BARRY ARRINDELL – NOTES
JEFF AUCOIN – TAPE
BRIAN BAILEY – TAPE

CORRECTIONS: $C_T = 0.0000116 (T-20)L_m$

COURSE	MEAS. DIST.	CORR. DIST.
AB	52.580	52.576
BC	28.945	28.943
CD	40.005	40.002
DE	38.513	38.510
EA	22.287	22.285

Checked:

FIGURE 2.21 Taping field notes for a closed traverse.

4. Recording the values incorrectly in the notes. Sometimes the note keeper hears the rear surveyor's callout correctly, but then transposes the figures when he or she enters it into the notes. This mistake can be eliminated if the note keeper calls out each value as it is recorded. The rear surveyor listens for these callouts to ensure that the numbers called out are the same as the data originally given.

5. Calling out figures ambiguously. The rear surveyor can call out 20.27 as "twenty (pause) two seven." This might be interpreted as 22.7. To avoid mistakes, this number should be called out as "twenty, decimal (or point), two, seven."

6. Not identifying correctly the zero point of the tape when a cloth or fiberglass tape is used. This mistake can be avoided if the surveyor checks unfamiliar tapes before use. The tape itself can be used to verify the zero mark.

7. Making arithmetic mistakes in sums of dimensions and in error corrections (e.g., temperature). These mistakes can be identified and corrected if each member of the crew is responsible for checking (and initialing) all computations.

Chap. 2   Distance Measurement

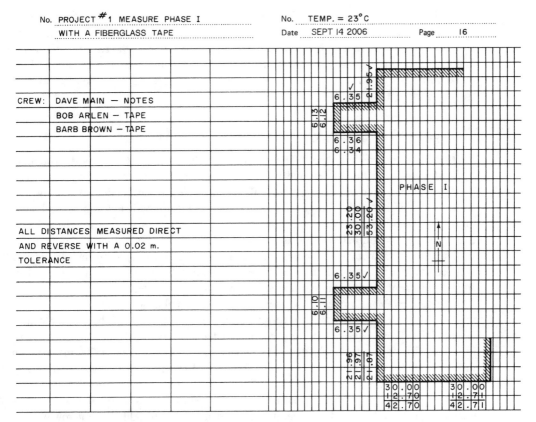

CREW: DAVE MAIN — NOTES

BOB ARLEN — TAPE

BARB BROWN — TAPE

PHASE I

N

ALL DISTANCES MEASURED DIRECT

AND REVERSE WITH A 0.02 m.

TOLERANCE

6.35 ✓

6.35 ✓

6.35 ✓

6.35 ✓

6.36
6.34

6.13
6.12

6.10
6.11

1.95 ✓

20.20
20.00
23.20
30.53

21.96
21.97
21.01
21.01

30.00
12.70
42.70

30.00
12.71
42.71

FIGURE 2.22   Taping field notes for building dimensions.

2.12   Field Notes for Taping

Section 1.15 introduced field notes and stressed the importance of neatness, legibility, completeness, and clarity of presentation. Sample field notes will be included in this text for all typical surveying operations. Figures 2.21 and 2.22 represent typical field notes from taping surveys.

      Figure 2.21 shows the taping notes for a traverse survey. The sides of the traverse have been measured forward and back, with the results being averaged (mean) if the discrepancy is within acceptable limits (0.015 in this example). Observe that the notes are clear and complete and generally satisfy the requirements listed in Section 1.15.

      Figure 2.22 shows taping notes for a building dimension survey. In this example, the measurements are entered right on the sketch of the building. If the sketch has lines that are too short to show the appropriate measured distance, those distances can be entered neatly in an uncrowded portion of the sketch, with arrows joining the measurements to the correct lines on the sketch. In this example, the required building walls have been measured, with the results entered on the sketch. Each wall is remeasured, and if the result is identical to the first measurement, a check mark is placed beside that measurement; if the result varies, the new measurement is also entered beside the original. If the two measurements do not agree within the specified tolerance (0.02 m), the wall is measured again until the tolerance has been met. Compare the page of field notes in Figure 2.22 to the requirements listed in Section 1.15.

2.13   Electronic Distance Measurement (EDM)

Electronic distance measurement (EDM), first introduced in the 1950s by the founders of Geodimeter Inc., has undergone continual refinement since those early days. The early instruments, which were capable of very precise measurements over long distances, were large, heavy, complicated, and expensive. Rapid advances in related technologies have provided lighter, simpler, and less expensive instruments. These EDM instruments are manufactured for use with theodolites and as modular components of total station instruments. Technological advances in electronics continue at a rapid rate—as evidenced by recent market surveys indicating that most new models of electronic instruments have been on the market for less than two years.

Current EDM instruments use infrared light, laser light, or microwaves. The once-popular microwave systems use a receiver/transmitter at both ends of the measured line, whereas infrared and laser systems utilize a transmitter at one end of the measured line and a reflecting prism at the other end. EDM instruments come in long range (10–20 km), medium range (3–10 km) and short range (0.5–3 km). Some laser EDM instruments measure relatively shorter distances (100–1,200 m) without a reflecting prism, reflecting the light directly off the feature (e.g., building wall) being measured. Microwave instruments are often used in hydrographic surveys and have a usual upper measuring range of 50 km. Although microwave systems can be used in poorer weather conditions (fog, rain, etc.) than can infrared and laser systems, the uncertainties caused by varying humidity conditions over the length of the measured line may result in lower accuracy expectations. Hydrographic EDM measuring and positioning techniques have largely been supplanted, in a few short years, by global positioning system (GPS) techniques (see Chapter 8).

EDM devices can be mounted on the standards or the telescope of most theodolites; they can also be mounted directly in a tribrach. When used with an electronic theodolite, the combined instruments can provide both the horizontal and the vertical position of one point relative to another. The slope distance provided by an add-on EDM device can be reduced to its horizontal and vertical equivalents by utilizing the slope angle provided by the theodolite. In total station instruments, this reduction is accomplished automatically.

2.14   Electronic Angle Measurement

The electronic digital theodolite, first introduced in the late 1960s by Carl Zeiss Inc., set the stage for modern field data collection and processing. (See Figure 4.7, which shows electronic angle measurement using a rotary encoder and photoelectric converters.) When the electronic theodolite is used with a built-in EDM device, (e.g., Trimble 3300 series; see Figure 2.23) or an add-on and interfaced EDM device (e.g., Wild T-1000; see Figure 2.24),

FIGURE 2.23 Trimble total stations. The Trimble 3303 and 3305 total stations incorporate DR technology (no prism required) and include the choice of the integrated Zeiss Elta control unit, detachable Geodimeter control unit or handheld TSCe data collector, and a wide range of software options. (Courtesy of Trimble Geomatics & Engineering Division, Dayton, Ohio)

FIGURE 2.24 Wild T-1000 electronic theodolite, shown with DI 1000 Distomat EDM and the GRE 3 data collector. (Courtesy of Leica Geosystems)

the surveyor has a very powerful instrument. Add to that instrument an on-board microprocessor that automatically monitors the instrument's operating status and manages built-in surveying programs, and a data collector (built-in or interfaced) that stores and processes measurements and attribute data, and you have what is known as a total station. (Total stations are described in detail in Chapter 5.)

2.15 Principles of Electronic Distance Measurement (EDM)

Figure 2.25 shows a wave of wavelength λ. The wave is traveling along the x axis with a velocity of 299,792.458 km/s (in vacuum). The frequency of the wave is the time taken for one complete wavelength:

$$\lambda = \frac{c}{f} \qquad (2.4)$$

where λ = wavelength in meters
c = velocity in km/s
f = frequency in hertz (one cycle per second)

Figure 2.26 shows the modulated electromagnetic wave leaving the EDM device and being reflected (light waves) or retransmitted (microwaves) back to the EDM device. You can see that the double distance ($2L$) is equal to a whole number of wavelengths ($n\lambda$), plus the partial wavelength (ϕ) occurring at the EDM instrument:

$$L = \frac{(n\lambda + \phi)}{2} \text{ meters} \qquad (2.5)$$

The partial wavelength (ϕ) is determined in the instrument by noting the phase delay required to match the transmitted and reflected or retransmitted waves precisely. Some

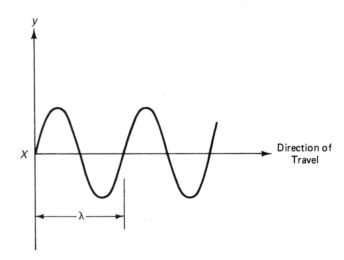

FIGURE 2.25 Light wave.

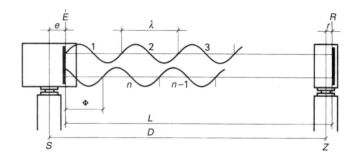

S   Station
Z   Target
E   Reference plane within the distance
    meter for phase comparison
    between transmitted and received
    wave
R   Reference plane for the reflection
    of the wave transmitted by the
    distance meter
a   Addition constant
e   Distance meter component of
    addition constant
r   Reflector component of addition
    constant
λ   Modulation wavelength
Φ   Fraction to be measured of a whole
    wavelength of modulation $(\triangle \lambda)$

The addition constant *a* applies to
a measuring equipment consisting of a
distance meter and reflector. The
components *e* and *r* are only auxiliary
quantities.

FIGURE 2.26 Principles of EDM measurement. (Courtesy of Leica Geosystems)

instruments can count the number of full wavelengths ($n\lambda$) or, instead, the instrument can send out a series (three or four) of modulated waves at different frequencies. (The frequency is typically reduced each time by a factor of 10 and, of course, the wavelength is increased each time also by a factor of 10.) By substituting the resulting values of λ and ϕ into Equation 2.5, the value of n can be found. The instruments are designed to carry out this procedure in a matter of seconds and then to display the value of L in digital form.

The velocity of light (including infrared) through the atmosphere can be affected by (1) temperature, (2) atmospheric pressure, and (3) water vapor content. In practice, the corrections for temperature and pressure can be performed manually by consulting nomographs similar to that shown in Figure 2.27, or the corrections can be performed automatically on some EDM devices by the on-board processor/calculator after the values for temperature and pressure have been entered.

For short distances using light-wave EDM, atmospheric corrections have a relatively small significance. For long distances using light-wave instruments and especially microwave instruments, atmospheric corrections can become quite important. Table 2.2 shows the comparative effects of the atmosphere on both light waves and microwaves.

At this point, it is also worth noting that several studies of general EDM use show that more than 90 percent of all distance determinations involve distances of 1,000 m or less and that more than 95 percent of all layout measurements involve distances of 400 m or less. The values in Table 2.2 seem to indicate that, for the type of measurements normally encountered in the construction and civil engineering fields, instrumental errors and centering errors hold much more significance than do atmosphere-related errors.

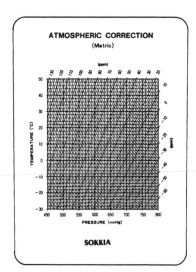

FIGURE 2.27 Atmospheric correction graph. (Courtesy of SOKKIA Corp.)

Table 2.2 ATMOSPHERIC ERRORS

Parameter	Error	Error (Parts per Million)	
		Light Wave	Microwave
t, temperature	+1°C	−1.0	−1.25
p, pressure	+1 mm Hg	+0.4	+0.4
e, partial water-vapor pressure	1 mm Hg	−0.05	+ 7 at 20°C
			+17 at 45°C

2.16 EDM Characteristics

Following are the characteristics of recent models of add-on EDM devices. Generally, the more expensive instruments have longer distance ranges and higher precision.

Distance range: 800 m to 1 km (single prism with average atmospheric conditions).

Short-range EDM can be extended to 1,300 m using 3 prisms.

Long-range EDM can be extended to 15 km using 11 prisms (Leica Geosystems).

Accuracy range:

± (5 mm + 5 ppm) for short-range EDM.

± (2 mm + 1 ppm) for long-range EDM.

Measuring time: 1.5 seconds for short-range EDM to 3.5 seconds for long-range EDM. (Both accuracy and time are considerably reduced for tracking mode measurements.)

Slope reduction: manual or automatic on some models.

Average of repeated measurements: available on some models.

Battery capability: 1,400 to 4,200 measurements, depending on the size and condition of the battery and the temperature.

FIGURE 2.28 Various target and reflector systems in tribrach mounts. (Courtesy of Topcon Positioning Systems, Inc.)

Temperature range: −20°C to +50°C. (*Note*: In the northern United States and Canada, temperatures can easily drop below −20°C during the winter months.)

Nonprism measurements: available on some models; distances from 100 to 1,200 m (see Section 2.21).

2.17 Prisms

Prisms are used with electro-optical EDM (light, laser, and infrared) to reflect the transmitted signals (see Figure 2.28). A single reflector is a cube corner prism that has the characteristic of reflecting light rays back precisely in the same direction as they are received. This retro-direct capability means that the prism can be somewhat misaligned with respect to the EDM instrument and still be effective. Cutting the corners off a solid glass cube forms a cube corner prism; the quality of the prism is determined by the flatness of the surfaces and the perpendicularity of the 90° surfaces. Prisms can be tribrach-mounted on a tripod, centered by optical or laser plummet, or attached to a prism pole held vertical on a point with the aid of a bull's-eye level. Prisms must be tribrach-mounted, however, if a higher level of accuracy is required.

In control surveys, tribrach-mounted prisms can be detached from their tribrachs and then interchanged with a theodolite/total station similarly mounted at the other end of the line being measured. This interchangeability of prism and theodolite/total station (also targets) speeds up the work because the tribrach mounted on the tripod is centered and leveled only once. Equipment that can be interchanged and mounted on tribrachs already set up is known as **forced-centering** equipment.

Prisms mounted on adjustable-length prism poles are portable and thus are particularly suited for stakeout and topographic surveys. Figure 2.29 shows the prism pole being steadied with the aid of an additional target pole. The height of the prism is normally set to equal the height of the instrument (hi). It is particularly important that prisms mounted on

FIGURE 2.29 Steadying the EDM reflector with the aid of a second target pole.

poles or tribrachs be permitted to tilt up and down so that they can be perpendicular to the signals that are being sent from much higher or lower positions.

2.18 EDM Instrument Accuracies

EDM accuracies are stated in terms of a constant instrumental error and a measuring error proportional to the distance being measured. Typically, accuracy is claimed as ±[5 mm + 5 parts per million (ppm)] or ±(0.02 ft + 5 ppm). The 5 mm (0.02 ft) is the instrument error that is independent of the length of the measurement, whereas the 5 ppm (5 mm/km) denotes the distance-related error.

Most instruments now on the market have claimed accuracies in the range of ±(2 mm + 1 ppm) to ±(10 mm + 10 ppm). The proportional part error (ppm) is insignificant for most work, and the constant part of the error assumes less significance as the distances being measured lengthen. At 10 m, an error of 5 mm represents an accuracy of 1:2,000; at 100 m, an error of 5 mm represents an accuracy of 1/20,000. At 1,000 m, the same instrumental error represents an accuracy of 1/200,000.

When dealing with accuracy, note that both the EDM and the prism reflectors must be corrected for their off-center characteristics. The measurement being recorded goes from the electrical center of the EDM device to the back of the prism (allowing for refraction through glass) and then back to the electrical center of the EDM device. The EDM device manufacturer at the factory compensates for the difference between the electrical center of the EDM device and the plumb line through the tribrach center. The prism constant (–30 to –40 mm) is eliminated either by the EDM device manufacturer at the factory or in the field.

The EDM/prism constant value can be field-checked in the following manner. A long line (>1 km) is laid out with end stations and an intermediate station (see Figure 2.30). The overall distance AC is measured, along with partial lengths AB and BC. The constant value will be present in all measurements; therefore,

$$AC - AB - BC = \text{instrument/prism constant} \qquad (2.6)$$

the constant can also be determined by measuring a known baseline, if one can be conveniently accessed.

2.19 EDM Operation

Figures 2.31 and 2.32 shows first-generation short- to medium-range EDM devices. Although these devices have been largely replaced by total stations (Chapter 5), they are included here as an introduction to the more advanced instruments. The operation of all EDM devices involves the following basic steps: (1) set up, (2) aim, (3) measure, and (4) record.

FIGURE 2.30 Method of determining the instrument-reflector constant.

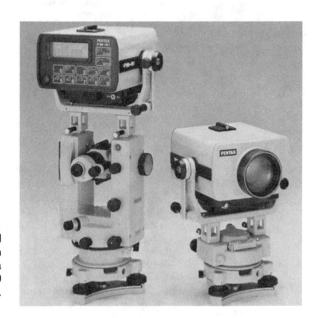

FIGURE 2.31 Pentax PM 81 EDM mounted on a 6-second Pentax theodolite and also shown as tribrach-mounted. EDM has a triple-prism range of 2 km (6,600 ft) with SE = ±(5 mm + 5 ppm). (Courtesy of Pentax Corp.)

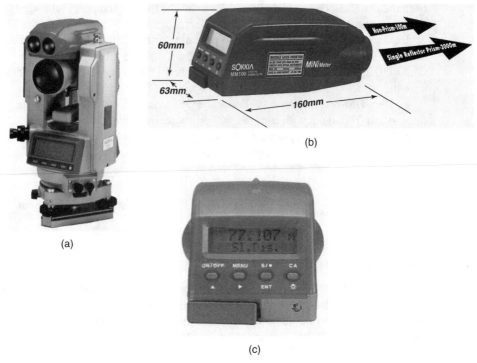

(a)

(b)

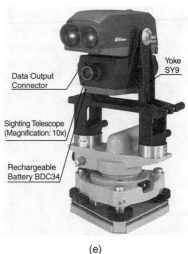

(c)

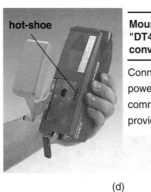

hot-shoe

**Mounting on the
"DT4F" is by
convenient hot-shoe**

Connection via which
power supply and data
communication are
provided.

(d)

Yoke
SY9

Data Output
Connector

Sighting Telescope
(Magnification: 10x)

Rechargeable
Battery BDC34

(e)

FIGURE 2.32 Sokkia MiNi Meter MM100 laser add-on EDM. (a) EDM shown mounted on the telescope of Sokkia DT4F electronic 5" theodolite. (b) MiNi Meter dimensions. (c) Display screen showing menu button, which provides access to programs permitting input for atmospheric corrections, height measurements, horizontal and vertical distances, self-diagnostics, etc. (d) Illustration showing "hot shoe" electronics connection and counterweight. (e) MiNi Meter shown mounted directly into a tribrach—with attached sighting telescope, data output connector, and battery. (Courtesy of SOKKIA Corp.)

2.19.1 Set Up

Tribrach-mounted EDM devices are simply inserted into the tribrach (forced centering)—after the tribrach has been set over the point—by means of the optical or laser plummet. Telescope or theodolite yoke-mounted EDM devices are simply attached to the theodolite either before or after the theodolite has been set over the point. Prisms are set over the remote station point either by inserting the prism into a tribrach (forced centering) that has already been set up or by holding the prism vertically over the point on a prism pole. The EDM is turned on, and a quick check is made (of the battery, display, and the like) to ensure that it is in good working order. The height of the instrument (telescope axis) and the height of the prism (center) are measured and recorded; the prism, when mounted on an adjustable prism pole, is usually set to the height of the theodolite or the EDM. It is important to check the accuracy of the optical/laser plummet regularly. If the prism is mounted on an adjustable pole, it is important that the prism can be tilted up and down for short sights and that the circular bubble be checked regularly for proper adjustment.

2.19.2 Aim

The EDM device is aimed at the prism by using either the built-in sighting devices on the EDM or the theodolite telescope. Telescope or yoke-mounted EDM devices have the optical line of sight a bit lower than the electronic signal. Some EDM devices have a sighting telescope mounted on top of the instrument; in those cases, the optical line of sight is a bit higher than the electronic signal.

Most instrument manufacturers provide prism/target assemblies, which permit fast optical sightings for both optical and electronic alignment (see Figure 2.28). That is, when the telescope cross hair is on target, the electronic signal is maximized at the center of the prism. The surveyor can (if necessary) set the electronic signal precisely on the prism center by adjusting the appropriate horizontal and vertical slow-motion screws until a maximum signal intensity is indicated on the display (this display is not available on all EDM devices). Some older EDMs have an attenuator that must be adjusted for varying distances—the signal strength is reduced for short distances so that the receiving electronics are not overloaded. Newer EDMs have automatic signal attenuation.

2.19.3 Measure

The slope distance measurement is accomplished by simply pressing the measure button and waiting a few seconds for the result to appear in the display. The displays are either LCD (most) or LED. The measurement is shown to two decimals of a foot or three decimals of a meter; a foot/meter switch readily switches from one system to the other. If no measurement appears in the display, the surveyor should check the switch position, battery status, attenuation, and cross-hair location (sometimes the stadia hair is mistakenly centered instead of the main cross hair).

Add-on EDM devices with built-in calculators or microprocessors can be used to compute horizontal and vertical distances; coordinates; and atmospheric, curvature, and prism constant corrections. The required input data (vertical angle, ppm, prism constant, etc.) are entered via the keyboard. Most EDM devices have a tracking mode (which is very useful in layout surveys). The tracking mode permits continuous distance updates as the prism is moved ever closer to its final layout position.

Handheld radios and cell phones are useful for all EDM work because the long distances put a halt to normal voice communications. In layout work, clear communications are essential if the points are to be located properly. All microwave EDMs permit voice communication, which is carried right on the measuring signal.

2.19.4 Record

The measured data can be recorded conventionally in field-note format, or they can be entered manually into an electronic data collector. The distance data must be accompanied by all relevant atmospheric and instrumental correction factors. (Total station instruments, which have automatic data acquisition capabilities, are discussed in Chapter 5.)

2.20 Geometry of Electronic Distance Measurements

Figure 2.33 illustrates the use of EDM when the optical target and the reflecting prism are at the same height (see Figure 2.28 for a single prism assembly). The EDM device is used to measure the slope distance (S), and the accompanying theodolite is used to measure the slope angle (α). The heights of the EDM device and theodolite (hi) are measured with a steel tape or by a graduated tripod-centering rod; the height of the reflector/target is measured in a similar fashion. As noted earlier, adjustable-length prism poles permit the surveyor to set the height of the prism (HR) equal to the height of the instrument (hi), thus simplifying the computations. From Figure 2.33, if the elevation of station A is known and the elevation of station B is required:

$$\text{Elev. STA. B} = \text{elev. STA. A} + \text{hi} \pm V - \text{HR} \tag{2.7}$$

When the EDM device is mounted on the theodolite and the target is located beneath the prism, the geometric relationship can be as shown in Figure 2.34.

The additional problem encountered in the situation depicted in Figure 2.34 is the computation of the correction to the vertical angle ($\Delta\alpha$) that occurs when Δhi and ΔHR are different. The precise size of the vertical angle is important because it is used in conjunction with the measured slope distance to compute the horizontal and vertical distances.

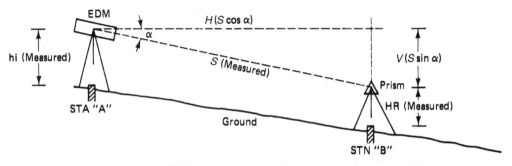

FIGURE 2.33 Geometry of an EDM calculation, general case.

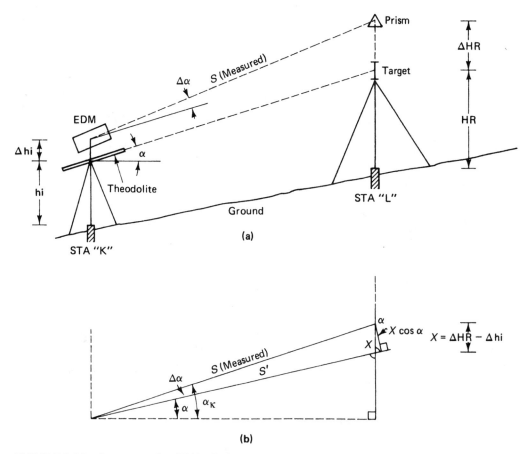

FIGURE 2.34 Geometry of an EDM calculation, usual case.

In Figure 2.34 the difference between ΔHR and Δhi is X (i.e., $\Delta HR - \Delta hi = X$). The small triangle formed by extending S' [see Figure 2.34(b)] has the hypotenuse equal to X and an angle of α. This permits computation of the side ($X \cos \alpha$), which can be used together with S to determine $\Delta \alpha$:

$$\frac{(X \cos \alpha)}{S} = \sin \Delta \alpha$$

■ **EXAMPLE 2.6**

An EDM slope distance AB is determined to be 561.276 m. The EDM device is 1.820 m above station A, and the prism is 1.986 m above station B. The EDM device is mounted on a theodolite whose optical center is 1.720 m above the station. The theodolite was used to measure the vertical angle (+6°21′38″) to a target on the prism pole; the target is 1.810 m above station B. Compute both the horizontal distance AB and the elevation of station B if the elevation of station A = 186.275 m.

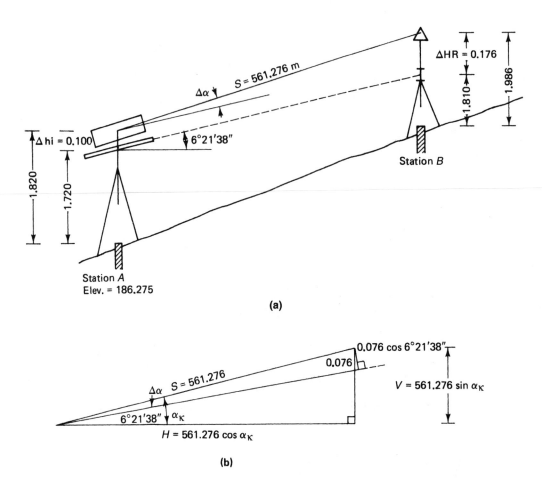

FIGURE 2.35 Illustration for Example 2.6. (a) Schematic of EDM measurement. (b) Deduced geometry for an EDM measurement.

Solution

The given data are shown in Figure 2.35(a), and the resultant figure is shown in Figure 2.35(b). The X value introduced in Figure 2.34(b) is determined, in this case, as follows:

$$X = (1.986 - 1.810) - (1.820 - 1.720)$$
$$= 0.176 - 0.100$$
$$= 0.076 \text{ m}$$
$$\sin \Delta\alpha = \frac{(0.076 \cos 6°21'38'')}{561.276}$$

$$\Delta\alpha = 28''$$
$$\alpha_\kappa = 6°22'06''$$
$$H = 561.276 \cos 6°22'06''$$
$$= 557.813 \text{ m}$$

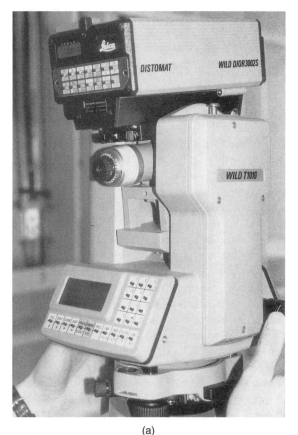

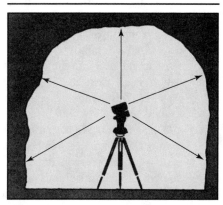

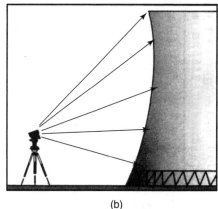

(a)   (b)

FIGURE 2.36 Distance measurement without reflectors. (a) Wild T1010 electronic theodolite, together with an interfaced DIOR 3002S prismless EDM (angle accuracy is 3 seconds). (b) Illustrations of two possible uses for this technique. Upper: tunnel cross sections. Lower: profiling a difficult-access feature. (Courtesy of Leica Geosystems)

$$\notin \alpha = 28-$$
$$\alpha_\kappa = 6°22\ 06-$$
$$H = 561.276 \cos 6°22\ 06-$$
$$= 557.813 \text{ m}$$

If H had been computed using the field vertical angle of $6°21'38''$, the result would have been 557.821 m, not a significant difference in this example.

$$\text{Elev. B} = \text{elev. A} + 1.820 + 561.276 \sin 6°22\ 06-- 1.986$$
$$= 186.275 + 1.820 + 62.257 - 1.986$$
$$= 248.366 \text{ m}$$

If V had been computed using $6°21'38''$, the result would have been 62.181 m instead of 62.257 m, a more significant discrepancy.

can be used for reflectorless measurement. When the reflecting surface is uneven or not at right angles to the measuring beam, varying amounts of the light pulses are not returned to the instrument. When using longer-range pulsed-laser techniques, as many as 20,000 pulses per second are employed to ensure that sufficient data is received. The pulsing of this somewhat strong laser emission still results in a safety designation of Class I—the safest designation. The use of Class II, III, and IV lasers require eye protection (see also Section 5.10). Another consideration is that various surfaces have different reflective properties; for example, a bright white surface at a right angle to the measuring beam may reflect almost 100 percent of the light, but most natural surfaces reflect light at a rate of only 18 percent.

EDM instruments can be used conventionally with the reflecting prisms for distances up to 4 km; when used without prisms, the range drops to 100 to 1,200 m, depending on the equipment, light conditions (cloudy days and night darkness provide better measuring distances), angle of reflection from the surface, and reflective properties of the measuring surface. With prisms, the available accuracy is about $\pm(1–3 \text{ mm} + 1 \text{ ppm})$; without prisms, the available accuracy ranges from $\pm(3 \text{ mm} + 3 \text{ ppm})$ up to about $\pm 10 \text{ mm}$. Targets with light-colored and flat surfaces perpendicular to the measuring beam (e.g., building walls) provide the best ranges and accuracies. Because of the different reflective capabilities of various surfaces, comparison of different survey equipment must be made based on standard surfaces; the Kodak Gray Card has been chosen as such a standard. This card is gray on one side and white on the other. The white side reflects 90 percent of white light and gray side reflects 18 percent of white light. An EDM range to a Kodak Gray Card (18 percent reflective) is considered a good indicator of typical surveying capabilities.

EDM instruments also provide quick results (0.8 seconds in rapid mode and 0.3 seconds in tracking mode), which means that applications for moving targets are possible. Applications are expected for near-shore hydrographic surveying and in many areas of heavy construction. This technique is already being used, with an interfaced data collector, to measure cross sections automatically in mining applications—with plotted cross sections and excavated volumes automatically generated by digital plotter and computer. Other applications include cross-sectioning above-ground excavated works and material stockpiles; measuring to dangerous or difficult access points, for example, bridge components, cooling towers, and dam faces; and automatically measuring liquid surfaces, for example, municipal water reservoirs and catchment ponds. These new techniques may have some potential in industrial surveying, where production line rates require this type of monitoring.

EDM instruments can be used with an attached visible laser, which helps to identify positively the feature being measured; that is, the visible laser beam is set on the desired feature so that the surveyor can be sure that the correct surface, not some feature just beside or behind it, is being measured. Some instruments require the surveyor to measure to possibly conflicting objects (for example, utility wires that may cross the line of sight closer to the instrument). In this situation, the surveyor first measures to the possibly conflicting wires and then directs the total station software only to show measurements beyond that range. Because the measurement is so fast, care must be taken not to measure mistakenly to an object that may temporarily intersect the measuring signal, for example, a truck or other traffic. See Section 5.10 for additional information on reflectorless distance measurement. Table 2.3 (the result of Trimble's research) shows ranges for reflectorless measurements using both time-of-flight (TOF) or pulsed-laser (DR300+) technology and phase-shift techniques (DR standard).

Table 2.3 TRIMBLE DR RANGE TO VARIOUS TARGET SURFACES

Surface	Dr300+	Dr Standard
Kodak Gray Card, 90%	>800 m (2,625 ft)	>240 m (787 ft)
Kodak Gray Card, 18%	>300 m (984 ft)	>120 m (393 ft)
Concrete	>400 m (1,312 ft)	>100 m (328 ft)
Wood	>400 m (1,312 ft)	>200 m (656 ft)
Light rock	>300 m (984 ft)	>150 m (492 ft)
Dark rock	>200 m (656 ft)	>80 m (262 ft)

Source: From *Direct Reflex EDM Technology for the Surveyor and Civil Engineer*, R. Hoglund and P. Large, 2005, Trimble Survey, Westminister, Colorado, USA.

Problems

2.1. Describe the relative advantages and disadvantages of measuring with steel tapes and with EDMs.

2.2. Give two examples of possible uses for each of the following field measurement techniques: (a) pacing, (b) odometer, (c) EDM, (d) subtense bar, (e) fiberglass tape, (f) steel tape.

2.3. The following measurements were taken from an early topographic survey where the measurements were made with a Gunter's chain. Convert each of these measurements to both feet and meters: (a) 23 chains, 92 links; (b) 48.71 chains; (c) 12 chains, 22 links; (d) 8 chains, 31 links.

2.4. You must determine the ground clearance of an overhead electrical cable. Surveyor B is positioned directly under the cable (surveyor B's position can be checked by his sighting past the string of a plumb bob, held in his outstretched hand, to the cable); surveyor A sets her clinometer to 45° and then backs away from surveyor B until the overhead electrical cable is on the cross hair of the leveled clinometer. At this point, surveyors A and B determine the distance between them to be 61.5 ft. Surveyor A then sets the clinometer to 0° and sights surveyor B; this horizontal line of sight cuts surveyor B at a distance of 4.3 ft above the ground. Determine the ground clearance of the electrical cable.

2.5. A 100-ft cut steel tape was used to check the layout of a building wall. The rear surveyor held 42 ft while the head surveyor cut 0.33 ft. What is the distance measured?

2.6. A 100-ft add steel tape was used to measure a partial baseline distance. The rear surveyor held 47 ft, while the head surveyor held 0.62 ft. What is the distance measured?

2.7. The slope distance between two points is 22.745 m, and the slope angle is 1°42′. Compute the horizontal distance.

2.8. The slope distance between two points is 73.79 ft, and the difference in elevation between them is 8.45 ft. Compute the horizontal distance.

2.9. A distance of 162.102 m was measured along a 2 percent slope. Compute the horizontal distance.

2.10. To verify the constant of a particular prism, a straight line *EFG* is laid out. The EDM instrument is first set up at *E*, with the following measurements recorded:

$$EG = 566.711 \text{ m} \qquad EF = 238.778 \text{ m}$$

2.9. A distance of 162.102 m was measured along a 2 percent slope. Compute the horizontal distance.

2.10. To verify the constant of a particular prism, a straight line *EFG* is laid out. The EDM instrument is first set up at *E*, with the following measurements recorded:

$$EG = 566.711 \text{ m} \qquad EF = 238.778 \text{ m}$$

The EDM instrument is then set up at *F*, where distance *FG* is recorded as 327.963 m. Determine the prism constant.

2.11. The EDM slope distance between two points is 3,964.37 ft, and the vertical angle is +2°45′30″ (the vertical angles were read at both ends of the line and then averaged). If the elevation of the instrument station is 285.69 ft and the heights of the instrument, EDM, target, and reflector are all equal to 5.08 ft, compute the elevation of the target station and the horizontal distance to that station.

2.12. A line *AB* is measured at both ends as follows:

Instrument at *A*, slope distance = 1458.777 m, zenith angle = 91°26′50″

Instrument at *B*, slope distance = 1458.757 m, zenith angle = 88°33′22″

The heights of the instrument, reflector, and target are equal for each observation.
(a) Compute the horizontal distance *AB*.
(b) If the elevation at *A* is 211.841 m, what is the elevation at *B*?

2.13. A coaxial EDM instrument at station K (elevation = 232.47 ft) is used to sight stations L, M, and N, with the heights of the instrument, target, and reflector equal for each sighting. The results are as follows:

Instrument at STA. L, zenith angle = 86°30′, EDM distance = 3,000.00 ft

Instrument at STA. M, zenith angle = 91°30′, EDM distance = 3,000.00 ft

Instrument at STA. N, zenith angle = 90°00′, EDM distance = 2,000.00 ft

Compute the elevations of L, M, and N

2.14. Refer to Figure 2.35. A top-mounted EDM instrument is set up at STA. *A* (elevation = 110.222 m). Using the following values, compute the horizontal distance from *A* to *B* and the elevation of *B*. The optical center of the theodolite is 1.601 m (hi) above station, and a vertical angle of +4°18′30″ is measured to the target, which is 1.915 (HR) above station. The EDM instrument center is 0.100 m (Δhi) above the theodolite, and the reflecting prism is 0.150 m (ΔHR) above the target. The slope distance is measured to be 387.603 m.

2.15. Refer to Figure 2.35. A top-mounted EDM instrument is set up at STA. *A* (elevation = 531.49 ft). Using the following values, compute the horizontal distance from *A* to *B* and the elevation of *B*. The optical center of the theodolite is 5.21 ft (hi) above station, and a vertical angle of +3°14′30″ is measured to the target, which is 5.78 ft (HR) above station. The EDM instrument center is 0.31 ft (Δhi) above the theodolite, and the reflecting prism is 0.39 ft (ΔHR) above the target. The slope distance is recorded as 536.88 ft.

Chapter 3

Leveling

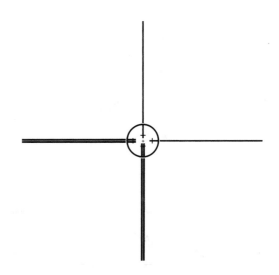

3.1 General Background

Leveling is the procedure used when one is determining differences in elevation between points that are some distance from each other. An **elevation** is a vertical distance above or below a reference datum. In surveying, the reference datum that is universally employed is **mean sea level (MSL)**. In North America, nineteen years of observations at tidal stations in twenty-six locations on the Atlantic, Pacific, and Gulf of Mexico shorelines were reduced and adjusted to provide the National Geodetic Vertical Datum (NGVD) of 1929. That datum has been further refined to reflect gravimetric and other anomalies in the 1988 general control re-adjustment called the North American Vertical Datum (NAVD88). Because of the inconsistencies found in widespread determinations of mean sea level, the NAVD88 datum has been tied to one representative tidal gage bench mark known as **Father Point**, which is located on the south shore of the mouth of the St. Lawrence River at Rimouski, Quebec. Although the NAVD does not precisely agree with mean sea level at some specific points on the earth's surface, the term *mean sea level* is often used to describe the datum. MSL is assigned a vertical value (elevation) of 0.000 ft or 0.000 m. See Figure 3.1. (Specifications for vertical control in the United States and Canada are shown in Tables 3.1 and 3.2.)

> A **vertical line** is a line from the surface of the earth to the earth's center. It is also referred to as a plumb line or a line of gravity.
>
> A **level line** is a line in a level surface. A level surface is a curved surface parallel to the mean surface of the earth. A level surface is best visualized as being the surface of a large body of water at rest.
>
> A **horizontal line** is a straight line perpendicular to a vertical line.

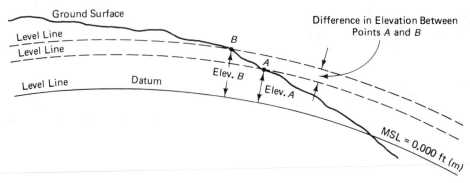

FIGURE 3.1   Leveling concepts.

Table 3.1   NATIONAL OCEAN SURVEY, U.S. COAST AND GEODETIC SURVEYS: CLASSIFICATION, STANDARDS OF ACCURACY, AND GENERAL SPECIFICATIONS FOR VERTICAL CONTROL

Classification	First Order Class I, Class II	Second Order Class I	Second Order Class II	Third Order
Principal uses Minimum standards; higher accuracies may be used for special purposes	Basic framework of the National Network and of metropolitan area control	Secondary control of the National Network and of metropolitan area control	Control densification, usually adjusted to the National Network	Miscellaneous local control; may not be adjusted to the National Network
	Extensive engineering projects	Large engineering projects	Local engineering projects	Small engineering projects
	Regional crustal movement investigations	Local crustal movement and subsidence investigations	Topographic mapping	Small-scale topographic mapping
	Determining geopotential values	Support for lower-order control	Studies of rapid subsidence	Drainage studies and gradient establishment in mountainous areas
			Support for local surveys	
Maximum closures[*] Section: forward and backward	$3 \text{ mm } \sqrt{K}$ (*Class I*) $4 \text{ mm } \sqrt{K}$ (*Class II*)	$6 \text{ mm } \sqrt{K}$	$8 \text{ mm } \sqrt{K}$	$12 \text{ mm } \sqrt{K}$
Loop or line	$4 \text{ mm } \sqrt{K}$ (*Class I*) $5 \text{ mm } \sqrt{K}$ (*Class II*)	$6 \text{ mm } \sqrt{K}$	$8 \text{ mm } \sqrt{K}$	$12 \text{ mm } \sqrt{K}$

[*]Check between forward and backward runnings where K is the distance in kilometers.

3.2   Theory of Differential Leveling

Differential leveling is used to determine differences in elevation between points (that are some distance from each other) by using a surveyors' level and a graduated measuring rod. The surveyors' level consists of a cross hair–equipped telescope and an attached spirit level, both of which are mounted on a sturdy tripod. The surveyor can sight through the telescope to a rod graduated in feet or meters and determine a measurement reading at the point where the cross hair intersects the rod.

In Figure 3.2, if the rod reading at $A = 6.27$ ft and the rod reading at $B = 4.69$ ft, the difference in elevation between A and B is $6.27 - 4.69 = 1.58$ ft. If the elevation of A is

Table 3.2 CLASSIFICATION, STANDARDS OF ACCURACY, AND GENERAL SPECIFICATIONS FOR VERTICAL CONTROL—CANADA

Classification	Special Order	First Order	Second Order (First-Order Procedures Recommended)	Third Order	Fourth Order
Allowable discrepancy between forward and backward levelings	± 3 mm $\sqrt{K}$ ± 0.012 ft $\sqrt{m}$	± 4 mm $\sqrt{K}$ ± 0.017 ft $\sqrt{m}$	± 8 mm $\sqrt{K}$ ± 0.035 ft $\sqrt{m}$	± 24 mm $\sqrt{K}$ ± 0.10 ft $\sqrt{m}$	± 120 mm $\sqrt{K}$ ± 0.5 ft $\sqrt{m}$
Instruments: Self-leveling high-speed compensator Level vial	Equivalent to 10″/2mm level vial 10″/2 mm	Equivalent to 10″/2 mm level vial 10″/2 mm	Equivalent to 20″/2 mm level vial 20″/2 mm	Equivalent to sensitivity below 40″ to 50″/2 mm	Equivalent to sensitivity below 40″ to 50″/2 mm
Telescopic magnification	40×	40×	40×		

Source: Adapted from "Specifications and Recommendations for Control Surveys and Survey Markers" (Surveys and Mapping Branch. Department of Energy. Mines and Resources, Ottawa, Canada 1973).

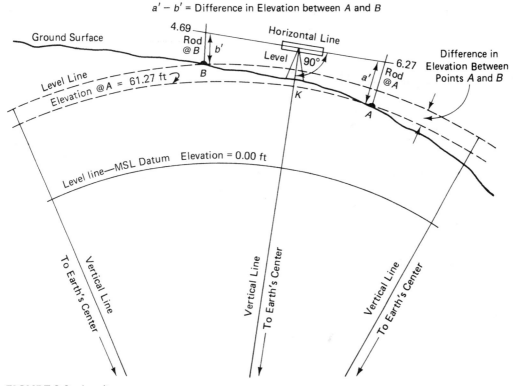

FIGURE 3.2 Leveling process.

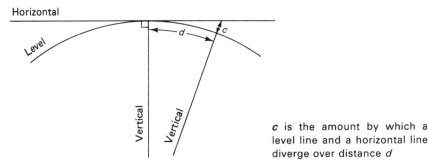

Horizontal

Level

Vertical

Vertical

d

c

c is the amount by which a
level line and a horizontal line
diverge over distance d

FIGURE 3.3 Relationship between a horizontal line and a level line.

61.27 ft (above MSL), then the elevation of B is 61.27 + 1.58 = 62.85 ft. That is, 61.27
(elev. A) + 6.27 (rod reading at A) − 4.69 (rod reading at B) = 62.85 (elev. B).

In Figure 3.3, you can see a potential problem. Whereas elevations are referenced to
level lines (surfaces), the line of sight through the telescope of a surveyors' level is, in fact,
almost a horizontal line. All rod readings taken with a surveyors' level will contain an error
c over a distance d. I have greatly exaggerated the curvature of the level lines shown in
Figures 3.1 through 3.3 for illustrative purposes. In fact, the divergence between a level
line and a horizontal line is quite small. For example, over a distance of 1,000 ft, the
divergence is 0.024 ft, and for a distance of 300 ft, the divergence is only 0.002 ft
(0.0008 m in 100 m).

3.3 Curvature and Refraction

The previous section introduced the concept of curvature error—that is, the divergence between a level line and a horizontal line over a specified distance. When considering the divergence between level and horizontal lines, you must also account for the fact that all
sight lines are refracted downward by the earth's atmosphere. Although the magnitude of
the refraction error depends on atmospheric conditions, it is generally considered to be
about one-seventh of the curvature error. You can see in Figure 3.4 that the refraction error of AB compensates for part of the curvature error of AE, resulting in a net error due to
curvature and refraction (c + r) of BE. From Figure 3.4, the curvature error can be computed as follows:

$$(R + c)^2 = R^2 + KA^2$$
$$R^2 + 2Rc + c^2 = R^2 + KA^2$$
$$c(2R + c) = KA^2$$
$$c = \frac{KA^2}{2R + c^*} \approx \frac{KA^2}{2R} \qquad (3.1)$$

---

*In the term $(2R + c)$, c is so small when compared to R that it can be safely ignored.

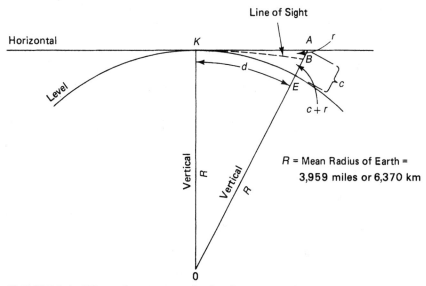

Line of Sight

Horizontal

K

A

r

B

d

E

c

Level

c + r

Vertical

R

Vertical

R

R = Mean Radius of Earth = 3,959 miles or 6,370 km

0

FIGURE 3.4   Effects of curvature and refraction.

Consider $R = 6370$ km:

$$c = \frac{KA^2}{2 \times 6370} = 0.0000785 \, KA^2 \, \text{km} = 0.0785 \, KA^2 \, \text{m}$$

Refraction (r) is affected by atmospheric pressure, temperature, and geographic location but, as noted earlier, it is usually considered to be about one-seventh of the curvature error (c). If $r = 0.14c$:

$$c + r = 0.0675 \, K^2$$

where $K = KA$ (Figure 3.4) and is the length of sight in kilometers. The combined effects of curvature and refraction ($c + r$) can be determined from the following formulas:

$(c + r)_m = 0.0675K^2$ $(c + r)_m$ in meters, K in kilometers   (3.2)

$(c + r)_{ft} = 0.0574K^2$ $(c + r)_{ft}$ in feet, K in miles   (3.3)

$(c + r)_{ft} = 0.0206M^2$ $(c + r)_{ft}$ in feet, M in thousands of feet   (3.4)

■   **EXAMPLE 3.1**

Calculate the error due to curvature and refraction for the following distances:

(a) 2,500 ft

(b) 400 ft

Table 3.3 SELECTED VALUES FOR $(C + R)$ AND DISTANCE

Distance (m)	30	60	100	120	150	300	1 km
$(c + r)_m$	0.0001	0.0002	0.0007	0.001	0.002	0.006	0.068
Distance (ft)	100	200	300	400	500	1,000	1 mi
$(c + r)_{ft}$	0.000	0.001	0.002	0.003	0.005	0.021	0.574

a.) 2,500 ft

b.) 400 ft

(c) 2.7 miles

(d) 1.8 km

Solution

(a) $(c + r) = 0.0206 \times 2.5^2 = 0.13$ ft

(b) $(c + r) = 0.0206 \times 0.4^2 = 0.003$ ft

(c) $(c + r) = 0.574 \times 2.7^2 = 4.18$ ft

(d) $(c + r) = 0.0675 \times 1.8^2 = 0.219$ m

You can see from the values in Table 3.3 that $(c + r)$ errors are relatively insignificant for differential leveling. Even for precise leveling, where distances of rod readings are seldom in excess of 200 ft (60 m), it would seem that this error is of only marginal importance. We will see in Section 3.11 that the field technique of balancing the distances of rod readings (from the instrument) effectively cancels out this type of error.

3.4   Types of Surveying Levels

3.4.1   Automatic Level

The automatic level [see Figure 3.5 and Figure 1.5(a)] employs a gravity-referenced prism or mirror compensator to orient the line of sight (line of collimation) automatically. The instrument is leveled quickly when a circular spirit level is used; when the bubble has been centered (or nearly so), the compensator takes over and maintains a horizontal line of sight, even if the telescope is slightly tilted. Automatic levels are extremely popular in present-day surveying operations and are available from most survey instrument manufacturers. They are quick to set up, easy to use, and can be obtained for use at almost any required precision.

A word of caution: all automatic levels employ a compensator referenced by gravity. This operation normally entails freely moving prisms or mirrors, some of which are hung by fine wires. If a wire or fulcrum breaks, the compensator will become inoperative, and all subsequent rod readings will be incorrect.

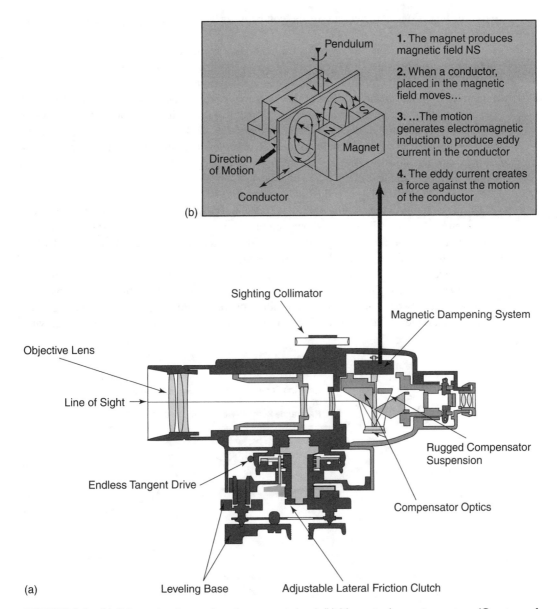

(b)

Pendulum

1. The magnet produces magnetic field NS

2. When a conductor, placed in the magnetic field moves...

3. ...The motion generates electromagnetic induction to produce eddy current in the conductor

4. The eddy current creates a force against the motion of the conductor

Magnet

Direction of Motion

Conductor

Sighting Collimator

Magnetic Dampening System

Objective Lens

Line of Sight →

Rugged Compensator Suspension

Endless Tangent Drive

Compensator Optics

(a)

Leveling Base

Adjustable Lateral Friction Clutch

FIGURE 3.5 (a) Schematic of an engineer's automatic level. (b) Magnetic dampening system. (Courtesy of SOKKIA Corp.)

The operating status of the compensator can be verified by tapping the end of the telescope or by slightly turning one of the leveling screws (one manufacturer provides a push button), causing the telescopic line of sight to veer from horizontal. If the compensator is operative, the cross hair will appear to deflect momentarily before returning to its original rod reading. Constant checking of the compensator will avoid costly mistakes caused by broken components.

Most levels (most surveying instruments) now come equipped with a three-screw leveling base. Whereas the support for a four-screw leveling base (see Appendix G) is the center bearing, the three-screw instruments are supported entirely by the foot screws themselves. Thus, the adjustment of the foot screws of a three-screw instrument effectively raises or lowers the height of the instrument line of sight. Adjustment of the foot screws of a four-screw instrument does not affect the height of the instrument line of sight because the center bearing supports the instrument. The surveyor should thus be aware that adjustments made to a three-screw level in the midst of a setup operation will effectively change the elevation of the line of sight and could cause significant errors on very precise surveys (e.g., benchmark leveling or industrial surveying).

The bubble in the circular spirit level is centered by adjusting one or more of the three independent screws. Figure 3.6 shows the positions for a telescope equipped with a tube level when using three leveling foot screws. If you keep this configuration in mind when leveling the circular spirit level, the movement of the bubble can be predicted easily.

Levels used to establish or densify vertical control are designed and manufactured to give precise results. The magnifying power, setting accuracy of the tubular level or compensator, quality of optics, and so on, are all improved to provide precise rod readings. The least count on leveling rods is 0.01 ft or 0.001 m. Precise levels are usually equipped with

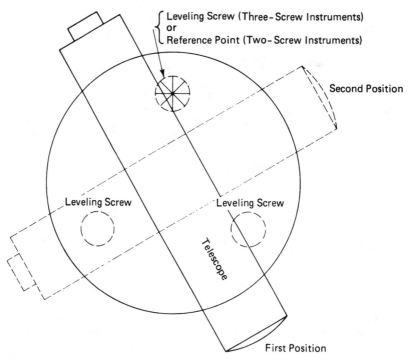

FIGURE 3.6 Telescope positions when leveling a three-screw instrument.

optical micrometers so that readings can be determined one or two places beyond the rod's least count.

Many automatic levels utilize a concave base so that, when the level is attached to its domed-head tripod top, it can be roughly leveled by sliding the instrument over the tripod top. This rough leveling can be accomplished in a few seconds. If the bull's-eye bubble is nearly centered by this maneuver and the compensator is activated, the leveling screws may not be needed at all to level the instrument.

3.4.2 Digital Level

Figure 3.7 shows a digital level and bar-code rod. This level features digital electronic image-processor that uses a charge-coupled device (CCD) for determining heights and distances with the automatic recording of data for later transfer to a computer. The digital level is an automatic level (pendulum compensator) capable of normal optical leveling with a rod graduated in feet or meters. When used in electronic mode and with the rod face graduated in bar code, this level can, with the press of a button, capture and process the image of the bar-code rod for distances in the range of about 1.8 m to about 100 m. This simple one-button operation initiates storage of the rod reading and distance measurement and the computation of the point elevation. The processed image of the rod reading is compared, using the instrument's central processing unit (CPU), with the image of the whole rod, which is

FIGURE 3.7 DNA03 digital level and bar-code rod. (Courtesy of Leica Geosystems)

stored permanently in the level's memory module, thus determining height and distance values. Data can be stored in internal on-board memory or on easily transferable PC cards and then transferred to a computer via an RS232 connection or by transferring the PC card from the digital level to the computer. Most levels can operate effectively with only 30 cm of the rod visible. Work can proceed in the dark by illuminating the rod face with a small flashlight. The rod shown in Figure 3.7 is 4.05 m long (others are 5 m long); it is graduated in bar code on one side and in either feet or meters on the other side.

After the instrument has been leveled, the image of the bar code must be focused properly by the operator. Next, the operator presses the measure button to begin the image processing, which takes about 3 s. Although the heights and distances are automatically determined and recorded (if desired), the horizontal angles are read and recorded manually.

Preprogrammed functions include level loop survey, two-peg test, self-test, set time, and set units. Coding can be the same as that used with total stations (see Chapter 5), which means that the processed leveling data can be transferred directly to the computer database. The bar code can be read in the range of 1.8 to 100 m away from the instrument; the rod can be read optically as close as 0.5 m. If the rod is not plumb or is held upside down, an error message flashes on the screen. Other error messages include "Instrument not level," "Low battery," and "Memory almost full." Rechargeable batteries are said to last for more than 2,000 measurements. Distance accuracy is in the range of 1/1,000, whereas leveling accuracy (for the more precise digital levels) is stated as having a standard deviation for a 1-km double run of 0.3 mm to 1.0 mm for electronic measurement and 2.0 mm for optical measurement. Manufacturers report that use of the digital level increases productivity by about 50 percent, with the added bonus of the elimination of field-book entry mistakes. The more precise digital levels can be used in first- and second-order leveling, whereas the less precise digital levels can be used in third-order leveling and construction surveys.

3.4.3   Tilting Level

The mostly obsolete tilting level is roughly leveled by observing the bubble in the circular spirit level. Just before each important rod reading is to be taken, and while the telescope is pointing at the rod, the telescope is leveled precisely by manipulating a tilting screw, which effectively raises or lowers the eyepiece end of the telescope. The level is equipped with a tube level, which is leveled precisely by operating the tilting screw. The bubble is viewed through a separate eyepiece or, as you can see in Figure 3.8, through the telescope. The image of the bubble is longitudinally split in two and viewed with the aid of prisms. One-half of each end of the bubble can be observed [see Figure 3.8(b)] and, after adjustment, the two half-ends are brought to coincidence and appear as a continuous curve. When coincidence has been achieved, the telescope has been leveled precisely. It has been estimated (by Leica Geosystems) that the accuracy of centering a level bubble with reference to the open tubular scale graduated at intervals of 2 mm is about one-fifth of a division, or 0.4 mm. With coincidence-type (split-bubble) levels, however, this accuracy increases to about one-fortieth of a division, or 0.05 mm. These levels are useful when a relatively high degree of precision is required; if tilting levels are used on ordinary work (e.g., earthwork), however, the time (expense) involved in setting the split bubble to coincidence for each rod reading can scarcely be justified. Automatic levels, digital levels, and total stations have mostly replaced these levels.

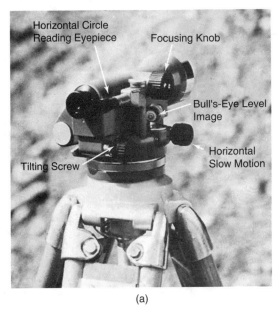

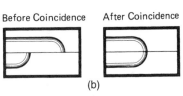

Before Coincidence   After Coincidence

(a)   (b)

FIGURE 3.8 (a) Leica engineering tilting level, GK 23 C. (b) Split bubble, before and after coincidence. (Courtesy of Leica Geosystems)

The level tube, used in many levels, is a sealed glass tube filled mostly with alcohol or a similar substance with a low freezing point. The upper (and sometimes lower) surface has been ground to form a circular arc. The degree of precision possessed by a surveyors' level is partly a function of the sensitivity of the level tube; the sensitivity of the level tube is directly related to the radius of curvature of the upper surface of the level tube. The larger the radius of curvature, the more sensitive the level tube. Sensitivity is usually expressed as the central angle subtending one division (usually 2 mm) marked on the surface of the level tube. The sensitivity of many engineers' levels is 30″; that is, for a 2-mm arc, the central angle is 30″ (R = 13.75 m, or 45 ft). See Figure 3.9. The sensitivity of levels used for precise work is usually 10″; that is, R = 41.25 m, or 135 ft.

Precise tilting levels, in addition to possessing more sensitive level tubes, have improved optics, including a greater magnification power. The relationship between the quality of the optical system and the sensitivity of the level tube can be simply stated: for any observable movement of the bubble in the level tube, there should be an observable movement of the cross hair on the leveling rod.

3.5 Leveling Rods

Leveling rods are manufactured from wood, metal, or fiberglass and are graduated in feet or meters. The foot rod can be read directly to 0.01 ft, whereas the metric rod can usually be read directly only to 0.01 m, with millimeters being estimated. Metric rod readings are normally booked to the closest 1/3 cm or 1/2 cm (i.e., 0.000, 0.003, 0.005, 0.007, and

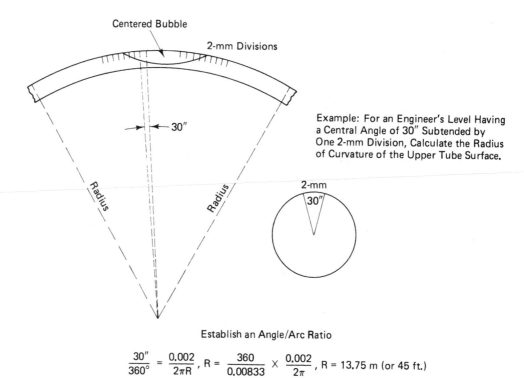

Example: For an Engineer's Level Having a Central Angle of 30″ Subtended by One 2-mm Division, Calculate the Radius of Curvature of the Upper Tube Surface.

Establish an Angle/Arc Ratio

$$\frac{30''}{360°} = \frac{0.002}{2\pi R}, \quad R = \frac{360}{0.00833} \times \frac{0.002}{2\pi}, \quad R = 13.75 \text{ m (or 45 ft.)}$$

FIGURE 3.9 Level tube showing the relationship between the central angle per division and the radius of curvature of the level tube upper surface. (Courtesy of Leica Geosystems)

0.010). More precise values can be obtained by using an optical micrometer. One-piece rods are used for more precise work. The most precise work requires the face of the rod to be an invar strip held in place under temperature-compensating spring tension (invar is a metal that has a very low rate of thermal expansion).

Most leveling surveys utilize two- or three-piece rods graduated in either feet or meters. The sole of the rod is a metal plate that withstands the constant wear and tear of leveling. The zero mark is at the bottom of the metal plate. The rods are graduated in a wide variety of patterns, all of which respond readily to logical analysis. The surveyor should study an unfamiliar rod at close quarters prior to leveling to ensure that he or she understands the graduations thoroughly. (See Figure 3.10 for a variety of graduation markings.)

The rectangular sectioned rods are of either the folding (hinged) or the sliding variety. Newer fiberglass rods have oval or circular cross sections and fit telescopically together for heights of 3, 5, and 7 m, from a stored height of 1.5 m (equivalent foot rods are also available). Benchmark leveling utilizes folding (one-piece) rods or invar rods, both of which have built-in handles and rod levels. All other rods can be plumbed by using a rod level (see Figure 3.11), or by waving the rod (see Figure 3.12).

Level Rod Faces

Rod faces pictured are approximately one-half actual size.

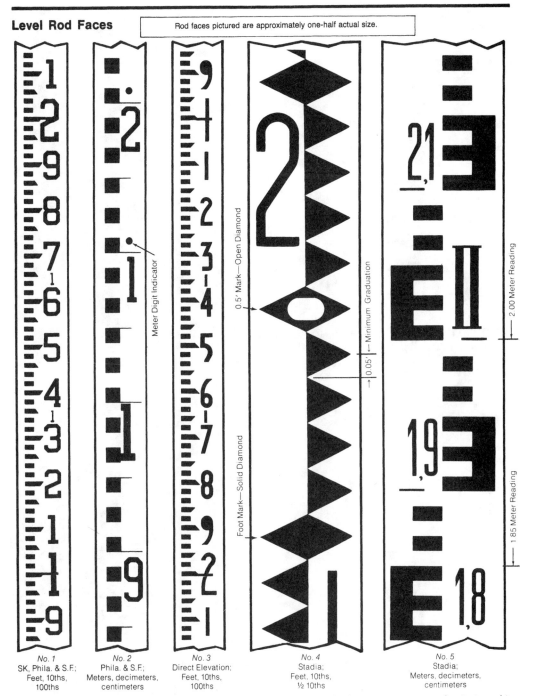

No. 1	No. 2	No. 3	No. 4	No. 5
SK, Phila. & S.F.; Feet, 10ths, 100ths	Phila. & S.F.; Meters, decimeters, centimeters	Direct Elevation; Feet, 10ths, 100ths	Stadia; Feet, 10ths, ½ 10ths	Stadia; Meters, decimeters, centimeters

FIGURE 3.10 Traditional rectangular cross-section leveling rods showing various graduation markings. (Courtesy of SOKKIA Corp.)

Sec. 3.5   Leveling Rods

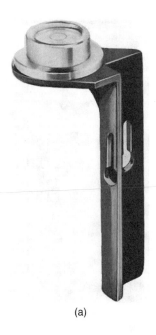

(a)

FIGURE 3.11 (a) Circular rod level. (Courtesy of Keuffel & Esser Co.) (b) Circular level, shown with a leveling rod.

(b)

FIGURE 3.12 (a) Waving the rod.

(continued)

(a)

Chap. 3   Leveling

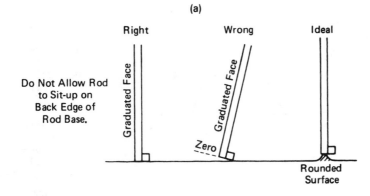

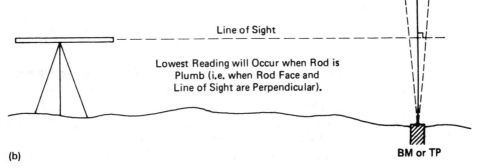

FIGURE 3.12 *(continued)*   (b) Waving the rod slightly to and from the instrument allows the instrument operator to take the most precise (lowest) reading.

3.6   Definitions for Differential Leveling

A **benchmark (BM)** is a permanent point of known elevation. Benchmarks are established by using precise leveling techniques and instrumentation; more recently, precise GPS techniques have been used. Benchmarks are bronze disks or plugs set usually into vertical wall faces. It is important that the benchmark be placed in a structure with substantial footings (at least below minimum frost depth penetration) that will resist vertical movement due to settling or upheaval. Benchmark elevations and locations are published by federal, state or provincial, and municipal agencies and are available to surveyors for a nominal fee. See Figure 3.13 for an illustration of the use of a BM.

A **temporary benchmark (TBM)** is a semipermanent point of known elevation. TBMs can be flange bolts on fire hydrants, nails in the roots of trees, top corners of concrete culvert head walls, and so on. The elevations of TBMs are not normally published, but they are available in the field notes of various surveying agencies.

A **turning point (TP)** is a point temporarily used to transfer an elevation (see Figure 3.14).

A **backsight (BS)** is a rod reading taken on a point of known elevation to establish the elevation of the instrument line of sight. See Figures 3.13 and 3.14.

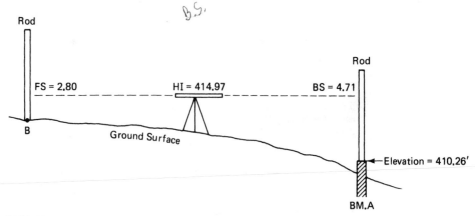

FIGURE 3.13 Leveling procedure: one setup.

The **height of instrument (HI)** is the elevation of the line of sight through the level (i.e., elevation of BM + BS = HI). See Figures 3.13 and 3.14.

A **foresight (FS)** is a rod reading taken on a turning point, benchmark, or temporary benchmark to determine its elevation [i.e., HI − FS = elevation of TP (or BM or TBM)]. See Figures 3.13 and 3.14.

An **intermediate foresight (IS)** is a rod reading taken at any other point where the elevation is required (see Figures 3.18 and 3.19), i.e., HI − IS = elevation.

Most engineering leveling projects are initiated to determine the elevations of intermediate points (as in profiles, cross sections, etc.). The addition of backsights to elevations to obtain heights of instrument and the subtraction of foresights from heights of instrument to obtain new elevations are known as note reductions.

3.7   Techniques of Leveling

In leveling (as opposed to theodolite work), the instrument can usually be set up in a relatively convenient location. If the level has to be set up on a hard surface, such as asphalt or concrete, the tripod legs are spread out to provide a more stable setup. When the level is to be set up on a soft surface (e.g., turf), the tripod is first set up so that the tripod top is nearly horizontal, and then the tripod legs are pushed firmly into the earth. The tripod legs are snugly tightened to the tripod top so that a leg, when raised, just falls back under the force of its own weight. Undertightening can cause an unsteady setup, just as overtightening can cause an unsteady setup due to torque strain. On hills, it is customary to place one leg uphill and two legs downhill; the instrument operator stands facing uphill while setting up the instrument. The tripod legs can be adjustable or straight-leg. The straight-leg tripod is recommended for leveling because it contributes to a more stable setup. After the tripod has been set roughly level, with the feet firmly pushed into the ground, the instrument can be leveled.

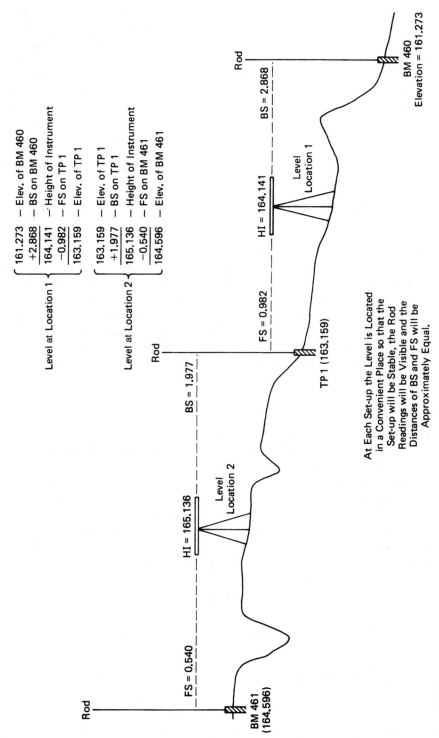

Level at Location 1 {
161.273 — Elev. of BM 460
+2.868 — BS on BM 460
164.141 — Height of Instrument
−0.982 — FS on TP 1
163.159 — Elev. of TP 1

Level at Location 2 {
163.159 — Elev. of TP 1
+1.977 — BS on TP 1
165.136 — Height of Instrument
−0.540 — FS on BM 461
164.596 — Elev. of BM 461

At Each Set-up the Level is Located in a Convenient Place so that the Set-up will be Stable, the Rod Readings will be Visible and the Distances of BS and FS will be Approximately Equal.

Rod

BS = 2.868

BM 460
Elevation = 161.273

Level Location 1

HI = 164.141

FS = 0.982

TP 1 (163.159)

Rod

BS = 1.977

Level Location 2

HI = 165.136

FS = 0.540

Rod

BM 461
(164.596)

FIGURE 3.14 Leveling procedure: more than one setup.

Four-screw instruments (see Appendix G) attach to the tripod via a threaded base and are leveled by rotating the telescope until it is directly over a pair of opposite leveling screws, and then by proceeding as described in Section G.2 and Figure G.7. Three-screw instruments are attached to the tripod via a threaded bolt that projects up from the tripod top into the leveling base of the instrument. The three-screw instrument, which usually has a circular bull's-eye bubble level, is leveled as described in Section 3.4.1 and Figure 3.6. Unlike the four-screw instrument, the three-screw instrument can be manipulated by moving the screws one at a time, although experienced surveyors usually manipulate at least two at a time. After the bubble has been centered, the instrument can be revolved to check that the circular bubble remains centered.

Once the level has been set up, preparation for the rod readings can take place. The eyepiece lenses are focused by turning the eyepiece focusing ring until the cross hairs are as black and as sharp as possible (it helps to have the telescope pointing to a light-colored background for this operation). Next, the rod is brought into focus by turning the telescope focusing screw until the rod graduations are as clear as possible. If both these focusing operations have been carried out correctly, the cross hairs appear to be superimposed on the leveling rod. If either focusing operation (cross hair or rod focus) has not been carried out properly, the cross hair appears to move slightly up and down as the observer's head moves slightly up and down. The apparent movement of the cross hair can cause incorrect readings. If one or both focus adjustments have been made improperly, the resultant reading error is known as **parallax**.

Figure 3.13 shows one complete leveling cycle; actual leveling operations are no more complicated than that shown. Leveling operations typically involve numerous repetitions of this leveling cycle. Some operations require that additional (intermediate) rod readings be taken at each instrument setup.

$$\text{Existing elevation} + \text{BS} = \text{HI} \qquad (3.5)$$

$$\text{HI} - \text{FS} = \text{new elevation} \qquad (3.6)$$

These two equations describe the differential leveling process completely.

When leveling between benchmarks or turning points, the level is set approximately midway between the BS and FS locations to eliminate (or minimize) errors due to curvature and refraction (Section 3.3) and errors due to a faulty line of sight (Section 3.11). To ensure that the rod is plumb, either a rod level (Figure 3.11) is used or the surveyor gently "waves" the rod toward and away from the instrument. The correct rod reading is the lowest reading observed. The surveyor must ensure that the rod does not sit up on the back edge of the base and thus effectively raises the zero mark on the rod off the BM (or TP). The instrument operator is sure that the rod has been waved properly if the readings decrease to a minimum value and then increase in value (see Figure 3.12).

To determine the elevation of B in Figure 3.13:

Elevation BM.A	410.26
Backsight rod reading at BM.A	+ 4.71 BS
Height (elevation) of instrument line of sight	414.97 HI
Foresight rod reading at TBM B	− 2.80 FS
Elevation TBM B	412.17

After the rod reading of 4.71 is taken at BM.A, the elevation of the line of sight of the instrument is known to be 414.97 (410.26 + 4.71). The elevation of TBM B can be determined by holding the rod at B, sighting the rod with the instrument, and reading the rod (2.80 ft). The elevation of TBM B is therefore $414.97 - 2.80 = 412.17$ ft. In addition to determining the elevation of TBM B, the elevations of any other points lower than the line of sight and visible from the level can be determined in a similar manner.

The situation depicted in Figure 3.14 shows the technique used when the point whose elevation is to be determined (BM 461) is too far from the point of known elevation (BM 460) for a single-setup solution. The elevation of an intermediate point (TP 1) is determined, thus allowing the surveyor to move the level to a location where BM 461 can be "seen." Real-life situations may require numerous setups and the determination of the elevation of many turning points before getting close enough to determine the elevation of the desired point. When the elevation of the desired point has been determined, the surveyor must then either continue the survey to a point (BM) of known elevation or return (loop) the survey to the point of commencement. The survey must be closed onto a point of known elevation so that the accuracy and acceptability of the survey can be determined. If the closure is not within allowable limits, the survey must be repeated.

The arithmetic can be verified by performing the **arithmetic check** (page check). All BSs are added, and all FSs are subtracted. When the sum of BS is added to the original elevation and then the sum of FS is subtracted from that total, the remainder should be the same as the final elevation calculated (see Figure 3.15).

BENCHMARK VERIFICATION

BM460 to BM461

Job SOMERVILLE-ROD, LEVEL #09 KANEN- λ 19°C, CLOUDY

Date MAY 11, 2006 Page 62

STA	B.S. (+)	H.I.		F.S. (−)	ELEV.	DESCRIPTION
BM 460	2.868	164.141			161.273	BRONZE PLATE SET IN ––– ETC.
T.P. 1	1.977	165.136		0.982	163.159	NAIL IN ROOT OF MAPLE ––– ETC
BM 461				0.540	164.596	BRONZE PLATE SET IN ––– ETC.
Σ	4.845			1.522		
ARITHMETIC CHECK:						
	161.273 + 4.845 − 1.522			=	164.596	

FIGURE 3.15 Leveling field notes and arithmetic check (data from Figure 3.14).

$$\text{Starting elevation} + \Sigma BS - \Sigma FS = \text{ending elevation}$$

In the 1800s and early 1900s, leveling procedures like the one described here were used to survey locations for railroads that traversed North America between the Atlantic Ocean and the Pacific Ocean.

3.8   Benchmark Leveling (Vertical Control Surveys)

Benchmark leveling is the type of leveling employed when a system of benchmarks is to be established or when an existing system of benchmarks is to be extended or densified. Perhaps a benchmark is required in a new location, or perhaps an existing benchmark has been destroyed and a suitable replacement is required. Benchmark leveling is typified by the relatively high level of precision specified, both for the instrumentation and for the technique itself.

The specifications shown in Tables 3.1 and 3.2 cover the techniques of precise leveling. Precise levels with coincidence tubular bubbles of a sensitivity of 10 seconds per 2-min division (or the equivalent for automatic levels) and with parallel-plate micrometers are used almost exclusively for this type of work. Invar-faced rods, together with a base plate (see Figure 3.16), rod level, and supports, are used in pairs to minimize the time required for successive readings.

Tripods for this type of work are longer than usual, thus enabling the surveyor to keep the line of sight farther above the ground to minimize interference and errors due to refraction. Ideally the work is performed on a cloudy, windless day, although work can proceed on a sunny day if the instrument is protected from the sun and its possible differential thermal effects.

At the national level, benchmarks are established by federal agencies utilizing first-order methods and first-order instruments. The same high requirements are also specified for state and provincial grids, but as work proceeds from the whole to the part (i.e., down to municipal or regional grids), the rigid specifications are relaxed somewhat. For most engineering works, benchmarks are established (from the municipal or regional grid) at third-order specifications. Benchmarks established to control isolated construction projects may be at even lower orders of accuracy.

It is customary in benchmark leveling at all orders of accuracy to verify first that the elevation of the starting benchmark is correct. Two-way leveling to the closest adjacent benchmark satisfies this verification requirement. This check is particularly important when the survey loops back to close on the starting benchmark and no other verification is planned.

With the ongoing improvements in GPS techniques and in geoid modeling (Geoid99), first- and second-order vertical control surveys are now performed routinely using GPS surveys (see Chapter 8). The GPS approach to vertical control surveys is reported to be much more productive (up to ten times) than conventional precise spirit leveling.[*]

In addition to GPS, another alternative to spirit leveling for the establishment of benchmarks is the use of total stations. Instrument manufacturers now refer angle

---

[*]*Professional Surveyor Magazine Newsletter*, December 12, 2002.

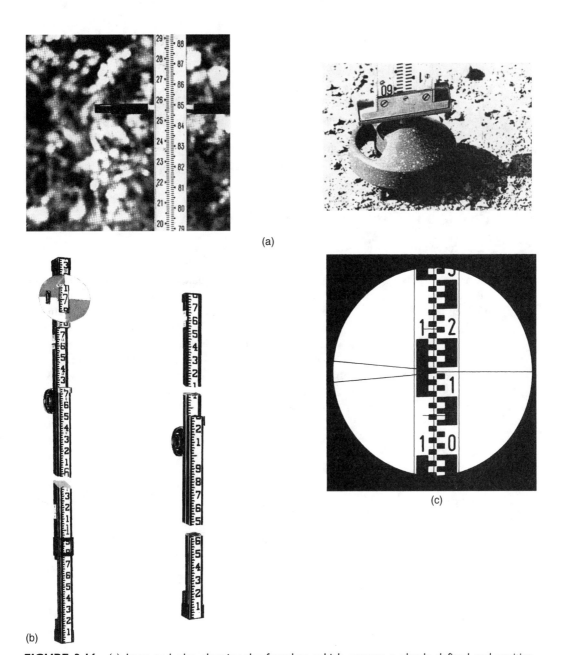

(a)

(b)

(c)

FIGURE 3.16 (a) Invar rod, also showing the footplate, which ensures a clearly defined rod position. (Courtesy of Leica Geosystems) (b) Philadelphia rod (*left*), which can be read directly by the instrument operator or the rod holder after the target has been set. (Courtesy of Keuffel & Esser Co.) Frisco rod (*right*) with two or three sliding sections having replaceable metal scales. (c) Metric rod; horizontal cross-hair reading on 1.143 m. (Courtesy of Leica Geosystems)

accuracies of theodolites and total stations to the Deutsches Institut für Normung (DIN); and to the International Organization of Standards (ISO): see the web sites for both organizations: www.din.de and www.iso.ch, respectively. Both organizations use the same procedures to determine the standard deviation (*So*) of a horizontal direction. Angle accuracies are covered in DIN 18723, part 3, and in ISO 12857, part 2. These standards indicate that a surveyor can obtain results consistent with stated instrument accuracies by using two sets of observations (direct and reverse) on each prism. In addition, the surveyor should try to set the total station midway between the two prisms being measured and apply relevant atmospheric corrections and prism constant corrections.

3.9   Profile and Cross-Section Leveling

In engineering surveying, we often consider a route (road, sewer pipeline, channel, etc.) from three distinct perspectives. The **plan** view of route location is the same as if we were in an aircraft looking straight down. The **profile** of the route is a side view or elevation (see Figure 3.17) in which the longitudinal surfaces are highlighted (e.g., road, top and bottom of pipelines). The **cross section** shows the end view of a section at a station (0 + 60 in Figure 3.17) and is at right angles to the centerline. These three views (plan, profile, and cross section) together completely define the route in *X*, *Y*, and *Z* coordinates.

Profile levels are taken along a path that holds interest for the designer. In roadwork, preliminary surveys often profile the proposed location of the centerline (₵) (see Figure 3.18). The proposed ₵ is staked out at an even interval (50 to 100 ft, or 20 to 30 m). The level is set up in a convenient location so that the benchmark—and as many intermediate points as possible—can be sighted. Rod readings are taken at the even station locations and at any other point where the ground surface has a significant change in slope. When the rod is moved to

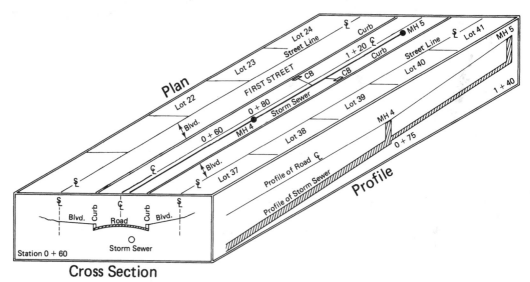

FIGURE 3.17   Relationship of plan, profile, and cross-section views.

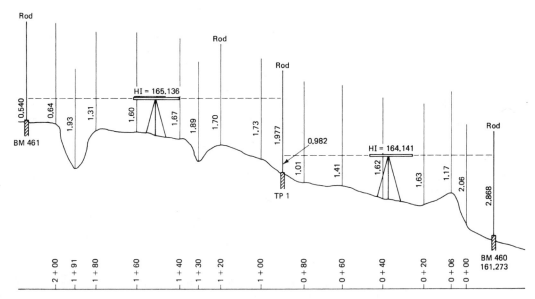

FIGURE 3.18 Example of profile leveling; see Figure 3.19 for survey notes.

a new location and it cannot be seen from the instrument, a turning point is called for so that the instrument can be moved ahead and the remaining stations leveled.

The turning point can be taken on a wood stake, the corner of a concrete step or concrete head wall, a lug on the flange of a hydrant, and so on. The turning point should be a solid, well-defined point that can be precisely described and—it is hoped—found again at a future date. In the case of leveling across fields, it usually is not possible to find turning point features of any permanence. In that case, stakes are driven in and then abandoned when the survey is finished. In the field notes shown in Figure 3.19, the survey was closed at an acceptable point: to BM 461. Had there been no benchmark at the end of the profile survey, the surveyor would have looped back and closed to the initial benchmark. The note on Figure 3.18 at the BM 461 elevation shows that the correct elevation of BM 461 is 164.591 m, which means that there was a survey error of 0.005 m. over a distance of 200 m. Using the standards shown in Tables 3.1 and 3.2, this result qualifies for consideration at the third level of accuracy in both the United States and Canada.

The intermediate sights (IS) in Figure 3.19 are shown in a separate column, and the elevations at the intermediate sights show the same number of decimals as are shown in the rod readings. Rod readings on turf, ground, and the like, are usually taken to the closest 0.1 ft or 0.01 m. Rod readings taken on concrete, steel, asphalt, and so on, are usually taken to the closest 0.01 ft or 0.003 m. It is a waste of time and money to read the rod more precisely than conditions warrant. (Refer to Chapter 7 for details on plotting the profile.)

When the final route of the facility has been selected, additional surveying is required. Once again, the ℄ is staked (if necessary), and cross sections are taken at all even stations. In roadwork, rod readings are taken along a line perpendicular to the ℄ at each even station. The rod is held at each significant change in surface slope and at the limits of the job. In

PROFILE OF PROPOSED

ROAD 0 + 00 to 2 + 00

Job 21 °C — SUNNY   LEVEL L—14

Date   AUG 3 2006   Page   72

STA.	B.S.	H.I.	I.S	F.S.	ELEV.	DESCRIPTION
BM 460	2.868	164.141			161.273	BRONZE PLATE SET IN --- ETC.
0 + 00			2.06		162.08	℄
0 + 06			1.17		162.97	℄ —TOP OF BERM
0 + 20			1.63		162.51	℄
0 + 40			1.62		162.52	℄
0 + 60			1.41		162.73	℄
0 + 80			1.01		163.13	℄
T.P. 1	1.977	165.136		0.982	163.159	NAIL IN ROOT OF MAPLE --- ETC.
1 + 00			1.73		163.41	℄
1 + 20			1.70		163.44	℄
1 + 30			1.89		163.25	℄ BOTTOM OF GULLY
1 + 40			1.67		163.47	℄
1 + 60			1.60		163.54	℄
1 + 80			1.31		163.83	℄
1 + 91			1.93		163.21	℄ BOTTOM OF GULLY
2 + 00			0.64		164.50	℄
BM 461				0.540	164.596	BRONZE PLATE SET IN --- ETC.
						164.591 — PUBLISHED ELEV.
	Σ=4.845			Σ=1.522		E = 164.596
ARITHMETIC CHECK: 161.273 + 4.845 −1.522						164.591
					= 164.596	0.005
						ALLOWABLE ERROR (3ᴿᴰ ORDER)
						= 12 mm √K, = .012√.2 = .0054 m
						ABOVE ERROR (.005) SATISFIES 3ᴿᴰ ORDER.

FIGURE 3.19 Profile field notes.

uniformly sloping land areas, the only rod readings required at each cross-sectioned station are often at the ℄ and the two street lines (for roadwork). Chapter 7 shows how the cross sections are plotted and then utilized to compute volumes of cut and fill.

Figure 3.20 illustrates the rod positions required to define the ground surface suitably at 2 + 60, at right angles to the ℄. Figure 3.21 shows the field notes for this typical cross section—in a format favored by municipal surveyors. Figure 3.22 shows the same field data entered in a cross-section note form favored by many highway agencies. Note that the HI (353.213) has been rounded to two decimals (353.21) in Figures 3.21 and 3.22 to facilitate reduction of the two-decimal rod readings. The rounded value is placed in brackets to distinguish it from the correct HI, from which the next FS will be subtracted (or from which any three-decimal intermediate rod readings are subtracted).

Road and highway construction often requires the location of granular (e.g., sand, gravel) deposits for use in the highway roadbed. These **borrow pits** (gravel pits) are surveyed to determine the volume of material "borrowed" and transported to the site. Before any excavation takes place, one or more reference baselines are established, and two benchmarks (at a minimum) are located in convenient locations. The reference lines are located in secure locations where neither the stripping and stockpiling of topsoil nor the actual

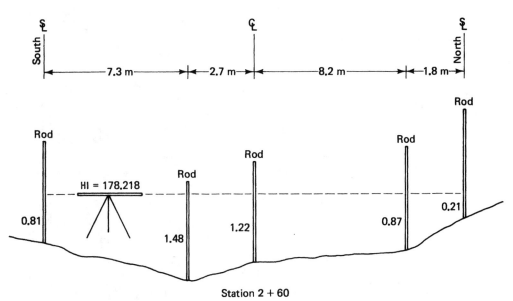

FIGURE 3.20 Cross-section surveying.

SMITH—NOTES JONES—ROD
BROWN—𝜋 TYLER—TAPE

CROSS—SECTIONS FOR PROPOSED

LOCATION OF DUNCAN ROAD

Job 14 °C CLOUDY LEVEL #6

Date NOV 24 2006 Page 23

STA.	B.S.	H.I.	I.S.	F.S.	ELEV.	DESCRIPTION
BM #28	2.011	178.218			176.207	BRONZE PLATE SET IN SOUTH WALL 0.50m
		(178.22)				ABOVE GROUND, CIVIC #2242, 23ᴿᴰ AVE.
2 + 60						
10m LT			0.81		177.41	S. ₵
2.7m LT			1.48		176.74	BOTTOM OF SWALE
₵			1.22		177.00	₵
8.2m RT			0.87		177.35	CHANGE IN SLOPE
10 m RT			0.21		178.01	N. ₵
2 + 80						
10 m LT			1.02		177.20	S. ₵
3.8m LT			1.64		176.58	BOTTOM OF SWALE
₵			1.51		176.71	₵
7.8m RT			1.10		177.12	CHANGE IN SLOPE
10 m RT			0.43		177.79	N. ₵

FIGURE 3.21 Cross-section notes (municipal format).

CROSS-SECTIONS FOR PROPOSED LOCATION OF DUNCAN HIGHWAY

Job 14 °C — SMITH–NOTES, BROWN–⊼, JONES–ROD, TYLER–TAPE — CLOUDY — LEVEL 6

Date NOV 24 2006 — Page 23

STA.	B.S.	H.I.	I.S	F.S.	ELEV.	
	2.011	178.218			176.207	BRONZE PLATE SET IN SOUTH WALL
		(178.22)				0.50 ABOVE GROUND, CIVIC #2242, 23RD AVE.
						LEFT ℄ RIGHT
						10.0 2.7 8.2 10.0
2+60						0.81 1.48 1.22 0.87 0.21
						177.41 176.74 177.00 177.35 178.01
						10.0 3.8 7.8 10.0
2+80						1.02 1.64 1.51 1.10 0.43
						177.20 176.58 176.71 177.12 177.79

FIGURE 3.22 Cross-section notes (highway format).

excavation of the granular material will endanger the stake (see Figure 3.23). Cross sections are taken over (and beyond) the area of proposed excavation. These original cross sections will be used as data against which interim and final survey measurements will be compared to determine total excavation. The volumes calculated from the cross sections (see Chapter 7) are often converted to tons (tonnes) for payment purposes. In many locations, weigh scales are located at the pit to aid in converting the volumes and as a check on the calculated quantities.

The original cross sections are taken over a grid at 50-ft (20-m) intervals. As the excavation proceeds, additional rod readings (in addition to 50-ft grid readings) for the top and bottom of the excavation are required. Permanent targets can be established to assist in the visual alignment of the cross-section lines running perpendicular to the baseline at each 50-ft station. If permanent targets have not been erected, a surveyor on the baseline can keep the rod on line by using a prism or estimated right angles.

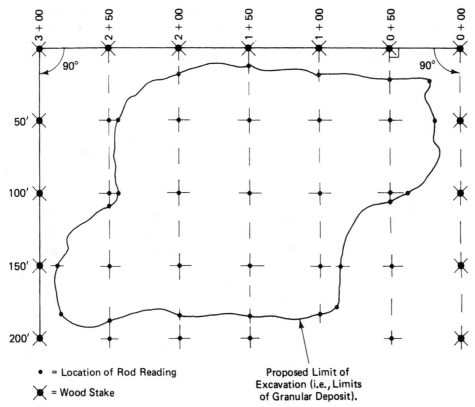

• = Location of Rod Reading

✖ = Wood Stake

Proposed Limit of
Excavation (i.e., Limits
of Granular Deposit).

FIGURE 3.23   Baseline control for a borrow pit survey.

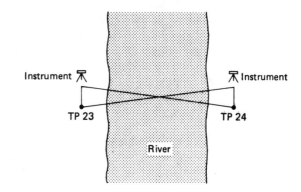

FIGURE 3.24   Reciprocal leveling.

3.10 Reciprocal Leveling

Section 3.7 advised the surveyor to keep BS and FS distances roughly equal so that instrumental and natural errors cancel out. In some situations, such as river or valley crossings, it is not always possible to balance BS and FS distances. The reciprocal leveling technique is illustrated in Figure 3.24. The level is set up and readings are taken on TP 23 and TP 24.

Precision can be improved by taking several readings on the far point (TP 24) and then averaging the results. The level is then moved to the far side of the river, and the process is repeated. The differences in elevation thus obtained are averaged to obtain the final result. The averaging process eliminates instrumental errors and natural errors, such as curvature. Errors due to refraction can be minimized by ensuring that the elapsed time for the process is kept to a minimum.

3.11 Peg Test

The purpose of the peg test is to check that the line of sight through the level is horizontal (i.e., parallel to the axis of the bubble tube). The line-of-sight axis is defined by the location of the horizontal cross hair (see Figure 3.25). Refer to Chapter 4 for a description of the horizontal cross-hair orientation adjustment.

To perform the peg test, the surveyor first places two stakes at a distance of 200 to 300 ft (60 to 90 m) apart. The level is set up midway (paced) between the two stakes, and rod readings are taken at both locations [see Figure 3.26(a)]. If the line of sight through the level is not horizontal, the errors in rod readings (Δe_1) at both points A and B are identical because the level is halfway between the points. The errors are identical, so the calculated difference in elevation between points A and B (difference in rod readings) is the *true* difference in elevation.

The level is then moved to one of the points (A) and set up so that a rod reading can be taken in one of two ways. The first technique requires that the level be set so that the eyepiece of the telescope just touches the rod as it is being held plumb at point A. The rod reading (a_2) can then be determined by sighting backward through the objective lens at a pencil point that is being moved slowly up and down the rod. When sighting backward through the telescope, the pencil point can be centered precisely, even though the cross hairs are not visible, because the circular field of view is relatively small. The second, more popular technique requires that the level be placed close (e.g., at the minimum focusing distance) to the rod, and then a normal sighting is taken. Any rod reading error introduced using this very short sight is relatively insignificant. Once the rod reading at A has been determined and booked, the rod is held at B and a normal rod reading is obtained.

■ **EXAMPLE 3.2**
Refer to Figure 3.26.

First setup: Rod reading at A, a_1 = 1.075
Rod reading at B, b_1 = <u>1.247</u>
True difference in elevations = 0.172

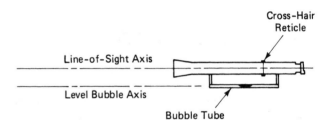

FIGURE 3.25 Optical axis and level tube axis.

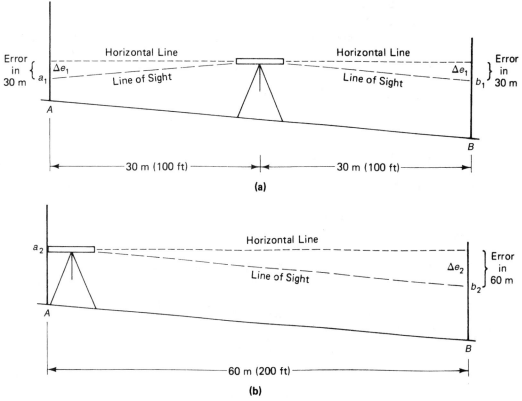

FIGURE 3.26 Peg test.

Second setup: Rod reading at A, a_2 = 1.783

Rod reading at B, b_1 = <u>1.946</u>*

Apparent difference in elevation = 0.163

Error(Δe_2)in 60 m = 0.009

This is an error of -0.00015 m/m. Therefore, the collimation correction (C factor) = $+0.00015$ m/m.

In Section 3.7, you were advised to keep the BS and FS distances equal; the peg test illustrates clearly the benefits to be gained from the use of this technique. If the BS and FS distances are kept roughly equal, errors due to a faulty line of sight simply do not have the opportunity to develop. The peg test can also be accomplished by using the techniques of reciprocal leveling (Section 3.10).

---

*Had there been no error in the instrument line of sight, the rod reading at b_2 would have been 1.955, i.e., 1.783 + 0.172.

■ EXAMPLE 3.3

If the level used in the peg test of Example 3.2 is used in the field with a BS distance of 80 m and an FS distance of 70 m, the net error in the rod readings would be $10 \times 0.00015 = 0.0015$ (0.002). That is, for a relatively large differential between BS and FS distances and a large line-of-sight error (0.009 for 60 m), the effect on the survey is negligible for ordinary work.

3.12 Three-Wire Leveling

Leveling can be performed by utilizing the stadia cross hairs found on most levels and on all theodolites (see Figure 3.27). Each backsight (BS) and foresight (FS) is recorded by reading the stadia hairs in addition to the horizontal cross hair. The stadia hairs (wires) are positioned an equal distance above and below the main cross hair and are spaced to give 1.00 ft (m) of interval for each 100 ft (m) of horizontal distance that the rod is away from the level. The three readings are averaged to obtain the desired value.

The recording of three readings at each sighting enables the surveyor to perform a relatively precise survey while utilizing ordinary levels. Readings to the closest thousandth of a foot (mm) are estimated and recorded. The leveling rod used for this type of work should be calibrated to ensure its integrity. Use of an invar rod is recommended.

Figure 3.28 shows typical notes for benchmark leveling. A realistic survey would include a completed loop or a check into another BM of known elevation. If the collimation correction as calculated in Section 3.11 (10.00015 m/m) is applied to the survey shown in Figure 3.28, the correction to the elevation is as follows:

$$C = +0.00015 \times (62.9 - 61.5) = +0.0002$$
$$\text{Sum of FS corrected to } 5.7196 + 0.0002 = 5.7198$$

The elevation of BM 201 in Figure 3.28 is calculated as follows:

$$
\begin{aligned}
\text{Elev. BM17} &= & 186.2830 \\
+ \ \Sigma \text{BS} &= + & \underline{2.4143} \\
& & 188.6973 \\
- \Sigma \text{FS (corrected)} &= - & 5.7198 \\
\text{Elev. BM 201} &= & 182.9775 \ \text{(corrected for collimation)}
\end{aligned}
$$

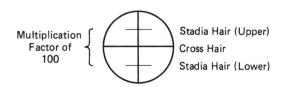

FIGURE 3.27 Reticle cross hairs.

Multiplication Factor of 100 { Stadia Hair (Upper) / Cross Hair / Stadia Hair (Lower)

100 X Stadia Hair Interval = Distance

B.M. LEVELING–3 WIRE
B.M. 17 to B.M. 201
(RETURN RUN ON P.48)

JONES–NOTES
SMITH–$\bar{x}$
BROWN–ROD
GREEN–ROD

Job ROD 19,   INST. L.33   8 °C, CLOUDY
Date   MAR 3, 2006   Page   47

STA.	B.S.	DIST.	F.S.	DIST.	ELEV.	DESCRIPTION
BM 17					186.2830	BRONZE PLATE SET IN WALL --- ETC.
	0.825		1.775			
	0.725	10.0	1.673	10.2	+0.7253	
	0.626	9.9	1.572	10.1	187.0083	
	2.176	19.9	5.020	20.3	−1.6733	
	+0.7253		−1.6733			
TP 1					185.3350	N.LUG TOP FLANGE FIRE HYD. N/S
	0.698		1.750			MAIN ST. OPP. CIVIC #181.
	0.571	12.7	1.620	13.0	+0.5710	
	0.444	12.7	1.490	13.0	185.9060	
	1.713	25.4	4.860	26.0	−1.6200	
	+0.5710		−1.6200			
TP 2					184.2860	N.LUG TOP FLANGE FIRE HYD. N/S
	1.199		2.509			MAIN ST. OPP. CIVIC #163.
	1.118	8.1	2.427	8.2	+1.1180	
	1.037	8.1	2.343	8.4	185.4040	
	3.354	16.2	7.279	16.6	−2.4263	
	+1.1180		−2.4263			
BM 201					182.9777	BRONZE PLATE SET IN ESTLY FACE
						OF RETAINING WALL --- ETC.
Σ	+2.4143	61.5m	−5.7196	62.9m		

ARITHMETIC CHECK: 186.283 + 2.4143 − 5.7196 = ✓
182.9777

FIGURE 3.28 Survey notes for three-wire leveling.

When levels are being used for precise purposes, it is customary to determine the collimation correction at least once each day.

3.13 Trigonometric Leveling

The difference in elevation between A and B in (Figure 3.29) can be determined if the vertical angle (a) and the slope distance (S) are measured. These measurements can be taken with total stations or with EDM/theodolite combinations.

$$V = S \sin \alpha \tag{3.7}$$

$$\text{Elevation at } \bar{\wedge} + \text{hi} \pm V - \text{rod reading (RR)} = \text{elevation at rod} \tag{3.8}$$

The hi in this case is not the elevation of the line of sight (HI), as it is in differential leveling. Instead, hi here refers to the distance from point A up to the optical center of the theodolite or total station, measured with a steel tape or rod. Modern practice involving the use of total stations routinely gives the elevations of sighted points by processing the differences in elevation between the total station point and the sighted points—along with the horizontal distances to those points. See also Section 5.3.1, which discusses total station techniques

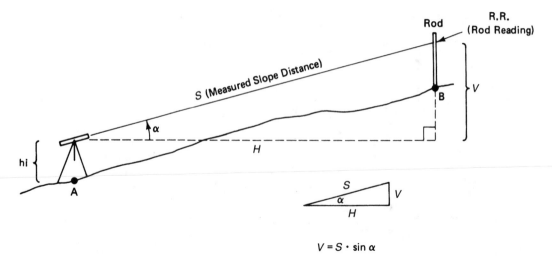

$$V = S \cdot \sin \alpha$$

FIGURE 3.29 Trigonometric leveling.

of trigonometric leveling. Instruments with dual-axis compensators can produce very accurate results.

■ **EXAMPLE 3.4**

See Figure 3.30. Using Equation 3.7, we have:

$$V = S \sin \alpha$$
$$= 82.18 \sin 30° \, 22'$$
$$= 41.54 \text{ ft}$$

Using Equation 3.8, we have:

$$\text{Elev. at } \overline{\wedge} + \text{hi} \pm V - \text{RR} = \text{elevation at rod}$$
$$361.29 + 4.72 - 41.54 - 4.00 = 320.47$$

The RR in Example 3.4 *could* have been 4.72, the value of the hi. If that reading had been visible, the surveyor would have sighted on it to eliminate + hi and −RR from the calculation; that is:

$$\text{Elev. at } \overline{\wedge} - V = \text{elevation at rod} \qquad (3.9)$$

In this example, the station (chainage) of the rod station can also be determined.

3.14 Level Loop Adjustments

In Section 3.7, we noted that level surveys had to be closed within acceptable tolerances or the survey would have to be repeated. The tolerances for various orders of surveys were shown in Tables 3.1 and 3.2.

If a level survey were performed to establish new benchmarks, it would be desirable to proportion any acceptable error throughout the length of the survey. Because the error

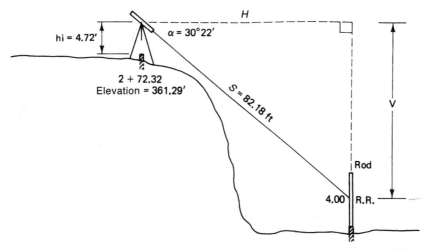

FIGURE 3.30 Example of trigonometric leveling (see Example 3.4 in Section 3.13).

tolerances shown in Tables 3.1 and 3.2 are based on the distances surveyed, adjustments to the level loop are based on the relevant distances, or on the number of instrument setups, which is a factor directly related to the distance surveyed.

■ EXAMPLE 3.5

A level circuit is shown in Figure 3.31. The survey, needed for local engineering projects, commenced at BM 20. The elevations of new benchmarks 201, 202, and 203 were determined. Then the level survey was looped back to BM 20, the point of commencement (the survey could have terminated at any established BM).

Solution

According to Table 3.1, the allowable error for a second-order, class II (local engineering projects) survey is $0.008 \sqrt{K}$; thus, $0.008 \sqrt{4.7} = 0.017$ m is the permissible error.

The error in the survey of this example was found to be -0.015 m over a total distance of 4.7 km—in this case, an acceptable error. It only remains for this acceptable error to be distributed over the length of the survey. The error is proportioned according to the fraction of cumulative distance over total distance, as shown in Table 3.4. More complex adjustments are normally performed by computer, using the adjustment method of least squares.

3.15 Suggestions for Rod Work

The following list provides some reminders when performing rod work:

1. The rod should be extended and clamped properly. Ensure that the bottom of the sole plate does not become encrusted with mud and the like, which could result in mistaken readings. If a rod target is being used, ensure that it is positioned properly and that it cannot slip.

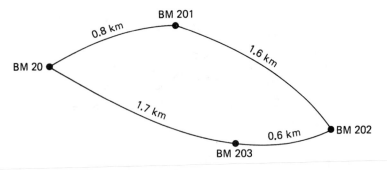

FIGURE 3.31 Level loop.

Total Distance Around Loop is 4.7 km.

Table 3.4 LEVEL LOOP ADJUSTMENTS

BM	Loop Distance, Cumulative (km)	Field Elevation	Correction, Cumulative Distance × E = Total Distance	Adjusted Elevation
20		186.273 (fixed)		186.273
201	0.8	184.242	+ 0.8/4.7 × 0.015 = + 0.003 =	184.245
202	2.4	182.297	+ 2.4/4.7 × 0.015 = + 0.008 =	182.305
203	3.0	184.227	+ 3.0/4.7 × 0.015 = + 0.010 =	184.237
20	4.7	186.258	+ 4.7/4.7 × 0.015 = + 0.015 =	186.273
			$E = 186.273 - 186.258 = -0.015$ m	

2. The rod should be held plumb for all rod readings. Use a rod level, or wave the rod to and from the instrument so that the lowest (indicating a plumb rod) reading can be determined. This practice is particularly important for all backsights and foresights.

3. Ensure that all points used as turning points are describable, identifiable, and capable of having the elevation determined to the closest 0.01 ft or 0.001 m. The TP should be nearly equidistant from the two proposed instrument locations.

4. Ensure that the rod is held in precisely the same position for the backsight as it was for the foresight for all turning points.

5. If the rod is being held temporarily near, but not on, a required location, the face of the rod should be turned away from the instrument so that the instrument operator cannot take a mistaken reading. This type of mistaken reading usually occurs when the distance between the two surveyors is too far to allow for voice communication and sometimes even for good visual contact.

3.16 Suggestions for Instrument Work

The following list of reminders will be useful when performing instrument work:

1. Use a straight-leg (nonadjustable) tripod, if possible.

2. Tripod legs should be tightened so that when one leg is extended horizontally, it falls slowly back to the ground under its own weight.

3. The instrument can be carried comfortably resting on one shoulder. If tree branches or other obstructions (e.g., door frames) threaten the safety of the instrument, it should be cradled under one arm, with the instrument forward, where it can be seen.

4. When setting up the instrument, gently force the legs into the ground by applying weight on the tripod shoe spurs. On rigid surfaces (e.g., concrete), the tripod legs should be spread farther apart to increase stability.

5. When the tripod is to be set up on a side-hill, two legs should be placed downhill and the third leg placed uphill. The instrument can be set up roughly level by careful manipulation of the third, uphill leg.

6. The location of the level setup should be chosen after considering the ability to see the maximum number of rod locations, particularly BS and FS locations.

7. Prior to taking rod readings, the cross hair should be sharply focused; it helps to point the instrument toward a light-colored background (e.g., the sky).

8. When the surveyor observes apparent movement of the cross hairs on the rod (parallax), he or she should carefully check the cross-hair focus adjustment and the objective focus adjustment for consistent results.

9. The surveyor should read the rod consistently at either the top or the bottom of the cross hair.

10. Never move the level before a foresight is taken; otherwise, all work done from that HI will have to be repeated.

11. Check to ensure that the level bubble remains centered or that the compensating device (in automatic levels) is operating.

12. Rod readings (and the line of sight) should be kept at least 18 in. (0.5 m) above the ground surface to help minimize refraction errors when performing a precise level survey.

3.17 Mistakes in Leveling

Mistakes in level loops can be detected by performing arithmetic checks and also by closing on the starting BM or on any other BM whose elevation is known. Mistakes in rod readings that do not form part of a level loop, such as in intermediate sights taken in profiles, cross sections, or construction grades, are a much more irksome problem. It is bad enough to discover that a level loop contains mistakes and must be repeated, but it is a far more serious problem to have to redesign a highway profile because a key elevation contains a mistake, or to have to break out a concrete bridge abutment (the day after the concrete was poured) because the grade stake elevation contains a mistake. Because intermediate rod readings cannot be checked (without releveling), it is essential that the opportunities for mistakes be minimized.

Common mistakes in leveling include the following: misreading the foot (meter) value, transposing figures, not holding the rod in the correct location, resting the hands on the tripod while reading the rod and causing the instrument to go off level, entering the rod readings incorrectly (i.e., switching BS and FS), giving the wrong station identification to a correct rod reading, and making mistakes in the note reduction arithmetic. Most mistakes in arithmetic can be eliminated by having the other crew members check the reductions and initial each page of notes checked. Mistakes in the leveling operation cannot be completely eliminated, but they can be minimized if the crew members are aware that mistakes can (and probably will) occur. All crew members should be constantly alert to the possible occurrence of mistakes, and all crew members should try to develop strict routines for doing their work so that mistakes, when they do eventually occur, will be all the more noticeable.

Problems

3.1. Compute the error due to curvature and refraction for the following distances:

    (a) 700 ft

    (b) 3,000 ft

    (c) 500 m

    (d) 1.75 miles

    (e) 3,500 m

    (f) 5 kilometers

3.2. Determine the rod readings indicated on the foot and metric rod illustrations shown in Figure 3.32. The foot readings are to the closest 0.01 ft, and the metric readings are to the closest 1/2 or 1/3 centimeter.

3.3. An offshore drilling rig is being towed out to sea. What is the maximum distance away that the navigation lights can still be seen by an observer standing at the shoreline? The observer's eye height is 5′9″, and the uppermost navigation light is 310 ft above the water.

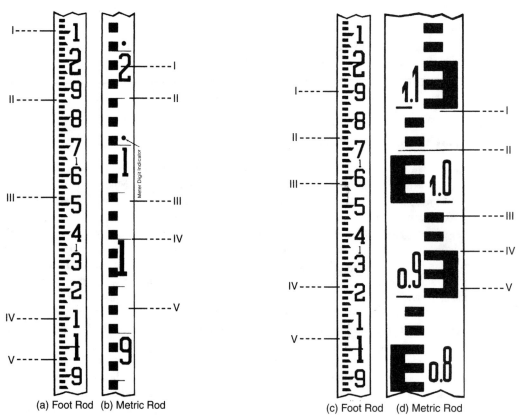

(a) Foot Rod (b) Metric Rod   (c) Foot Rod   (d) Metric Rod

FIGURE 3.32 (a) Foot rod. (b) Metric rod. (c) Foot rod. (d) Metric rod.
(Courtesy of Sokkia Co. Ltd.)

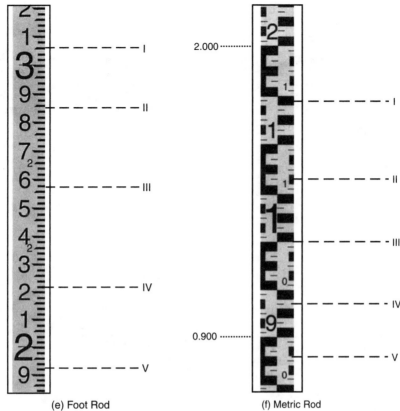

(e) Foot Rod   (f) Metric Rod

FIGURE 3.32 *(continued)* (e) Foot rod. (f) Metric rod.

3.4. Prepare a set of level notes for the survey in Figure 3.33. Show the arithmetic check.

3.5. Prepare a set of profile leveling notes for the survey in Figure 3.34. In addition to computing all elevations, show the arithmetic check and the resulting error in closure.

3.6. Complete the accompanying set of differential leveling notes, and perform the arithmetic check.

Station	BS	HI	FS	Elevation
BM 3	1.613			148.610
TP 1	1.425		1.927	
TP 2	1.307		1.710	
TP 3	1.340		1.273	
BM 3			0.780	

3.7. If the loop distance in Problem 3.6 is 700 m, at what order of survey do the results qualify? (Use Table 3.1 or Table 3.2.)

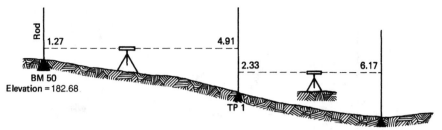

FIGURE 3.33

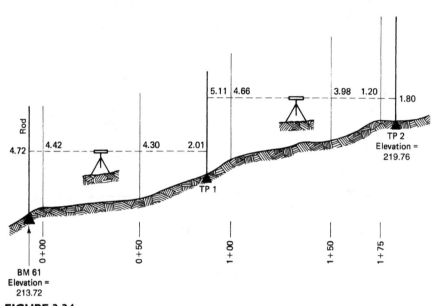

FIGURE 3.34

3.8. Reduce the accompanying set of differential leveling notes, and perform the arithmetic check.

Station	BS	HI	FS	Elevation
BM 100	2.71			267.09
TP 1	3.62		4.88	
TP 2	3.51		3.97	
TP 3	3.17		2.81	
TP 4	1.47		1.62	
BM 100			1.21	

3.9. If the distance leveled in Problem 3.8 is 1,000 ft, for what order of survey do the results qualify? (See Tables 3.1 and 3.2).

3.10 Reduce the accompanying set of profile notes, and perform the arithmetic check.

Station	BS	HI	IS	FS	Elevation
BM S.101	0.475				159.987
0 + 000			0.02		
0 + 020			0.41		
0 + 040			0.73		
0 + 060			0.70		
0 + 066.28			0.726		
0 + 080			1.38		
0 + 100			1.75		
0 + 120			2.47		
TP 1	0.666			2.993	
0 + 140			0.57		
0 + 143.78			0.634		
0 + 147.02			0.681		
0 + 160			0.71		
0 + 180			0.69		
0 + 200			1.37		
TP 2	0.033			1.705	
BM S. 102				2.891	

3.11. Reduce the following set of municipal cross-section notes.

Station	BS	HI	IS	FS	Elevation
BM 41	6.21				413.33
TP 13	4.10			0.89	
12 + 00					
50 ft left			3.9		
18.3 ft left			4.6		
₵			6.33		
20.1 ft right			7.9		
50 ft right			8.2		
13 + 00					
50 ft left			5.0		
19.6 ft left			5.7		
₵			7.54		
20.7 ft right			7.9		
50 ft right			8.4		
TP 14	7.39			1.12	
BM S.22				2.41	

3.12. Complete the accompanying set of highway cross-section notes.

Station	BS	HI	FS	Elevation	Left		₵	Right	
BM 107	7.71			256.71					
80 + 50					60′	28′		32′	60′
					9.7	8.0	5.7	4.3	4.0
81 + 00					60′	25′		30′	60′
					10.1	9.7	6.8	6.0	5.3
81 + 50					60′	27′		33′	60′
					11.7	11.0	9.2	8.3	8.0
TP 1			10.17						

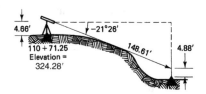

FIGURE 3.35

3.13. A level is set up midway between two wood stakes that are about 300′ apart. The rod reading on stake A is 8.72′ and it is 5.61′ on stake B. The level is then moved to point B and set up so that the eyepiece end of the telescope is just touching the rod as it is held plumb on the stake. A reading of 5.42′ is taken on the rod at B by sighting backward through the telescope. The level is then sighted on the rod held on stake A, where a reading of 8.57′ is noted.

(a) What is the correct difference in elevation between the tops of stakes A and B?

(b) If the level had been in perfect adjustment, what reading would have been observed at A from the second setup?

(c) What is the line-of-sight error in 300′?

(d) Describe how you would eliminate the line-of-sight error from the telescope.

3.14. A preengineering baseline was run down a very steep hill (see Figure 3.35). Rather than measure horizontally downhill with the steel tape, the surveyor measured the vertical angle with a theodolite and the slope distance with an EDM. The vertical angle was −21°26′, turned to a prism on a plumbed range pole 4.88′ above the ground. The slope distance from the theodolite to the prism was 148.61′. The theodolite's optical center was 4.669 above the upper baseline station at 110 + 71.25.

(a) If the elevation of the upper station is 324.28, what is the elevation of the lower station?

(b) What is the chainage of the lower station?

3.15. You must establish the elevation of point B from point A (elevation 187.298 m). Points A and B are on opposite sides of a twelve-lane highway. Reciprocal leveling is used, with the following results:

Setup at A side of highway:

Rod reading on A = 0.673 m

Rod readings on B = 2.416 and 2.418 m

Setup at B side of highway:

Rod reading on B = 2.992 m

Rod readings on A = 1.254 and 1.250 m

(a) What is the elevation of point B?

(b) What is the leveling error?

3.16. Reduce the following set of differential leveling notes and perform the arithmetic check.

Station	BS	HI	FS	Elevation
BM 130	0.702			168.213
TP #1	0.970		1.111	
TP #2	0.559		0.679	
TP #3	1.744		2.780	
BM K110	1.973		1.668	
TP #4	1.927		1.788	
BM 132			0.888	

(a) Determine the order of accuracy. Refer to Table 3.1 or 3.2.

(b) Adjust the elevation of BM K110. The length of the level run was 780 m, with setups equally spaced. The elevation of BM 132 is known to be 167.185 m.

Chapter 4

Angles and Theodolites

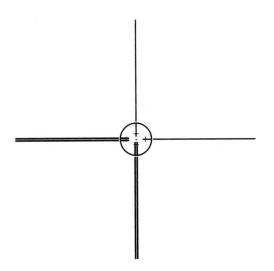

4.1 General Background

We noted in section 1.9 that the units of angular measurement employed in North American practice are degrees, minutes, and seconds. For the most part, angles in surveying are measured with a theodolite or total station, although angles can be measured with clinometers, sextants (hydrographic surveys), or compasses.

4.2 Reference Directions for Vertical Angles

Vertical angles, which are used in slope distance corrections (Section 2.8) or in height determination (Section 3.13), are referenced to (1) the horizon by plus (up) or minus (down) angles, (2) the zenith, or (3) the nadir (see Figure 4.1). **Zenith** and **nadir** are terms describing points on a celestial sphere (i.e., a sphere of infinitely large radius with its center at the center of the earth). The zenith is directly above the observer and the nadir is directly below the observer; the zenith, nadir, and observer are all on the same vertical line.

4.3 Meridians

A line on the mean surface of the earth joining the north and south poles is called a **meridian**. All lines of longitude are meridians. The term *meridian* can be defined more precisely by noting that it is the line formed by the intersection with the earth's surface of a plane that includes the earth's axis of rotation. The meridian, as described, is known as the **geographic meridian.**

    Magnetic meridians are meridians that are parallel to the directions taken by freely moving magnetized needles, as in a compass. Whereas geographic meridians are fixed, magnetic meridians vary with time and location. **Grid meridians** are lines that are parallel to a grid reference meridian (central meridian) and are described in detail in Chapter 9.

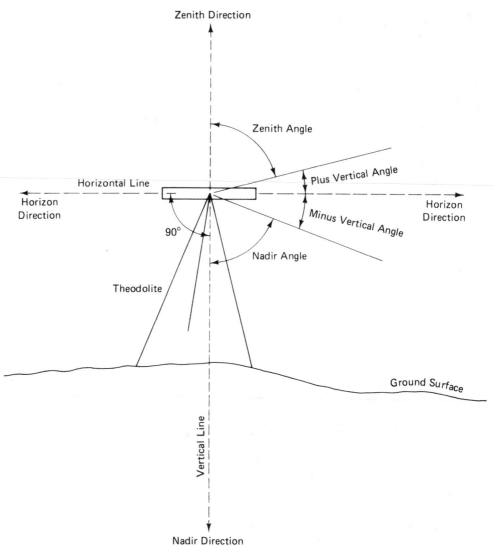

FIGURE 4.1 The three reference directions for vertical angles: horizontal, zenith, and nadir.

We saw in Section 4.2 that vertical angles were referenced to a horizontal line (plus or minus) or to a vertical line (from either the zenith or nadir direction); in contrast, we now see that all horizontal directions are referenced to meridians.

4.4 Horizontal Angles

Horizontal angles are usually measured with a theodolite or total station whose precision can range from 1 second to 20 seconds of arc. Angles can be measured between lines forming a closed traverse, between lines forming an open traverse, or between a line and a point to aid in the location of that point.

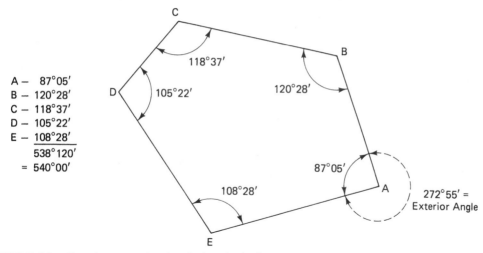

FIGURE 4.2 Closed traverse showing the interior angles.

4.4.1 Interior Angles

For all closed polygons of n sides, the sum of the interior angles will be $(n - 2)180°$; the sum of the exterior angles will be $(n + 2)180°$. In Figure 4.2, the interior angles of a five-sided closed polygon have been measured as shown. For a five-sided polygon, the sum of the interior angles must be $(5 - 2)180° = 540°$; the angles shown in Figure 4.2 do, in fact, total 540°. In practical field problems, however, the total is usually marginally more or less than $(n - 2)180°$, and it is then up to the surveyors to determine if the error of angular closure is within tolerances as specified for that survey. The adjustment of angular errors is discussed in Chapter 6.

Note that the exterior angles at each station in Figure 4.2 could have been measured instead of the interior angles as shown. (The exterior angle at A of $272°55''$ is shown.) Generally, exterior angles are measured only occasionally to serve as a check on the interior angle.

4.4.2 Deflection Angles

An open traverse is illustrated in Figure 4.3(a). The **deflection angles** shown are measured from the prolongation of the back survey line to the forward line. The angles are measured either to the left (L) or to the right (R) of the projected line. The direction (L or R) must be shown, along with the numerical value. It is also possible to measure the change in direction [see Figure 4.3(b)] by directly sighting the back line and turning the angle left or right to the forward line.

4.5 Theodolites

Theodolites are survey instruments designed to measure horizontal and vertical angles precisely. In addition to measuring horizontal and vertical angles, theodolites can also be used

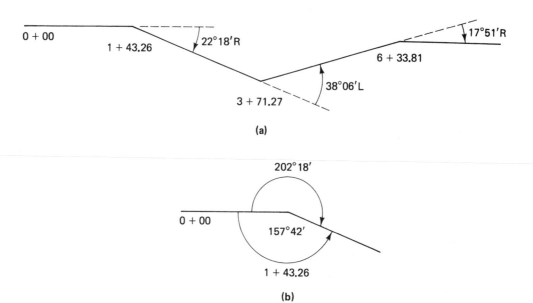

FIGURE 4.3 (a) Open traverse showing deflection angles. (b) Same traverse showing angle right (202° 18′) and angle left (157°42′).

to mark out straight and curved lines in the field. Theodolites (also called transits) have gone through three distinct evolutionary stages during the twentieth century:

- The open-face, vernier-equipped engineers' transit (American transit). See Figure G.8.
- The enclosed, optical readout theodolites with direct digital readouts, or micrometer-equipped readouts (for more precise readings). See Figure 4.4.
- The enclosed electronic theodolite with direct readouts. See Figure 4.5.

Most recent manufactured theodolites are electronic, but many of the earlier optical instruments and even a few vernier instruments still survive in the field (and in the classroom)—no doubt a tribute to the excellent craftsmanship of the instrument makers. In past editions of this text, the instruments were introduced chronologically, but in this edition, the vernier transits are introduced last (in Appendix G) in recognition of their fading importance.

The electronic theodolite will probably be the last in the line of transits/theodolites. Because of the versatility and lower costs of electronic components, future field instruments will be more like the total station (Chapter 5), which combines all the features of a theodolite with the additional capabilities of measuring horizontal and vertical distances (electronically), and storing all measurements, with relevant attribute data, for future transfer to the computer. By 2005, some total stations even included a global positioning system (GPS) receiver (see Figure 5.29) to permit precise horizontal and vertical positioning. The sections in this chapter that deal with theodolite setup, instrument adjustments, and general use, such as prolonging a straight line, bucking-in, and prolonging a line past an obstacle, are equally applicable to total stations; total station operations are described more fully in Chapter 5.

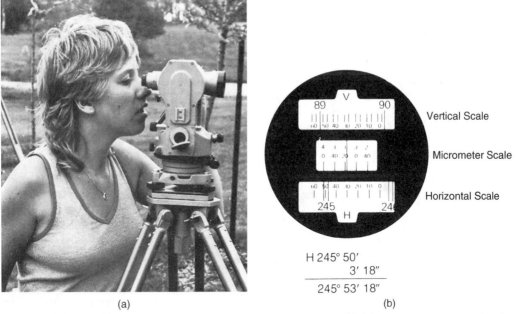

Vertical Scale

Micrometer Scale

Horizontal Scale

H 245° 50'
<u>3' 18"</u>
245° 53' 18"

(a)   (b)

FIGURE 4.4   (a) Twenty-second micrometer theodolite, the Sokkia TM 20. (b) Horizontal and vertical scales with micrometer scale.

FIGURE 4.5   Leica T-1600 digital electronic theodolite. This instrument has keyboards front and back, an angle accuracy of 1.5", automatic error monitoring, some surveying programs (e.g., "free stationing"), and interfaces that permit the addition of a data-storage module and an electronic distance measurement device. (Courtesy of Leica Geosystems)

4.6 Electronic Theodolites

Electronic theodolites operate similarly to optical theodolites (Section 4.8); one major difference is that these instruments usually have only one motion (upper) and accordingly have only one horizontal clamp and slow-motion screw. Angle readouts can be to 1″, with precision ranging from 0.5″ to 20″. The surveyor should check the specifications of new instruments to determine their precision, rather than simply accept the lowest readout as relevant (some instruments with 1″ readouts may be capable of only 5″ precision).

Digital readouts eliminate the uncertainty associated with the reading and interpolation of scale and micrometer settings. Electronic theodolites have zero-set buttons for quick instrument orientation after the backsight has been taken (any angular value can be set for the backsight). Horizontal angles can be turned left or right, and repeat-angle averaging is available on some models. Figures 4.5, 4.6, and 4.7 are typical of the more recently introduced theodolites. The display windows for horizontal and vertical angles are located at both the front and rear of many instruments for easy access.

The instruments shown in Figures 4.6 and 4.7 are basic electronic theodolites, whereas the instrument shown in Figure 4.5 has additional capabilities, including the expansion to total station capability with the inclusion of modular components such as a central processor, EDM, and data collection. Figure 4.6 shows the operation keys and display area typical of many of these instruments. After turning on some instruments, the operator must activate the vertical circle by turning the telescope slowly through the horizon; newer instruments don't require this referencing action. The vertical circle can be set with zero at the zenith or at the horizon; the factory setting of zenith can be changed by setting the appropriate dip switch as described in the instrument's manual. The status of the battery charge can be monitored on the display panel, giving the operator ample warning of the need to replace and/or recharge the battery.

4.6.1 Angle Measurement

Most surveying measurements are performed at least twice; this repetition improves precision and helps eliminate mistakes. Angles are usually measured at least twice: once with the telescope in its normal position and once with the telescope inverted. After the theodolite has been set over a station, the angle measurement begins by sighting the left hand (usually) target and clamping the instrument's horizontal motion. The target is then sighted precisely by using the fine adjustment, or slow-motion screw. The horizontal angle is set to zero by pressing the zero-set button. Then the horizontal clamp is loosened and the right-hand target sighted. The clamp is tightened and the telescope fine-adjusted onto the target. The hold button is pressed and the angle is read and booked (pressing the hold button ensures that the original angle stays on the instrument display until the surveyor is ready to measure the angle a second time).

To prepare to double the angle (i.e., measure the same angle a second time), the clamp is loosened and the telescope transited. The left-hand point is now resighted. To turn the double angle, simply press (release) the hold button, release the clamp, and resight the right-hand target a second time. After fine-adjusting on the target, the double angle is read and booked. To obtain the mean angle, divide the double angle by 2. See Figure 4.8 for typical field notes.

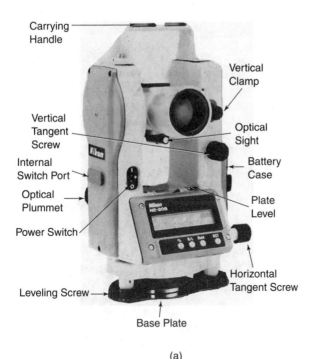

Carrying Handle

Vertical Clamp

Vertical Tangent Screw

Optical Sight

Internal Switch Port

Battery Case

Optical Plummet

Plate Level

Power Switch

Horizontal Tangent Screw

Leveling Screw

Base Plate

(a)

VA (Vertical Angle) or % (Percentage of Grade) Display Symbol

G (GON) Display Symbol (Available)

Battery Charge Level Indicator

HA (Clockwise Horizontal Angle) or HL (Counter-clockwise Horizontal Angle) Display Symbol

HOLD (Horizontal Angle Hold) Display (Available)

Vertical Angle/Grade Display Key

Horizontal Angle Zero Reset Key

Horizontal Angle Selection Key
   R: Clockwise
   L: Counterclockwise

Horizontal Angle HOLD Key

(b)

FIGURE 4.6 Nikon NE-20S electronic digital theodolite. (a) Theodolite. (b) Operation keys and display. (Courtesy of Nikon Instruments, Inc.)

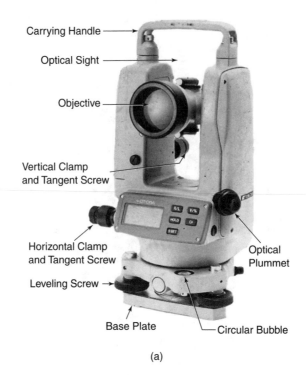

Carrying Handle

Optical Sight

Objective

Vertical Clamp
and Tangent Screw

Horizontal Clamp
and Tangent Screw

Leveling Screw

Optical
Plummet

Base Plate

Circular Bubble

(a)

ROTARY ENCODER SYSTEM FOR ELECTRONIC THEODOLITES AND TRANSITS

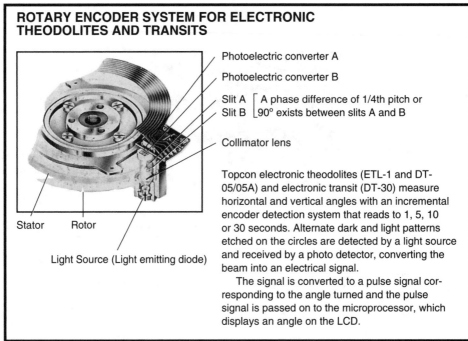

Photoelectric converter A

Photoelectric converter B

Slit A ⎡ A phase difference of 1/4th pitch or
Slit B ⎣ 90° exists between slits A and B

Collimator lens

Topcon electronic theodolites (ETL-1 and DT-05/05A) and electronic transit (DT-30) measure horizontal and vertical angles with an incremental encoder detection system that reads to 1, 5, 10 or 30 seconds. Alternate dark and light patterns etched on the circles are detected by a light source and received by a photo detector, converting the beam into an electrical signal.

The signal is converted to a pulse signal corresponding to the angle turned and the pulse signal is passed on to the microprocessor, which displays an angle on the LCD.

Stator   Rotor

Light Source (Light emitting diode)

FIGURE 4.7 Topcon DT-05 electronic digital theodolite. (a) Theodolite. (b) Encoder system for angle readouts. (Courtesy of Topcon Positioning Systems, Inc.)

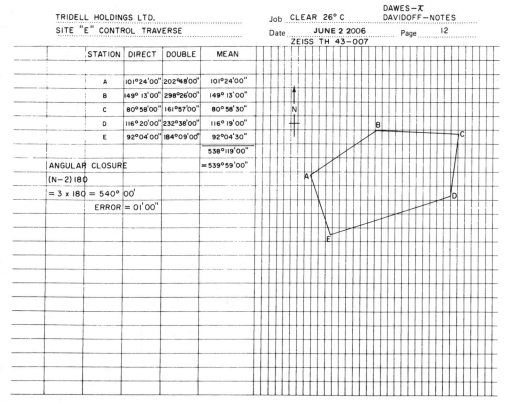

TRIDELL HOLDINGS LTD.
SITE "E" CONTROL TRAVERSE

Job CLEAR 26° C
Date JUNE 2 2006   Page 12

DAWES—π
DAVIDOFF—NOTES
ZEISS TH 43—007

STATION	DIRECT	DOUBLE	MEAN
A	101°24'00"	202°48'00"	101°24'00"
B	149°13'00"	298°26'00"	149°13'00"
C	80°58'00"	161°57'00"	80°58'30"
D	116°20'00"	232°38'00"	116°19'00"
E	92°04'00"	184°09'00"	92°04'30"
			538°119'00"
ANGULAR CLOSURE			= 539°59'00"

(N−2)180
= 3 × 180 = 540° 00'
ERROR = 01'00"

FIGURE 4.8 Field notes for angles by repetition (closed traverse).

4.6.2 Typical Specifications for Electronic Theodolites

The typical specifications for electronic theodolites are listed below:

Magnification: 26× to 30×

Field of view: 1.5°

Shortest viewing distance: 1.0 m

Angle readout, direct: 1″ to 20″; accuracy: 1″ to 20″*

Angle measurement, electronic and incremental: see Figure 4.7(b)

Level sensitivity:

Plate bubble vial—40″/2 mm

Circular bubble vial—10′/2 mm

---

*Accuracies are now specified by most surveying instrument manufacturers by reference to DIN 18723. DIN (Deutsches Institut für Normung) is known in English-speaking countries as the German Institute for Standards. These accuracies are tied directly to surveying practice. For example, to achieve a claimed accuracy (stated in terms of 1 standard deviation) of ±5 seconds, the surveyor must turn the angle four times (two on face 1 and two with the telescope inverted on face 2). This practice assumes that collimation errors, centering errors, and the like, have been eliminated before measuring the angles. DIN specifications can be purchased at www2.din.de/index.php?lang=en.

Electronic theodolites are quickly replacing optical theodolites (which replaced the vernier transit). They are simpler to use and less expensive to purchase and repair, and their use of electronic components seems to indicate a continuing drop in both purchase and repair costs. Some of these instruments have various built-in functions that enable the operator to perform other theodolite operations, such as remote object elevation, and distance between remote points (see Chapter 5). The instrumentation technology is evolving so rapidly that most new instruments now on the market have been in production for only a year or two, and this statement has been valid since the early 1990s.

4.7   Theodolite/Total Station Setup

Follow this procedure to set up a theodolite or total station:

1. Place the instrument over the point, with the tripod plate as level as possible and with two tripod legs on the downhill side, if applicable.
2. Stand back a pace or two and see if the instrument appears to be over the station. If it does not, adjust the location, and check again from a pace or two away.
3. Move to a position 90° opposed to the original inspection location and repeat step 2. (*Note:* This simple act of "eyeing-in" the instrument from two directions, 90° opposed, takes only seconds but could save a great deal of time in the long run.)
4. Check to see that the station point can now be seen through the optical plummet (or that the laser plummet spot is reasonably close to the setup mark) and then firmly push in the tripod legs by pressing down on the tripod shoe spurs. If the point is now not visible in the optical plumb sight, leave one leg in the ground, lift the other two legs, and rotate the instrument, all the while looking through the optical plumb sight. When the point is sighted, carefully lower the two legs to the ground, keeping the station point in view.
5. While looking through the optical plummet (or at the laser spot), manipulate the leveling screws until the cross hair (bull's-eye) of the optical plummet or the laser spot is directly on the station mark.
6. Level the theodolite circular bubble by adjusting the tripod legs up or down. This step is accomplished by noting which leg, when slid up or down, would move the circular bubble into the bull's-eye. Upon adjusting that leg, the bubble will either move into the circle (the instrument is level), or it will slide around until it is exactly opposite another tripod leg. That leg is then adjusted up or down until the bubble moves into the circle. If the bubble does not move into the circle, adjust the leg until the bubble is directly opposite another leg and repeat the process. If this manipulation has been done correctly, the bubble will be centered after the second leg has been adjusted; it is seldom necessary to adjust the legs more than three times. (Comfort can be taken from the fact that these manipulations take less time to perform than they do to read about!)
7. Perform a check through the optical plummet or note the location of the laser spot to confirm that it is still quite close to being over the station mark.
8. Turn one (or more) leveling screw(s) to ensure that the circular bubble is now centered exactly (if necessary).

9. Loosen the tripod clamp bolt a bit and slide the instrument on the flat tripod top (if necessary) until the optical plummet or laser spot is centered exactly on the station mark. Retighten the tripod clamp bolt and reset the circular bubble, if necessary. When sliding the instrument on the tripod top, do not twist the instrument, but move it in a rectangular fashion. This precaution ensures that the instrument will not go seriously off level if the tripod top itself is not close to being level.

10. The instrument can now be leveled precisely by centering the tubular bubble. Set the tubular bubble so that it is aligned in the same direction as two of the foot screws. Turn these two screws (together or independently) until the bubble is centered. Then turn the instrument 90°; at this point, the tubular bubble will be aligned with the third leveling screw. Next, turn that third screw to center the bubble. The instrument now should be level, although it is always checked by turning the instrument through 180° and noting the new bubble location. See Section 4.13.2 for adjustment procedures. On instruments with dual-axis compensation, final leveling can be achieved by viewing the electronic display [see Figure 5.29(b)] and then turning the appropriate leveling screws. This latter technique is faster because the instrument does not have to be rotated repeatedly.

4.8 Repeating Optical Theodolites

Figures 4.4 and 4.9 show repeating optical micrometer theodolites. Optical theodolites are characterized by three-screw leveling heads, optical plummets, light weight, and glass circles being read either directly or with the aid of a micrometer; angles (0° to 360°) are normally read in the clockwise direction. In contrast with the American engineers' transit (see Appendix G), most theodolites do not come equipped with compasses or telescope levels. Most theodolites

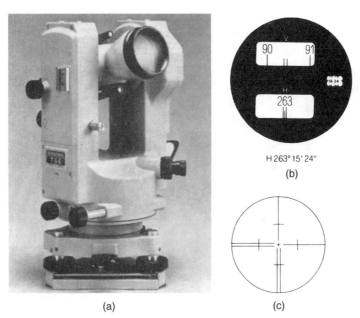

H 263° 15′ 24″

(b)

(a)   (c)

FIGURE 4.9 (a) Six-second repeating micrometer theodolite. (b) Horizontal circle and micrometer reading. (c) Cross-hair reticule pattern. (Courtesy of SOKKIA Corp.)

are now equipped with a compensating device that automatically indexes the horizontal position when the vertical circle has been set to the horizontal setting of 90° (or 270°).

The horizontal position angular setting (in the vertical plane) for theodolites is 90° (270°), whereas for the transit it is 0° (see Appendix G). *A word of caution:* Although all theodolites have a horizontal setting of 90° direct or 270° inverted, some theodolites have their zero set at the nadir, but most have the zero set at the zenith. The method of graduation can be ascertained quickly in the field simply by setting the telescope in an upward (positive) direction and noting the scale reading. If the reading is less than 90°, the zero has been referenced to the zenith direction; if the reading is more than 90°, the zero has been referenced to the nadir direction.

By means of prisms and lenses, the graduations of both the vertical and horizontal circles are projected to one location just adjacent to the telescope eyepiece, where they are read by means of a microscope. An adjustable mirror is located on one of the standards and it controls the illumination needed for the circle-reading procedure. Light required for underground or night work is directed through the mirror window by attached lamps powered by battery packs.

Optical theodolites have two horizontal motions. The alidade (see Figure G.5) can be revolved on the circle assembly, with the circle assembly locked to the leveling head. This arrangement allows you to turn and read horizontal angles. Alternatively, the alidade and circle assembly can also be clamped together but revolve freely on the leveling head. This setup permits you to keep a set angle value on the circle while turning the instrument to sight another point.

The theodolite tripod has a flat base through which a bolt is threaded up into the three-screw leveling base (tribrach), thus securing the instrument to the tripod. Most theodolites have a tribrach release feature that permits the alidade and circle assemblies to be lifted from the tribrach and interchanged with a target or prism (see Figures 2.28 and 4.10). When a theodolite, minus its tribrach, is placed on a tribrach vacated by the target or prism, it will be instantly over the point and nearly level. This system, called **forced centering**, speeds up the work and reduces the centering errors associated with multiple setups.

FIGURE 4.10 A variety of tribrach-mounted traverse targets. Targets and theodolites can be interchanged easily to save setup time (forced-centering system). (Courtesy of SOKKIA Corp.)

Optical plummets can be mounted in the alidade or in the tribrach. Alidade-mounted optical plummets can be checked for accuracy simply by revolving the alidade around its vertical axis and noting the location of the optical plummet cross hairs (bull's-eye) with respect to the station mark. Tribrach-mounted optical plummets can be checked by means of a plumb bob. Adjustments can be made by manipulating the appropriate adjusting screws, or the instrument can be sent out for shop analysis and adjustment.

Like the engineers' transit (see Appendix G), many repeating theodolites have two independent motions (upper and lower), which necessitates upper and lower clamps with their attendant tangent screws.

When some tribrach-equipped lasers are used in mining surveys for zenith and nadir plumbing, the alidade (upper portion of the theodolite) can be released from the tribrach, with the laser plummet then able to project upward to ceiling stations. After the tribrach is centered properly under the ceiling mark, the alidade is then replaced in the tribrach, and the instrument is ready for final settings. This visual plumb line helps the surveyor to position the instrument over (under) the station mark more quickly than with the use of an optical plummet. Some manufacturers place the laser plummet in the alidade portion of the instrument, which permits an easy check on the beam's accuracy because the operator can simply rotate the instrument and observe whether the beam stays on the point. (Upward plumbing is not possible with alidade-mounted laser plummets.)

4.9 Angle Measurement with an Optical Theodolite

The technique for turning and doubling (repeating) an angle is the same as that described for a transit (Section G.3.8). The only difference in the procedure is that of zeroing and reading the scales. In the case of the direct reading optical scale, zeroing the circle is simply a matter of turning the circle until the zero-degree mark lines up approximately with the zero-minute mark on the scale. Once the upper clamp has been tightened, the setting can be accomplished precisely by manipulation of the upper tangent (slow-motion) screw.

In the case of the optical micrometer instruments (see Figures 4.4 and 4.9), it is important first to set the micrometer to zero and then to set the horizontal circle to zero. When the first angle has been turned, note that the horizontal (or vertical) circle index mark is not directly over a degree mark (the micrometer scale is still set to zero). The micrometer knob is turned until the circle index mark is set precisely to a degree mark, and the micrometer scale then records how far the index mark was moved. The micrometer scale does not have to be reset to zero for subsequent angle readings unless a new reference backsight is taken.

The vertical circle is read in the same way. Use the same micrometer scale when vertical angles are being read.

4.10 Direction Optical Theodolites

The essential differences between a direction optical theodolite and a repeating optical theodolite is that the direction theodolite's circles are divided more precisely and the instrument has only one motion (upper), whereas the optical repeating theodolite has two motions (upper and lower). Because it is difficult to set angle values precisely on this type of instrument, angles are usually determined by reading the initial direction and the final direction, and then by determining the difference between the two to obtain the angle.

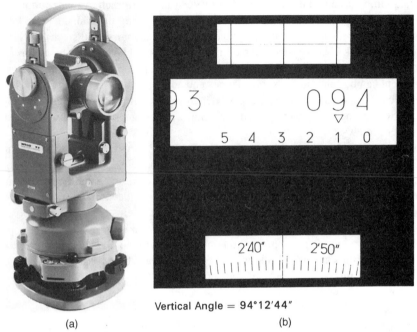

FIGURE 4.11 (a) Wild T-2, a 1-second optical direction theodolite. (b) Vertical circle reading. (Courtesy of Leica Geosystems)

Direction optical theodolites are generally more precise; for example, the Wild T-2 shown in Figure 4.11 reads directly to 01″ and by estimation to 0.5″, whereas the Wild T-3 shown in Figure 4.12 reads directly to 0.2″ and by estimation to 0.1″. In the case of the Wild T-2 and the other 1-second theodolites, the micrometer is turned to force the index to read an even 10′ (the grid lines shown above [beside] the scale are brought to coincidence), and then the micrometer scale reading (02′44″) is added to the circle reading (94°10′) to give a result of 94°12′44″. In the case of the Wild T-3, both sides of the circle are viewed simultaneously; one reading is shown erect, the other inverted. The micrometer knob is used to align the erect and inverted circle markings precisely. Each division on the circle is 04′. If the lower scale is moved half a division, however, the upper also moves half a division (that is, a movement of only 02′), once again causing the markings to align. The circle index line is between the 73° and 74° mark, indicating that the value being read is 73°. Minutes can be read on the circle by counting the number of divisions from the erect 73° to the inverted value, that is, 180° different than 73° (i.e., 253°). In this case, the number of divisions between these two numbers is 13, each having a value of 02′ (i.e., 26′). Precise electronic theodolites have replaced these optical instruments.

4.11 Angles Measured with a Direction Theodolite

As noted earlier, it is not always possible to set angles on a direction theodolite scale precisely, so directions are observed and then subtracted one from the other to determine

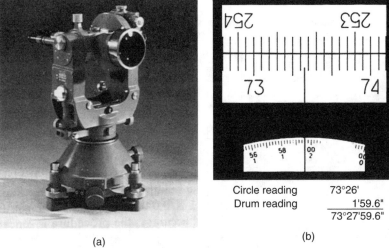

Circle reading	73°26'
Drum reading	1'59.6"
	73°27'59.6"

(a) (b)

FIGURE 4.12 (a) Wild T-3 precise theodolite for first-order surveying. (b) Circle reading (least graduation is 4 minutes) and micrometer reading (least graduation is 0.2 seconds). (Courtesy of Leica Geosystems) On the micrometer, a value of 01'59.6" can be read. The reading is, therefore, 73°27'59.6".

angles. Furthermore, if several sightings are required for precision purposes, it is customary to distribute the initial settings around the circle to minimize the effect of circle graduation distortions. For example, if you are using a directional theodolite where both sides of the circle are viewed simultaneously, the initial settings (positions) would be distributed per $180/n$, where n is the number of settings required by the precision specifications. (Specifications for precise surveys are published by the National Geodetic Survey in the United States and by the Geodetic Surveys of Canada.) To be consistent, not only should the initial settings be distributed uniformly around the circle, but the range of the micrometer scale should also be noted and appropriately apportioned.

The direct readings are taken first in a clockwise direction; the telescope is then transited (plunged), and the reverse readings are taken counterclockwise. In Figure 4.13, the last entry at position 1 is 180°00'12" (R). If the angles (shown in the abstract) do not meet the required accuracy, the procedure is repeated while the instrument still occupies that station.

4.12 Geometry of the Theodolite and Total Station

The vertical axis of the theodolite goes up through the center of the spindles and is oriented over a specific point on the earth's surface. The circle assembly and alidade revolve about this axis. The horizontal axis of the telescope is perpendicular to the vertical axis, and the telescope and vertical circle revolve about it. The line of sight (line of collimation) is a line joining the intersection of the reticule cross hairs and the center of the objective lens. The line of sight is perpendicular to the horizontal axis and should be truly horizontal when the telescope level bubble (if there is one) is centered and when the vertical circle is set at 90° or 270° (or 0° for vernier transits).

FOXLEA SUBDIVISION

DIRECTIONS FOR CONTROL EXTENSION

Job CLEAR, 17 °C HABKIRK—NOTES

Date MAR 13, 2006 Page 28

WILD T-2 #4128-B

π @ STATION #481

STATION SIGHTED	D/R	READING	MEAN D/R	REDUCED DIRECTION
POSITION 1				
1001	D	0° 00' 08"		
	R	180° 00' 12"	10"	0° 00' 00"
778	D	40° 37' 44"		
	R	220° 37' 47"	46"	40° 37' 36"
779	D	78° 52' 19"		
	R	258° 52' 13"	16"	78° 52' 06"
POSITION 2				
1001	D	45° 02' 22"		
	R	225° 02' 26"	24"	0° 00' 00"
778	D	85° 40' 02"		
	R	265° 40' 05"	04"	40° 37' 40"
779	D	123° 54' 30"		
	R	303° 54' 36"	33"	78° 52' 09"
POSITION 3				
1001	D	90° 05' 03"		
	R	270° 05' 07"	05"	0° 00' 00"
778	D	130° 42' 44"		
	R	310° 42' 45"	44"	40° 37' 39"
779	D	168° 57' 10"		
	R	348° 57' 14"	12"	78° 52' 07"
POSITION 4				
	ETC.			

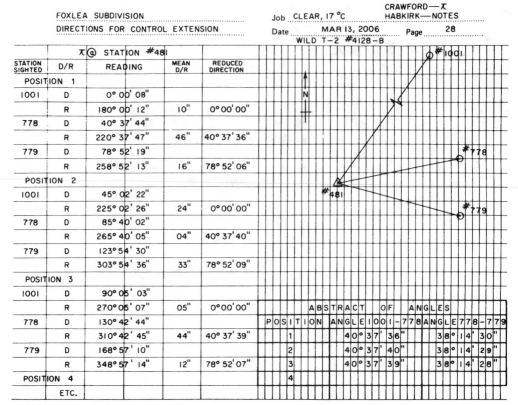

ABSTRACT OF ANGLES

POSITION	ANGLE 1001-778	ANGLE 778-779
1	40° 37' 36"	38° 14' 30"
2	40° 37' 40"	38° 14' 29"
3	40° 37' 39"	38° 14' 28"
4		

FIGURE 4.13 Field notes for angles by direction.

4.13 Adjustment of the Theodolite and Total Station

Figure 4.14 shows the geometric features of the theodolite. The most important relationships are as follows:

1. The axis of the plate bubble should be in a plane perpendicular to the vertical axis.
2. The vertical cross hair should be perpendicular to the horizontal axis (tilting axis).
3. The line of sight should be perpendicular to the horizontal axis.
4. The horizontal axis should be perpendicular to the vertical axis (standards adjustment)

In addition, the following secondary feature must be considered:

5. The vertical circle should read 90° (or 270°) when the telescope is level.

These features are discussed in the following paragraphs. See Appendix G for additional adjustments applicable to the engineers' transit.

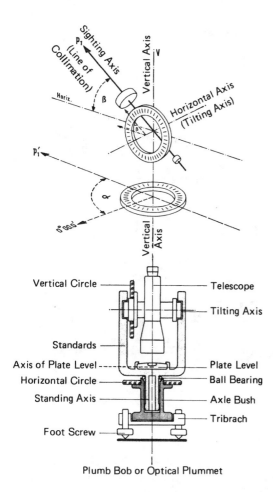

Vertical Axis

Sighting Axis
(Line of Collimation)

P_1

Horiz.

β

Horizontal Axis
(Tilting Axis)

P_1'

α

0° 00.0'

Vertical Axis

Vertical Circle — — Telescope

Tilting Axis

Standards

Axis of Plate Level — — Plate Level

Horizontal Circle — — Ball Bearing

Standing Axis — — Axle Bush

Foot Screw — — Tribrach

Plumb Bob or Optical Plummet

FIGURE 4.14 Geometry of the theodolite and total station. (Courtesy of Leica Geosystems)

4.13.1   Plate Bubbles

We have noted that, after a bubble has been centered, its adjustment is checked by rotating the instrument through 180°. If the bubble does not remain centered, it can be set properly by bringing the bubble halfway back using the foot screws. For example, if you are checking a tubular bubble accuracy position and it is out by four division marks, the bubble can now be set properly by turning the foot screws until the bubble is only two division marks off center. The bubble should remain in this off-center position as the telescope is rotated, indicating that the instrument is, in fact, level. Although the instrument can now be safely used, it is customary to remove the error by adjusting the bubble tube.

The bubble tube can now be adjusted by turning the capstan screws at one end of the bubble tube until the bubble becomes precisely centered. The entire leveling and adjusting procedure is repeated until the bubble remains centered as the instrument is rotated and checked in all positions. All capstan screw adjustments are best done in small increments. That is, if the end of the bubble tube is to be lowered, first loosen the lower capstan screw

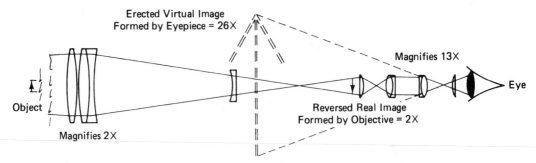

FIGURE 4.15 Diagram of a telescope optical system. (Courtesy of SOKKIA Corp.)

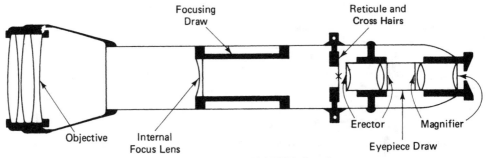

FIGURE 4.16 Theodolite telescope. (Courtesy of SOKKIA Corp.)

a slight turn (say, one-eighth), then tighten (snug) the top capstan screw to close the gap. This incremental adjustment is continued until the bubble is centered precisely.

4.13.2 Vertical Cross Hair

If the vertical cross hair is perpendicular to the horizontal axis, all parts of the vertical cross hair can be used for line and angle sightings. This adjustment can be checked by first sighting a well-defined distant point and then clamping the horizontal movements. The telescope (see Figures 4.15 and 4.16) is now moved up and down so that the point sighted appears to move on the vertical cross hair. If the point appears to move off the vertical cross hair, an error exists and the cross-hair reticule must be rotated slightly until the sighted point appears to stay on the vertical cross hair as it is being revolved. The reticule can be adjusted slightly by loosening two adjacent capstan screws, rotating the reticule, and then retightening the same two capstan screws.

This same cross-hair orientation adjustment is performed on the level, but in the case of the level, the horizontal cross hair is of prime importance. The horizontal cross hair is checked by first sighting a distant point on the horizontal cross hair with the vertical clamp set. Then move the telescope left and right, checking to see that the sighted point remains on the horizontal cross hair. The adjustment for any maladjustment of the reticule is performed as described previously.

4.13.3 Line-of-Sight Collimation Corrections

4.13.3.1 Vertical Cross Hair.
The vertical line of sight should be perpendicular to the horizontal axis so that a vertical plane is formed by the complete revolution of the telescope on its axis. The technique for this testing and adjustment is very similar to the surveying technique of double centering (described in Section 4.15). The testing and adjustment procedure is described in the following steps:

TEST

1. Set up the instrument at point A [see Figure 4.17(a)]
2. Take a backsight on any well-defined point B (a well-defined point on the horizon is best).
3. Transit (plunge) the telescope and set a point C_1 on the opposite side of the transit 300 to 400 ft away, at roughly the same elevation as the instrument station.
4. After loosening the upper or lower horizontal plate clamp (if the instrument has two clamps), and with the telescope still inverted from step 3, sight point B again.
5. After sighting on point B, transit the telescope and sight in the direction of the previously set point C_1. It is highly probable that the line of sight will not fall precisely on point C_1, so set a new point—point C_2— adjacent to point C_1.
6. With a steel tape, measure between point C_1 and point C_2 and set point C midway between them. Point C will be the correct point—the point established precisely on the projection of line BA. These six steps describe double centering (a surveying technique of producing a straight line — see Section 4.15).

ADJUSTMENT

7. Because the distance C_1C or CC_2 is double the sighting error, the line-of-sight correction is accomplished by first setting point E midway between C and C_2 (or one-quarter of the way from C_1 to C_2).
8. After sighting the vertical cross hair on point E, the vertical cross-hair correction adjustment is performed by adjusting the left and right capstan screws (see Figure 4.16) until the vertical cross hair is positioned directly on point C. The capstan screws are adjusted by first loosening one screw a small turn and then immediately tightening the opposite capstan screw to take up the slack. This procedure is repeated until the vertical cross hair is positioned precisely on point C.

Note: On instruments with dual-axis compensation, the components of the standing axis tilt are measured, thus enabling the instrument to correct angles automatically for the standing axis tilt.

4.13.3.2 Horizontal Cross Hair.
The horizontal cross hair must be adjusted so that it lies on the optical axis of the telescope. To test this relationship, set up the instrument

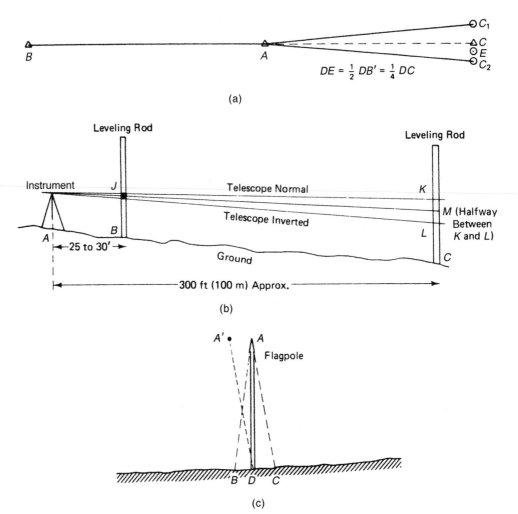

$$DE = \tfrac{1}{2} DB' = \tfrac{1}{4} DC$$

(a)

(b)

(c)

FIGURE 4.17 Instrument adjustments. (a) Line of sight perpendicular to horizontal axis. (b) Horizontal cross-hair adjustment. (c) Standards adjustment.

at point A [see Figure 5.13(b)], and place two stakes B and C in a straight line, B about 25 ft away and C about 300 ft away. With the vertical motion clamped, take a reading first on $C(K)$ and then on $B(J)$. Transit (plunge) the telescope and set the horizontal cross hair on the previous rod reading (J) at B and then take a reading at point C. If the theodolite is in perfect adjustment, the two rod readings at C will be the same; if the cross hair is out of adjustment, the rod reading at C will be at some reading L instead of K. To adjust the cross hair, adjust the cross-hair reticule up or down by first loosening and then tightening the appropriate opposing capstan screws until the cross hair lines up with the average of the two readings (M).

Note: Total station software can address both these errors in the following manner. After making sure the instrument is leveled precisely and after sighting a distant backsight, transit the telescope forward to sight a point on a rod (face 1); the horizontal and vertical angles are recorded. When sighting the point on the rod again in face 2, the horizontal and vertical angles are noted again. The horizontal angle should be different from the face 1 reading by exactly 180°—any discrepancy is double the horizontal collimation error. Similarly, if the two vertical angles (face 1 and face 2) do not equal 360°, the discrepancy is twice the vertical collimation error. If this procedure is repeated several times, the total station software can estimate the collimation errors closely and proceed to remove these errors in future single-face measurement work.

4.13.4 Circular Level

Optical and electronic theodolites and total stations use a circular level for rough leveling as well as plate level(s) for fine leveling. (See Chapter 5 for electronic leveling techniques.) After the plate level has been set and adjusted, as described in Section 4.13.2, the circular bubble can be adjusted (centered) by turning one or more of the three adjusting screws around the bubble.

4.13.5 Optical and Laser Plummets

The optical axis of these plummets is aligned with the vertical axis of the instrument if the reticule of the optical plummet or laser dot stays superimposed on the ground mark when the instrument is revolved through 180°. If the reticule or laser dot does not stay on the mark, the plummet can be adjusted in the following manner. The reticule or laser dot is put over the ground mark by adjusting the leveling screws [see Figure 4.18(a)]. If the plummet is not in adjustment, the reticule (laser dot) appears to be in a new location (*X*) after the theodolite is turned about 180°; point *P* is marked halfway between the two locations. The adjusting screws [see Figure 4.18(b)] are turned until the reticule image or laser dot is over *P*, indicating that the plummet is now in adjustment.

4.14 Laying Off Angles

4.14.1 Case 1

The angle is to be laid out no more precisely than the least count of the theodolite or total station. Assume that you must complete a route survey where a deflection angle (31°12′ R) has been determined from aerial photos to position the ℄ (centerline) clear of natural obstructions [see Figure 4.19(a)]. The surveyor sets the instrument at the *PI* (point of intersection of the tangents), and then sights the back line with the telescope reversed and the horizontal circle set to zero. Next, the surveyor transits (plunges) the telescope, turns off the required deflection angle, and sets a point on line. The deflection angle is 31°12′ R and a point is set at *a'*. The surveyor then loosens the lower motion (repeating instruments), sights again at the back line, transits the telescope, and turns off the required value (31°12′ × 2 = 62°24′). It is very likely that this line of sight will not precisely match the first sighting at *a'*; if the line does not cross *a'*, a new mark *a''* is made and the instrument operator gives a sighting on the correct point *a*, which is midway between *a'* and *a''*.

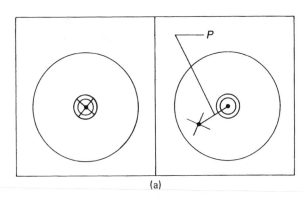

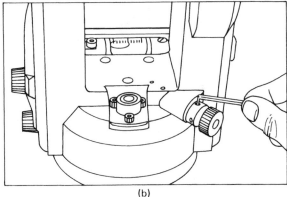

FIGURE 4.18 Optical plummet adjustment. (a) Point P is marked halfway between the original position (·) and the 180° position (X). (b) Plummet adjusting screws are turned until X becomes superimposed on P. (Courtesy of Nikon Instruments, Inc.)

4.14.2 Case 2

The angle is to be laid out more precisely than the least count of the instrument permits. Assume that an angle of 27°30′45″ is required in a heavy construction layout, and that a 01-minute theodolite is being used. In Figure 4.19(b), the theodolite is set up at A zeroed on B with an angle of 27°31′ turned to set point C'. The angle is then repeated to point C' a suitable number of times so that an accurate value of that angle can be determined. Let's assume that the scale reading after four repetitions is 110°05′, giving a mean angle value of 27°31′15″ for angle BAC'. If the layout distance of AC is 250.00 ft, point C can be located precisely by measuring from C' a distance $C'C$:

$$C'C = 250.00 \tan 0°00′30″$$
$$= 0.036 \text{ ft}$$

After point C has been located, its position can be verified by repeating angles to it from point B [see Figure 4.19(b)].

4.15 Prolonging a Straight Line (Double Centering)

Prolonging a straight line (also known as **double centering**) is a common survey procedure used every time a straight line must be prolonged. The best example of this requirement is

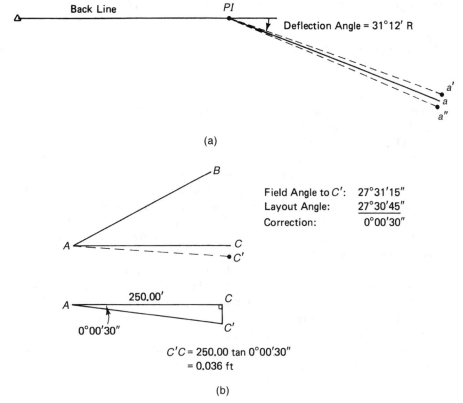

(a)

Field Angle to C': 27°31'15"
Layout Angle: 27°30'45"
Correction: 0°00'30"

$C'C = 250.00 \tan 0°00'30"$
$= 0.036$ ft

(b)

FIGURE 4.19 Laying off an angle. (a) Case 1, angle laid out to the least count of the instrument. (b) Case 2, angle to be laid out more precisely than the least count permitted by the instrument.

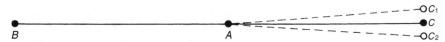

FIGURE 4.20 Double centering to prolong a straight line.

in route surveying, where straight lines are routinely prolonged over long distances and often over very difficult terrain. The technique of reversion (the same technique as used in repeating, or doubling, angles) is used to ensure that the straight line is prolonged properly.

In Figure 4.20, the straight line BA is to be prolonged to C [see also Figure 4.17(a)]. With the instrument at A, a sight is made carefully on station B. The telescope is transited, and a temporary point is set at C_1. The theodolite is revolved back to station B (the telescope is in a position reversed to the original sighting), and a new sighting is made. The telescope is transited, and a temporary mark is made at C_2, adjacent to C. (*Note:* Over short distances, well-adjusted theodolites will show no appreciable displacement between points C_1 and C_2. However, over the longer distances normally encountered in this type of work, all theodolites and total stations will display a displacement between direct and reversed sightings.

The longer the forward sighting, the greater the displacement.) The correct location of station C is established midway between C_1 and C_2 by measuring with a steel tape.

4.16 Bucking-In (Interlining)

It is sometimes necessary to establish a straight line between two points that themselves are not intervisible (i.e., a theodolite or total station set up at one point cannot, because of an intervening hill, be sighted at the other required point). It is usually possible to find an intermediate position from which both points can be seen.

In Figure 4.21, points A and B are not intervisible, but point C is in an area from which both A and B can be seen. The bucking-in procedure is as follows. The instrument is set up in the area of C (at C_1) and as close to line AB as is possible to estimate. The instrument is roughly leveled and a sight is taken on point A; then the telescope is transited and a sight is taken toward B. The line of sight will, of course, not be on B but on point B_1 some distance away. Noting roughly the distance B_1B and the position of the instrument between A and B (e.g., halfway, one-third, or one-quarter of the distance AB), the surveyor estimates proportionately how far the instrument is to be moved to be on the line AB. The instrument is once again roughly leveled (position C_2), and the sighting procedure is repeated.

This trial-and-error technique is repeated until, after sighting A, the transited line of sight falls on point B or close enough to point B so that it can be set precisely by shifting the theodolite on the leveling head shifting plate. When the line has been established, a point is set at or near point C so that the position can be saved for future use.

The entire procedure of bucking-in can be accomplished in a surprisingly short period of time. All but the final instrument setups are only roughly leveled, and at no time does the instrument have to be set up over a point.

4.17 Intersection of Two Straight Lines

The intersection of two straight lines is also a very common survey operation. In municipal surveying, street surveys usually begin $(0 + 00)$ at the intersection of the centerlines ₵ of two streets, and the station and angle of the intersections of all subsequent street's ₵ are routinely determined. Figure 4.22(a) illustrates the need for intersecting points on a municipal survey, and Figure 4.22(b) illustrates how the intersection point is located. In Figure 4.22(b), with the instrument set on a Main Street station and a sight taken also on the

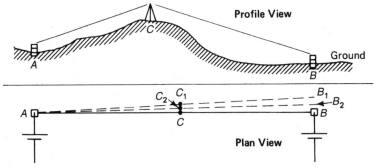

FIGURE 4.21 Bucking-in, or interlining.

Chap. 4 Angles and Theodolites

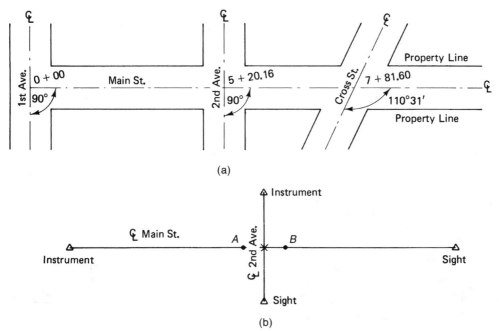

FIGURE 4.22 Intersection of two straight lines. (a) Example of the intersection of the centerlines of streets. (b) Intersecting technique.

Main Street ₵ (the longer the sight, the more precise the sighting), two points (2 to 4 ft apart) are established on the Main Street ₵, one point on either side of where the surveyor estimates that the 2nd Avenue ₵ will intersect. The instrument is then moved to a 2nd Avenue station, and a sight is taken some distance away, on the far side of the Main Street ₵. The surveyor can stretch a plumb bob string over the two points (A and B) established on the Main Street ₵, and the instrument operator can note where on the string the 2nd Avenue ₵ intersects. If the two points (A and B) are reasonably close together (2 to 3 ft), the surveyor can use the plumb bob itself to take a line from the instrument operator on the plumb bob string; otherwise, the instrument operator can take a line with a pencil or any other suitable sighting target. The intersection point is then suitably marked (e.g., nail and flagging on asphalt, wood stake with tack on ground), and then the angle of intersection and the station (chainage) of the point can be determined. After marking and referencing the intersection point, the surveyors remove the temporary markers A and B.

4.18 Prolonging a Measured Line by Triangulation over an Obstacle

In route surveying, obstacles such as rivers or chasms must be traversed. Whereas double centering can conveniently prolong the alignment, the station (chainage) may be deduced from the construction of a geometric figure. In Figure 4.23(a), the distance from 1 + 121.271 to the station established on the far side of the river can be determined by solving

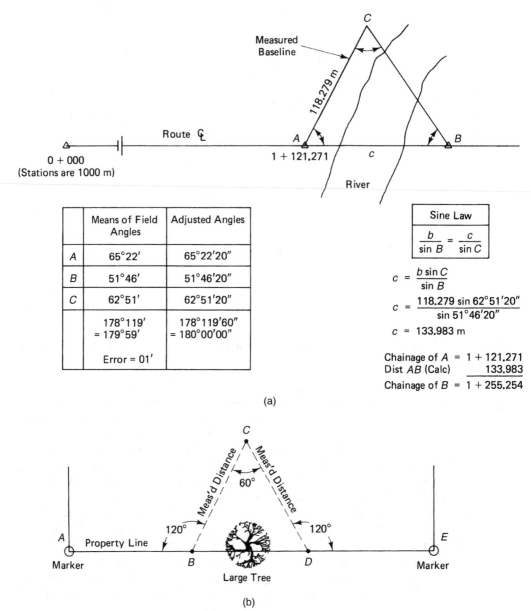

FIGURE 4.23 (a) Prolonging a measured line over an obstacle by triangulation. (b) Prolonging a line past an obstacle by triangulation.

the constructed triangle (triangulation). The ideal (geometrically strongest—see Section 9.8) triangle is one with angles close to 60° (equilateral), although angles as small as 20° may be acceptable. Rugged terrain and heavy tree cover adjacent to the river often result in a less than optimal geometric figure. The baseline and a minimum of two angles are measured so that the missing distance can be calculated. The third angle (on the far side of the river) should also be measured to check for mistakes and to reduce errors.

See Figure 4.23(b) for another illustration of triangulation. In Figure 4.23(b), the line must be prolonged past an obstacle: a large tree. In this case, a triangle is constructed with the three angles and two distances measured, as shown. As we noted earlier, the closer the constructed triangle is to equilateral, the stronger is the calculated distance (*BD*). Also, the optimal equilateral figure cannot always be constructed due to topographic constraints, and angles as small as 20° are acceptable for many surveys.

4.19 Prolonging a Line Past an Obstacle

In property surveying, obstacles such as trees often block the path of the survey. In route surveying, it is customary for the surveyor to cut down the offending trees (later construction will require them to be removed in any case), but in property surveying, the property owner would be quite upset to find valuable trees destroyed just so the surveyor could establish a boundary line. Thus, the surveyor must find an alternative method of providing distances and/or locations for blocked survey lines.

Figure 4.24(a) illustrates the technique of right-angle offset. Boundary line *AF* cannot be run because of the wooded area. The survey continues normally to point *B*, just clear of the wooded area. At *B*, a right angle is turned (and doubled), and point *C* is located a sufficient distance away from *B* to provide a clear parallel line to the boundary line. The instrument is set at *C* and sighted at *B* (great care must be exercised because of the short sighting distance); an angle of 90° is turned to locate point *D*. Point *E* is located on the

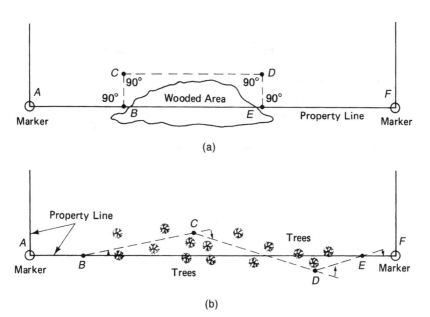

FIGURE 4.24 Prolonging a line past an obstacle. (a) Right-angle offset method. (b) Random-line method.

boundary line using a right angle and the offset distance used for *BC*. The survey can then continue to *F*. If distance *CD* is measured, then the required boundary distance (*AF*) is *AB* + *CD* + *EF*. If intermediate points are required on the boundary line between *B* and *E* (e.g., fencing layout), a right angle can be turned from a convenient location on *CD*, and the offset distance (*BC*) is used to measure back to the boundary line. Use of this technique minimizes the destruction of trees and other obstructions.

In Figure 4.24(b), trees are scattered over the area, preventing the establishment of a right-angle offset line. In this case, a random line (open traverse) is run (by deflection angles) through the scattered trees. The distance *AF* is the sum of *AB*, *EF*, and the resultant of *BE*. (See Chapter 6 for appropriate computation techniques for these type of problems.)

The technique of right-angle offsets has a larger potential for error because of the weaknesses associated with several short sightings; at the same time, however, this technique gives a simple and direct method for establishing intermediate points on the boundary line. In contrast, the random-line and triangulation methods provide for stronger geometric solutions to the missing property line distances, but they also require less direct and much more cumbersome calculations for the placement of intermediate line points (e.g., fence layout).

Review Questions

4.1 How does an electronic theodolite differ from a total station?

4.2 Describe, in point form, how to set up a theodolite or total station over a station point.

4.3 What are the difference between repeating and direction theodolites?

4.4 Describe, in point form, how you can use a theodolite or a total station to prolong a straight line *XY* to a new point *Z*. (Assume the instrument is set up over point *Y* and is level.)

4.5 How can you check to see that the plate bubbles of a theodolite or total station are adjusted correctly?

4.6 How can you determine the exact point of intersection of the centerlines of two cross streets?

4.7 What are some techniques of establishing a property line when trees block the instrument's line of sight?

4.8 Why do route surveyors prefer to use deflection angles instead of angles turned to the right or left when they are producing the survey line forward?

4.9 Describe how the introduction of electronic theodolites changed the field of surveying.

Chapter 5

Total Stations

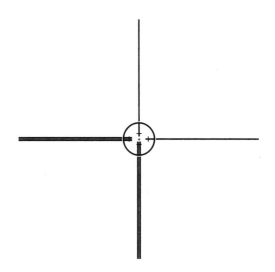

5.1   General Background

The electronic digital theodolite, first introduced in the late 1960s by Carl Zeiss Inc., helped set the stage for modern field data collection and processing. (See Figure 4.6 for a typical electronic theodolite.) When the electronic theodolite is used with a built-in electronic data measurement (EDM) or an add-on and interfaced EDM, the surveyor has a very powerful instrument. Add to that instrument an on-board microprocessor that automatically monitors the instrument's operating status and manages built-in surveying programs and a data collector (built-in or interfaced) that stores and processes measurements and attribute data, and you have what is known as a total station (see Figures 5.1 to 5.4).

5.2   Total Station Capabilities

5.2.1   General Background

Total stations can measure and record horizontal and vertical angles together with slope distances. The microprocessors in the total stations can perform several different mathematical operations, for example, averaging multiple angle measurements; averaging multiple distance measurements; determining horizontal and vertical distances; determining X, Y, and Z coordinates; and determining remote object elevations (that is, heights of sighted features) and distances between remote points. Some total stations also have the on-board capability of measuring atmospheric conditions and then applying corrections to field measurements. Some total stations are equipped to store collected data in on-board data cards (see Figures 5.1 and 5.2), whereas other total stations come equipped with data collectors, also known as field controllers, that are connected by cable (or wirelessly) to the instrument (see Figures 5.3 and 5.4).

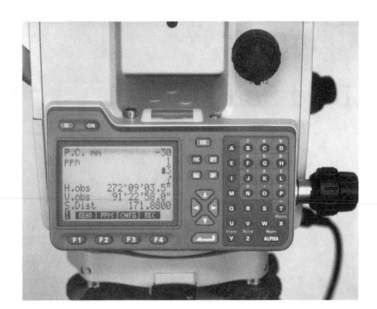

(a) (b)

■ Graphic "Bull's-Eye" Level

A graphically displayed "bull's-eye" lets you quickly and efficiently level the instrument.

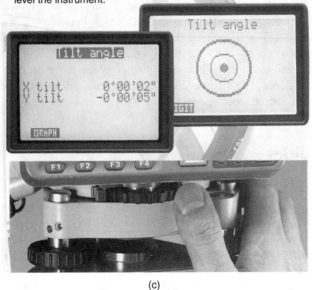

(c)

FIGURE 5.1 (a) Sokkia SET 1000 total station having angle display of 0.5 seconds (1″ accuracy) and a distance range to 2,400 m using one prism. The instrument comes with a complete complement of surveying programs, dual-axis compensation, and the ability to measure (to 120 m) to reflective sheet targets—a feature suited to industrial surveying measurements. (b) Keyboard and LCD display. (c) Graphics bull's-eye level allows you to level the instrument while observing the graphics display. (Courtesy of SOKKIA Corp.)

(continued)

Simultaneous Detection of Inclination in Two Directions and Automatic Compensation

The built-in dual-axis tilt sensor constantly monitors the inclination of the vertical axis in two directions. It calculates the compensation value and automatically corrects the horizontal and vertical angles. (The compensation range is +3'.)

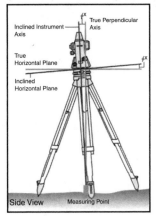
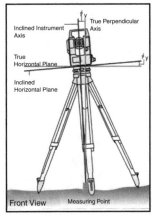

(d)

FIGURE 5.1 *(continued)*　(d) Dual-axis compensation illustration. (Courtesy of SOKKIA Corp.)

Figure 5.1 shows a typical total station, one of a series of instruments that have angle accuracies from 0.5 to 5 seconds, and distance ranges (to one prism) from 1,600 to 3,000 m. We noted in Chapter 3 that modern levels employ compensation devices to correct the line of sight automatically to its horizontal position; total stations come equipped with single-axis or dual-axis compensation [see Figure 5.1(d)]. With total stations, dual-axis compensation measures and corrects for left/right (lateral) tilt and forward/back (longitudinal) tilt. Tilt errors affect the accuracies of vertical angle measurements. Vertical angle errors can be minimized by averaging the readings of two-face measurements; with dual-axis compensation, however, the instrument's processor can determine tilt errors and remove their effects from the measurements, thus much improving the surveyor's efficiency. Dual-axis compensation also permits the inclusion of electronic instrument leveling [see Figures 5.1(c) and 5.25(b)], a technique that is much faster than the repetitive leveling of plate bubbles in the two positions (180° opposed) required by instruments without dual-axis compensation.

Modern instruments have a wide variety of built-in instrument monitoring, data collection, and layout programs and rapid battery charging that can charge the battery in 70 minutes. Data are stored on-board in internal memory (about 1,300 points) and/or on memory cards (about 2,000 points per card). Data can be transferred directly to the computer from the total station via an RS232 cable, or data can be transferred from the data-storage cards first to a card reader-writer and from there to the computer. Newer total stations have the capability of downloading and uploading data wirelessly, via the Internet, using the latest in cell-phone technology.

Figure 5.2 shows another typical total station with card driver for application program cards (upper drive) and for data-storage cards (lower drive). The data-storage cards can be

Upper Card Drive
Holds Program Cards

Lower Card Drive
Holds Data Memory
Cards

(a)

Main Menu

FNC
Switches
View/Edit
Total Station
Help
Notes
Setting
Status
Check Level
Others

FILE MANAGEMENT
1. File Manager
2. File Settings
3. Data Transfer

SURVEY I
1. Station Setup
2. Collection
3. Stakeout
4. Cogo

CONFIGURATION
1. Calibration
2. Settings

STATION SETUP
1. Known Station
2. Resection

COGO
1. Inverse
2. Calc. Pt.

(b)

FIGURE 5.2 (a) Nikon DTM 750 total station featuring on-board storage on PCMCIA computer cards; applications software on PCMCIA cards (upper drive): guidelight for layout work (to 100 m); dual-axis tilt sensor; EDM range of 2,700 m (8,900 ft) to one prism; angle precision of 1 second to 5 seconds; distance precision of ± (2 + 2 ppm) mm. The operating system is MS-DOS compatible. (b) Menu schematic for the Nikon DTM 750. (Courtesy of Nikon Instruments, Inc.)

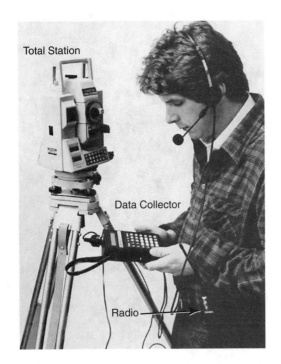

Total Station

Data Collector

Radio

FIGURE 5.3 Sokkia total station with cable-connected electronic field book. Also shown is a two-way radio (two-mile range) with push-to-talk headset.

removed when they are full and read into a computer using standard PCMCIA card drives—now standard on most notebook computers. The operating system for this and some other total stations is MS-DOS compatible, which permits the addition of user-defined applications software.

All the data collectors (built-in and handheld) described here are capable of doing much more than just collecting data. The capabilities vary a great deal from one model to another. The computational characteristics of the main total stations themselves also vary widely. Some total stations (without the attached data collector) can compute remote elevations and distances between remote points, whereas others require the interfaced data collector to perform these functions.

Many recently introduced data collectors are really handheld computers, ranging in price from $600 to $2,000. Costs have moderated recently because the surveying field has seen an influx of various less expensive handheld or pocket PC's (for example, from Compaq and Panasonic) for which total station data-collection software has been written. Some data collectors have graphics touch screens (stylus or finger) that permits the surveyor to input or edit data rapidly. For example, some collectors allow the surveyor to select (by touching) the line-work function that enables displayed points to be joined into lines—without the need for special line-work coding (see Section 5.5). If the total station is used alone, the capability of performing all survey computations—including traverse closures and adjustments—is highly desirable. If the total station is used in topographic surveys as a part of a system (field data collection, data processing, and digital plotting), however, the extended computational capacity of some data collectors (e.g., traverse closure computations) becomes less important.

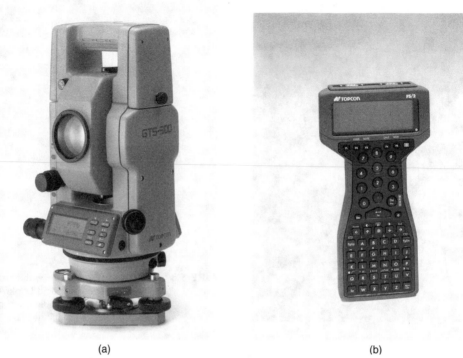

| (a) | (b) |

FIGURE 5.4 (a) Topcon GTS 300 total station. The 300 series instruments have angle accuracies from 1 second to 10 seconds and single prism distances from 1,200 m (3,900 ft) to 2,400 m (7,900 ft) at 2 mm + 2 ppm accuracy. (b) Topcon FS2 data collector for use with GTS 300 series total stations. (Courtesy of Topcon Positioning Systems, Inc.)

If the total station is used as part of a topographic surveying system (field data collection—data processing—digital plotting), the data collector need collect only the basic information, that is, slope distance, horizontal angle, and vertical angle, giving coordinates, and the sighted points' attribute data, such as point number, point description, and any required secondary tagged data. Computations and adjustments to the field data are then performed by one of the many coordinate geometry programs now available for surveyors and engineers. However, more sophisticated data collectors are needed in many applications, for example, machine guidance, construction layouts, GIS surveys, real-time GPS surveys (see Chapter 8), etc.

Most early total stations (and some current models) use the absolute method for reading angles. These instruments are essentially optical coincidence instruments with photoelectric sensors that scan and read the circles, which are divided into preassigned values, from 0° to 360° (or 0 to 400 grad or gon). Other models employ a rotary encoder technique of angle measurement (see Figure 4.7). These instruments have two (one stationary and one rotating) glass circles divided into many graduations (5,000 to 20,000). Light is projected through the circles and is converted to electronic signals, which the instrument's processor converts to the angle measurement between the fixed and rotating circles. Accuracies can be improved by averaging the results of the measurements on opposite sides of the circles. Some manufacturers determine the errors in the finished instrument and then install a processor program to remove such errors during field use, thus giving one-face measurements the accuracy of two-face measurements. Both systems enable the surveyor to assign 0° (or any other value) conveniently to an instrument setting after the instrument has been sighted-in and clamped.

Total stations have on-board microprocessors that monitor the instrument status (e.g., level and plumb orientation, battery status, return signal strength) and make corrections to measured data for the first of these conditions when warranted. In addition, the microprocessor controls the acquisition of angles and distances and then computes horizontal distances, vertical distances, coordinates, and the like. Many total stations are designed so that the data stored in the data collector can be downloaded quickly to a computer via an RS232 interface. The manufacturer usually supplies the download program; a second program is required to translate the raw data into a format that is compatible with the surveyor's coordinate geometry (i.e., processing) programs.

Most total stations also enable the topographic surveyor to capture the slope distance and the horizontal and vertical angles to a point simply by pressing one button. The point number (most total stations have automatic point-number incrementation) and point description for that point can then be entered and recorded. In addition, the wise surveyor prepares field notes showing the overall detail and the individual point locations. These field notes help the surveyor to keep track of the completeness of the work, and they are invaluable during data editing and preparation of the plot file.

5.2.2 Total Station Instrument Errors

Section 4.13 explained the basic tests for checking theodolite and total station errors. One advantage of total stations over electronic theodolites is that the total station has built-in programs designed to monitor the operation of the instrument and to recalibrate as needed. One error, the line-of-sight error (see Section 4.13.3), can be removed by using two-face measurements (double centering), or simply by using the instrument's on-board calibration program. This line-of-sight error affects the horizontal angle readings more seriously as the vertical angle increases. The tilting axis error [Figure 5.1(d)], which has no impact on perfectly horizontal sightings, also affects the horizontal angle reading more seriously as the vertical angle increases. The effect of the error can also be removed from measurements by utilizing the total station's built-in calibration procedures or by using two-face measurements. The automatic target recognition (ATR) collimation error reflects the angular divergence between the line of sight and the digital camera axis. This ATR is calibrated using on-board software (see Section 5.7.1).

5.3 Total Station Field Techniques

Total stations and/or their attached data collectors have been programmed to perform a wide variety of surveying functions. All total station programs require that the instrument station and at least one reference station be identified so that all subsequent tied-in stations can be defined by their X (easting), Y (northing), and Z (elevation) coordinates. The instrument station's coordinates and elevation, together with the azimuth to the backsight reference station (or its coordinates), can be entered in the field or uploaded prior to going out to the field. After setup and before the instrument has been oriented for surveying as described above, the hi (height of instrument above the instrument station) and prism heights must be measured and recorded.

Typical total station programs include point location, missing line measurement, resection, azimuth determination, remote object elevation, offset measurements, layout or setting out positions, and area computations. All these topics are discussed in the following sections.

5.3.1 Point Location

5.3.1.1 General Background.
After the instrument has been properly oriented, the coordinates (northing, easting, and elevation) of any sighted point can be determined, displayed, and recorded in the following format: N. E. Z. or E. N. Z.; the order you choose reflects the format needs of the software program chosen to process the field data. At this time, the sighted point is numbered and coded for attribute data (point description)—all of which is recorded with the location data. This program is used extensively in topographic surveys. See Figure 5.5.

5.3.1.2 Trigonometric Leveling.
Chapter 3 explained how elevations were determined using automatic and digital levels, and Sections 3.8 and 3.13 briefly introduced the concept of trigonometric leveling. Field practice has shown that total station observations can produce elevations at a level of accuracy suitable for the majority of topographic and engineering applications. For elevation work, the total station should be equipped with dual-axis compensation to ensure that angle errors remain tolerable, and for more precise work, the total station should have the ability to measure to 1 second of arc.

In addition to the uncertainties caused by errors in the zenith angle, the surveyor should be mindful of the errors associated with telescope collimation; other instrument and

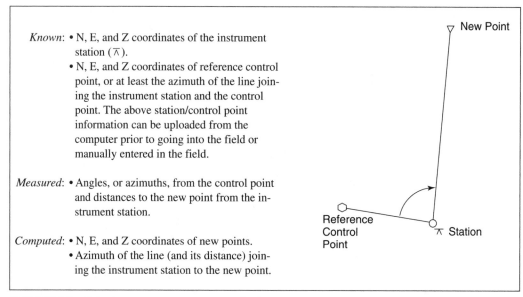

Known: • N, E, and Z coordinates of the instrument station (⊼).
• N, E, and Z coordinates of reference control point, or at least the azimuth of the line joining the instrument station and the control point. The above station/control point information can be uploaded from the computer prior to going into the field or manually entered in the field.

Measured: • Angles, or azimuths, from the control point and distances to the new point from the instrument station.

Computed: • N, E, and Z coordinates of new points.
• Azimuth of the line (and its distance) joining the instrument station to the new point.

New Point

Reference Control Point

⊼ Station

FIGURE 5.5 Point location program inputs and solutions.

prism imperfections; measurement of the height of the instrument (hi); measurement of the height of the prism (HR); curvature, refraction, and other natural effects; and errors associated with the use of a handheld prism pole. Also to improve precision, multiple readings can be averaged, readings can be taken from both face 1 and face 2 of the instrument, and reciprocal readings can be taken from either end of the line, with the results averaged. If the accuracy of a total station is unknown or suspect and if the effects of the measuring conditions are unknown, the surveyor should compare his or her trigonometric leveling results with differential leveling results (using the same control points) under similar measuring conditions.

Table 5.1 shows the effect of angle uncertainty on various distances. The basis for the tabulated results is the cosine function of ($90° \pm$ the angle uncertainty) times the distance. Analyze the errors shown in Table 5.1 with respect to the closure requirements shown in Tables 3.1 and 3.2.

5.3.2   Missing Line Measurement

This program enables the surveyor to determine the horizontal and slope distances between any two sighted points, as well as the directions of the lines joining those sighted points (see Figure 5.6). On-board programs first determine the N, E, and Z coordinates of the sighted points and then compute (inverse) the adjoining distances and directions.

5.3.3   Resection

This technique permits the surveyor to set up the total station at any convenient position (sometimes referred to as a **free station**) and then to determine the coordinates and elevation of that instrument position by sighting previously coordinated reference stations (see Figure 5.7). When sighting only two points of known position, it is necessary to measure and record both the distances and the angle between the reference points. When sighting several points (three or more) of known position, it is necessary only to measure the angles between the points. It is important to stress that most surveyors take more readings than are minimally necessary to obtain a solution; these redundant measurements give the surveyor increased precision and a check on the accuracy of the results. Once the instrument station's coordinates have been determined, the instrument can now be oriented, and the surveyor can continue to survey using any of the other techniques described in this section.

Table 5.1 ELEVATION ERRORS

| Distance (ft/m) | Zenith Angle Uncertainty | | | | |
	1 Second	5 Seconds	10 Seconds	20 Seconds	60 Seconds
100	0.0005	0.0024	0.005	0.0097	0.029
200	0.0010	0.0048	0.010	0.0194	0.058
300	0.0015	0.0072	0.015	0.029	0.087
400	0.0019	0.0097	0.019	0.039	0.116
500	0.0024	0.0121	0.024	0.049	0.145
800	0.0039	0.0194	0.039	0.078	0.233
1,000	0.0048	0.0242	0.0485	0.097	0.291

Known:	• N, E, and Z coordinates of the instrument station ($\overline{\wedge}$). • N, E, and Z coordinates of a reference control point, or at least the azimuth of the line joining the instrument station and the control point.	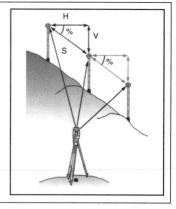
Measured:	• Angles, or azimuths, from the control point and distances to the new points from the instrument station.	
Computed:	• N, E, and Z (elevation) coordinates of new points. • Azimuths and horizontal, slope and vertical distances between the sighted points. • Slopes of the lines joining any sighted points.	

FIGURE 5.6 Missing line measurement program inputs and solutions. (Graphics courtesy of SOKKIA Corp.)

Known:	• N, E, and Z coordinates of control point # 1. • N, E, and Z coordinates of control point # 2. • N, E, and Z coordinates of additional sighted control stations (up to a total of ten control points). These can be entered manually or uploaded from the computer— depending on the instrument's capabilities.	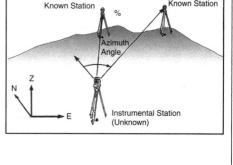
Measured:	• Angles between the sighted control points. • Distances are also required from the instrument station to the control points, if only two control points are sighted. • The more measurements (angles and distances), the better the resection solution.	
Computed:	• N, E, and Z (elevation) coordinates of the instrument station ($\overline{\wedge}$).	

FIGURE 5.7 Resection program inputs and solutions. (Graphics courtesy of SOKKIA Corp.)

5.3.4 Azimuth

When the coordinates of the instrument station and a backsight reference station have been entered into the instrument processor, the azimuth of a line joining any sighted points can be readily displayed and recorded (see Figure 5.8). When using a motorized total station equipped with automatic target recognition, the automatic angle measurement program permits you to repeat angles automatically to as many as ten previously sighted stations (including on face 1 and face 2). Just specify the required accuracy (standard deviation) and the sets of angles to be measured (up to twenty angles sets at each point), and the motorized instrument performs the necessary operations automatically.

Known: • N, E, and Z coordinates of the instrument station (π).
   • N, E, and Z coordinates of a reference control point,
      or at least the azimuth of the line joining the instru-
      ment station and the control point.

Measured: • Angles from the control point and distances to the
      new points from the instrument station.

Computed: • Azimuth of the lines joining the new points to the in-
      strument station.
   • Slopes of the lines joining any sighted points.
   • Elevations of any sighted points.
   • Coordinates of sighted points.

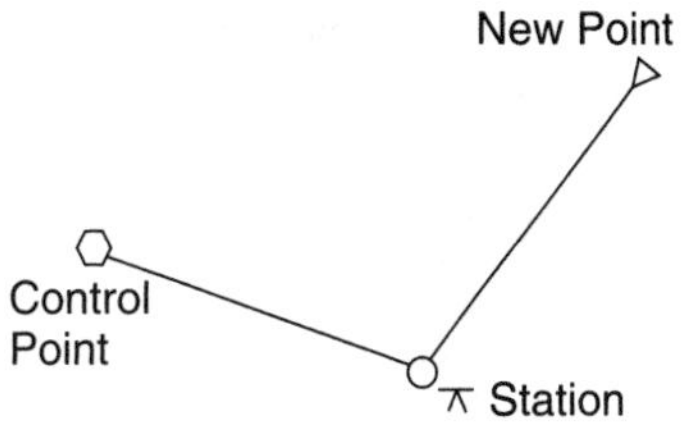

FIGURE 5.8 Azimuth program inputs and solutions.

Known: • N, E, and Z coordinates of the instrument
      station ($\overline{\lambda}$).
   • N, E, and Z coordinates of a reference
      control point, or at least the azimuth of
      the line joining the instrument station
      and the control point.

Measured: • Horizontal angle, or azimuth, from the
      reference control point (optional) and
      distance from the instrument station
      to the prism held directly below
      (or above) the target point.
   • Vertical angles to the prism and to
      the target point hi and height of the prism.

Computed: • Distance from the ground to the target
      point (and its coordinates if required).

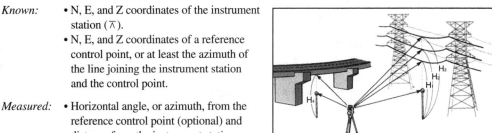

FIGURE 5.9 Remote object elevation program inputs and solutions. (Graphics courtesy of SOKKIA Corp.)

5.3.5 Remote Object Elevation

The surveyor can determine the heights of inaccessible points (e.g., electrical conductors, bridge components, etc.) by simply sighting the pole-mounted prism while it is held directly under the object. When the object itself is then sighted, the object height can be promptly displayed (the prism height must first be entered into the total station). See Figure 5.9.

5.3.6 Distance Offset Measurements

When an object is hidden from the total station, a measurement can be taken to the prism held out in view of the total station. The offset distance is measured from the prism (see Figure 5.10). The angle (usually 90°) or direction to the hidden object, along with the measured distance, is entered into the total station, enabling it to compute the position of the hidden object.

5.3.7 Angle Offset Measurements

When the center of a solid object (e.g., a concrete column, tree, etc.) is to be located, its position can be ascertained by turning angles from each side to the centerpoint (see Figure 5.11). The software then computes the center location of the measured solid object.

5.3.8 Layout or Setting-Out Positions

After the station numbers, coordinates, and elevations of the layout points have been uploaded into the total station, the layout/setting-out software enables the surveyor to locate any layout point by simply entering that point's number when prompted by the layout software. The instrument's display shows the left/right, forward/back, and up/down movements needed to place the prism in each of the desired position locations (see Figure 5.12). This capability is a great aid in property and construction layouts.

Known:	• N, E, and Z coordinates of the instrument station ($\overline{\wedge}$). • N, E, and Z coordinates (or the azimuth) for the reference control point.
Measured:	• Distance from the instrument station to the offset point. • Distance (l) from the offset point (at a right angle to the instrument line of sight) to the measuring point.
Computed:	• N, E, and Z (elevation) coordinates of the hidden measuring point. • Azimuths and distance from the instrument station to the hidden measuring point.

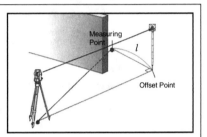

FIGURE 5.10 Offset measurements (distance) program solutions and inputs. (Graphics courtesy of SOKKIA Corp.)

Known:	• N, E, and Z coordinates of the instrument station ($\overline{\wedge}$). • N, E, and Z coordinates (or the azimuth) of the reference control point.
Measured:	• Angles from the prism being held on either side of the measuring point, to the target centerpoint. (The prism must be held such that both readings are the same distance from the instrument station—if both sides are measured).
Computed:	• N, E, and Z coordinates of the hidden measuring points. • Azimuths and distance from the instrument station to the hidden measuring point.

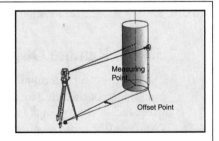

FIGURE 5.11 Offset measurements (angles) program inputs and solutions. (Graphics courtesy of SOKKIA Corp.)

Known:	• N, E, and Z coordinates of the instrument station ($\bar{\lambda}$).
	• N, E, and Z coordinates of a reference control point, or at least the azimuth of the line joining the instrument station and the control point.
	• N, E, and Z coordinates of the proposed layout points (entered manually or previously uploaded from the computer).
Measured:	• Indicated angles (on the instrument display), or azimuths, from the control point and distances to each layout point.
	• The angles may be turned manually or automatically if a servomotor-driven instrument is being used.
	• The distances (horizontal and vertical) are continually remeasured as the prism is eventually moved to the required layout position.

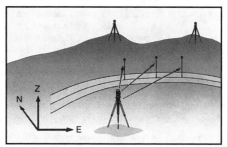

FIGURE 5.12 Laying out or setting out program inputs and solutions. (Graphics courtesy of SOKKIA Corp.)

5.3.9 Area Computation

While this program has been selected, the processor computes the area enclosed by a series of measured points. The processor first determines the coordinates of each station as described earlier and then, using those coordinates, computes the area in a manner similar to that described in Section 6.12. See Figure 5.13.

5.3.10 Summary of Typical Total Station Characteristics

The following lists summarize the characteristics of a typical total station.

PARAMETER INPUT

1. Angle units: degrees or gon
2. Distance units: feet or meters
3. Pressure units: inches HG or mm HG[*]
4. Temperature units: °F or °C[*]
5. Prism constant (usually −0.03 m). This constant is normally entered at the factory. When measuring to reflective paper, this prism constant must be corrected.
6. Offset distance (used when the prism cannot be held at the center of the object)
7. Face 1 or face 2 selection
8. Automatic point number incrementation (yes or no)
9. Height of instrument (hi)
10. Height of reflector (HR)
11. Point numbers and code numbers for occupied and sighted stations
12. Date and time settings (for total stations with on-board clocks)

---

[*]Some newer instruments have built-in sensors that detect atmospheric effects and automatically correct readings for these natural errors.

Known:	• N, E, and Z coordinates of the instrument station ($\overline{\wedge}$).
	• N, E, and Z coordinates of a reference control point, or at least the azimuth of the line joining the instrument station and the control point.
	• N, E, and Z coordinates of the proposed layout points (entered manually or previously uploaded from the computer).
Measured:	• Angles, or azimuths, from the control point and distances to the new points from the instrument station.
Computed:	• N, E, and Z coordinates of the area boundary points.
	• Area enclosed by the coordinated points.

FIGURE 5.13 Area computation program inputs and solutions. (Graphics courtesy of SOKKIA Corp.)

CAPABILITIES (COMMON TO MANY TOTAL STATIONS)

1. Telescope magnification power of 26× to 33× (most common is 30×)
2. Capable of orienting the horizontal circle to zero (or any other value) after sighting a backsight
3. Electronic leveling
4. Out-of-level warning
5. Dual-axis compensation
6. Automatic focus
7. Typical distance accuracy of ± (3 mm + 2 ppm)
8. Typical angle accuracy (and display) ranges from 1 to 10 seconds
9. Prismless measurement (70 to 1,200 m measurement range). This range depends on both the instrument's capabilities and on the reflective properties of the target's measuring surface.
10. Guide lights to permit prism-holder's self alignment
11. Horizontal and vertical collimation corrections—stored on-board and automatically applied
12. Vertical circle indexing and other error corrections—stored on-board and automatically applied
13. Central processor monitors and/or controls:
    * Battery status, signal attenuation, horizontal and vertical axes status, collimation factors
    * Computes coordinates: northing, easting, and elevation, or easting, northing, and elevation
    * Traverse closure adjustments and traverse areas
    * Topography reductions
    * Remote object elevation (that is, object heights)
    * Distances between remote points (missing line measurement)
    * Inversing
    * Resection
    * Layout (setting out; for example 3D and 2D highway layout programs

- Records search and review
- Programmable features—that is, load external programs
- Transfer of data to the computer (downloading)
- Transfer of computer files to the data collector (uploading) for layout and reference purposes

14. Robotic capabilities (700 to 1,200 m range for control):
- Servomotors—one to six speeds
- Automatic target recognition
- Target tracking/searching

15. Digital imaging (e.g., Topcon GPT 7000i)

5.4 Field Procedures for Total Stations in Topographic Surveys

Total stations can be used in any type of preliminary survey, control survey, or layout survey. They are particularly well suited for topographic surveys, in which the surveyor can capture the northings, eastings, and elevations of a large number of points—700 to 1,000 points per day. This significant increase in productivity means that in some medium-density areas, ground surveys are once again competitive in cost with aerial surveys. Although the increase in efficiency in the field survey is notable, an even more significant increase in efficiency can occur when the total station is part of a computerized surveying system—data collection, data processing (reductions and adjustments), and plotting (see Figure 7.1).

One of the notable advantages of using electronic surveying techniques is that data are recorded in electronic field books; this cuts down considerably on the time required to record data and on the opportunity of making transcription mistakes. Does this mean that manual field notes are a thing of the past? The answer is no. Even in electronic surveys, there is a need for neat, comprehensive field notes. At the very least, such notes will contain the project title, field personnel, equipment description, date, weather conditions, instrument setup station identification, and backsight station(s) identification. In addition, for topographic surveys, many surveyors include in the manual notes a sketch showing the area's selected (in some cases, all) details—individual details such as trees, catch basins, and poles, and stringed detail such as the water's edge, curbs, walks, and building outlines. As the survey proceeds, the surveyor places all or selected point identification numbers directly on the sketch feature, thus clearly showing the type of detail being recorded. Later, after the data have been transferred to the computer, and as the graphics features are being edited, manual field notes are invaluable in clearing up problems associated with any incorrect or ambiguous labeling and numbering of survey points. It seems that surveyors who use modern data collectors with touch-screen graphics have a reduced need for manual field notes.

5.4.1 Initial Data Entry

Most data collectors are designed to prompt for, or accept, some or all of the following initial configuration and operation data:

- Project description
- Date and crew

- Temperature
- Pressure (some data collectors require a ppm correction input, which is read from a temperature/pressure graph—see Figure 2.27)
- Prism constant (-0.03 m is a typical value—check the specifications)
- Curvature and refraction settings (see Chapter 3)
- Sea-level corrections (see Chapter 9)
- Number of measurement repetitions—angle or distance (the average value is computed)
- Choice of face 1 and face 2 positions
- Choice of automatic point-number increments (yes/no), and the value of the first point number in a series of measurements
- Choice of Imperial units or SI units for all data

Many of these prompts can be bypassed, causing the microprocessor to use previously selected, or default, values in its computations. After the initial data have been entered, and the operation mode has been selected, most data collectors prompt the operator for all station and measurement entries.

5.4.2 Survey Station Descriptors

Each survey station, or shot location (point), must be described with respect to surveying activity (e.g., backsight, intermediate sight, foresight), station identification, and descriptive attribute data. In many cases, total stations prompt for the data entry [e.g., occupied station, backsight reference station(s)] and then automatically assign appropriate labels, which will then show up on the survey printout. Point description data can be entered as alphabet or numeric codes (see Figure 5.14). This descriptive data will also show up on the printout and can (if desired) be tagged to show up on the plotted drawing. Some data collectors can be equipped with bar-code readers that permit instantaneous entry of descriptive data when they are used with prepared code sheets.

Some data collectors are designed to work with all (or most) total stations on the market. Figure 5.15 shows two such data collectors. Each of these data collectors has its own unique routine and coding requirements. As this technology continues to evolve and continues to take over many surveying functions, more standardized procedures will likely develop. Newer data collectors (and computer programs) permit the surveyor to enter not only the point code, but also various levels of attribute data for each coded point. For example, a tie-in to a utility pole can also tag the pole number, the use (e.g., electricity, telephone), the material (e.g., wood, concrete, steel), connecting poles, year of installation, and so on. This type of expanded attribute data is typical of the data collected for a municipal geographic information system (GIS). See Chapter 7.

Figure 5.15(a) shows a Trimble TSCe collector/control unit. This data collector can be used with both robotic total stations and GPS receivers. When used with TDS SurveyPro robotics software, this collector/controller features real-time maps, 3D design stakeout and interactive digital terrain modeling (DTM) with real-time cut-and-fill computations. When used with TDS SurveyPro GPS software, this collector/controller can be used for general GPS work as well as for real time (RTK) measurements at centimeter-level

Point Identification Codes
(shown is part of Seneca dictionary)

Survey Points

01	BM	Bench Mark
02	CM	Concrete Monument
03	SIB	Standard Iron Bar
04	IB	Iron Bar
05	RIB	Round Iron Bar
06	IP	Iron Pipe
07	WS	Wooden Stake
08	MTR	Coordinate Monument
09	CC	Cut Cross
10	N&W	Nail and Washer
11	ROA	Roadway
12	SL	Street Line
13	EL	Easement Line
14	ROW	Right of Way
15	CL	Centerline

Topography

16	EW	Edge Walk
17	ESHLD	Edge Shoulder
18	C&G	Curb and Gutter
19	EWAT	Edge of Water
20	EP	Edge of Pavement
21	RD	CL Road
22	TS	Top of Slope
23	BS	Bottom of Slope
24	CSW	Concrete Sidewalk
25	ASW	Asphalt Sidewalk
26	RW	Retaining Wall
27	DECT	Deciduous Tree
28	CONT	Coniferous Tree
29	HDGE	Hedge
30	GDR	Guide Rail
31	DW	Driveway
32	CLF	Chain Link Fence
33	PWF	Post and Wire Fence
34	WDF	Wooden Fence

Code Sheet for Field Use

1 BM	2 CM	3 SIB	4 IB	5 RIB
6 IP	7 WS	8 MTR	9 CC	10 N W
11 ROA	12 SL	13 EL	14 ROW	15 CL
16 EW	17 ESHL	18 C G	19 EWAT	20 EP
21 RD	22 TS	23 BS	24 CSW	25 ASW
26 RW	27 DECT	28 CONT	29 HDGE	30 GDR
31 DW	32 CLF	33 PWF	34 WDF	35 SIGN
36 MB	37 STM	38 HDW	39 CULV	40 SWLE
41 PSTA	42 SAN	43 BRTH	44 CB	45 DCB
46 HYD	47 V	48 V CH	49 M CH	50 ARV
51 WKEY	52 HP	53 UTV	54 LS	55 TP
56 PED	57 TMH	58 TB	59 BCM	60 GUY
61 TLG	62 BLDG	63 GAR	64 FDN	65 RWYX
66 RAIL	67 GASV	68 GSMH	69 G	70 GMRK
71 TL	72 PKMR	73 TSS	74 SCT	75 BR
76 ABUT	77 PIER	78 FTG	79 EDB	80 POR
81 SLS	82 WTT	83 STR	84 BUS	85 PLY
86 TEN	0	0	0	0
0	0	0	0	0
0	0	0	0	0

(a)   (b)

FIGURE 5.14 (a) Alphanumeric codes for sighted point descriptions. (b) Code sheet for field use.

accuracy. Figures 5.15(b), (c), and (d) show a handheld computer, together with a few screen displays, which has software designed for both data collection and stakeout activities. Although this data collector has fewer options than the one shown in Figure 5.15(a), its functionality and much lower cost make it attractive to some surveyors.

All total station/GPS manufacturers produce data collectors and software to execute topographic and layout surveys. Independent companies are also in the market. Micro-Survey Software Inc. has a touch-screen data collector, using total station/GPS software called FieldGenius, that works with most surveying hardware and sells for less than

(a)

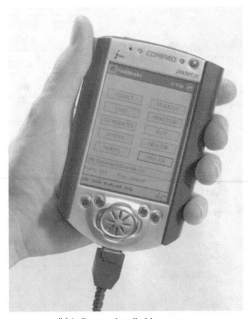

(b) i. Compaq handheld computer.

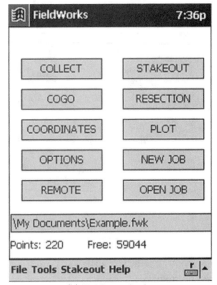

(b) ii. Menu default settings.

(b) iii. Menu selection.

FIGURE 5.15 (a) Trimble's TSCe data collector for use with robotic total stations and GPS receivers. When used with TDS Survey Pro Robotics software, this collector features real-time maps, 3D design stakeout, and interactive DTM with real-time cut-and-fill computations. When used with TDS Survey Pro GPS software, this collector can be used for general GPS work, as well as for RTK measurements at centimeter-level accuracy. (Portions © 2001 Trimble Navigation Limited. All Rights Reserved) (b) Handheld computer and surveying software. (Courtesy of XYZ Works)

$2,000 (including a collector). Another example is Penmap PC, marketed by Strata Software, using a GPS receiver and collecting data in a large-format, touch-screen tablet PC (see Figure 8.22).

5.4.3 AASHTO's Survey Data Management System (SDMS)

The American Association of State Highway and Transportation Officials (AASHTO) developed the Survey Data Management System (SDMS) in 1991 to aid in the nationwide standardization of field coding and computer processing procedures used in highway surveys. The field data can be captured automatically in the data collector, as in the case of total stations, or the data can be entered manually into the data collector for a wide variety of theodolite and level surveys. AASHTO's software is compatible with most recently designed total stations and some third-party data collectors, and data processing can be accomplished using most MS-DOS–based computers. For additional information on these coding standards, contact AASHTO at 444 N. Capitol St., N.W., Suite 225, Washington, DC 20001 (202-624-5800).

5.4.4 Occupied Point (Instrument Station) Entries

The following information is entered into the instrument to describe the setup station:

- Height of instrument (measured value is entered)
- Station number, for example, 111 (see the example in Figure 5.16)
- Station identification code [see the code sheet in Figure 5.14(b)]

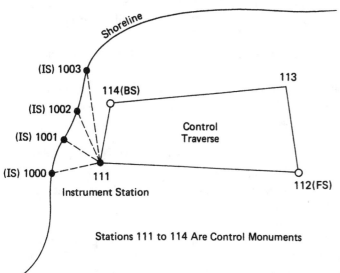

FIGURE 5.16 Sketch showing intermediate shoreline ties to a control traverse.

- Coordinates of occupied station—the coordinates can be assumed, state plane, or universal transverse Mercator (UTM). Such coordinates may have previously been uploaded from a computer.
- Coordinates of backsight (BS) station, or reference azimuth to BS station

Note: With some data collectors, the coordinates of the above stations may instead be uploaded from computer files. Once the surveyor identifies by point number specific points as being instrument station and backsight reference station(s), the coordinates of those stations then become active in the processor.

5.4.5 Sighted Point Entries

The following information is entered into the instrument to describe the sighted points:

- Height of prism/reflector (HR)—measured value is entered
- Station number—for example, 114 (BS); see Figure 5.16
- Station identification code (see Figure 5.14)

5.4.6 Procedures for the Example Shown in Figure 5.16

1. Set the total station at station 111 and enter the initial data and instrument station data, as shown in Sections 5.4.1 and 5.4.4.
2. Measure the height of the instrument, or adjust the height of the reflector to equal the height of the instrument.
3. Sight at station 114; zero the horizontal circle (any other value can be set instead of zero, if desired). Most total stations have a zero-set button.
4. Enter code (e.g., BS), or respond to the data collector prompt.
5. Measure and enter the height of the prism/reflector (HR). If the height of the reflector has been adjusted to equal the height of the instrument, the value of 1.000 is often entered for both values because these hi and HR values usually cancel each other in computations.
6. Press the appropriate measure buttons, for example, slope distance, horizontal angle, vertical angle.
7. Press the record button after each measurement; most instruments measure and record slope, horizontal, and vertical data after the pressing of just one button (when they are in the automatic mode).
8. After the station measurements have been recorded, the data collector program prompts for the station point number (e.g., 114) and the station identification code (e.g., 02 from Figure 5.14, which identifies "concrete monument").
9. If appropriate, as in traverse surveys, the next sight is to the FS station; repeat steps 5 through 8, using relevant data.
10. While at station 111, any number of intermediate sights (IS) can be taken to define the topographic features being surveyed. Most instruments have the option of speeding up the work by employing "automatic point number incrementation." For example, if the topographic readings are to begin with point number 1000, the surveyor will be

prompted to accept the next number in sequence (e.g., 1001, 1002, 1003, etc.) at each new reading; if the prism is later being held at a previously numbered point (e.g., control point #107), the prompted value can easily be temporarily overridden with 107 being entered.

In topographic work, the prism/reflector is usually mounted on an adjustable-length prism pole, with the height of the prism (HR) being set equal to the height of the total station (hi). The prism pole can be steadied with a brace pole, as shown in Figure 2.29, to improve the accuracy for more precise sightings.

Some software permits the surveyor to identify, by attribute name or an additional code number, points that will be connected (stringed) on the resultant plan (e.g., shoreline points in this example). This connect (on and off) feature permits the field surveyor to prepare the plan (for graphics terminal or digital plotter) while performing the actual field survey. Also, the surveyor can later connect the points while in edit mode on the computer, depending on the software in use; clear field notes are essential for this activity. Data collectors with touch-screen graphics can have points joined by screen touches—without any need for special stringed coding.

11. When all the topographic detail in the area of the occupied station (111) has been collected, the total station can be moved* to the next traverse station (e.g., 112 in this example), and the data collection can proceed in the same manner as that already described, that is, BS @ STA.111, FS @ STA.113, and take all relevant IS readings.

5.4.7 Data Download and Processing

In the example shown in Figure 5.16, the collected data now must be downloaded to a computer. The download computer program is normally supplied by the total station manufacturer, and the actual transfer can be cabled through an RS232 interface cable. Once the data are in the computer, the data must be sorted into a format that is compatible with the computer program that will process the data; this translation program is usually written or purchased separately by the surveyor.

Many modern total stations have data stored on-board, which eliminates the handheld data collectors. Some instruments store data on a module that can be transferred to a computer-connected reading device. Some manufacturers use PCMCIA cards, which can be read directly into a computer through a PCMCIA reader (see Figure 5.2). Other total stations, including the geodimeter (see Figure 5.25), can be downloaded by connecting the instrument (or its keyboard) directly to the computer; as noted earlier, some total stations can download wirelessly using modern cell-phone technology.

If the topographic data have been tied to a closed traverse, the traverse closure is calculated, and then all adjusted values for northings, eastings, and elevations (Y, X, Z, respectively) are computed. Some total stations have sophisticated data collectors (which are actually small computers) that can perform preliminary analysis, adjustments, and coordinate computations, whereas others require the computer program to perform these functions.

---

*When the total station is to be moved to another setup station, the instrument is always removed from the tripod and carried separately by its handle or in its case.

Once the field data have been stored in coordinate files, the data required for plotting by digital plotters can be assembled, and the survey can be plotted quickly at any desired scale. The survey can also be plotted at an interactive graphics terminal for graphics editing with the aid of one of the many available computer-aided design (CAD) programs.

5.5 Field-Generated Graphics

Many surveying software programs permit the field surveyor to identify field data shots so that subsequent processing will produce appropriate computer graphics. For example, MicroSurvey Software, Inc., has a typical "description to graphics" feature that enables the surveyor to join field shots, such as curb-line shots (see Figure 5.17) by adding a "Z" prefix to all but the last "CURB" descriptor. When the program first encounters the "Z" prefix, it begins joining points with the same descriptors; when the program encounters the first curb descriptor without the "Z" prefix, the joining of points is terminated. Rounding (e.g., curved curbs) can be introduced by substituting an "X" prefix for the "Z" prefix (see Figure 5.17). Other typical MicroSurvey graphics prefixes include the following:

- "Y" joins the last identical descriptor by drawing a line at a right angle to the established line (see the fence line in Figure 5.17).
- "-" causes a dot to be created in the drawing file, which is later transferred to the plan. The dot on the plan can itself be replaced by inserting a previously created symbol block, for example, a tree, a manhole, a hydrant (see MH in Figure 5.17). If a second dash follows the first prefix dash, the ground elevation will not be transferred to the graphics file (see HYD in Figure 5.17). This capability is useful because some feature elevations may not be required.
- "." instructs the system to close back on the first point of the string of descriptors with the same characters (see BLDG and BUS shelter in Figure 5.17 and POND in Figure 5.19).

Other software programs create graphic stringing by different techniques. Some manufacturers, for example, Sokkia, give the software code itself a stringing capability (e.g., fence1, curb1, curb2, ₵), which the surveyor can easily turn on and off (see Figure 5.18). In some cases, it may be more efficient to assign the point descriptors from the computer keyboard after the survey has been completed. For example, the entry of descriptors is time consuming on some data collectors, particularly in automatic mode. In addition, some topographic features (e.g., the edge of the water in a pond or lake) can be captured in sequence, thus permitting the surveyor to edit these descriptors efficiently from the computer in the processing stage instead of repeatedly entering dozens or even hundreds of identical attribute codes (some data collectors prompt for the last entry, thus avoiding the rekeying of identical attributes or descriptors). If the point descriptors are to be added at the computer, clear field notes are indispensable.

See Figure 5.19 for an illustration of this stringing technique. The pond edge has been picked up (defined) by ten shots, beginning with #58 and ending with #67. Using the MicroSurvey program introduced earlier, the point description edit feature is selected from the pull-down menu, and the following steps occur:

1. "Points to be described?" 58..66 (*enter*)
2. "Description?" ZPOND (*enter*)

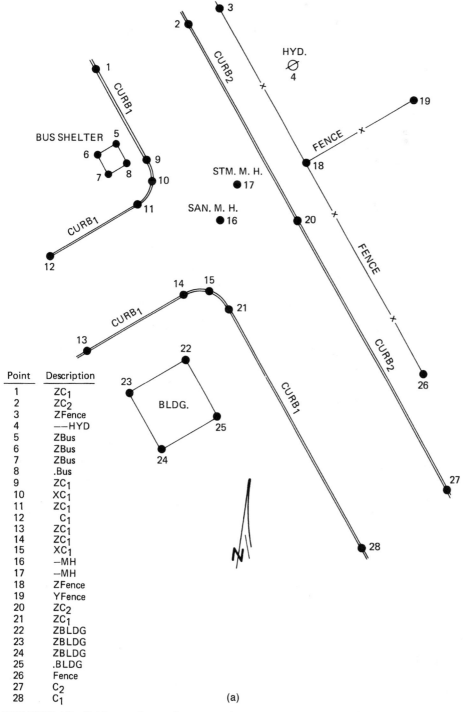

Point	Description
1	ZC_1
2	ZC_2
3	ZFence
4	——HYD
5	ZBus
6	ZBus
7	ZBus
8	.Bus
9	ZC_1
10	XC_1
11	ZC_1
12	C_1
13	ZC_1
14	ZC_1
15	XC_1
16	—MH
17	—MH
18	ZFence
19	YFence
20	ZC_2
21	ZC_1
22	ZBLDG
23	ZBLDG
24	ZBLDG
25	.BLDG
26	Fence
27	C_2
28	C_1

(a)

FIGURE 5.17 Field notes for total station graphics descriptors—MicroSurvey Software, Inc. codes.

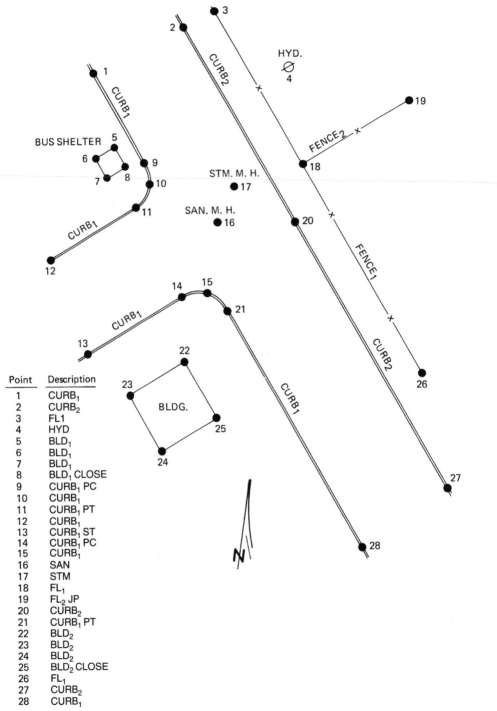

Point	Description
1	$CURB_1$
2	$CURB_2$
3	FL1
4	HYD
5	BLD_1
6	BLD_1
7	BLD_1
8	BLD_1 CLOSE
9	$CURB_1$ PC
10	$CURB_1$
11	$CURB_1$ PT
12	$CURB_1$
13	$CURB_1$ ST
14	$CURB_1$ PC
15	$CURB_1$
16	SAN
17	STM
18	FL_1
19	FL_2 JP
20	$CURB_2$
21	$CURB_1$ PT
22	BLD_2
23	BLD_2
24	BLD_2
25	BLD_2 CLOSE
26	FL_1
27	$CURB_2$
28	$CURB_1$

FIGURE 5.18 Field notes for total station graphics descriptors—SOKKIA codes.

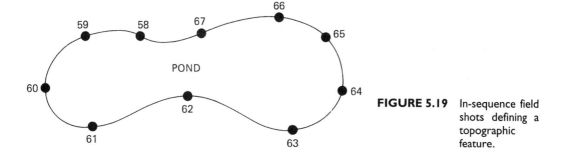

FIGURE 5.19 In-sequence field shots defining a topographic feature.

3. "Points to be described?" 67 *(enter)*

4. "Description?" .POND *(enter)*

After the point descriptions have been suitably coded (either by direct field coding or by postsurvey editing), a second command is accessed from another pull-down menu that simply (in one operation) converts the coded point description file so that the shape of the pond is produced in graphics. The four descriptor operations described here are less work than would be required to describe each point using field entries for the ten points shown in Figure 5.19. Larger features requiring many more field shots would be even more conducive to this type of postsurvey editing of descriptors.

It is safe to say that most projects requiring that graphics be developed from total station surveys utilize a combination of point description field coding and postsurvey point description editing. Note that the success of some of these modern surveys still depends to a significant degree on old-fashioned, reliable survey field notes. It is becoming clear that the "drafting" of the plan of survey is increasingly becoming the responsibility of the surveyor, either through direct field coding techniques or through postsurvey data processing. All recently introduced surveying software programs enable the surveyor to produce a complete plan of survey. Data collector software is now capable of exporting files in DXF format for AutoCad postprocessing or shape files for further processing in ESRI GIS programs. Also, alignments and cross sections are created on desktop computers and then exported to data collectors in the recently established standard Land XML file format.

5.6 Construction Layout Using Total Stations

We saw in the previous section that total stations are particularly well suited for collecting data in topographic surveys; we also noted that the collected data could be readily downloaded to a computer and processed into point coordinates—northing, easting, elevation (Y, X, Z)—along with point attribute data. The significant increases in efficiency made possible with total station topographic surveys can also be realized in layout surveys when the original point coordinates exist in computer memory or computer disks, together with the coordinates of all the key design points. To illustrate, consider the example of a road construction project.

First, the topographic detail is collected using total stations set up at various control points (preliminary survey). The detail is then transferred to the computer (and adjusted, if necessary), and converted into Y, X, and Z coordinates. Various coordinate geometry and road design programs can then be used to design the proposed road. When the proposed horizontal, cross-section, and profile alignments have been established, the proposed coordinates (Y, X, and Z) for all key horizontal and vertical (elevation) features can be computed and stored in computer files. The points coordinated include top-of-curb and centerline (℄) positions at regular stations, as well as all changes in direction or slope. Catch basins, traffic islands, and the like, are also included, as are all curved or irregular road components.

The computer files now include coordinates of all control stations, all topographic detail, and (finally) all design component points. Control point and layout point coordinates (X, Y, and Z) can then be uploaded into the total station. The layout is accomplished by setting the instrument to layout mode, then setting up at an identified control point, and properly orienting the sighted instrument toward another identified control point. Second, while still in layout mode, and after the first layout point number is entered into the instrument, the required layout angle or azimuth and layout distance are displayed. To set the selected layout point from the instrument station, the layout angle (or azimuth) is then turned (automatically by motorized total stations) and the distance is set by following the display instructions (backward/forward, left/right, up/down) to locate the desired layout point. When the total station is set to tracking mode, the surveyor can first set the prism close to target by rapid trial-and-error measurements.

Figure 5.20 illustrates a computer printout for a road construction project. The total station is set at control monument CX-80 with a reference backsight on RAP (reference azimuth point) 2 (point 957 on the printout). The surveyor has a choice of (1) setting the actual azimuth (213°57′01″, line 957) on the backsight and then turning the horizontal circle to the printed azimuths of the desired layout points or (2) setting 0° for the backsight and then turning the clockwise angle listed for each desired layout point. As a safeguard, after sighting the reference backsight, the surveyor usually sights a second or third control monument (see the top twelve points on the printout) to check the azimuth or angle setting. The computer printout also lists the point coordinates and baseline offsets for each layout point.

Figure 5.21 shows a portion of the construction drawing that accompanies the computer printout shown in Figure 5.20. The drawing (usually drawn by digital plotter from computer files) is an aid to the surveyor in the field for laying out the works correctly. All the layout points listed on the printout are also shown on the drawing, together with curve data and other explanatory notes. Modern total stations offer an even more efficient technique. Instead of having the layout data only on a printout similar to that shown in Figure 5.20, as noted earlier, the coordinates for all layout points can be uploaded into the total station microprocessor. The surveyor in the field can then identify the occupied control point and the reference backsight point(s), thus orienting the total station.

The desired layout point number is then entered, with the required layout angle and distance being inversed from the stored coordinates, and displayed. The layout can proceed by setting the correct azimuth. Then, by trial and error, the prism is moved to the layout distance (the total station is set to tracking mode for all but the final measurements). With some total stations, the prism is simply tracked with the remaining left/right (±) distance being displayed alongside the remaining near/far (±) distance. When the correct location has been

THE MUNICIPALITY OF METROPOLITAN TORONTO - DEPARTMENT OF ROADS AND TRAFFIC

W.R. ALLEN ROAD FROM SHEPPARD AVENUE TO STANSTEAD DRIVE

ENGINEERING STAKEOUT

FROM STATION   269+80.00   TO STATION   271+30.00

BASE LINE

INSTRUMENT ON   CONST CONTROL MON CX-80

                              AZIMUTH   213-57- 1

SIGHTING   RAP #2 - ANTENNA C.F.B.

POINT NO.	STATION	DESCRIPTION	OFFSET FROM BASELINE		AZIMUTH DEG-MIN-SEC	DISTANCE	CLOCKWISE TURN ANGLE	ELEVATION	COORDINATES NORTH	EAST
876	269+89.355	CONST CONTROL MON CX-76	22.862	LEFT	170-49-16	80.430	316-52-15	0.0	4845374.460	307710.370
885	271+29.785	CONST CONTROL MON CX-85	22.857	LEFT	350-49- 8	60.000	136-52- 7	0.0	4845513.091	307687.967
877	269+95.164	CONST CONTROL MON CX-77	22.861	RIGHT	139-19-24	87.513	285-22-24	0.0	4845387.490	307754.580
878	269+97.098	CONST CONTROL MON CX-78	38.095	RIGHT	130-50-11	94.861	276-53-11	0.0	4845391.830	307769.310
879	269+27.530	CONST CONTROL MON CX-79	38.081	RIGHT	115-33-22	74.155	261-36-21	0.0	4845421.870	307764.440
881	270+69.932	CONST CONTROL MON CX-81	22.862	RIGHT	80-38- 4	45.719	226-41- 3	0.0	4845461.300	307742.650
958	290+51.899	RAP #3 - RADIO TOWER	294.749	LEFT	344-19-40	1990.850	130-22-40	0.0	4847370.697	307159.747
884	271+29.932	CONST CONTROL MON CX-84	22.862	RIGHT	28- 3-30	75.551	174- 6-29	0.0	4845520.531	307733.077
959	0+00.000	RAP #4 - CN TOWER	0.0		152-46-14	********	298-49-14	0.0	4833410.793	313894.638
933	270+16.535	CONST CONTROL MON CX-133	61.272	RIGHT	113- 9- 1	99.566	259-12- 0	0.0	4845633.810	307789.088
956	271+72.134	RAP #1 - BILLBOARD FRAME	471.682	RIGHT	69- 7-32	505.019	215-10-31	0.0	4845358.277	308169.411
960	269+67.759	CONTROL MON MTR77-6119	9.329	RIGHT	153-18-33	106.983	299-21-32	196.768	4845259.249	307745.594
957	268+98.575	RAP #2 - ANTENNA C.F.B.	183.253	LEFT	213-57- 1	234.606	0- 0- 0	0.0	4845368.291	307566.519
483	269+83.555	BC CORNER ROUND	25.637	LEFT	172-39-51	86.275	318-42-51	0.0	4845368.437	307708.556
753	269+83.622	CATCH BASIN GUTTER	25.142	LEFT	172-20-12	86.193	318-23-11	0.0	4845371.643	307709.034
485	269+84.988	PI CORNER ROUND	13.500	LEFT	164-31-15	85.312	310-34-15	0.0	4845374.136	307720.309
486	269+88.075	MP CORNER ROUND	16.973	LEFT	166-41-56	81.922	312-44-55	0.0	4845379.569	307716.388
446	269+88.654	PI CORNER ROUND	13.500	RIGHT	146-40-47	88.906	292-43-46	0.0	4845378.744	307746.378
2103	269+90.000	C/L OF CONSTRUCTION	0.0		154-49-55	82.995	300-52-54	196.352	4845383.986	307732.836
444	269+90.625	BC CORNER ROUND	28.986	RIGHT	137-35-48	94.626	283-38-47	0.0	4845385.613	307761.351
754	269+91.351	CATCH BASIN GUTTER	34.690	RIGHT	134-33- 3	97.281	280-36- 2	0.0	4845386.179	307766.866
461	269+91.604	BC CORNER ROUND	36.674	RIGHT	133-31-51	98.267	279-34-50	0.0	4845381.308	307768.784
434	269+92.355	BULLNOSE TOP OF CURB	1.500	RIGHT	153-21-22	81.172	299-24-21	196.425	4845382.168	307733.941
437	269+93.105	BC BULLNOSE TOP OF CURB	2.250	RIGHT	152-41-18	80.687	298-44-17	196.395	4845382.048	307734.562
435	269+93.105	CP BULLNOSE TOP OF ISLAND	1.500	RIGHT	153-11-45	80.456	299-14-44	196.410	4845381.929	307733.821
436	269+93.105	EC BULLNOSE TOP OF CURB	0.750	RIGHT	153-42-23	80.233	299-45-22	196.425	4845385.538	307733.081
447	269+93.947	MP CORNER ROUND	18.162	RIGHT	142-24-37	86.221	288-27-36	0.0	4845381.513	307750.135
482	269+97.210	CENTER PT CORNER ROUND	27.250	LEFT	174-16-54	72.708	320-19-53	0.0	4845383.707	307704.785
484	269+97.210	EC CORNER ROUND-TOP CURB	13.500	LEFT	163-28-17	73.176	309-31-16	196.290	4845384.488	307718.358
755	269+98.000	CATCH BASIN TOP OF CURB	13.500	LEFT	163-23-29	72.392	309-26-28	196.270	4845386.462	307718.232
2107	270+00.000	TOP OF CURB	13.500	LEFT	163-10-51	70.410	309-13-51	196.036	4845388.616	307717.913
2106	270+00.000	C/L OF CONSTRUCTION	0.0		152-40-57	73.433	298-43-56	196.156	4845397.175	307731.240
443	270+04.265	CENTER PT CORNER ROUND	27.250	RIGHT	133-24-39	82.485	279-27-38	0.0	4845394.981	307757.460
445	270+04.265	EC CORNER ROUND-TOP CURB	13.500	RIGHT	141-47-32	74.932	287-50-31	196.050	4845395.968	307743.887
756	270+05.265	CATCH BASIN TOP OF CURB	13.500	RIGHT	141-25- 0	74.059	287-28- 0	196.110	4845400.642	307743.727
2111	270+10.000	TOP OF CURB	13.500	RIGHT	139-30-47	69.973	285-33-46	195.854	4845396.334	307742.971
2110	270+10.000	TOP OF CURB	13.500	LEFT	161-55-21	60.513	307-58-20	195.854	4845398.488	307716.317
2109	270+10.000	C/L OF CONSTRUCTION	0.0		149-56-42	64.006	295-56-41	195.974	4845406.206	307729.644
2113	270+20.000	TOP OF CURB	13.500	LEFT	160-10-24	50.657	306-13-23	195.686		307714.722

FIGURE 5.20 Computer printout of layout data for a road construction project. (Courtesy of Department of Roads and Traffic, City of Toronto)

reached, both displays show 0.000 m (0.00 ft). When a motorized total station is used (see Section 5.7.2), the instrument itself turns the appropriate angle after the layout point number has been entered.

If the instrument is set up at an unknown position (free station), its coordinates can be determined by sighting control stations whose coordinates have been previously uploaded into the total station microprocessor. This technique, known as resection, is available on all modern total stations (see Section 5.3.3). Sightings on two control points can locate the instrument station, although additional sightings (up to a total of four) are recommended to provide a stronger solution and an indication of the accuracy level achieved.

Some theodolites and total stations come equipped with a guide light, which can help move the prism-holder on line very quickly (see Figures 5.22 and 5.23). The TC 800 is a total station that can be turned on and used immediately (no initialization procedure). It comes equipped with internal storage for 2,000 points and an electronic guide light, which is very useful in layout surveys because prism-holders can quickly place themselves on-line by noting the colored lights sent from the total station. The flashing lights (yellow on the

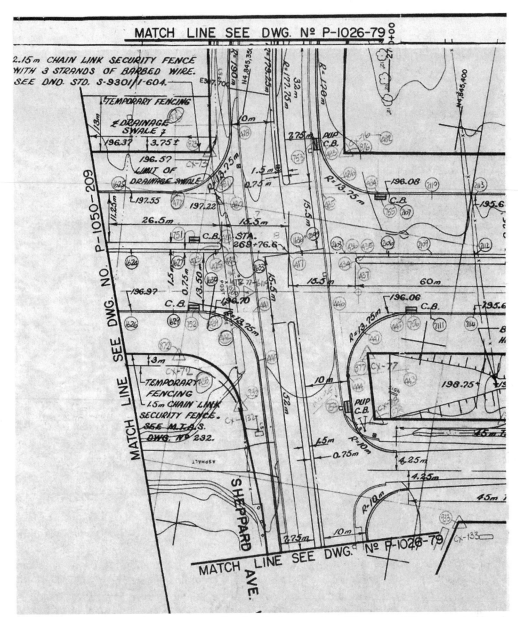

FIGURE 5.21 Portion of a construction plan showing layout points. (Courtesy of Department of Roads and Traffic, City of Toronto)

left and red on the right, as viewed by the prism-holder), which are 12 m wide at a distance of 100 m, enable the prism-holder to place the prism on-line, with final adjustments as given by the instrument operator. With automatic target recognition (ATR, see Section 5.7.1), the sighting-in process is completed automatically.

FIGURE 5.22 Leica TC 800 total station with electronic guide light. (Courtesy of Leica Geosystems Inc.)

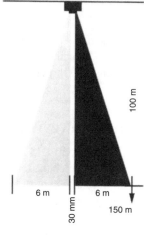

6 m · 6 m

30 mm

100 m

150 m

FIGURE 5.23 Electronic guide light. (Courtesy of Leica Geosystems Inc.)

5.7 Motorized Total Stations

One significant improvement to the total station has been the addition of servomotors to drive both the horizontal and the vertical motions of these instruments. Motorized instruments have been designed to search automatically for prism targets and then lock onto them precisely, to turn angles automatically to designated points using the uploaded coordinates of those points, and to repeat angles by automatically double-centering. These instruments, when combined with a remote controller held by the prism surveyor, enable the survey to proceed with a reduced need for field personnel.

5.7.1 Automatic Target Recognition (ATR)

Some instruments are designed with automatic target recognition, which utilizes an infrared light bundle or laser beam that is sent coaxially through the telescope to a prism. The return signal is received by an internal charge-coupled device (CCD) camera. First, the telescope must be pointed roughly at the target prism, either manually or under software control. Then the motorized instrument places the cross hairs almost on the prism center (within two seconds of arc). Any residual offset error is measured and applied to the horizontal and vertical angles. The ATR module is a digital camera that notes the offset of the reflected laser beam, permitting the instrument then to move automatically until the cross hairs have been set on the point electronically. After the point has thus been precisely sighted, the instrument can then read and record the angle and distance. Reports indicate that the time required for this process (which eliminates manual sighting and focusing) is only one-third to one-half the time required to obtain the same results using conventional total station sighting techniques. ATR comes with a lock-on mode, where the instrument, once sighted on the prism, continues to follow the prism as it is moved from station to station. (Automatic target recognition is also referred to as autolock and autotracking.) To ensure that the prism is always pointed at the instrument, a 360° prism (see Figures 5.22 and 5.24) assists the surveyor in keeping the lock-on for a period of time. If lock-on is lost because of intervening obstacles, it is reestablished after manually and roughly pointing at the prism. Automatic target recognition recognizes targets up to 2,200 m away, functions in darkness, requires no focusing or fine pointing, works with all types of prisms, and maintains a lock on prisms moving up to speeds of 11 mph, or 5 mps (at a distance of 100 m).

5.7.2 Remote-Controlled Surveying

Geodimeter, the company that first introduced EDM equipment in the early 1950s, introduced in the late 1980s a survey system in which the total station (Geodimeter series 4000; see Figure 5.25) was equipped with motors to control both horizontal and vertical movements. This total station can be used as a conventional instrument, but when it is interfaced with a controller located with the prism, the surveyor at the prism station, by means of radio telemetry, can remotely control the station instrument. Some instruments introduced more recently use optical controls instead of radio control.

When the remote control button on the total station is activated, control of the station instrument is transferred to the remote controller. See Figure 5.26. The remote controller consists of the pole (with circular bubble), the prism, a data collector (up to 10,000 points), telemetry equipment for communicating with the station instrument, and a sighting telescope that, when aimed back at the station instrument, permits a sensing of the angle of inclination, which is then transmitted to the station instrument via radio communication. As a result, the instrument can move its telescope automatically to the proper angle of

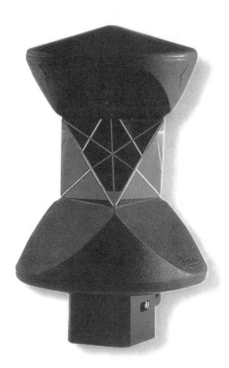

FIGURE 5.24 Leica 360° prism—used with remote-controlled total stations and with automatic target recognition (ATR) total stations. ATR eliminates fine pointing and focusing. The 360° feature means that the prism is always facing the instrument. (Courtesy of Leica Geosystems Inc.)

inclination, thus enabling the instrument to commence an automatic horizontal sweeping that results in the station instrument being locked precisely onto the prism. Other manufacturers employ a vertical laser fan to help locate the prism.

A typical operation requires that the station unit be placed over a control station or over a free station whose coordinates can be determined using resection techniques (see Section 5.3.3) and that a backsight be taken to another control point, thus fixing the location and orientation of the total station. The operation then begins with both units being activated at the remote controller. The remote controller sighting telescope is aimed at the station unit, and the sensed vertical angle is sent via telemetry to the station unit. The station unit then sets its telescope automatically at the correct angle in the vertical plane and begins a horizontal search for the remote controller prism. The search area can be limited to a specific sector (e.g., 70°), thus reducing search time. The limiting range of this instrument is said to be about 700 m. When the measurements (angle and distance) have been completed, the point number and attribute data codes can be entered into the data collector attached to the prism pole.

When used for setting out, the desired point number is entered at the remote controller keyboard. The total station instrument then automatically turns the required angle, which it computes from previously uploaded coordinates held in storage (both the total station and the remote controller have the points in storage). The remote controller operator can position the prism roughly on-line by noting the Track-Light®, which shows as red or green (for this instrument), depending on whether the operator is left or right, respectively, of the line,

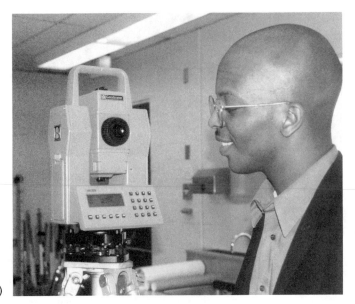

(a)

(b)

FIGURE 5.25 (a) Geodimeter 4400 base station. A total station equipped with servomotors controlling both the horizontal and vertical circle movements. Can be used alone as a conventional total station or as a robotic base station controlled by the remote positioning unit (RPU) operator. (b) Geodimeter keyboard showing in-process electronic leveling. Upper cursor can also be centered by finally adjusting the third leveling screw.

and as white when the operator is on the line (see also the electronic guide light shown in Figures 5.22 and 5.23).

Distance and angle readouts are then observed by the operator to position the prism pole precisely at the layout point location. Because the unit can fast-track (0.4 s) precise measurements and because it is also capable of averaging multiple measurement readings, very precise results can be obtained when using the prism pole by slightly "waving" the pole left and right and back and forth in a deliberate pattern. All but the backsight reference are obtained using infrared and telemetry, so the system can be used effectively after dark, thus

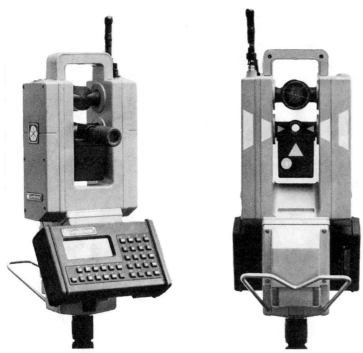

FIGURE 5.26 Remote positioning unit (RPU)—a combination of prism, data collector, and radio communicator (with the base station) that permits the operator to engage in one-person surveys.

FIGURE 5.27 Leica TPS System 1000, used for roadway stakeout. The surveyor is controlling the remote-controlled total station (TCA 1100) at the prism pole using the RCS 1000 controller together with a radio modem. The assistant is placing the steel bar marker at the previous set-out point. (Courtesy of Leica Geosystems Inc.)

permitting nighttime layouts for next-day construction and for surveys in high-volume traffic areas that can be accomplished efficiently only in low-volume time periods.

Figure 5.27 shows a remotely controlled total station manufactured by Leica Geosystems Incorporated. This system utilizes ATR and the electronic guide light to search for and

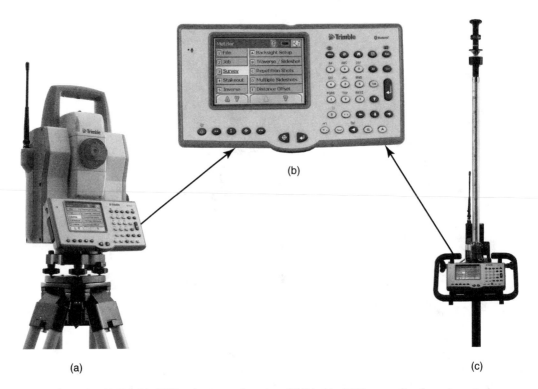

(b)

(a)   (c)

FIGURE 5.28 (a) Trimble 5600 robotic total station. (b) Trimble ACU controller, featuring wireless communications and a choice of software for all robotic total station applications as well as all GPS applications. (c) Remote pole unit containing an ACU controller, 360° prism, and radio. (Courtesy of Trimble)

then position the prism on the correct layout line—where the operator then notes the angle and distance readouts to determine the precise layout location. Figure 5.28 shows a motorized total station (Trimble 5600totalstation). It has many features, including a four-speed servo; Autolock, a coaxial prism sensor that locks quickly and precisely on the target prism and then follows it as it moves; robotic operation (one-person operation) utilizing radio control from the remote roving station instrument [see Figure 5.28(c)]; direct reflex (DR) EDM, which permits measurements to features without the use of reflecting prisms up to 600 m away, aided by a laser marking device; Tracklight®, a multicolored beam that provides a fast technique of positioning the prism-holder in setting-out activities; and a full complement of computational, point acquisition, and setting-out software.

5.8 Summary of Modern Total Station Characteristics and Capabilities

Modern total stations have some or all of the capabilities listed below:

- Some instruments require that horizontal and vertical circles be revolved through 360° to initialize angle-measuring operations, whereas some newer instruments require no such initialization for angle-measuring readiness.

- They can be equipped with servomotors to drive the horizontal and vertical motions (the basis for robotic operation).
- A robotic instrument can be controlled at a distance of up to 1,200 m from the prism pole.
- Telescope magnification for most total stations is at 26× or 30×.
- The minimum focus distance is in the range of 0.5 to 1.7 m.
- Distance measurement to a single prism is in the range of 2 to 5 km (10 km for a triple prism).
- A coaxial visible laser has a reflectorless measuring range of 250 to 1,200 m.
- Angle accuracies are in the range of 1 to 10 seconds. Accuracy is defined as the standard deviation, direct and reverse DIN specification 18723.
- Distance accuracies, using reflectors, are as follows: normal, 2 mm + 2 ppm; tracking, 5 mm + 2 ppm; reflectorless: 3 mm + 2 ppm; longer-range reflectorless: 5 mm + 2 ppm.
- Eye-safe lasers are class 1 and class 2 (class 2 requires some caution).
- Reflectorless ranges are up to 800 m to a 90 percent (reflection) Kodak Gray Card, and up to 300 m to a 18 percent (reflection) Kodak Gray Card.
- Auto focusing.
- Automatic environment sensing (temperature and pressure) with ppm corrections applied to measured distances, or manual entry for measured corrections.
- Wireless Blue Tooth communications, or cable connections.
- Standard data collectors (on-board or attached), or collectors with full graphics (four to eight lines of text) with touch-screen command capabilities for quick editing and line-work creation, in addition to keyboard commands.
- On-board calibration programs for line-of-sight errors, tilting axis errors, compensator index errors, vertical index errors, and automatic target lock-on calibration. Corrections for sensed errors can be applied automatically to all measurements.
- Endless drives for horizontal and vertical motions (nonmotorized total stations).
- Laser plummets.
- Guide lights for prism placements, with a range of 5 to 150 m, used in layout surveys.
- Automatic target recognition (ATR), also known as autolock and autotracking, with an upper range of 2,000 m for standard prisms, and about half that value for 360° prisms.
- Single-axis or dual-axis compensation.

The capabilities of total stations are constantly being improved. Thus, this list is not meant to be comprehensive.

5.9 Instruments Combining Total Station Capabilities and GPS Receiver Capabilities

Leica Geosystems Inc. was the first in early 2005 to market a total station equipped with an integrated dual-channel GPS receiver (see Figure 5.29). All TPS 1200 total stations can be upgraded to achieve total station/GPS receiver capability. The GPS receiver, together with

FIGURE 5.29 Leica SmartStation total station with integrated GPS. (Courtesy of Leica Geosystems, Inc.)

a communications device, are mounted directly on top of the total station yoke so that they are on the vertical axis of the total station and thus correctly positioned over the station point. The total station keyboard can be used to control all total station and GPS receiver operations. RTK accuracies are said to be 10 mm + 1 ppm horizontal, and 20 mm + 1 ppm vertical. The GPS receiver can identify the instrument's first uncoordinated position (free station) to high accuracy when working differentially with a GPS base station (located within 50 km). Once the instrument station has been coordinated, a second station can be similarly established, and the total station can then be oriented while you are backsighting the first established station. All necessary GPS software is included in the total station processor. Early comparative trials indicate that, for some applications, this new technology can be 30 to 40 percent faster than traditional total station surveying. Because nearby ground control points may not have to be occupied, the surveyor saves the time it takes to set up precisely over a point and to bring control into the project area from control stations that are too far away to be of immediate use. If the instrument is set up in a convenient un-coordinated location where you can see the maximum number of potential survey points, you need only to turn on the GPS receiver and then level and orient the total station to a backsight reference. By the time the surveyor is ready to begin the survey and assuming a GPS reference base station is within 50 km, the GPS receiver is functioning in RTK mode. The survey can commence with the points requiring positioning being determined using to-tal station techniques and/or by using the GPS receiver after it has been transferred from the total station and placed on a rover antenna pole for roving positioning. The GPS data, collected at the total station, are stored along with all the data collected using total station techniques on compact memory flash cards. Data can be transferred to cell phones using wireless (e.g., Bluetooth) technology before being transmitted to the project computer for downloading. This new technology, which represents a new era in surveying, will change the way we look at traditional traverses and control surveying in general, as well as topo-graphic surveys and layout surveys.

5.10 Portable/Handheld Total Stations

Several instrument manufacturers produce lower-precision reflectorless total stations that can be handheld, like a camcorder, or pole/tripod-mounted, like typical surveying instru-ments. These instruments have three integral components:

1. The pulse laser (Food and Drug Administration [FDA] Class 1[*]) distance meter meas-ures reflectorlessly and/or to reflective papers and prisms. When used reflectorlessly, the distance range of 300 to 1,200 m varies with different manufacturers and model types; the type of sighted surface (masonry, solid trees, bushes, etc.) determines the distance range because some surfaces are much more reflective than others. Masonry and concrete surfaces reflect light well, while trees, bushes, sod, etc., reflect light to lesser degrees (see Table 2.3). When used with prisms, the distance range increases to 5 to 8 km. Some instruments can be limited to an expected distance-range enve-lope. This measuring restriction instructs the instrument not to measure and record distances outside the envelope. This feature permits the surveyor to sight more dis-tant points through nearby clutter such as intervening electrical wires, branches, and other foliage. Distance accuracies are in the range of 1 to 10 cm, and ±(5 cm + 20 ppm) is typical. Power is supplied by two to six AA batteries or two C-cell batteries; battery type depends on the model. See Figure 5.30.

2. Angle encoders and built-in inclinometers determine the horizontal and vertical an-gles, much like the total stations described earlier in this chapter, and can produce ac-curacies up to greater than 1 minute of arc. The angle accuracy range of instruments on the market in 2005 is 0.01° to 0.2°. Obviously, mounted instruments produce bet-ter accuracies than do handheld units.

3. Data collection can be provided by many of the collectors on the market. These hand-held total stations can be used in a wide variety of mapping and GIS surveys or they can be used in GPS surveys. For example, GPS surveys can be extended into areas

---

[*]FDA Class 1 lasers are considered safe for the eyes. Class 2 lasers should be used with caution and should not be operated at eye level. Class 3A, 3B, and 4 lasers should be used only with appropriate eye protection.

FIGURE 5.30 The reflectorless total station is equipped with integrating distance measurement, angle encoding, and data collection. (Courtesy of Riegl USA Inc., Orlando, Florida)

under tree canopy or structural obstructions (where GPS signals are blocked), with the positional data collected directly into the GPS data collector.

Review Questions

5.1 What are the advantages of using a total station rather than an electronic theodolite and steel tape?

5.2 When would you use a steel tape rather than a total station to measure a distance?

5.3 What impact did the creation of electronic angle measurement have on surveying procedures?

5.4 With the ability to record field measurements and point descriptions in an instrument controller or electronic field book, why are manual field notes still important?

5.5 Explain the importance of electronic surveying in the extended field of surveying and data processing, now often referred to as the science of geomatics.

5.6 Using a programmed total station, how would you tie in (locate) a water valve "hidden" behind a building corner?

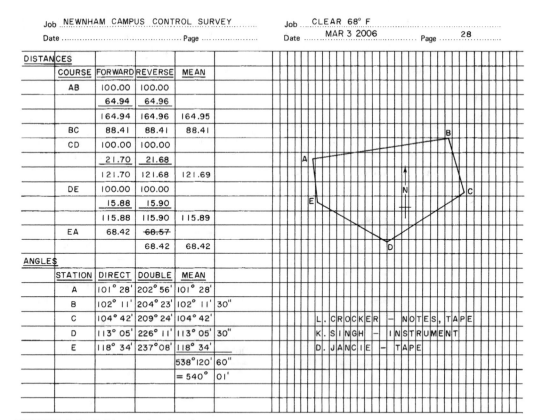

FIGURE 6.4 Field notes for closed traverse.

next and verified, using a steel tape or EDM instrument. The interior angle is measured at each station, and each angle is measured at least twice. Figure 6.4 illustrates typical field notes for a loop traverse survey. In this type of survey, distances are booked simply as dimensions, not as stations or chainages.

6.2 Balancing Field Angles

For a closed polygon of n sides, the sum of the interior angles is $(n - 2)180°$. In Figure 6.2, the interior angles of a five-sided polygon have been measured as shown in the field notes in Figure 6.4. For a five-sided closed figure, the sum of the interior angles must be $(5 - 2)180° = 540°$. You can see in Figure 6.4 that the interior angles add to 540°01'—an excess of one minute.

Before mathematical analysis can begin—that is, before the bearings or azimuths can be computed—the field angles must be adjusted so that their sum exactly equals the correct geometric total. The angles can be balanced by distributing the error evenly to each angle, or one or more angles can be arbitrarily adjusted to force the closure. The total allowable error of angular closure is quite small (see Chapter 9); if the field results exceed the allowable error, the survey must be repeated.

The angles for the traverse in Figure 6.4 are shown in Table 6.1. Also shown are the results of equally balanced angles and arbitrarily balanced angles. The angles can be arbitrarily balanced if the required precision will not be affected or if one or two setups are suspect (e.g., due to unstable ground, or very short sightings).

6.3 Meridians

A line on the surface of the earth joining the north and south poles is called a geographic, astronomic, or "true" meridian. Figure 6.5 illustrates that **geographic meridian** is another term for a line of longitude. The figure also illustrates that all geographic meridians converge at the poles. **Grid meridians** are lines that are parallel to a grid reference meridian (a central meridian—see Figure 6.5). Rectangular coordinate grids are discussed further in

Table 6.1 TWO METHODS OF ADJUSTING FIELD ANGLES

Station	Field Angle	Arbitrarily Balanced	Equally Balanced
A	101°28′	101°28′	101°27′48″
B	102°11′30″	102°11′	102°11′18″
C	104°42′	104°42′	104°41′48″
D	113°05′30″	113°05′	113°05′18″
E	118°34′	118°34′	118°33′48″
	= 538°120′60″	= 538°120′	= 538°117′180″
	= 540°01′00″	= 540°00′	= 540°00′00″

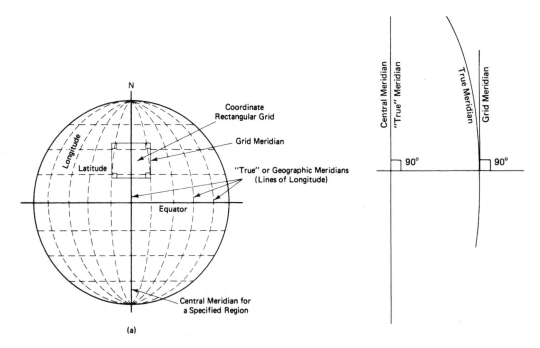

(a)

FIGURE 6.5 Relationship between "true" meridians and grid meridians. (a) Illustration of geographic ("true") meridians and grid meridians. (b) Illustration of geographic north and grid north.

Chapter 9. **Magnetic meridians** are lines parallel to the directions taken by freely moving magnetized needles, as in a compass. Whereas geographic and grid meridians are fixed, magnetic meridians can vary with time and location.

Geographic meridians can be established by tying into an existing survey line whose geographic direction is known or whose direction can be established by observations on the sun or on Polaris (the North Star) or through GPS surveys. Grid meridians can be established by tying into an existing survey line whose grid direction is known or whose direction can be established by tying into coordinate grid monuments whose rectangular coordinates are known. On small or isolated surveys of only limited importance, meridians are sometimes assumed (e.g., one of the survey lines is simply designated as being "due north"), and the whole survey is referenced to that assumed direction.

Meridians are important to the surveyor because they are used as reference directions for surveys. All survey lines can be related to each other and to the real world by angles measured from meridians. These angles are called bearings and azimuths.

6.4 Bearings

A **bearing** is the direction of a line given by the acute angle between the line and a meridian. The bearing angle, which can be measured clockwise or counterclockwise from the north or south end of a meridian, is always accompanied by the letters that describe the quadrant in which the line is located (NE, SE, SW, and NW). Figure 6.6 illustrates the concepts of bearings. The given angles for lines K1 and K4 are acute angles measured from the meridian and, as such, are bearing angles. The given angles for lines K2 and K3 are both measured

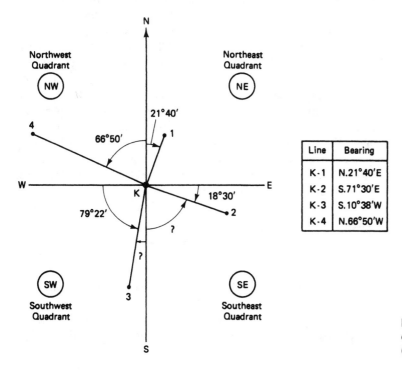

Line	Bearing
K-1	N.21°40′E
K-2	S.71°30′E
K-3	S.10°38′W
K-4	N.66°50′W

FIGURE 6.6 Bearings calculated from given data (answers in box).

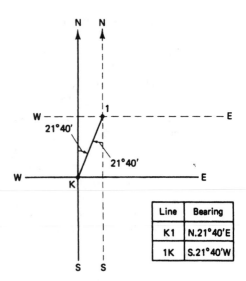

FIGURE 6.7 Illustration of reversed bearings.

Line	Bearing
K1	N.21°40′E
1K	S.21°40′W

from the E/W axis and therefore are not bearing angles; here, the given angles must be subtracted from 90° to determine the bearing angle.

All lines have two directions: forward and reverse. Figure 6.7 shows that to reverse a bearing, the letters are simply switched, for example, N to S and E to W. To illustrate, consider walking with a compass along a line in a northeasterly direction. If you were to stop and return along the same line, the compass would indicate that you would then be walking in a southwesterly direction. In Figure 6.7, when the meridian is drawn through point K, the line K1 is being considered, and the bearing is NE. If the meridian is drawn through point 1, however, the line 1K is being considered, and the bearing is SW. In computations, the direction of a line therefore depends on which end of the line the meridian is placed. When computing the bearings of adjacent sides (as in a closed traverse) the surveyor must routinely reverse bearings as the computations proceed around the traverse (see Figure 6.8).

Figure 6.8 shows a five-sided traverse with geometrically closed angles and a given bearing for side AE of S 7°21′ E. The problem here is to compute the bearings of the remaining sides. To begin, the surveyor must decide whether to solve the problem going clockwise or counterclockwise. In this example, the surveyor decided to work clockwise, and the bearing of side AB is computed first using the angle at A. Had the surveyor decided to work the computations counterclockwise, the first computation would be for the bearing of side ED, using the angle at E.

To compute the bearing of side AB, the two sides AB and AE are drawn, together with a meridian through point A. The known data are then placed on the sketch, and a question mark is placed in the location of the required bearing angle. Step 1 in Figure 6.8 shows the bearing angle of 7°21′ and the interior angle at A of 101°28′. A question mark is placed in the position of the required bearing angle for side AB. If the sketch has been drawn and labeled properly, the procedure for the solution will become apparent. Here, it is apparent that the required bearing angle (AB) + the interior angle (A) + the bearing angle (AE) = 180°; that is, the bearing of AB = N 71°11′ E.

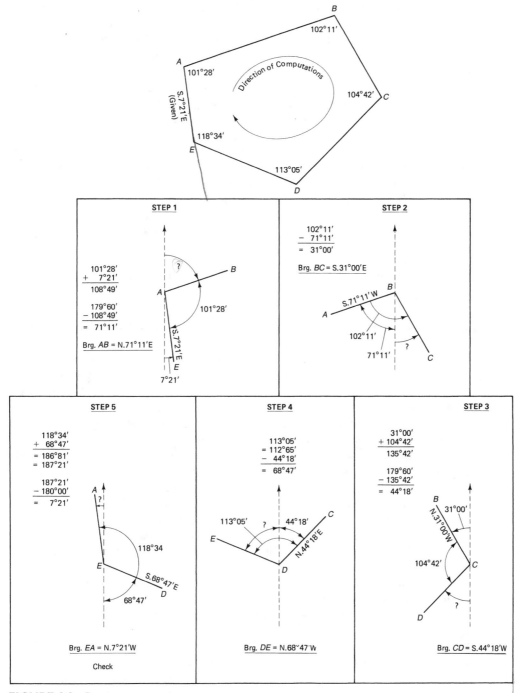

FIGURE 6.8 Bearing computations.

To compute the bearing of the next side, BC, the lines BA and BC are drawn, together with a meridian through point B. Once again, the known data are placed on the sketch, and a question mark is placed in the position of the required bearing angle for side BC. In step 1, the bearing of AB was computed as N 71°11′ E. When this information is to be shown on the sketch for step 2, it is obvious that something must be done to make these data comply with the new location of the meridian—that is, the bearing must be reversed. With the meridian through point B, the line being considered is BA (not AB) and the bearing becomes S 71°11′ W. It is apparent from the sketch and data shown in step 2 that the required bearing angle (BC) = the interior angle (B) − the bearing angle (AB), a value of S 31°00′ E.

The remaining bearings are computed in steps 3, 4, and 5. Step 5 involves the computation of the bearing for side EA. This last step provides the surveyor with a check on all the computations; that is, if the computed bearing turns out to be the same as the starting bearing (7°21′), the work is correct. If the computed bearing does not agree with the starting value, the work must be checked and the mistake found and corrected. If the work has been done with neat, well-labeled sketches similar to that shown in Figure 6.8, any mistake(s) will be found quickly and corrected.

6.5   Azimuths

An **azimuth** is the direction of a line given by an angle measured clockwise from the north end of a meridian. In some astronomic, geodetic, and state plane grid projects, azimuths are measured clockwise from the south end of the meridian. In this text, azimuths are referenced from north only. Azimuths can range in magnitude from 0° to 360°. Values in excess of 360°, which are sometimes encountered in computations, are valid but are usually reduced by 360° before final listing.

Figure 6.9 illustrates the concepts of azimuths. The given angle for K1 is a clockwise angle measured from the north end of the meridian and, as such, is the azimuth of K1. The given angle for K2 is measured clockwise from the easterly axis, which itself is already measured 90° from the north end of the meridian. The azimuth of K2 is therefore 90° + 18°46′ = 108°46′.

The given angle for K3 is measured clockwise from the south end of the meridian, which itself is 180° from the north end; the required azimuth of K3 is therefore 180° + 38°07′ = 218°07′. The given angle for K4 is measured counterclockwise from the north end of the meridian; the required azimuth is therefore 360° (359°60′) − 25°25′ = 334°35′.

We noted in the previous section that each line has two directions: forward and reverse. In Figure 6.10, when the reference meridian is at point K, the line K1 is being considered, and its azimuth is 10°20′. When the reference meridian is at point 1, however, the line 1K is being considered, and the original azimuth must be reversed by 180°; its azimuth is now 190°20′.

To summarize, bearings are reversed by simply switching the direction letters, for example, N to S and E to W; azimuths are reversed by numerically changing the value by 180°.

Figure 6.11 shows a five-sided traverse with geometrically closed angles and a given azimuth for side AE of 172°39′. Note that this azimuth of 172°39′ for AE is identical in direction to the bearing of S 7°21′ E for AE in Figure 6.8. In fact, the problems illustrated in

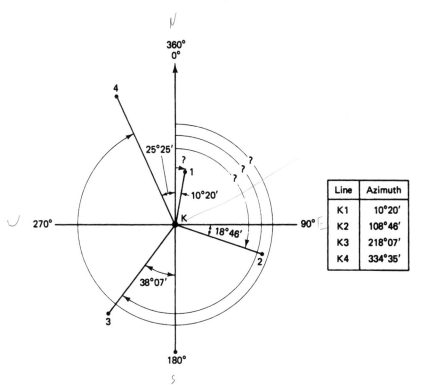

Line	Azimuth
K1	10°20′
K2	108°46′
K3	218°07′
K4	334°35′

FIGURE 6.9 Azimuths calculated from given data (answers in box).

Figures 6.8 and 6.11 are identical, except that the directions are given in bearings in Figure 6.8 and in azimuths in Figure 6.11.

If we wish to compute the azimuths proceeding in a clockwise manner, we must first compute the azimuth of side AB. To compute the azimuth of AB, the two sides AB and AE are drawn, together with a meridian drawn through point A. The known data are then placed on the sketch along with a question mark in the position of the required azimuth. In step 1, it is apparent that the required azimuth (AB) = [the given azimuth (AE) − the interior angle (A)], a value of 71°11′.

To compute the azimuth of the next side, BC, the lines BC and BA are drawn, together with a meridian through point B. Once again, the known data are placed on the sketch, and a question mark is placed in the position of the required azimuth angle for side BC.

In step 1, the azimuth of AB was computed to be 71°11′; when this result is transferred to the sketch for step 2, it is obvious that it must be altered to comply with the new location of the meridian at point B. That is, with the meridian at B, the line BA is being considered, and the azimuth of line AB must be changed by 180°, resulting in a value of 251°11′; this "back azimuth" computation is shown in the boxes in the sketches for steps 2 through 5 in Figure 6.11.

The remaining azimuths are computed in steps 3, 4, and 5. Step 5 provides the azimuth for side EA. This last step gives the surveyor a check on all the computations because the final azimuth (when reversed by 180°) should agree with the azimuth originally given for that line (AE). If the check does not work, all the computations must be reworked to find the mistake(s).

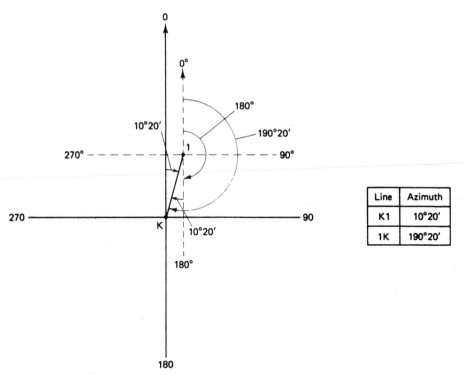

Line	Azimuth
K1	10°20'
1K	190°20'

FIGURE 6.10 Illustration of reversed azimuths.

Scrutinizing the five steps in Figure 6.11 reveals that the same procedure is involved in all the steps; that is, *in each case, when working clockwise, the desired azimuth is determined by subtracting the interior angle from the back azimuth of the previous course. This relationship is true every time the computations proceed in a clockwise direction.* If this problem had been solved by proceeding in a counterclockwise direction—that is, the azimuth of ED is computed first—you would have noted that, *in each case, when working counterclockwise, the desired azimuth was computed by adding the interior angle to the back azimuth of the previous course.*

The counterclockwise solution for azimuth computation is shown in Figure 6.12. Note that the computation follows a very systematic routine; it is so systematic that sketches are not required for each stage of the computation. A sketch is required at the beginning of the computation to give the overall sense of the problem and to ensure that the given azimuth is properly recognized as a forward or a back azimuth. (The terms *back azimuth* and *forward azimuth* are usually found only in computations; their directions depend entirely on the choice of clockwise or counterclockwise for the direction of the computation stages.)

Normally surveying problems are worked out by using either bearings or azimuths for the directions of survey lines. The surveyor must be prepared to convert readily from one system to the other. Figure 6.13 illustrates and defines the relationships of bearings and azimuths in all four quadrants.

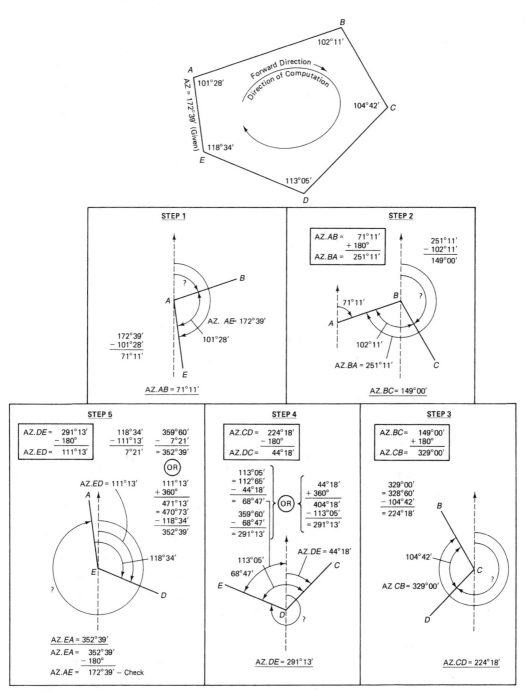

FIGURE 6.11 Azimuth computations.

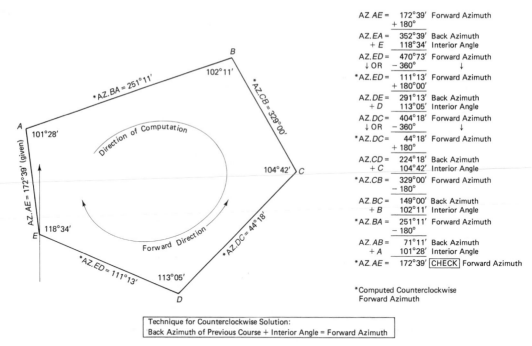

FIGURE 6.12 Azimuth computations, counterclockwise solution.

6.6   Latitudes and Departures

In Chapter 1, we spoke of the need for the surveyor to check the survey measurements to ensure that the required accuracies were achieved and to ensure that mistakes were eliminated. Checking can consist of repeating the measurements in the field, and/or checking can be accomplished using mathematical techniques. One such mathematical technique involves the computation and analysis of latitudes and departures.

In Section 1.4, we noted that a point can be located by polar ties (direction and distance) or by rectangular ties (two distances at 90°). In Figure 6.14(a), point B is located by polar ties from point A by direction (N 71°11′ E) and distance (164.95′). In Figure 6.14(b), point B is located by rectangular ties from point A by distance north ($\Delta N = 53.20′$) and distance east ($\Delta E = 156.13′$).

- By definition, **latitude** is the north/south rectangular component of a line (ΔN). To differentiate direction, north is considered positive ($+$), and south is considered negative ($-$).

- **Departure** is the east/west rectangular component of a line (ΔE). To differentiate direction, east is considered positive ($+$), and west is considered negative ($-$).

When working with azimuths, the plus/minus designation of the latitude and departure is given directly by the appropriate trigonometric function:

$$\text{Latitude } [\Delta N] = \text{distance } [S] \cos \text{bearing} \qquad (6.1)$$

or

$$\text{Latitude } [\Delta N] = \text{distance } [S] \cos \text{azimuth} \qquad (6.2)$$

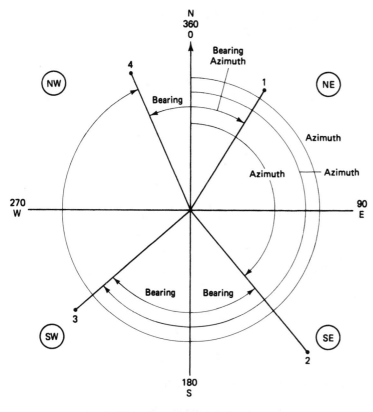

1. NE quadrant: bearing = azimuth
2. SE quadrant: bearing = 180° − azimuth
3. SW quadrant: bearing = azimuth − 180°
4. NW quadrant: bearing = 360° − azimuth

FIGURE 6.13
Relationships between bearings and azimuths.

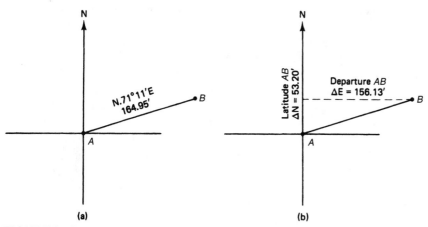

FIGURE 6.14   Location of a point. (a) Polar tie. (b) Rectangular tie.

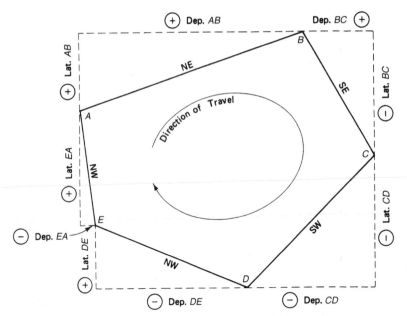

FIGURE 6.15 Closure of latitudes and departures (clockwise solution).

$$\text{Departure } [\Delta E] = \text{distance } [S] \sin \text{bearing} \qquad (6.3)$$

or

$$\text{Departure } [\Delta E] = \text{distance } [S] \sin \text{azimuth} \qquad (6.4)$$

Latitudes (lats) and departures (deps) can be used to compute the precision of a traverse survey by noting the plus/minus closure of both latitudes and departures. If the survey has been perfectly measured (angles and distances), the plus latitudes will equal the minus latitudes, and the plus departures will equal the minus departures.

In Figure 6.15, the survey has been analyzed in a clockwise manner (all algebraic signs and letter pairs are simply reversed for a counterclockwise approach). Latitudes DE, EA, and AB are all positive and should precisely equal (if the survey measurements were perfect) the latitudes of BC and CD, which are negative. Departures AB and BC are positive and ideally should equal the departures CD, DE, and EA, which are negative.

The following sections in this chapter involve traverse computations that use trigonometric functions of direction angles—either bearings or azimuths—to compute latitudes and departures. When bearings are used, the algebraic signs of the lats and deps are assigned according to the N/S and E/W directions of the bearings. When azimuths are used, the algebraic signs of the lats and deps are given directly by the calculator or computer, according to the trigonometric conventions illustrated in Figure 6.16.

■ **EXAMPLE 6.1** *Computation of Latitudes and Departures to Determine the Error of Closure and the Precision Ratio of a Traverse Survey*

The survey data shown in Figure 6.2 are used for this illustrative example. Following are all the steps required to adjust the field data and perform the necessary computations.

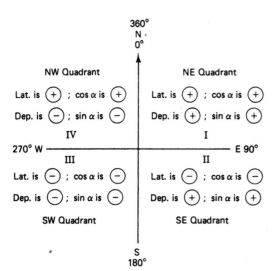

NW Quadrant

Lat. is (+) ; cos α is (+)

Dep. is (−) ; sin α is (−)

IV

NE Quadrant

Lat. is (+) ; cos α is (+)

Dep. is (+) ; sin α is (+)

I

270° W

III

II

E 90°

Lat. is (−) ; cos α is (−)

Dep. is (−) ; sin α is (−)

SW Quadrant

Lat. is (−) ; cos α is (−)

Dep. is (+) ; sin α is (+)

SE Quadrant

FIGURE 6.16 Algebraic signs of latitudes and departures by trigonometric functions where α is the azimuth.

Table 6.2 CLOSED TRAVERSE COMPUTATIONS

Course	Distance	Bearing	Azimuth	Latitude	Departure
AB	164.95′	N 71°11′ E	71°11′	+53.20	+156.13
BC	88.41′	S 31°00′ E	149°00′	−75.78	+45.53
CD	121.69′	S 44°18′ W	224°18′	−87.09	−84.99
DE	115.89′	N 68°47′ W	291°13′	+41.94	−108.03
EA	68.42′	N 7°21′ W	352°39′	+67.86	−8.75
	P = 559.36′			Σ lat = +0.13	Σ dep = −0.11

$$E = \sqrt{\Sigma \text{lat}^2 + \Sigma \text{dep}^2} \qquad E = \sqrt{0.13^2 + 0.11^2} \qquad E = 0.17'$$

where E is the linear error of closure

Solution

Step 1. *Balance the angles.* See Table 6.1; the arbitrarily balanced angles are used here.

Step 2. *Compute the bearings or azimuths.* See Figure 6.8 for bearings and Figure 6.11 or Figure 6.12 for azimuths.

Step 3. *Compute the latitudes and departures.* Table 6.2 shows a typical format for a closed traverse computation. Columns are included for both bearings and azimuths, although only one of those direction angles is needed for computations. Table 6.2 also shows that the latitudes fail to close by +0.13′ and that the departures fail to close by −0.11′.

Step 4. *Compute the linear error of closure.* See Figure 6.17 and Table 6.2. The linear error of closure is the net accumulation of the random errors associated with the traverse measurements. In Figure 6.17, the total error is showing up at A simply because the computation started at A. If the computation had started at any other station, the identical error of closure would have shown up at that station:

$$\text{Precision ratio} = \frac{E}{P}$$

Traverse Computation Begins at A, and Terminates at A′

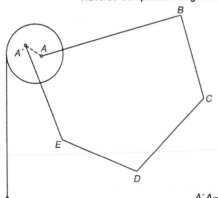

Closure Error = AA′
Closure Correction = A′A

$$A'A = \sqrt{C_{Lat.}2\,1\,C_{Dep.}2} = 0.17$$

Bearing of A′A Can Be Computed from the Relationship:

$$\text{Tan Bearing} = \frac{C_{Dep.}}{C_{Lat.}} = \frac{0.11}{-0.13}$$

Bearing Angle = 40.23636°
Bearing A′A = S.40°14′11″E

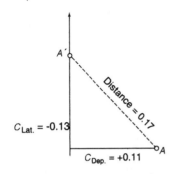

Σ lat. = error in latitudes
Σ dep. = error in departures
C lat. = required correction in latitudes
C dep. = required correction in departures

FIGURE 6.17 Closure error and closure correction.

where P = the perimeter of the traverse.

$$\text{Precision ratio} = \frac{0.17}{559.36} = \frac{1}{3{,}290} = \frac{1}{3{,}300}$$

Line A′A in Figure 6.17 is a graphical representation of the linear error of closure. The length of A′A is the square root of the sums of the squares of the C_{lat} and the C_{dep}:

$$A'A = \sqrt{C_{lat^2} + C_{dep^2}} = \sqrt{0.13^2 + 0.11^2} = 0.17$$

C_{lat} and C_{dep} are equal to and opposite in sign to Σ_{lat} and Σ_{dep} and reflect the consistent direction (in this example) with the clockwise approach to this problem. Sometimes it is advantageous to know the bearing of the linear error of closure. Figure 6.17 shows that C_{dep}/C_{lat} = tan bearing angle:

$$\text{Brg } A'A = S\ 40°14'11''\ E$$

Step 5. *Compute the precision ratio of the survey.* Table 6.2 shows that the precision ratio is the linear error of closure (E) divided by traverse perimeter (P). The resultant fraction (E/P) is always expressed with a numerator of 1 and with the denominator rounded to the closest 100 units.

$$\text{Precision ratio} \left(\frac{E}{P}\right) = \frac{0.17}{559.36} = \frac{1}{3,290} = \frac{1}{3,300}.$$

The concept of an accuracy ratio was introduced in Section 1.12. Most states and provinces have these ratios legislated for minimally acceptable surveys for property boundaries. These values vary from one area to another, but they are usually in the range of 1/5,000 to 1/7,500. In some cases, higher ratios (e.g., 1/10,000) are stipulated for high-cost downtown urban areas.

Engineering and construction surveys are performed at levels of 1/3,000 to 1/10,000, depending on the importance of the work and the materials being used. For example, a ditched highway could well be surveyed at 1/3,000, whereas overhead rails for a monorail transit system may require accuracies in the range of 1/7,500 to 1/10,000. Control surveys for both engineering and legal projects must be executed at higher levels of accuracy than are necessary for the actual location of the engineering or legal markers that are to be surveyed from those control surveys.

6.6.1 Summary of Initial Traverse Computations

The initial steps for computing the traverse are:

1. Balance the angles.
2. Compute the bearings and/or the azimuths.
3. Compute the latitudes and the departures.
4. Compute the linear error of closure.
5. Compute the precision ratio of the survey.

If the precision ratio is satisfactory, further treatment of the traverse data (for example, coordinate and area computations) is possible. If the precision ratio is unsatisfactory (for example, a precision ratio of only 1/2,500 when 1/3,000 was specified), complete the following steps:

1. Double-check all computations.
2. Double-check all field-book entries.
3. Compute the bearing of the linear error of closure, and check to see if it is similar to one of the course bearings (±5°). [*]

---

[*]If a large error (or mistake) has been made on the measurement of one course, it will significantly affect the bearing of the linear error of closure. Thus, if a check on the field work is necessary, the surveyor first computes the bearing of the linear error of closure and checks that bearing against the course bearings. If a similarity exists (±5°), that course is the first course remeasured in the field.

4. Remeasure the sides of the traverse, beginning with a course having a bearing similar to the bearing of the linear error of closure (if there is one).

5. When a mistake (or error) is found, try the corrected value in the traverse computation to determine the new precision ratio.

The search for mistakes and errors is normally confined only to the distance measurements. The angle measurements are initially checked by doubling the angles (Sections 4.6.1 and 4.9), and the angular geometric closure is checked at the conclusion of the survey for compliance with the survey specifications, that is, within a given tolerance of $(n - 2)180°$ (see Section 6.2).

6.7 Traverse Precision and Accuracy

The actual accuracy of a survey, as given by the precision ratio, can be misleading. The opportunity exists for significant errors to cancel each other out, which results in "high precision" closures from relatively imprecise field techniques. Many new surveying students are introduced to taped traverses when they are asked to survey a traverse at the 1/3,000 or 1/5,000 levels of precision. Whereas most new student crews will struggle to achieve 1/3,000, there always seems to be one crew (using the same equipment and techniques) that reports back with a substantially higher precision ratio, say, 1/8,000. In this case, it is safe to assume that the higher precision ratio obtained by the one crew is probably due to compensating errors rather than to superior skill.

For example, in a square-shaped traverse, systematic taping errors (e.g., long or short tape; see Section F.1) will be completely balanced and beyond mathematical detection. In fact, if a traverse has any two courses that are close to being parallel, identical systematic or random errors made on those courses will likely cancel out: the bearings will be reversed, and the nearly parallel latitudes and departures will have opposite algebraic signs.

To be sure that the computed precision ratio truly reflects the field work, the traverse survey must be performed to specifications that will produce the desired results. For example, if a taped survey is to result in an accuracy of 1/5,000, then the field distance-measuring techniques should resemble those specified in Table 2.1. Also, the angle-measuring techniques should be consistent with the distance-measuring techniques. Figure 6.18 illustrates the relationship between angular and linear measurements; the survey specifications should be designed so that the maximum allowable error in angle (E_a) should be roughly equivalent to the maximum allowable error in distance (E_d). If the linear accuracy is restricted to 1/5,000, the angular error (θ) should be consistent:

$$1/5,000 = \tan \theta$$
$$\theta = 0;00'41''$$

See Table 6.3 for additional linear and angular error relationships. For a five-sided closed traverse with a specification for precision of 1/3,000, the maximum angular error of closure is $01' \sqrt{5} = 02'$ (to the closest minute), and for a specified precision of 1/5,000, the maximum error of closure of the field angles is $30'' \sqrt{5} = 01'$ (to the closest 30 seconds).

Point Y is to Be Set Out from Fixed Points X and Z

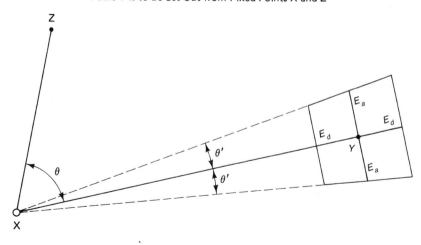

For the Line XY

E_d is the Possible Error in Distance Measurement and E_a is the Position Error Resulting from a Possible Angle Error of θ' in an Angle of θ.

FIGURE 6.18   Relationship between errors in linear and angular measurements.

Table 6.3   LINEAR AND ANGULAR ERROR RELATIONSHIPS*

Linear Accuracy Ratio	Maximum Angular Error (E)	Least Count of Angle Scale
1/1,000	0°03′26″	01′
1/3,000	0°01′09″	01′
1/5,000	0°00′41″	30″
1/7,500	0°00′28″	20″
1/10,000	0°00′21″	20″
1/20,000	0°00′10″	10″

*The overall allowable angular error in a closed traverse of n angles is $E_a \sqrt{n}$. Random errors accumulate as the square root of the number of observations.

6.8   Compass Rule Adjustment

The compass rule is used in many survey computations. The compass rule distributes the errors in latitude and departure for each traverse course in the same proportion as the course distance is to the traverse perimeter; that is, generally:

$$C_{lat}\,AB/\Sigma\,lat = \frac{AB}{P} \qquad \text{or} \qquad C_{lat}\,AB = \Sigma\,lat \times \frac{AB}{P} \qquad (6.5)$$

where $C_{lat}\,AB$ = correction in latitude AB
  $\Sigma\,lat$ = error of closure in latitude

Table 6.4 EXAMPLE 6.1: COMPASS RULE ADJUSTMENTS

Course	Distance	Bearing	Latitude	Departure	C_{lat}	C_{dep}	Latitudes (Balanced)	Departures (Balanced)
AB	164.95'	N71°11'E	+53.20	+156.13	−0.04	+0.03	+53.16	+156.16
BC	88.41'	S31°00'E	−75.78	+45.53	−0.02	+0.02	−75.80	+45.55
CD	121.69'	S44°18'W	−87.09	−84.99	−0.03	+0.03	−87.12	−84.96
DE	115.89'	N68°47'W	+41.94	−108.03	−0.03	+0.02	+41.91	−108.01
EA	68.42'	N 7°21'W	+67.86	−8.75	−0.01	+10.01	+67.85	−8.74
$P = 559.36'$			Σ lat = +0.13	Σ dep = −0.11	C_{lat} = −0.13	C_{dep} = +0.11	0.00	0.00

$$AB = \text{distance AB}$$
$$P = \text{perimeter of traverse}$$

and

$$C_{dep} \frac{AB}{\text{depB}} = \frac{AB}{P} \qquad \text{or} \qquad C_{dep} \, AB = \Sigma dep \times \frac{AB}{P} \tag{6.6}$$

where $C_{dep} \, AB$ = correction in departure AB
Σ dep = error of closure in departure
$\quad$ AB = distance AB
$\quad$ P = perimeter of traverse

Refer to Example 6.1. Table 6.2 has been expanded in Table 6.4 to provide space for traverse adjustments. The magnitudes of the individual corrections are shown next:

$$C_{lat} \, AB = 0.13 \times \frac{164.95}{559.36} = 0.04 \qquad C_{dep} \, AB = 0.11 \times \frac{164.95}{559.36} = 0.03$$

$$C_{lat} \, BC = 0.13 \times \frac{88.41}{559.36} = 0.02 \qquad C_{dep} \, BC = 0.11 \times \frac{88.41}{559.36} = 0.02$$

$$C_{lat} \, CD = 0.13 \times \frac{121.69}{559.36} = 0.03 \qquad C_{dep} \, CD = 0.11 \times \frac{121.69}{559} = 0.03$$

$$C_{lat} \, DE = 0.13 \times \frac{115.89}{559.36} = 0.03 \qquad C_{dep} \, DE = 0.11 \times \frac{115.89}{559.36} = 0.02$$

$$C_{lat} \, EA = 0.13 \times \frac{68.42}{559.36} = 0.01 \qquad C_{dep} \, EA = 0.11 \times \frac{68.42}{559.36} = 0.11$$

$$\text{Check: } C_{lat} = 0.13 \qquad\qquad \text{Check: } C_{dep} = 0.11$$

When these computations are performed on a handheld calculator, the constants 0.13/559.36 and 0.11/559.36 can be entered into storage for easy retrieval and thus quick computations.

The next step is to determine the algebraic sign to be used with the corrections. Quite simply, the corrections are opposite in sign to the errors. Therefore, for this example, the latitude corrections are all negative, and the departure corrections are all positive. The corrections are now added algebraically to arrive at the balanced values. For example, in Table 6.4, the correction for latitude BC is -0.02, which is to be "added" to latitude BC, -75.78. Because the correction is the same sign as the latitude, the two values are added to get the answer. In the case of course AB, the latitude correction (-0.04) and the latitude correction ($+53.20$) have opposite signs, indicating that the difference between the two values is the desired value (i.e., subtract to get the answer).

To check the work, the balanced latitudes and balanced departures are totaled to see if their respective sums are zero. The balanced latitude or balanced departure totals sometimes fail to equal zero by one last-place unit (0.01 in this example). This discrepancy is probably caused by rounding off and is normally of no consequence; the discrepancy is removed by arbitrarily changing one of the values to force the total to zero.

Note that when the error (in latitude or departure) to be distributed is quite small, the corrections can be assigned arbitrarily to appropriate courses. For example, if the error in latitude (or departure) is only 0.03 ft in a five-sided traverse, it is appropriate to apply corrections of 0.01 ft to the latitude of each of the three longest courses. Similarly, when the error in latitude for a five-sided traverse is $+0.06$, it is appropriate to apply a correction of -0.02 to the longest course latitude and a correction of -0.01 to each of the remaining four latitudes, the same solution provided by the compass rule.

Although the computations shown here are not tedious, you will be pleased to know that the computation of latitudes and departures and all the adjustments (usually using the least squares methods) to those computations are routinely performed on computers using different computer programs based on coordinate geometry (COGO). Some total station instruments also have these computational capabilities on-board.

6.9 Effects of Traverse Adjustments on Measured Angles and Distances

Once the latitudes and departures have been adjusted, the original polar coordinates (distance and direction) are no longer valid. In most cases, the adjustment required for polar coordinates is too small to warrant consideration. If the data are to be used for construction layout purposes, however, the corrected distances and directions should be used.

By way of example, consider the traverse data summarized in Table 6.4. We can use the corrected lats and deps to compute distances and bearings (azimuths) consistent with those corrected lats and deps. Figures 6.14 and 6.17 illustrate the trigonometric relationships among bearings, course distances, lats, and deps. It is clear that the distance $= \sqrt{\text{lat}^2 + \text{dep}^2}$ and that the tangent of the course bearing (azimuth) $=$ dep/lat. In Example 6.1:

$$\text{Adjusted distance AB} = \sqrt{53.16^2 + 156.16^2} = 164.96'$$

$$\text{Tan of adjusted bearing AB} = \frac{156.16}{53.16}$$

$$\text{Adjusted bearing AB} = \text{N71° 12'01" E}$$

The remaining corrected distances and bearings can be found in Table 6.5.

Table 6.5 ADJUSTMENT OF ORIGINAL DISTANCES AND BEARINGS

Course	Balanced Latitude	Balanced Departure	Adjusted Distance	Adjusted Bearing	Original Distance	Original Bearing
AB	+53.16	+156.16	164.969'	N 71°12'01" E	164.95'	N 71°11' E
BC	−75.80	+45.55	88.43'	S 31°00'10" E	88.41'	S 31°00' E
CD	−87.12	−84.96	121.69'	S 44°16'51" W	121.69'	S 44°18' W
DE	+41.91	−108.01	115.86'	N 68°47'34" W	115.89'	N 68°47' W
EA	+67.85	−8.74	68.41'	N 7°20'24" W	68.42'	N 7°21' W
	0.00	0.00	P = 559.35'		P = 559.36'	

6.10 Omitted Measurement Computations

The techniques developed in the computation of latitudes and departures can be used to solve for missing course information on a closed traverse. These techniques can also be utilized to solve any surveying problem that can be arranged in the form of a closed traverse. The case of one missing course is illustrated in Example 6.2 and in Section 14.7. Such a case requires the solution of a problem in which the bearing (azimuth) and distance for one course in a closed traverse are missing. Other variations of this problem include the case where the distance of one course and the bearing of another course are unknown. These cases can be solved by using missing course techniques (as outlined below), together with the sine law and/or cosine law (see Appendix A), which may be required for intermediate steps or cutoff lines.

■ EXAMPLE 6.2

In Figure 6.19, data for three of the four sides of the closed traverse are shown. In the field, the distances for AB, BC, and CD were measured. The interior angles at B and C were also measured. The bearing (azimuth) of AB was available from a previous survey. The bearings of BC and CD were computed from the given bearing and the measured angles at B and C.

Solution

Required are the distance DA and the bearing (azimuth) of DA. The problem is set up in the same manner as a closed traverse; see Table 6.6. When the latitudes and departures of AB, BC, and CD are computed and summed, the results indicate that the traverse failed to close by a line having a latitude of −202.82 and a departure of +276.13—i.e., line AD, in Figure 6.19. To be consistent with direction (clockwise in this example), we can say that the missing course is line DA with a latitude of +202.82 and a departure of −276.13 (i.e., C_{lat} and C_{dep}, Figure 6.19).

$$\text{Distance DA} = \sqrt{\text{lat DA}^2 + \text{dep DA}^2}$$
$$= \sqrt{202.82^2 + 276.13^2}$$
$$= 342.61'$$

$$\text{Tan brg. DA} = \frac{\text{dep AD}}{\text{lat AD}}$$
$$= \frac{-276.13}{+202.82}$$
$$= -1.3614535$$

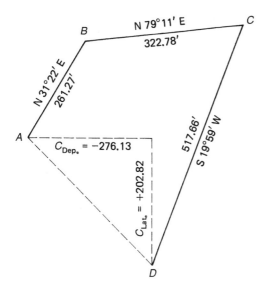

FIGURE 6.19 Example 6.2, missing course computation.

Table 6.6 EXAMPLE 6.2: MISSING COURSE

Course	Distance	Bearing	Latitude	Departure
AB	261.27	N 31°22′ E	+223.09	+135.99
BC	322.78	N 79°11′ E	+60.58	+317.05
CD	517.66	S 19°59′ W	−486.49	−176.91
			Σ lat = −202.82	Σ dep = +276.13
DA			C_{lat} = +202.82	C_{dep} = −276.13

Brg. DA = N53°42′W (to the closest minute)

Note that this technique does not permit a check on the precision ratio of the field work. Because DA is the closure course, its computed value will also contain all the accumulated errors in the field work (see Example 6.1, step 4).

6.11 Rectangular Coordinates of Traverse Stations

6.11.1 Coordinates Computed from Balanced Latitudes and Departures

Rectangular coordinates define the position of a point with respect to two perpendicular axes. Analytic geometry uses the concepts of a y axis (north-south) and an x axis (east-west), concepts that are obviously quite useful in surveying applications. In universal transverse Mercator (UTM) grid systems, the equator is used as the x axis, and the y axis is a central meridian through the middle of the zone in which the grid is located (see

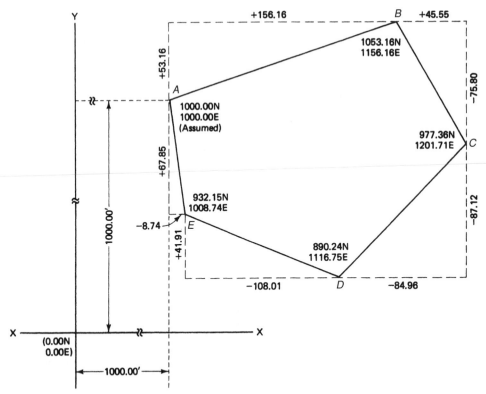

FIGURE 6.20 Station coordinates using balanced latitudes and departures.

Chapter 9). For surveys of a limited nature, where a coordinate grid has not been established, the coordinate axes can be assumed. If the axes are to be assumed, values are chosen so that the coordinates of all stations will be positive (i.e., stations will be in the northeast quadrant).

The traverse tabulated in Table 6.4 will be used for illustrative purposes. Values for the coordinates of station A are assumed to be 1000.00 north and 1000.00 east. To calculate the coordinates of the other traverse stations, it is simply a matter of applying the balanced latitudes and departures to the previously calculated coordinates. In Figure 6.20, the balanced latitude and departure of course AB are applied to the assumed coordinates of station A to determine the coordinates of station B, and so on. These simple computations are shown in Table 6.7. A check on the computation is possible by using the last latitude and departure (EA) to recalculate the coordinates of station A.

If a scale drawing of the traverse is required, it can be accomplished by methods of rectangular coordinates (where each station is located independently of the other stations by scaling the appropriate north and east distances from the axes) or the traverse can be drawn by the direction and distance method of polar coordinates (i.e., by scaling the interior angle between courses and by scaling the course distances; the stations are located in counterclockwise or clockwise sequence). In the rectangular coordinate system, plotting errors are isolated at each station, but in the polar coordinate layout method, all angle and

Table 6.7 COMPUTATION OF COORDINATES USING BALANCED LATITUDES AND DEPARTURES

Course	Balanced Latitude	Balanced Departure	Station	Northing	Easting
			A	1000.00	1000.00
AB	+53.16	+156.16		+53.16	+156.16
			B	1053.16	1156.16
BC	−75.80	+45.55		−75.80	+45.55
			C	977.36	1201.71
CD	−87.12	−84.96		−87.12	−84.96
			D	890.24	1116.75
DE	+41.91	−108.01		+41.91	−108.01
			E	932.15	1008.74
EA	+67.85	−8.74		+67.85	−8.74
			A	1000.00	1000.00
				check	check

distance scale errors are accumulated and show up only when the last distance and angle are scaled to relocate the starting point theoretically. The resultant plotting error is similar in type to the linear error of traverse closure (AA′) illustrated in Figure 6.17.

The use of coordinates to define the positions of boundary markers has been steadily increasing over the years. The storage of property-corner coordinates in large-memory civic computers permits lawyers, municipal authorities, and others to retrieve instantly current land registration assessment and other municipal information, for example, information concerning census and the level of municipal services. Although such use is important, the truly impressive impact of coordinate use results from the coordination of topographic detail (digitization) so that plans can be prepared by computer-assisted plotters (see Chapter 7), and from the coordination of all legal and engineering details. Not only are the plans produced by computer-assisted plotters, but the survey layout is also accomplished by sets of computer-generated coordinates (rectangular and polar) fed either manually or automatically through total stations (see Chapter 5). Complex layouts can then be accomplished quickly by a few surveyors from one or two centrally located control points, with a higher level of precision and a lower incidence of mistakes.

6.11.2 Adjusted Coordinates Computed from Raw-Data Coordinates

Section 6.8 demonstrated the adjustment of traverse errors by the adjustment of the individual ΔY's (northings) and ΔX's (eastings) for each traverse course using the compass rule. This traditional technique has been favored for many years. Because of the wide use of the computer in surveying solutions, however, coordinates are now often first computed from raw (unadjusted) bearing/distance data and then adjusted using the compass rule or the least squares technique. Because we are now working with coordinates and not individual ΔY's and ΔX's, the distance factor to be used in the compass rule must be cumulative. This technique will be illustrated using the same field data from Example 6.1. Corrections (C) to raw-data coordinates are shown in Tables 6.8 and 6.9.

Table 6.8 COMPUTATION OF RAW-DATA COORDINATES

Station/Course	Bearing	Distance	ΔY	ΔX	Coordinates (Raw-Data) Northing	Coordinates (Raw-Data) Easting
A					1000.00	1000.00
AB	N 71°11′ E	164.95	+53.20	+156.13		
B					1053.20	1156.13
BC	S 31°00′ E	88.41	−75.78	+45.53		
C					977.42	1201.66
CD	S 44°18′ W	121.69	−87.09	−84.99		
D					890.33	1116.67
DE	N 68°47′ W	115.89	+41.94	−108.03		
E					932.27	1008.64
EA	N 7°21′ W	68.42	+67.86	−8.75		
A					1000.13	999.89
			ΣΔY = +0.13	ΣΔX = −0.11		

Table 6.9 COMPUTATION OF ADJUSTED COORDINATES

Station	Coordinates (Raw-Data) Northing	Coordinates (Raw-Data) Easting	CΔY	CΔX	Coordinates (Adjusted) Northing	Coordinates (Adjusted) Easting
A	1000.00	1000.00			1000.00	1000.00
B	1053.20	1156.13	−0.04	+0.03	1053.16	1156.16
C	977.42	1201.66	−0.06	+0.05	977.36	1201.71
D	890.33	1116.67	−0.09	+0.07	890.24	1116.74
E	932.27	1008.64	−0.11	+0.10	932.16	1008.74
A	1000.13	999.89	20.13	10.11	1000.00	1000.00

$C\Delta Y$ = correction in northing (latitude) and $C\Delta X$ = correction in easting (departure). Correction C is opposite in sign to errors ΣY and ΣX.

Station B

$$C\Delta Y \text{ is } [AB/P] \ \Sigma \Delta Y = [164.95/559.36] \ 0.13 = -0.04$$
$$C\Delta X \text{ is } [AB/P] \ \Sigma \Delta X = [164.95/559.36] \ 0.11 = +0.03$$

Station C

$$C\Delta Y \text{ is } [(AB + BC)/P] \ \Sigma \Delta Y = [253.36/559.36] \ 0.13 = -0.06$$
$$C\Delta X \text{ is } [(AB + BC)/P] \ \Sigma \Delta X = [253.36/559.36] \ 0.11 = +0.05$$

Station D

$$C\Delta Y \text{ is } [(AB + BC \ 1 \ CD)/P] \ \Sigma \Delta Y = [375.05/559.36] \ 0.13 = -0.09$$
$$C\Delta X \text{ is } [(AB + BC \ 1 \ CD)/P] \ \Sigma \Delta Y = [375.05/559.36] \ 0.11 = +0.07$$

Station E

$$C\Delta Y \text{ is } [(AB + BC + CD + DE)/P] \Sigma\Delta Y = [490.94/559.36]\,0.13 = -0.11$$
$$C\Delta X \text{ is } [(AB + BC + CD + DE)/P] \Sigma\Delta X = [490.94/559.36]\,0.11 = +0.10$$

Station A

$$C\Delta Y \text{ is } [(AB + BC + CD + DE + EA)/P]\Sigma\Delta Y = [559.36/559.36]\,0.13 = -0.13$$
$$C\Delta X \text{ is } [(AB + BC + CD + DE + EA)/P]\Sigma\Delta X = [559.36/559.36]\,0.11 = +0.11$$

6.12 Area of a Closed Traverse by the Coordinate Method

When the coordinates of the stations of a closed traverse are known, it is a simple matter then to compute the area within the traverse, either by computer or by handheld calculator. Figure 6.21(a) shows a closed traverse 1, 2, 3, 4 with the appropriate X and Y coordinate distances. Figure 6.21(b) illustrates the technique used to compute the traverse area.

In Figure 6.21(b), you can see that the desired area of the traverse is, in effect, area 2 minus area 1. Area 2 is the sum of the areas of trapezoids 4′433′ and 3′322′. Area 1 is the sum of trapezoids 4′411′ and 1′122′:

$$\text{Area 2} = \frac{1}{2}(X_4 + X_3)(Y_4 - Y_3) + 1/2\,(X_3 + X_2)(Y_3 - Y_2)$$

$$\text{Area 1} = \frac{1}{2}(X_4 + X_1)(Y_4 - Y_1) + 1/2\,(X_1 + X_2)(Y_1 - Y_2)$$

$$2A = [(X_4 + X_3)(Y_4 - Y_3) + (X_3 + X_2)(Y_3 - Y_2)]$$
$$- [(X_4 + X_1)(Y_4 - Y_1) + (X_1 + X_2)(Y_1 - Y_2)]$$

Expand this expression, and collect the remaining terms:

$$2A = X_1(Y_2 - Y_4) + X_2(Y_3 - Y_1) + X_3(Y_4 - Y_2) + X_4(Y_1 - Y_3) \qquad (6.7)$$

Stated simply, the double area of a closed traverse is the algebraic sum of each X coordinate multiplied by the difference between the Y values of the adjacent stations. This double area is divided by 2 to determine the final area. The final area can be positive or negative, the algebraic sign reflecting only the direction of computation approach (clockwise or counterclockwise). The area is, of course, positive.

■ **EXAMPLE 6.3** *Area Computation by Coordinates*

Refer to the traverse example (Example 6.1) in Section 6.6. As illustrated in Figure 6.20, the station coordinates are summarized below:

Station	Northing	Easting
A	1000.00	1000.00
B	1053.16	1156.16
C	977.36	1201.71
D	890.24	1116.75
E	932.15	1008.74

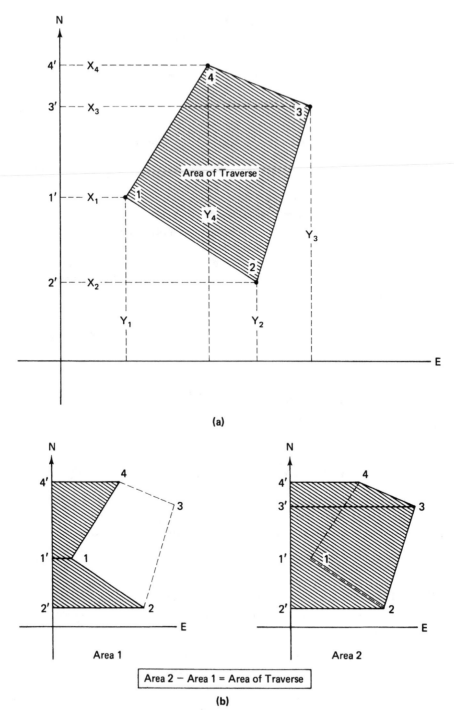

FIGURE 6.21 Area by rectangular coordinates.

The double area computation uses the relationships developed earlier in this section. Equation (6.7) must be expanded from the four-sided traverse shown in the illustrative example to the five-sided traverse of Example 6.1:

$$2A = X_1(Y_5 - Y_2) + X_2(Y_1 - Y_3) + X_3(Y_2 - Y_4) + X_4(Y_3 - Y_5) + X_5(Y_4 - Y_1)$$

that is, for the double area, each X coordinate is multiplied by the difference between the Y coordinates of the adjacent stations.

Using the station letters instead of the general case numbers shown above, we find the solution:

$$
\begin{aligned}
X_A(Y_E - Y_B) &= 1000.00(932.15 - 1053.16) = -121,010 \\
X_B(Y_A - Y_C) &= 1156.16(1000.00 - 977.36) = +26,175 \\
X_C(Y_B - Y_D) &= 1201.71(1053.16 - 890.24) = +195,783 \\
X_D(Y_C - Y_E) &= 1116.75(977.36 - 932.15) = +50,488 \\
X_E(Y_D - Y_A) &= 1008.74(890.24 - 1000.00) = -110,719 \\
2A &= 40,677 \text{ ft}^2 \\
A &= 20,338 \text{ ft}^2 \\
A &= \frac{20,358}{43,560} = 0.47 \text{ acre} \\
1 \text{ acre} &= 43,560 \text{ ft}^2
\end{aligned}
$$

Problems

6.1. A closed, five-sided traverse has the following interior angles: A = 125°25′, B = 93°03′, C = 88°38′, D = 116°16′, E = ? Find the angle at E.

6.2. A five-sided, closed traverse has the following interior angles:

$$
\begin{aligned}
A &= 75°10'30'' \\
B &= 137°43'00'' \\
C &= 88°49'30'' \\
D &= 113°27'30'' \\
E &= 124°47'00''
\end{aligned}
$$

Determine the angular error, and balance the angles by applying equal corrections to each angle.

6.3. Convert the following azimuths to bearings:
(a) 188°23′　(b) 167°52′　(c) 330°09′50″
(d) 29°33′　(e) 261°50′　(f) 166°59′

6.4. Convert the following bearings to azimuths:
(a) N 7°51′ W　(b) N 76°14′ E　(c) S 33°33′ E
(d) S 4°39′ W　(e) N 37°18′ W　(f) S 0°38′ E

6.5. Convert each of the azimuths given in Problem 6.3 to reverse (back) azimuths.

6.6. Convert each of the bearings given in Problem 6.4 to reverse (back) bearings.

6.7. An open traverse that runs from A to H has the following deflection angles: B = 1°03′R; C = 2°58′ R; D = 7°24′ L; E = 6°31′ L; F = 1°31′ R; G = 8°09′ L. If the bearing of AB is N 19°24′ E, compute the bearings of the remaining sides.

6.8. Closed traverse ABCD has the following bearings: AB = N 59°18′ E, BC = S 75°15′ E, CD = S 2°41′ W, DA = N 69°37′ W. Compute the interior angles, and show a geometric check for your work.

Use the sketch in Figure 6.22 and the following interior angles for Problems 6.9 through 6.11.

Angles

A = 101°28′
B = 102°11′
C = 104°42′
D = 113°05′
E = 118°34′

538°120′ = 540°00′

FIGURE 6.22 Sketch for Problem 6.9 through 6.11.

6.9. If bearing AB is N 53°55′ E, compute the bearings of the remaining sides. Provide two solutions, one working clockwise and one working counterclockwise.

6.10. If the azimuth of AB is 63°22′, compute the azimuths of the remaining sides. Provide two solutions, one working clockwise and one working counterclockwise.

6.11. If the azimuth of AB is 56°56′, compute the azimuths of the remaining sides. Provide two solutions, one working clockwise and one working counterclockwise.

6.12. See Figure 6.23. The four-sided, closed traverse has the following angles and distances:

A = 51°23′   AB = 713.93 ft
B = 105°39′   BC = 606.06 ft
C = 78°11′   CD = 391.27 ft
D = 124°47′   DA = 781.18 ft

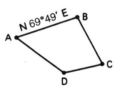

The bearing of AB is N 69°49′ E.
(a) Perform a check for angular closure.
(b) Compute both bearings and azimuths for all sides.
(c) Compute the latitudes and departures.
(d) Compute the linear error of closure and the precision ratio.

FIGURE 6.23 Sketch for Problem 6.12.

6.13. Using the data from Problem 6.12, and the compass rule, balance the latitudes and departures. Compute corrected distances and directions.

6.14. Using the data from Problem 6.13, compute the coordinates of stations B, C, and D. Assume that the coordinates of station A are 1000.00 ft N and 1000.00 ft E.

6.15. Using the data from Problem 6.14, compute the area (in acres) enclosed by the traverse.

6.16. See Figure 6.24. The five-sided, closed traverse has the following angles and distances:

A = 38°30′   AB = 371.006 m
B = 100°38′   BC = 110.222 m
C = 149°50′   CD = 139.872 m
D = 85°59′   DE = 103.119 m
E = 165°03′   EA = 319.860 m

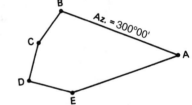

FIGURE 6.24 Sketch for Problem 6.16.

Side AB has an azimuth of 300°00′.

(a) Perform a check for angular closure.

(b) Compute both bearings and azimuths for all sides.

(c) Compute the latitudes and departures.

(d) Compute the linear error of closure and the precision ratio.

6.17. Using the data from Problem 6.16 and the compass rule, balance the latitudes and departures.

6.18. Using the data from Problem 6.17, compute the coordinates of stations C, D, E, and A. Assume that the coordinates of station B are 1000.000 m N and 1000.000 m E.

6.19. Using the data from Problem 6.18, compute the area (in hectares) enclosed by the traverse.

6.20. See Figure 6.25. The two frontage corners (A and D) of a large tract of land are joined by the following open traverse:

Course	Distance (ft)	Bearing
AB	80.32	N 70°10′07″ E
BC	953.83	N 74°29′00″ E
CD	818.49	N 70°22′45″ E

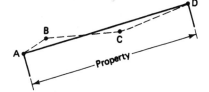

FIGURE 6.25 Sketch for Problem 6.20.

Compute the distance and bearing of the property frontage AD.

6.21. Given the following data for a closed property traverse, compute the missing data (i.e., distance CD and bearing DE).

Course	Distance (m)	Bearing
AB	537.144	N 37°10′49″ E
BC	1109.301	N 79°29′49″ E
CD	?	S 18°56′31″ W
DE	953.829	?
EA	483.669	N 26°58′31″ W

6.22. A six-sided traverse has the following station coordinates:

A: 559.319 N, 207.453 E; B: 738.562 N, 666.737 E; C: 541.742 N, 688.350 E;

D: 379.861 N, 839.008 E; E: 296.099 N, 604.048 E; F: 218.330 N, 323.936 E

Compute the distance and bearing of each side.

6.23. Using the data from Problem 6.22, compute the area (hectares) enclosed by the traverse.

6.24. A total station was set up at control station K, which is within the limits of a five-sided property (see Figure 6.26). The coordinates of station K are 2,000.000 N, 2,000.000 E. Azimuth angles and polar distances to the five property corners are as follows:

Direction	Azimuth	Horizontal Distance (m)
KA	286°51′30″	34.482
KB	37°35′28″	31.892
KC	90°27′56″	38.286
KD	166°26′49″	30.916
KE	247°28′43″	32.585

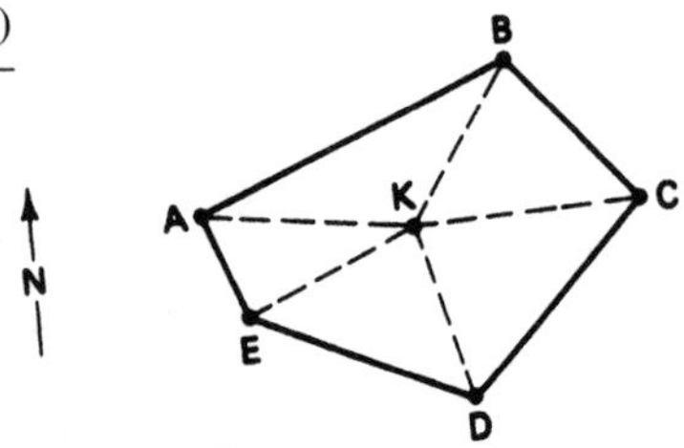

FIGURE 6.26 Sketch for Problem 6.24.

Compute the coordinates of the property corners A, B, C, D, and E.

6.25. Using the data from Problem 6.24, compute the area (hectares) of the property.

6.26. Using the data from Problem 6.24, compute the bearings (to the closest second) and distances (to three decimals) of the five sides of the property.

Chapter 7

An Introduction to Geomatics

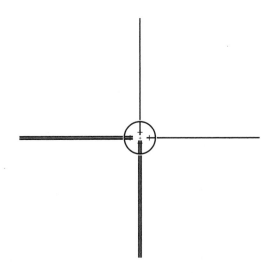

7.1   Geomatics Defined

Geomatics is a term used to describe the science and technology of dealing with earth measurement data. It includes field data collection, processing, and presentation. It has applications in all disciplines and professions that use earth-related spatial data. Some examples of these disciplines and professions include planning, geography, geodesy, infrastructure engineering, agriculture, natural resources, environment, land division and registration, project engineering, and mapping.

7.2   Branches of Geomatics

Figure 7.1 is a model of the science of geomatics and shows how all the branches and specializations are tied together by their common interest in earth measurement data and in their common dependence on computer science and information technology (IT). This computerized technology has changed the way field data is collected and processed. To appreciate the full impact of this new technology, you must view the overall operation, that is, from field to computer, computer processing, and data portrayal in the form of maps and plans. Data-collection techniques include field surveying, satellite positioning, and remotely sensed imagery obtained through aerial photography/imaging and satellite imagery. It also includes the acquisition of database material scanned from older maps and plans and data collected by related agencies, for example, census data such as TIGER. Data processing is handled through various computer programs designed to process the measurements and their attribute data, such as coordinate geometry (COGO), field data processing, and the processing of remotely sensed data from aerial photos (photogrammetry, including soft-copy photogrammetry) and satellite imagery analysis.

Processed data can then be used in computer programs for engineering design, relational database management, and geographic information systems (GISs). Data presentation

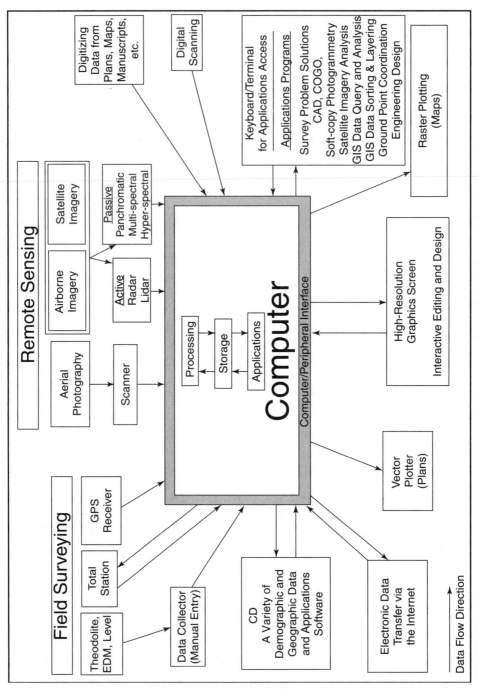

FIGURE 7.1 Geomatics data model, showing the collection, processing, analysis, design, and plotting of geodata.

is handled through the use of mapping and other illustrative computer programs; the presentations are displayed on computer screens (where interactive editing can occur) and are output from digital vector and raster plotting devices.

Once the positions and attributes of geographic entities have been digitized and stored in computer memory, they are available to a wide variety of users. Through the use of modern information technology (IT), geomatics brings together professionals in all related disciplines and professions.

7.3 Data-Collection Branch

Advances in computer science have had a tremendous impact on all aspects of modern technology, and the effects on the collection and processing of data in both field surveying and remotely sensed imagery have been significant. Chapters 1 to 5 described ground-based techniques for data collection. Survey data once laboriously collected with tapes, theodolites/transits, and levels can now be collected quickly and efficiently using total stations (Chapter 5) and GPS receivers (Chapter 8)—both featuring on-board (or attached) data collection. These latter techniques can provide the high accuracy results usually required in control surveys (see Chapter 9) and engineering surveys. When high accuracy is not a prime requirement, as in some GIS surveys (see Section 7.9) and many mapping surveys, data can be collected efficiently from airborne and satellite platforms (see Sections 7.6 and 7.7). The broad picture encompassing all aspects of data collection, as well as the storage, processing, analysis, planning and design, and presentation of the data, is now referred to as the field of geomatics.

The upper portion of the schematic in Figure 7.1 shows the various ways that data can be collected and transferred to the computer. In addition to total station techniques, field surveys can be performed using conventional surveying instruments (theodolites, EDMs, and levels), with the field data entered into a data collector instead of conventional field books. This manual entry of field data lacks the speed associated with interfaced equipment, but after the data have been entered, all the advantages of electronic techniques are available to the surveyor. The raw field data, collected and stored by the total station, are transferred to the computer through a standard RS232 interface connection; via a memory card; and/or, more recently, via wireless cellular connections. When the collected terrain data has been downloaded into a computer, coordinate geometry programs (COGO) and/or imagery analysis programs can be used to determine the positions (northing, easting, and elevation values) of all survey points. Also at this stage, additional data points (e.g., inaccessible ground points) can be computed and added to the data file.

Existing maps and plans have a wealth of lower-precision data that may be relevant for an area survey database. If such maps and plans are available, the data can be digitized on a digitizing table or by digital scanners and added to the Y, X, and Z coordinates files. Ideally, such files would show the level of precision at which this data was collected—to distinguish this data from data that may have been precisely collected. One of the more important features of the digitizer is its ability to digitize maps and plans drawn at various scales and store the distances and elevations in the computer at their ground (or grid) values.

The stereo analysis of aerial photos (see Section 7.6) is a very effective method of generating topographic ground data, particularly in high-density areas, where the costs for conventional surveys would be high. Many municipalities routinely fly all major roads at regular intervals (for example, every five years in developing areas) and create plans and profiles that can be used for design and construction (see Part II). With the advent of computerized (soft-copy) photogrammetric procedures, stereo-analyzers can coordinate all horizontal and vertical features from aerial images and transfer these Y (north), X (east), and Z (height) coordinates to computer storage.

You will see in Section 7.7 that satellite imagery is received from U.S. (e.g., EOS and Landsat), French, European, Japanese, Canadian, Chinese, and South American satellites, and that it can be processed by a digital image analysis system that classifies terrain into categories of soil and rock types and vegetative cover. These and other data can be digitized (georeferenced) and added to the spatial database.

7.3.1   General Background

Data-collection surveys, including topographic surveys, are used to determine the positions of built and natural features (for example, roads, buildings, trees, and shorelines). These built and natural features can then be plotted to scale on a map or plan. In addition, topographic surveys include the determination of ground elevations, which can later be plotted in the form of contours, cross sections, profiles, or simply spot elevations. In engineering and construction work, topographic surveys are often called preliminary or preengineering surveys. Large-area topographic surveys are usually performed using aerial photography/ imaging, with the resultant distances and elevations being derived from the photographs through the use of photogrammetric principles. The survey plan is normally drawn on a digital plotter.

For small-area topographic surveys, various types of ground survey techniques can be employed. Ground surveys can be accomplished by using (1) theodolite and tape, (2) total station, and (3) global positioning system (GPS) (see Chapter 8). Surveys executed using theodolite/tape and level/rod are usually plotted using conventional scale/protractor techniques. On the other hand, if the survey has been executed using a total station or GPS receiver, the survey drawing is plotted on a digital plotter. When total stations are used for topographic surveys, the horizontal (X and Y) and the vertical location (elevation) can be captured easily with just one sighting, with point descriptions and other attribute data entered into electronic storage for later transfer to the computer. (Electronic surveying techniques were discussed in detail in Chapter 5.)

7.3.1.1   Precision Required for Topographic Surveys.   If we consider plotting requirements only, the survey detail need only be located at a precision level consistent with standard plotting precision. Many municipal plans (including plan and profile) are drawn at 1 in. = 50 ft or 1 in. = 40 ft (1:500 metric). If we assume that points can be plotted to the closest 1/50 in. (0.5 mm), then location ties need only be to the closest 1 ft or 0.8 ft (0.25 m). For smaller-scale plans, the location precision can be relaxed even further.

In addition to providing plotting data, topographic surveys provide the designer with field dimensions that must be considered for related engineering design. For example, when you are designing an extension to an existing storm sewer, the topographic survey must include the location and elevations of all connecting pipe inverts (see Chapter 13). These values are determined more precisely (0.01 ft or 0.005 m) because of design requirements.

The following points should be considered with respect to levels of precision:

1. Some detail, for example, building corners, railway tracks, bridge beam seats, and pipe and culvert inverts, can be precisely defined and located.
2. Some detail cannot be defined or located precisely. Examples include stream banks, edges of gravel roads, limits of wooded areas, rock outcrops, and tops and bottoms of slopes.
3. Some detail can be located with only moderate precision with normal techniques. Examples are large single trees, manhole covers, and walkways.

When a topographic survey requires all three of the above levels of precision, the items in level 1 are located at a precision dictated by the design requirements; the items in levels 2 and 3 are usually located at the precision of level 3 (for example, 0.1 ft or 0.01 m). Because most natural features are themselves not precisely defined, topographic surveys in areas having only natural features (for example, stream or watercourse surveys, site development surveys, or large-scale mapping surveys) can be accomplished by using relatively imprecise survey methods (see aerial surveying in Section 7.6).

7.3.1.2 Traditional Rectangular Surveying Methods.

We focus here on the rectangular techniques, first introduced in Section 1.4, employing preelectronic field techniques. The rectangular technique discussed here utilizes right-angle offsets for detail location, and cross sections (level and rod) for elevations and profiles. Polar techniques (for example, the use of total stations) are also used for both horizontal positioning and elevations.

Ground surveys are based on survey lines or stations that are part of, or tied in to, the survey control. Horizontal survey control can consist of boundary lines, or offsets to boundary lines, such as centerlines (₵s); coordinate grid monuments; route survey traverses; or arbitrarily placed baselines or control monuments. Vertical survey control is based on benchmarks that already exist in the survey area or benchmarks that are established through differential leveling from other areas.

Surveyors are conscious of the need for accurate and well-referenced survey control. If the control is inaccurate, the survey and any resultant design will also be inaccurate. If the control is not well referenced, it will be costly (perhaps impossible) to relocate the control points precisely in the field once they are lost. In addition to providing control for the original survey, the same survey control must be used if additional survey work is required to complete the preengineering project and, of course, the original survey control must be used for subsequent construction layout surveys that may result from designs based on the original surveys. It is not unusual to have one or more years pass between the preliminary survey and the related construction layout.

7.3.1.3 Tie-Ins at Right Angles to Baselines.

Many ground-based topographic surveys (excluding mapping surveys but including many preengineering surveys) utilize the right-angle offset technique to locate detail. This technique not only provides the location of plan detail but also provides location for area elevations taken by cross sections.

Plan detail is located by measuring the distance perpendicularly from the baseline to the object and, in addition, measuring along the baseline to the point of perpendicularity.

FIGURE 7.2   Double right-angle prism. (Courtesy of Keuffel & Esser Co.)

The baseline is laid out in the field with stakes (nails in pavement) placed at appropriate intervals, usually 100 ft, or 20 to 30 m. A sketch is entered in the field book before the measuring commences. If the terrain is smooth, a tape can be laid on the ground between the station marks. This technique permits the surveyor to move along the tape (toward the forward station), noting and booking the stations of the sketched detail on both sides of the baseline. The right angle for each location tie can be established by using a pentaprism (Figure 7.2), or a right angle can be established approximately in the following manner. The surveyor stands on the baseline facing the detail to be tied in. He then points one arm down the baseline in one direction and the other arm down the baseline in the opposite direction; after checking both arms (pointed index fingers) for proper alignment, the surveyor closes his eyes while he swings his arms together in front of him, pointing (presumably) at the detail. If he is not pointing at the detail, the surveyor moves slightly along the baseline and repeats the procedure until the detail has been correctly sighted in. The station is then read off the tape and booked in the field notes. This approximate method is used a great deal in route surveys and municipal surveys. This swung-arm technique provides good results over short offset distances (50 ft, or 15 m). For longer offset distances or for very important detail, a pentaprism or even a theodolite can be used to determine the station.

Once all the stations have been booked for the interval (100 ft, or 20 to 30 m), only the offsets left and right of the baseline are left to be measured. If the steel tape has been left lying on the ground during the determination of the stations, it is usually left in place to mark the baseline while the offsets are measured from it with another tape (e.g., a cloth tape).

Figure 7.3(a) illustrates topographic field notes that have been booked when a single baseline was used, and Figure 7.3(b) illustrates such notes when a split baseline was used. In Figure 7.3(a), the offsets are shown on the dimension lines, and the stations are shown opposite the dimension line or as close as possible to the actual tie point on the baseline. In Figure 7.3(b), the baseline has been "split"; that is, two lines are drawn representing the baseline, leaving a space of zero dimensions between them for the inclusion of stations. The split baseline technique is particularly valuable in densely detailed areas where single baseline notes would become crowded and difficult to decipher. The earliest topographic surveyors in North America used the split-baseline method of note keeping (see Figure 7.4).

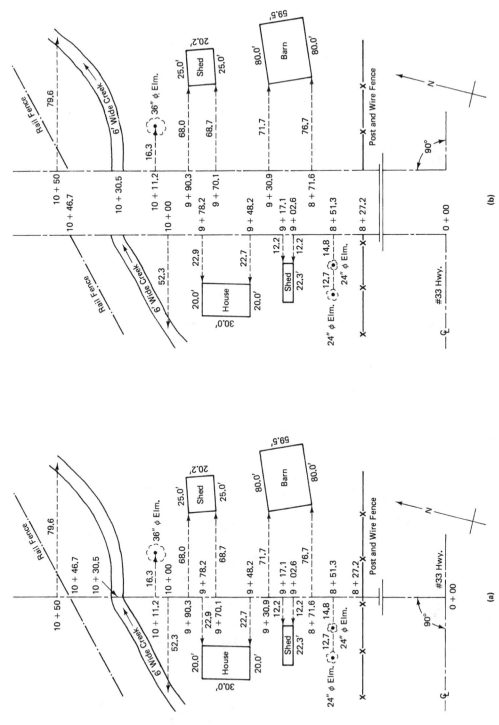

FIGURE 7.3 Topographic field notes. (a) Single baseline. (b) Split baseline.

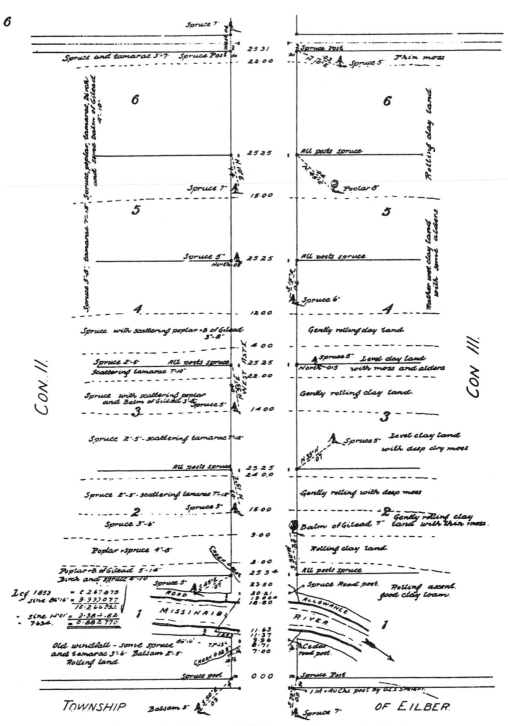

FIGURE 7.4 Original topographic field notes, 1907 (distances shown are in chains).

7.3.1.4 Cross Sections and Profiles.

Cross sections form series of elevations taken at right angles to a baseline at specific stations, whereas **profiles** are a series of elevations taken along a baseline at some specified repetitive station interval. The elevations thus determined can be plotted on maps and plans either as spot elevations or as contours, or they can be plotted as end areas for construction quantity estimating.

As in offset ties, the baseline interval is usually 100 ft (20 to 30 m), although in rapidly changing terrain, the interval is usually smaller (e.g., 50 ft or 10 to 15 m). In addition to the regular intervals, cross sections are taken at each abrupt change in the terrain (top and bottom of slopes, etc.).

Figure 7.5 illustrates how the rod readings are used to define the ground surface. In Figure 7.5(a), the uniform slope permits a minimum number of rod readings. In Figure 7.5(b), the varied slope requires several more (than the minimum) rod readings to define the ground surface adequately. Figure 7.5(c) illustrates how cross sections are taken before and after construction. Chapter 17 covers the calculation of the end area (lined section) at each station and then the volumes of cut and fill.

The profile consists of a series of elevations along the baseline. If cross sections have been taken, the necessary data for plotting a profile will also have been taken. If cross sections are not planned for an area for which a profile is required, the profile elevations can be determined by simply taking rod readings along the line at regular intervals and at all points where the ground slope changes (see Figure 3.19).

Typical field notes for profile leveling are shown in Figure 3.20. Cross sections are booked in two different formats. Figure 3.21 showed cross sections booked in standard level note format. All the rod readings for one station (that can be "seen" from the HI) are booked together. In Figure 3.22, the same data were entered in a format popular with highways agencies. The latter format is more compact and thus takes up less space in the field book; the former format takes up more space in the field book, but it allows for a description for each rod reading, an important consideration for municipal surveyors.

In cases where the terrain is very rugged, thus making the level and rod work very time consuming (many instrument setups), the surveys should be performed using total stations. Chapter 5 describe total station instruments. With polar techniques, these instruments can measure distances and differences in elevation very quickly. The roving surveyor holds a reflecting prism mounted on a range pole instead of holding a rod. Many of these instruments have the distance and elevation data recorded automatically for future computer processing, while others require that the data be entered manually into the data recorder.

7.4 Design and Plotting

Historically, map and plan preparation depended on measurements taken directly in the field, whereas modern data-collection practice has been expanded to include both previously referenced and remotely sensed data. The concept of maps and plans has also been expanded to include electronic images (electronic charts are used a great deal in navigation and marine surveying). Maps and plans (hard-copy and electronic) are generally prepared for one of two reasons: (1) when measured data is displayed to scale on a map or plan, the ground data is presented as an inventory or record of the features surveyed, (2) when the presented data is used to facilitate the design of infrastructure projects, private projects, or

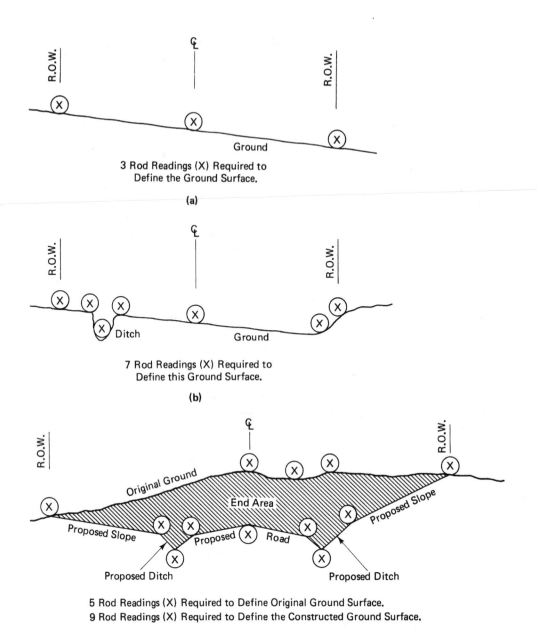

FIGURE 7.5 Cross sections used to define ground surface. (a) Uniform slope. (b) Varied slope. (c) Ground surface before and after construction.

land division, the plan or map is used as a design and layout tool. If the data are to be digitally plotted, a plot file may be created that contains point plot commands (including symbols) and join commands for straight and curved lines. Labels and other attribute data are also included.

Design programs are available for all engineering and construction endeavors. These software programs can work with the stored coordinates to provide different possible designs, which can then be analyzed quickly with respect to cost and other factors. Some design programs incorporate interactive graphics, which permit a plot of the survey to be shown to scale on a high-resolution graphics screen. On this screen, points and lines can be moved, created, edited, and so forth, with the final positions coordinated right on the screen and the new coordinates captured and added to the coordinates files. Once the coordinates of all field data have been determined, the design software can then compute the coordinates of all key points in the new facility (e.g., roads, site developments, etc.). The surveyor can take this design information back to the field and perform a construction layout survey (see Part II) showing the contractor how much cut and fill, for example, is required to bring the land to the proposed elevations and locations. Also, the design data can now be transferred directly (sometimes wirelessly) to construction machine controllers for machine guidance and control functions.

Three types of basic geospatial data are collected and coordinated:

- The horizontal position of natural and constructed features or entities.
- The vertical position (elevation) of the ground surface or built features.
- Attribute data describing the features or entities being surveyed.

Survey drafting is a term that covers a wide range of scale graphics, including both manual and automatic plotting. If the data has been collected electronically, computer programs can digitize the data, thus enabling the surveyor to process and selectively plot areas of interest. Figure 7.6 shows a digital workstation, which includes the computer, printer and digitizing table. Figure 7.7 shows a digital screen plot, which can be edited, and Figure 7.8 shows a digital plotter that can produce the plot at any scale using eight different pens; a variety of pens permits the use of different colors (or line widths).

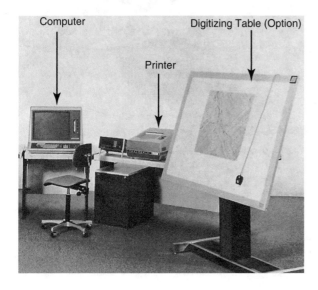

FIGURE 7.6 Wild Geomap workstation. (Courtesy of Leica Geosystems Inc.)

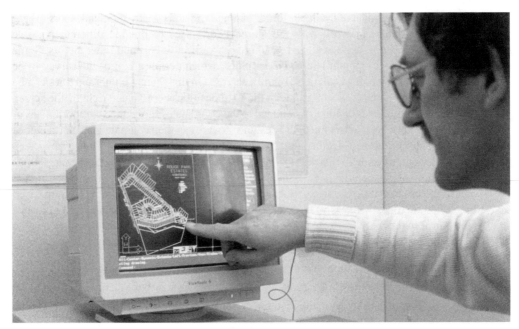

FIGURE 7.7 Land division design and editing on a desktop computer.

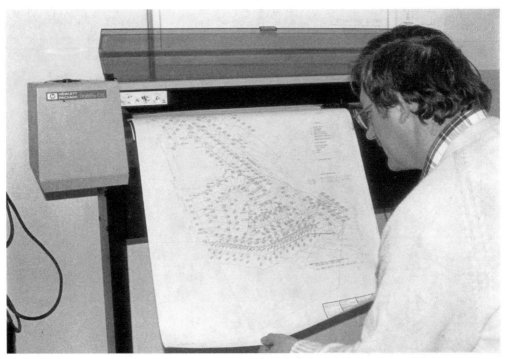

FIGURE 7.8 Land division plot on a Hewlett-Packard eight-pen digital plotter.

7.4.1 Digital Plotting

Data can be plotted onto a high-resolution graphics screen, where the plot can be checked for completeness and accuracy. When interactive graphics are available, the plotted features can be deleted, enhanced, corrected, cross-hatched, labeled, dimensioned, and so on. At this stage, a hard copy of the screen display can be printed either on a simple printer or on an ink-jet color printer. Plot files can be plotted directly on a digital plotter similar to that shown in Figure 7.8. The resultant plan can be plotted to any desired scale, limited only by the paper size. Some plotters have only one or two pens, although plotters are available with four to eight pens; a variety of pens permits colored plotting or plotting using various line weights. Plans, and plan and profiles, drawn on digital plotters are becoming more common on construction sites.

Automatic plotting (with a digital plotter) can be used when the field data have been coordinated and stored in computer memory. Coordinated field data are by-products of data collection by total stations; air photo stereo analysis; satellite imagery; and digitized, or scanned, data from existing plans and maps (see Figure 7.1).

A plot file can be created that includes the following typical commands:

- Title
- Scale
- Limits (e.g., can be defined by the southwesterly coordinates, northerly range, and easterly range)
- Plot all points?
- Connect specific points (through feature coding or CAD commands)
- Pen number (various pens can have different line weights or colors; two to four pens are common)
- Symbol (symbols are predesigned and stored by identification number in a symbol library)
- Height of characters (the heights of labels, letters and numbers, coordinates, and symbols can be defined)

The actual plotting can be performed simply by keying in the plot command required by the specific software and then by keying in the name of the plot file to be plotted. Also, the coordinated field point files can be transferred to an interactive graphics terminal (see Figure 7.7), with the survey plot being created and edited graphically right on the high-resolution graphics screen. Some surveying software programs have this graphics capability, whereas others permit coordinate files to be transferred easily to an independent graphics program (e.g., Autocad) for editing and plotting. Once the plot has been completed on the graphics screen (and all the point coordinates have been stored in the computer), the plot can be transferred to a digital plotter for final presentation.

Maps, which are usually drawn at a small scale, portray an inventory of all the topographic detail included in the survey specifications; on the other hand, plans, which are usually drawn at a much larger scale, not only show the existing terrain conditions but can also contain proposed locations for newly designed construction works. Table 7.1 shows typical scales for maps and plans, and Table 7.2 shows standard drawing sizes for both the foot and the metric systems.

The size of the drafting paper required can be determined by knowing the scale to be used and the area or length of the survey. Standard paper sizes are shown in Table 7.2.

Table 7.1 SUMMARY OF MAP SCALES AND CONTOUR INTERVALS

	Metric Scale		Foot/Inch Scale Equivalents	Contour Interval for Average Terrain*	Typical Uses
Large scale	1:10		1″ = 1′		Detail
	1:50		1/4″ = 1′, 1″ = 5′		Detail
	1:100		1/8″ = 1′, 1″ = 10′, 1″ = 8″		Detail, profiles
	1:200		1″ = 20′		Profiles
	1:500		1″ = 40′, 1″ = 50′	0.5 m, 1 ft	Municipal design plans
	1:1000		1″ = 80′, 1″ = 100′	1 m, 2 ft	Municipal services and site engineering
Intermediate scale	1:2000		1″ = 200′	2 m, 5 ft	Engineering studies and planning (e.g., drainage areas, route planning)
	1:5000		1″ = 400′	5 m, 10 ft	
	1:10 000		1″ = 800′	10 m, 20 ft	
Small scale	1:20 000	1:25 000	2 1/2″ = 1 mi		
	1:25 000				
	1:50 000	1:63 360	1″ = 1 mi		Topographic maps, Canada and the United States
	1:100 000	1:126 720	1/2″ = 1 mi		Geological maps, Canada and the United States
	1:200 000				
	1:250 000	1:250 000	1/4″ = 1 mi		Special-purpose maps and atlases (e.g., climate, minerals)
	1:500 000	1:625 000	1/10″ = 1 mi		
	1:1 000 000		1/16″ = 1 mi		

*The contour interval chosen must reflect the scale of the plan or map, but, additionally, the terrain (flat or steeply inclined) and the intended use of the plan are factors in choosing the appropriate contour interval.

The title block is often a standard size and has a format similar to that shown in Figure 7.9. The title block is usually placed in the lower right corner of the plan, but its location ultimately depends on the filing system used. Many consulting firms and engineering departments attempt to limit the variety of their drawing sizes so that plan filing can be standardized. Some vertical-hold filing cabinets are designed so that title blocks in the upper right corner can be seen more easily. Revisions to the plan are usually referenced immediately above the title block and show the date and a brief description of the revision.

7.4.2 Manual Plotting

Manual plotting begins by first plotting the survey control (e.g., ℄, traverse line, coordinate grid) on the drawing. The control is plotted so that the data plot will be centered suitably on the available paper. Sometimes the data outline is first plotted roughly on tracing paper so

Table 7.2 STANDARD DRAWING SIZES

International Standards Organization (ISO)						ACSM* Recommendations (metric)	
Inch Drawing Sizes			Metric Drawing Sizes (Millimeters)				
Drawing Size	Border Size	Overall Paper Size	Drawing Size	Border Size	Overall Paper Size	Drawing Size	Paper Size
						—	150×200
A	8.00×10.50	8.50×11.00	A4	195×282	210×297	A4	200×300
B	10.50×16.50	11.00×17.00	A3	277×400	297×420	A3	300×400
C	16.00×21.00	17.00×22.00	A2	400×574	420×594	A2	400×600
D	21.00×33.00	22.00×34.00	A1	574×821	594×841	A1	600×800
E	33.00×43.00	34.00×44.00	A0	811×1159	$841 \times 1,189$	A0	$800 \times 1,200$

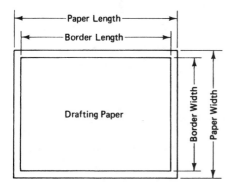

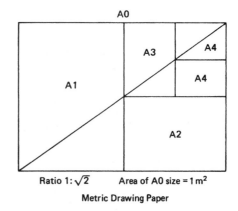

Metric Drawing Paper

*American Congress on Surveying and Mapping Metric Workshop, March 14, 1975. Paper sizes rounded off for simplicity, still have cut-in-half characteristic.

that the plan's dimension requirements can be oriented properly on the available drafting paper. It is customary to orient the data plot so that north is toward the top of the plan; a north arrow is included on all property survey plans and many engineering drawings. The north direction on maps is indicated clearly by lines of longitude or by N-S, E-W grid lines. The plan portion of the plan and profile usually does not have a north indication; instead, local practice dictates the direction of increasing chainage (e.g., chainage increasing left to right for west to east and south to north directions). After the control has been plotted and checked, the surveyor can plot the ground features using either rectangular (X, Y coordinates) or polar (r, θ coordinates) methods.

Rectangular plots (X, Y coordinates) can be laid out with a T-square and set square (right-angle triangle), although the parallel rule has now largely replaced the T-square. When either the parallel rule or the T-square is used, the paper is first set properly on the drawing board and then held in place with masking tape. Once the paper is square and secure, the parallel rule, together with a set square and scale, can be used to lay out and measure rectangular dimensions.

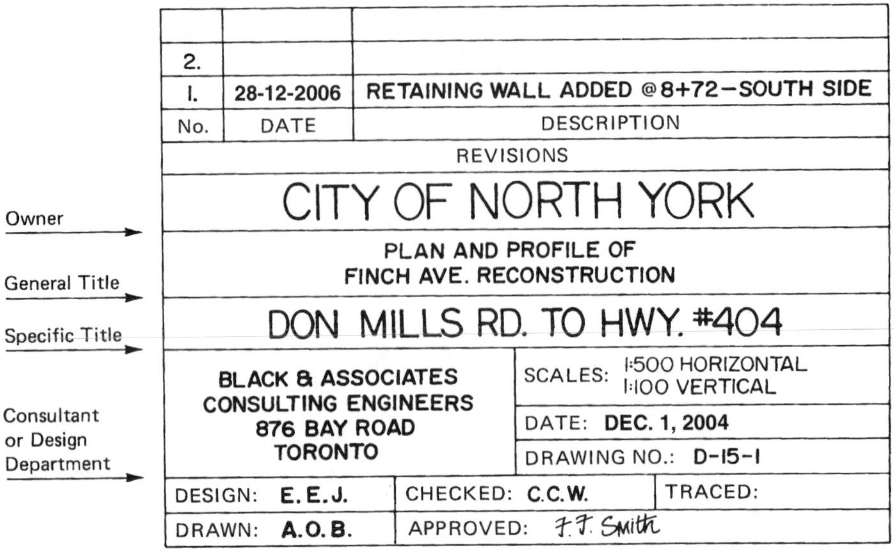

2.		
I.	28-12-2006	RETAINING WALL ADDED @ 8+72—SOUTH SIDE
No.	DATE	DESCRIPTION
		REVISIONS

Owner →

CITY OF NORTH YORK

General Title →

**PLAN AND PROFILE OF
FINCH AVE. RECONSTRUCTION**

Specific Title →

DON MILLS RD. TO HWY. #404

Consultant
or Design
Department →

BLACK & ASSOCIATES CONSULTING ENGINEERS 876 BAY ROAD TORONTO	SCALES: 1:500 HORIZONTAL 1:100 VERTICAL
	DATE: **DEC. 1, 2004**
	DRAWING NO.: **D-15-1**

DESIGN: **E.E.J.**	CHECKED: **C.C.W.**	TRACED:
DRAWN: **A.O.B.**	APPROVED: *F.F. Smith*	

FIGURE 7.9 Typical title block.

Polar plots are drawn with a protractor and scale. The protractor can be a plastic, graduated circle or half-circle with various size diameters (the larger the diameter, the more precise the protractor). A paper, full-circle protractor can also be used under or on the drafting paper. Finally, a flexible-arm drafting machine, complete with right angle–mounted graduated scales, can plot the data using polar techniques. See Figures 7.10 and 7.11 for standard map and plan symbols.

7.4.3   Computerized Surveying Computations and Drawing Preparation

Survey data can be transferred to the computer either as discrete plot points (northing, easting, and elevation) or as already joined graphic entities. If the data have been referenced to a control traverse for horizontal control, various software programs can quickly determine the acceptability of the traverse closure. Feature coding can produce graphics labels, or they can be created right in the CAD programs. Computer-generated models of surface elevations are called digital elevation models (DEMs) or digital terrain models (DTMs). Some agencies use these two terms interchangeably, while others regard a DTM as a DEM that includes the location of break lines. **Break lines** are joined coordinated points that define changes in slope, such as valley lines, ridge lines, the tops and bottoms of slopes, ditch lines, the tops and bottoms of curbs, etc. They are necessary for the generation of realistic surface contours.

Computer programs are also available to help create designs in digital terrain modeling (together with contour production), land division and road layout, highway layout, etc. The designer can quickly assemble a database of coordinated points reflecting both the existing surveyed ground points and the proposed key points created through the various design programs. Some surveying software includes drawing capabilities, whereas others

Primary highway, hard surface	Boundaries: National	
Secondary highway, hard surface	State	
Light-duty road, hard or improved surface	County, parish, municipio	
Unimproved road	Civil township, precinct, town, barrio	
Road under construction, alignment known	Incorporated city, village, town, hamlet	
Proposed road	Reservation, National or State	
Dual highway, dividing strip 25 feet or less	Small park, cemetery, airport, etc.	
Dual highway, dividing strip exceeding 25 feet	Land grant	
Trail	Township or range line, United States land survey	

Township or range line, approximate location

Railroad: single track and multiple track	Section line, United States land survey
Railroads in juxtaposition	Section line, approximate location
Narrow gage: single track and multiple track	Township line, not United States land survey
Railroad in street and carline	Section line, not United States land survey
Bridge: road and railroad	Found corner: section and closing
Drawbridge: road and railroad	Boundary monument: land grant and other
Footbridge	Fence or field line
Tunnel: road and railroad	
Overpass and underpass	Index contour Intermediate contour
Small masonry or concrete dam	Supplementary contour Depression contours
Dam with lock	Fill Cut
Dam with road	Levee Levee with road
Canal with lock	Mine dump Wash
	Tailings Tailings pond
Buildings (dwelling, place of employment, etc.)	Shifting sand or dunes Intricate surface
School, church, and cemetery Cem	Sand area Gravel beach
Buildings (barn, warehouse, etc.)	
Power transmission line with located metal tower	Perennial streams Intermittent streams
Telephone line, pipeline, etc. (labeled as to type)	Elevated aqueduct Aqueduct tunnel
Wells other than water (labeled as to type) oOil oGas	Water well and spring o Glacier
Tanks: oil, water, etc. (labeled only if water) Water	Small rapids Small falls
Located or landmark object; windmill o	Large rapids Large falls
Open pit, mine, or quarry; prospect x x	Intermittent lake Dry lake bed
Shaft and tunnel entrance	Foreshore flat Rock or coral reef
	Sounding, depth curve 10 Piling or dolphin
Horizontal and vertical control station:	Exposed wreck Sunken wreck
Tablet, spirit level elevation BM△5653	Rock, bare or awash; dangerous to navigation
Other recoverable mark, spirit level elevation △5455	
Horizontal control station: tablet, vertical angle elevation VABM△95/9	Marsh (swamp) Submerged marsh
Any recoverable mark, vertical angle or checked elevation △3775	Wooded marsh Mangrove
Vertical control station: tablet, spirit level elevation BM×957	Woods or brushwood Orchard
Other recoverable mark, spirit level elevation ×954	Vineyard Scrub
Spot elevation ×7369 ×7369	Land subject to
Water elevation 670 670	controlled inundation Urban area

FIGURE 7.10 Topographic map symbols. (Courtesy of U.S. Department of Interior, Geological Survey)

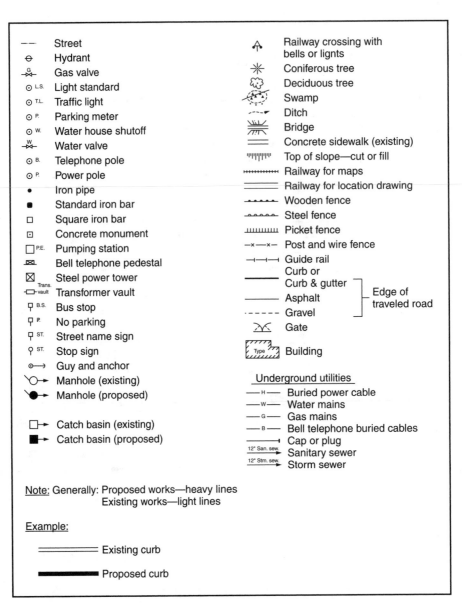

FIGURE 7.11 Municipal works plan symbols.

create DXF or DWG files designed for transfer to CAD programs; Autocad (ACAD) is the most widely used CAD program. Surveying, design, and drawing programs have some or all of the following capabilities:

- Survey data import
- Project definition with respect to map projection, horizontal and vertical datums, and ellipsoid and coordinate system

- Coordinate geometry (COGO) routines for accuracy determination and for the creation of auxiliary points
- Graphics creation
- Feature coding and labeling
- Digital terrain modeling and contouring, including break-line identification and contour smoothing
- Earthwork computations
- Design of land division, road (and other alignment) design, horizontal and vertical curves, site grading, etc.
- Creation of plot files
- File exports in DXF, DWG, and XML; compatibility with GIS through ESRI (e.g., Arcinfo) files and shape files

Engineering project data files include an original ground DTM, a finished project DTM, and design alignment data. The creation of these files enables construction companies to utilize machine guidance and control techniques to construct a facility (see Chapter 10).

7.5   Contours

Contours are lines drawn on a plan; they connect points having the same elevation. Contour lines represent an even value (see Table 7.1), with the selected contour interval kept consistent with the terrain, scale, and intended use of the plan. It is commonly accepted that elevations can be determined to half the contour interval; this convention permits, for example, a 10-ft contour interval on a plan where it is required to know elevations to the closest 5 ft.

Scaling between two adjacent points of known elevation enables the surveyor to plot a contour. Any scale can be used. In Figure 7.12(a), the scaled distance between points 1 and 2 is 0.75 unit, and the difference in elevation is 5.4 ft. The difference in elevation between point 1 and contour line 565 is 2.7 ft; therefore, the distance from point 1 to contour line 565 is:

$$\frac{2.7}{5.4} \times 0.75 = 0.38 \text{ units}$$

To verify this computation, the distance from contour line 565 to point 2 is:

$$\frac{2.7}{5.4} \times 0.75 = 0.38 \text{ unit}$$

$$0.38 + 0.38 \approx 0.75 \text{ Check}$$

The scaled distance between points 3 and 4 is 0.86 units, and their difference in elevation is 5.2 ft. The difference in elevation between point 3 and contour line 565 is 1.7 ft; therefore, the distance from point 3 to contour line 565 is:

$$\frac{1.7}{5.2} \times 0.86 = 0.28 \text{ units}$$

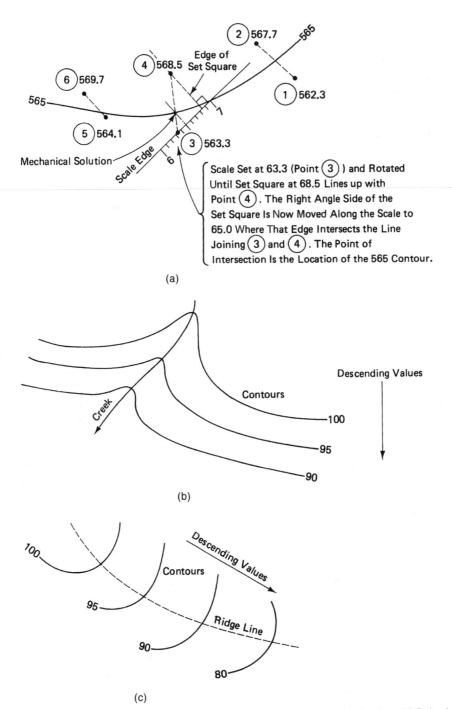

Scale Set at 63.3 (Point ③) and Rotated Until Set Square at 68.5 Lines up with Point ④ . The Right Angle Side of the Set Square Is Now Moved Along the Scale to 65.0 Where That Edge Intersects the Line Joining ③ and ④ . The Point of Intersection Is the Location of the 565 Contour.

(a)

(b)

(c)

FIGURE 7.12 Contours. (a) Plotting contours by interpolation, (b) Valley line. (c) Ridge line.

This computation can be verified by computing the distance from contour line 565 to point 4:

$$\frac{3.5}{5.2} \times 0.86 = 0.58 \text{ units}$$
$$0.58 + 0.28 = 0.86 \quad \text{Check}$$

The scaled distance between points 5 and 6 is 0.49 units, and the difference in elevation is 5.6 ft. The difference in elevation between point 5 and contour line 565 is 0.9 ft; therefore, the distance from point 5 to contour line 565 is:

$$\frac{0.9}{5.6} \times 0.49 = 0.08 \text{ units}$$

and from line 565 to point 6 the distance is:

$$\frac{4.7}{6.6} \times 0.49 = 0.41 \text{ units}$$

In addition to the foregoing arithmetic solution, contours can be interpolated by using mechanical techniques. It is possible to scale off units on a barely taut elastic band and then stretch the elastic so that the marked-off units fit the interval being analyzed. The problem can also be solved by rotating a scale, while a set square is used to line up the appropriate divisions with the field points. In Figure 7.12(a), a scale is set up at 63.3 on point 3 and then rotated until the 68.5 mark lines up with point 4, a set square being used on the scale. The set square is then slid along the scale until it lines up with 65.0; the intersection of the set square edge (90° to the scale) with the straight line joining points 3 and 4 yields the solution (i.e., the location of elevation at 565 ft). This latter technique is faster than the arithmetic technique.

Surveyors plot contours by analyzing adjacent field points, so it is essential that the ground slope be uniform between those points. An experienced survey crew ensures that enough rod readings are taken to define the ground surface suitably. The survey crew can further define the terrain if care is taken in identifying and tying in valley lines, ridge lines, and other break lines. Figure 7.12(b) shows how contour lines bend uphill as they cross a valley; the steeper the valley, the more the line diverges uphill. Figure 7.12(c) shows how contour lines bend downhill as they cross ridge lines. Figure 7.13 shows the plot of control, elevations, and valley and ridge lines. Figure 7.14 shows contours interpolated from the data in Figure 7.13. Figure 7.15 shows a plan with a derived profile line (AB).

If the contours are to be plotted using computer software, as is usually the case, additional information is required to permit the computer/plotter to produce contours that truly represent the surveyed ground surface. In addition to defining ridge and valley lines, as noted earlier, it is necessary for the surveyor to identify the break lines—lines that join points that define significant changes in slope, such as toe of slope, top and bottom of ditches and swales, ₵s, and the like. When the point numbers defining break lines are appropriately tagged—as required by the individual software—truly representative contours will be produced on the graphics screen and by the digital plotter.

Contours are now almost exclusively produced using any of the current software programs. Most programs generate a triangulated irregular network (TIN); the sides of the triangles are analyzed so that contour crossings can be interpolated. The field surveyor must

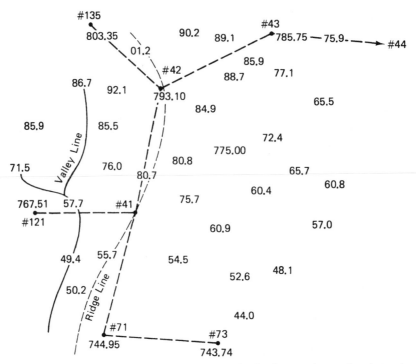

FIGURE 7.13 Plot of survey control, ridge and valley lines, and spot elevations.

also note and mark break lines that define significant changes in topography (e.g., top and bottom of hills, channels, depressions, creeks, etc.) so that the software program can generate contours that accurately reflect actual field conditions.

Figure 7.16 shows the steps in a typical contour production process. Two basic methods are used: the uniform grid approach and the more popular TIN approach (used here). First the triangulated irregular network (TIN) is created from the plotted points and defined break lines [Figure 7.16(a)], and then the raw contours [Figure 7.16(b)] are computer-generated from that data. To make the contours pleasing to the eye, as well as representative of the ground surface, some form of ground-true smoothing techniques must be used to soften the sharp angles that occur when contours are generated from TINs (as opposed to the less angular lines resulting in contours derived from a uniform grid approach) [Figure 7.16(c)]. For more on this topic, refer to A. H. J. Christensen, "Contour Smoothing by an Eclectic Procedure," *Photogrammetric Engineering and Remote Sensing (RE & RS)*, April 2001: 516.

7.5.1   Summary of Contour Characteristics

The following list provides a summary of contours:

1. Closely spaced contours indicate steep slopes.
2. Widely spaced contours indicate moderate slopes (spacing here is a relative relationship).

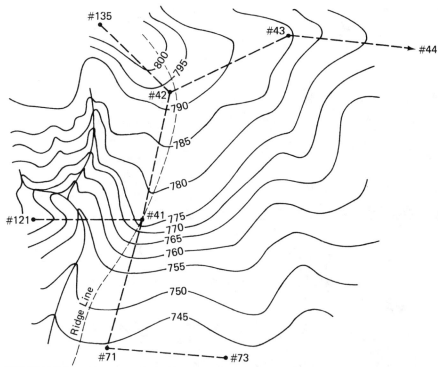

FIGURE 7.14 Contours plotted by interpolating between spot elevations, with additional plotting information given when the locations of ridge and valley lines are known.

CONTOUR LINES

These are drawn through points having the same elevation. They show the height of ground above sea level (M.S.L.) in either feet or metres and can be drawn at any desired interval.

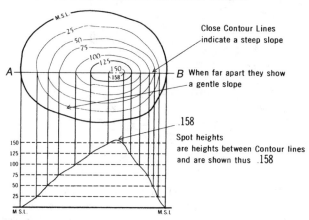

Close Contour Lines indicate a steep slope

When far apart they show a gentle slope

.158

Spot heights are heights between Contour lines and are shown thus .158

FIGURE 7.15 Contour plan with derived profile (line AB). (Courtesy of Department of Energy, Mines, and Resources, Canada)

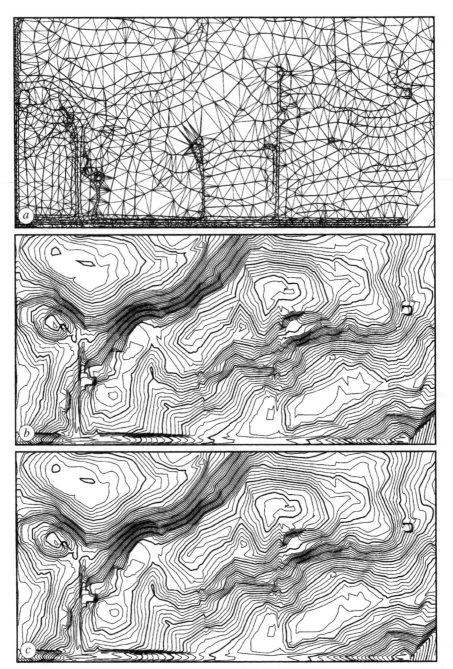

FIGURE 7.16 Contouring. (a) Break lines and TIN. (b) Raw contours. (c) Contour smoothing with the eclectic procedure. (Reproduced with permission, the American Society for Photogrammetry and Remote Sensing. A. H. J. Christensen, "Contour Smoothing by an Eclectic Procedure." *Photogrammetric Engineering and Remote Sensing [RE&RS].* April 2001:516.)

3. Contours must be labeled to give the elevation value. Either each line is labeled, or every fifth line is drawn darker (wider) and labeled.

4. Contours are not shown going through buildings.

5. Contours crossing a built horizontal surface (roads, railroads) are straight parallel lines as they cross the facility.

6. Because contours join points of equal elevation, contour lines cannot cross. (Caves present an exception.)

7. Contour lines cannot begin or end on the plan.

8. Depressions and hills look the same; one must note the contour value to distinguish the terrain (some agencies use hachures or shading to identify depressions).

9. Contours deflect uphill at valley lines and downhill at ridge lines. Contour line crossings are perpendicular: U-shaped for ridge crossings, V-shaped for valley crossings.

10. Contour lines must close on themselves, either on the plan or in locations off the plan.

11. The ground slope between contour lines is uniform. If the ground slope is not uniform between the points, additional readings (by total station or level) are taken at the time of the survey.

12. Important points can be further defined by including a spot elevation (height elevation).

13. Contour lines tend to parallel each other on uniform slopes.

By now, you have probably determined that the manual plotting of contours involves a great deal of time-consuming scaling operations. Fortunately, all the advantages of computer-based digital plotting are also available for the production of contour plans. In addition to contours, some software programs provide a three-dimensional perspective plot, which can portray the topography as viewed from any azimuth position and from various altitudes. See Figure 7.17 for a perspective view of a proposed road.

FIGURE 7.17 QuickSurf TGRID showing smoothed, rolling terrain coupled with road definition. (Courtesy of MicroSurvey Software, Inc.)

7.6 Aerial Photography

7.6.1 General Background

Airborne imagery introduces the reader to the topics of aerial photography (this section) and the more recent (2000) topic of airborne digital imagery (next section). Aerial photography has a history dating back to the mid-1800s, when balloons and even kites were used as camera platforms from which photos could be taken. About fifty years later, in 1908, photographs were taken from early aircraft. During World Wars I and II, the use of aerial photography mushroomed in the support of military reconnaissance. From the 1930s to the present, aerial photography became an accepted technique for collecting mapping and other ground data in North America. Most airborne imagery is still collected using aerial photographs, although it has been predicted that much of data capture and analysis will be accomplished using digital imagery techniques (see Section 7.7).

Under the proper conditions, cost savings for survey projects using aerial surveys rather than ground surveys can be enormous. Consequently, it is critical that the surveyor can identify the situations in which the use of aerial imagery is most appropriate. The first part of this section discusses the basic principles required to use aerial photographs intelligently, the terminology involved, the limitations of its use, and specific applications to various projects.

Photogrammetry is the science of making measurements from aerial photographs. Measurements of horizontal distances and elevations form the backbone of this science. These capabilities result in the compilation of **planimetric maps** or **orthophoto maps** showing the scaled horizontal locations of both natural and cultural features and in **topographic maps** showing spot elevations and contour lines.

Both black and white panchromatic and color film are used in aerial photography. Color film has three emulsions: blue, green, and red light sensitive. In color infrared (IR), the three emulsion layers are sensitive to green, red, and the photographic portion of near-IR—which are processed, in false color, to appear as blue, green, and red, respectively.

7.6.2 Aerial Camera Systems

The introduction of digital cameras has revolutionized photography. Digital and film-based cameras both use optical lenses. Whereas film-based cameras use photographic film to record an image, digital cameras record image data with electronic sensors: charge-coupled devices (CCDs) or complementary metal-oxide-semiconductor (CMOS) devices. One chief advantage to digital cameras is that the image data can be stored, transmitted, and analyzed electronically. Cameras on-board satellites can capture photographic data and then have this data, along with other sensed data, transmitted back to earth for additional electronic processing.

Although airborne digital imagery is already making a significant impact in the remote sensing field, we will begin by discussing film-based photography, a technology that still accounts for much of aerial imaging. The 9″ by 9″ format used for most film-based photographic cameras captures a wealth of topographic detail, and the photos, or the film itself, can be scanned efficiently (see Figure 7.18), thus preparing the image data for electronic processing.

The camera used for most aerial photography is shown in Figure 7.19; it uses a fixed focal length, large-format negative, usually 9 in. (230 mm) by 9 in. (230 mm). The drive

FIGURE 7.18 DSW300 scanner with Sun Ultra 60® host computer. This high-precision photogrammetric photo and film scanner has four elements: (1) movable *xy* cross carriage stage with flat film platen; (2) fixed array charge-coupled device (CCD) camera and image optics; (3) Xenon light source and color scanning software; and (4) computer hardware, including storage. (Courtesy of Leica Geosystems Inc.)

mechanism is housed in the camera body, as shown in Figure 7.19. It is motor-driven, and the time between exposures to achieve the required overlap is set based on the photographic scale and the ground speed of the aircraft. The film is thus advanced from the feed spool to the take-up spool at automatic intervals. The focal plane is equipped with a vacuum device to hold the film flat at the instant of exposure. The camera also has four fiducial marks built in so that each exposure can be oriented properly to the camera calibration. Most cameras also record the frame number, time of exposure, and height of the aircraft on each exposure. Some include the image of a level bubble.

The camera mount permits flexible movement of the camera for leveling purposes. Using the level bubble mounted on the top of the camera body as the indicator, the operator should make every attempt to have the camera as level as possible at the instant of exposure. This requires constant attention by the operator because the aircraft is subject to pitching and rolling, resulting in a tilt when the photographs are taken. The viewfinder is mounted vertically to show the area being photographed at any time.

Two points, N_F and N_R, are shown on the optical axis in Figure 7.19. These are the front and rear nodal points of the lens system, respectively. When light rays strike the front

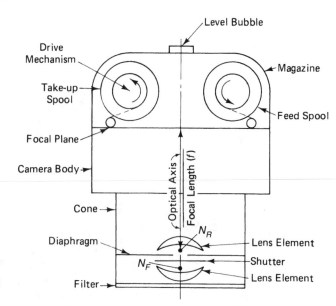

FIGURE 7.19 Components of aerial survey camera (large format).

nodal point (N_F), they are refracted by the lens so that they emerge from the rear nodal point (N_R) parallel with their original direction. The focal length (f) of the lens is the distance between the rear nodal point and the focal plane along the optical axis, as shown in Figure 7.19. The value of the focal length is accurately determined through calibration for each camera. The most common focal length for aerial cameras is 6 in. (152 mm).

Because atmospheric haze contains an excessive amount of blue light, a filter is used in front of the lens to absorb some of the blue light, thus reducing the haze on the actual photograph. A yellow, orange, or red filter is used, depending on atmospheric conditions and the flying height of the aircraft above mean ground level.

The shutter of a modern aerial camera is capable of speeds ranging from 1/50 s to 1/2,000 s. The range is commonly between 1/100 s and 1/1,000 s. A fast shutter speed minimizes blurring of the photograph, known as image motion, which is caused by the ground speed of the aircraft.

7.6.3 Photographic Scale

The scale of a photograph is the ratio between a distance measured on the photograph and the ground distance between the same two points. The features, both natural and cultural, shown on a photograph are similar to those on a planimetric map. There is one important difference. The planimetric map has been rectified through ground control so that the horizontal scale is consistent among any points on the map. The air photo contains scale variations unless the camera was perfectly level at the instant of exposure and the terrain being photographed was also level. Because the aircraft is subject to tip, tilt, and changes in altitude due to updrafts and downdrafts, the chances of the focal plane being level at the instant of exposure are minimal. In addition, the terrain is seldom flat. As illustrated in Figure 7.20,

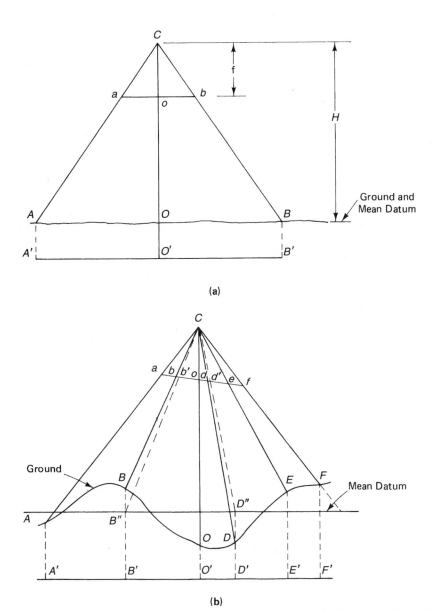

FIGURE 7.20 Scale differences caused by tilt and topography. (a) Level focal plane and level ground. (b) Tilted focal plane and hilly topography.

any change in elevation causes scale variations. The basic problem is transferring an uneven surface like the ground to the flat focal plane of the camera.

In Figure 7.20(a), points A, O, and B are at the same elevation and the focal plane is level. Therefore, all scales are true on the photograph because the distance $AO = A'O'$ and $OB = O'B'$. A, O, and B are points on a level reference datum that is comparable to the surface of a planimetric map. Therefore, under these unusual circumstances, the scale of the photograph is uniform because the ratio *ao:AO* is the same as the ratio *ob:OB*.

Figure 7.20(b) illustrates a more realistic situation. The focal plane is tilted and the topographic relief is variable. Points *A, B, O, D, E,* and *F* are at different elevations. You can see that, although *A'B'* equals *B'O'*, the ratio *ab:bo* is far from equal. Therefore, the photographic scale for points between *a* and *b* is significantly different than that for points between *b* and *o*. The same variations in scale can also be seen for the points between *d* and *e* and between *e* and *f.*

The overall average scale of the photograph is based partially on the elevations of the mean datum shown in Figure 7.20(b). The mean datum elevation is intended to be the average ground elevation. Examining the most accurate contour maps of the area available permits the selection of the apparent average elevation. Distances between points on the photograph that are situated at the elevations of the mean datum will be at the intended scale. Distances between points having elevations above or below the mean datum will be at a different photograph scale, depending on the magnitude of the local relief.

The scale of a vertical photograph can be calculated from the focal length of the camera and the flying height above the mean datum. Note that the flying height and the altitude are different elevations having the following relationship:

$$\text{altitude} = \text{flying height} + \text{mean datum}$$

By similar triangles, as shown in Figure 7.20(a), we have the following relationship:

$$\frac{ao}{AO} = \frac{Co}{CO} = \frac{f}{H}$$

where *AO/ao* = scale ratio between the ground and the photograph
f = focal length
H = flying height above mean datum

Therefore, the scale ratio is:

$$SR = \frac{H}{f} \tag{7.1}$$

For example, if H = 1,500 m and f = 150 mm, then:

$$SR = \frac{1,500}{0.150} = 10,000$$

Therefore, the average scale of the photograph is 1:10,000. In the foot system, the scale would be stated as 1 in. = 10,000 /12, or 1 in. = 833 ft. The conversion factor of 12 is required to convert both sides of the equation to the same unit.

7.6.4 Flying Heights and Altitude

When planning an air photo acquisition mission, the flying height and altitude must be determined, particularly if the surveyor is acquiring supplementary aerial photography using a small-format camera. The flying height is determined using the same relationships discussed in Section 7.6.3 and illustrated in Figure 7.20. Using the relationship in Equation 7.1, we have $H = SR \times f$.

■ **EXAMPLE 7.1**

If the desired scale ratio (SR) is 1:10,000 and the focal length of the lens (f) is 150 mm, find H.

Solution
Use Equation 7.1:

$$H = 10,000 \times 0.150 = 1,500 \text{ m}$$

■ **EXAMPLE 7.2**

If the desired scale ratio (SR) is 1:5,000 and the focal length of the lens is 50 mm, find H.

Solution
Use Equation 7.1:

$$H = 5,000 \times 0.050 = 250 \text{ m}$$

The flying heights calculated in Examples 7.1 and 7.2 are the vertical distances that the aircraft must fly above the mean datum, illustrated in Figure 7.20. Therefore, the altitude at which the plane must fly is calculated by adding the elevation of the mean datum to the flying height. If the elevation of the mean datum were 330 ft (100 m), the altitudes for Examples 7.1 and 7.2 would be 1,600 m (5,250 ft) and 350 m (1,150 ft), respectively. These values are the readings for the aircraft altimeter throughout the flight to achieve the desired average photographic scale.

If the scale of existing photographs is unknown, it can be determined by comparing a distance measured on the photograph with the corresponding distance measured on the ground or on a map of known scale. The points used for this comparison must be easily identifiable on both the photograph and the map. Examples include road intersections, building corners, and river or stream intersections. The photographic scale is found using the following relationship:

$$\frac{\text{photo scale}}{\text{map scale}} = \frac{\text{photo distance}}{\text{map distance}}$$

Because this relationship is based on ratios, the scales on the left side must be expressed in the same units. The same applies to the measured distances on the right side of the equation. For example, if the distance between two identifiable points on the photograph is 5.75 in. (14.38 cm), on the map, it is 1.42 in. (3.55 cm), and the map scale is 1:50,000, then the photo scale is:

$$\frac{\text{photo scale}}{1:50,000} = \frac{5.75 \text{ in.}}{1.42 \text{ in.}}$$

$$\text{photo scale} = \frac{5.75}{50,000 \times 1.42}$$

$$= 1:12,348$$

If the scale is required in inches and feet, it is calculated by dividing 12,348 by 12 (number of inches per foot), which yields 1,029. Therefore, the photo scale is 1 in. = 1,029 ft between these points only. The scale will be different in areas that have different ground elevations.

7.6.5 Relief (Radial) Displacement

Relief displacement occurs when the point being photographed is not at the elevation of the mean datum. As previously explained and as illustrated in Figure 7.20(a), when all ground points are at the same elevation, no relief displacement occurs. However, the displacement of point b on the focal plane (photograph) in Figure 7.20(b) is illustrated. Because point B is above the mean datum, it appears at point b on the photograph rather than at point b', which would be the location on the photograph for point B', and point B' on the mean datum.

Fiducial marks are placed precisely on the camera back plate so that they reproduce in exactly the same position on each air photo negative. These marks are located either in the corners, as illustrated in Figure 7.21, or in the middle of each side, as illustrated in Figure 7.22. Their primary function is the location of the principal point, which is located at the intersection of straight lines drawn between each set of opposite fiducial marks (as illustrated in Figure 7.22).

Relief displacement depends on the position of the point on the photograph and the elevation of the ground point above or below the mean datum. Note the following in Figure 7.20(b):

- The displacement at the center (or principal point), represented by O on the photograph, is zero.
- The farther that the ground point is located from the principal point, the greater the relief displacement. The displacement dd' is less than bb', even though the ground point D is farther below the mean datum than point B is above it.

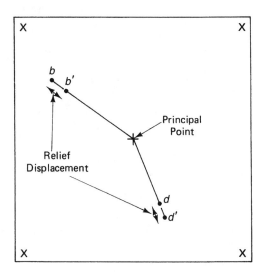

FIGURE 7.21 Direction of relief displacement [compare with Figure 7.20(b)]; X denotes fiducial marks.

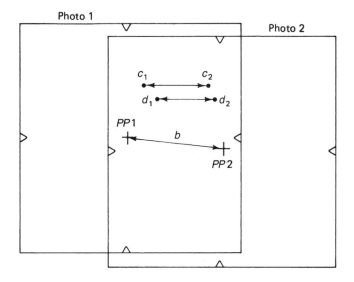

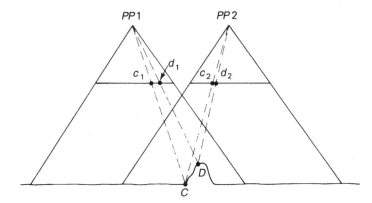

FIGURE 7.22 X parallax along the flight line.

- The greater the elevation of the ground point above or below the mean datum, the greater is the displacement. If the ground elevation of point B were increased so that it was farther above the mean datum, the displacement bb' would increase correspondingly.

Relief displacement is radial to the principal point of the photograph, as illustrated in Figure 7.21. The direction of the relief distortion on the photograph is shown for photo points b and d in Figure 7.20(b).

7.6.6 Flight Lines and Photograph Overlap

It is important for you to understand the techniques by which aerial photographs are taken. Once the photograph scale, flying height, and altitude have been calculated, the details of implementing the mission are carefully planned. Although the planning process is beyond the scope of this text, the most significant factors include the following:

- A suitable aircraft and technical personnel must be arranged, including their availability if the time period for acquiring the photography is critical to the project. The costs of having the aircraft and personnel available, known as mobilization, are extremely high.
- The study area must be outlined carefully and the means of navigating the aircraft along each flight line, using either ground features or magnetic bearings, must be determined. More recently, GPS is used in aircraft to maintain proper flight line alignment.
- The photographs must be taken under cloudless skies. The presence of high clouds above the aircraft altitude is unacceptable because of the shadows cast on the ground by the clouds. Therefore, the aircraft personnel are often required to wait for suitable weather conditions. This downtime can be very expensive. Most aerial photographs are taken between 10 A.M. and 2 P.M. to minimize the effect of long shadows, which obscure terrain features. Consequently, the weather has to be suitable at the right time.

To achieve photogrammetric mapping and to examine the terrain for air photo interpretation purposes, it is essential that each point on the ground appear in two adjacent photographs along a flight line so that all points can be viewed stereoscopically. Figure 7.23 illustrates the relative locations of flight lines and photograph overlaps, both along the flight line and between adjacent flight lines. An area over which it has been decided to acquire air photo coverage is called a block. The block is outlined on the most accurate available topographic map. The locations of the flight lines required to cover the area properly are then plotted. The flight lines *A* and *B* in Figure 7.23(a) are two examples. The aircraft proceeds along flight line *A*, and air photos are taken at time intervals calculated to provide 60 percent forward overlap between adjacent photographs. As illustrated in Figure 7.23(a), the format of each photograph is square. Therefore, the hatched area represents the forward overlap between air photos 2 and 3, flight line *A*. The minimum overlap to ensure that all ground points show on two adjacent photographs is 50 percent. However, at least 60 percent forward overlap is standard because the aircraft is subject to altitude variations, tip, and tilt as the flight proceeds. The extra 10 percent forward overlap allows for these variables. The air photo coverage of flight line *B* overlaps that of *A* by 25 percent, as illustrated in Figure 7.23(a). This overlap not only ensures that no gaps of unphotographed ground exist, but it is also necessary to extend control between flight lines for photogrammetric methods. (This technique is often called sidelap.)

The flight line in profile view is illustrated in Figure 7.23(b). The single-hatched areas represent the forward overlap between air photos 1 and 2 and photos 2 and 3. Because of the forward overlap of 60 percent, the ground points in the double-hatched area will appear on each of the three photographs, thus permitting full photogrammetric treatment.

The number of air photos required to cover a block or study area is a very important consideration. Keep in mind that each air photo has to be cataloged and stored, or scanned into a digital file. Most important, these photographs have to be used individually and collectively and/or examined for photogrammetric mapping and/or air photo interpretation purposes. All other factors being equal, such as focal length and format size, the photographic scale is the controlling factor regarding the number of air photos required.

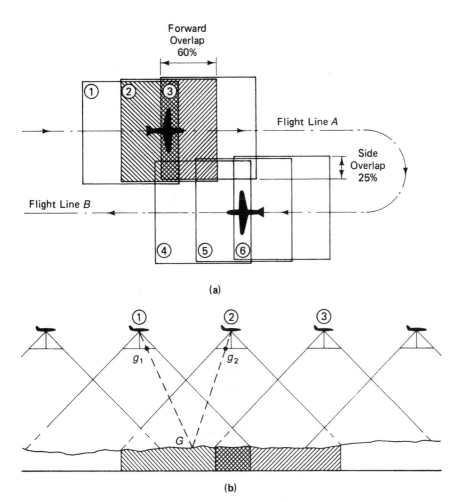

FIGURE 7.23 Flight lines and photographic overlap. (a) Photographic overlap. (b) Overlap along flight line.

The approximate number of air photos required to cover a given area stereoscopically (every ground point shows on at least two adjacent photos along the flight line) can be easily calculated. The basic relationships required for this computation are set out next for a forward overlap of 60 percent and a side overlap of 25 percent. For a photographic scale of 1:10,000, the area covered by one photograph, accounting for loss of effective area through overlaps, is 0.4 square miles, or 1 km². Therefore, the number of air photos required to cover a 200-square-mile (500 km²) area is 200/0.4 = 500, or 500/1 = 500.

It is important to realize that the approximate number of photographs varies as the square of the photographic scale. For example, if the scale were 1:5,000 versus 1:10,000 in Figure 7.22, the aircraft would be flying at one-half the altitude. Consequently, the ground area covered by each air photo would be reduced by half in *both* directions. Thus, twice the number of air photos would be required along each flight line, and the number of flight lines required to cover the same area would be doubled. The following list items illustrate the effect on the total number of air photos required, based on the coverage required for a 200-square-mile (500 km^2) area:

1. For a scale of 1:5,000, the number of photographs is $500 \times (10,000/5,000)^2 = 2,000$.
2. For a scale of 1:2,000, the number of photographs is $500 \times (10,000/2,000)^2 = 12,500$.
3. For a scale of 1:20,000, the number of photographs is $500 \times (10,000/20,000)^2 = 125$.

Thus, you can see that the proper selection of scale for the mapping or air photo interpretation purposes intended is critical. The scale requirements for photogrammetric mapping depend on the accuracies of the analytical equipment to be used in producing the planimetric maps. For general air photo interpretation purposes, including survey boundary line evidence (cut lines, land-use changes), photographic scales of between 1:10,000 and 1:20,000 are optimal.

7.6.7 Ground Control for Mapping

As stated previously, the aerial photograph is not perfectly level at the instant of exposure and the ground surface is seldom flat. As a result, ground control points are required to manipulate the air photos physically or mathematically before mapping can be done. Recent advances in GPS and inertial measurement systems (IMS) permit the collection of aerial and airborne imaging to be georeferenced without the need for time-consuming extensive horizontal control surveys.

Ground control is required for each data point positioning. The accuracy with which the measurements are made varies depending on the following final product requirements:

- Measurements of distances and elevations, such as building dimensions, highway or road locations, and cross-sectional information for quantity payments for cut and fill in construction projects
- Preparation of planimetric and topographic maps, usually including contour lines at a fixed interval
- Construction of controlled mosaics
- Construction of orthophotos and rectified photographs

7.6.8 Photogrammetric Stereoscopic Plotting Techniques

Stereoplotters have traditionally been used for image rectification, that is, to extract planimetric and elevation data from stereo-paired aerial photographs for the preparation of topographic maps. The photogrammetric process includes the following steps:

1. Establish ground control for aerial photos.
2. Obtain aerial photographs.
3. Orient adjacent photos so that ground control matches.
4. Use aerotriangulation to reduce the number of ground points needed.
5. Generate DEM/DTM.
6. Produce the orthophoto.
7. Collect data using photogrammetric techniques.

Steps 3 to 7 are accomplished using stereoplotting equipment and techniques. Essentially, stereoplotters incorporate two adjustable projectors that are used to duplicate the attitude of the camera at the time the film was exposed. Camera tilt and differences in flying height and flying speed can be noted and rectified. A floating mark can be made to rest on the ground surface when viewing is stereoscopic, thus enabling the skilled operator to trace planimetric detail and to deduce both elevations and contours.

In the past fifty years, aerial photo stereoplotting has undergone four distinct evolutions. The original stereoplotter (e.g., Kelsh Plotter) was a heavy and delicate mechanical device. Then came the analog stereoplotter and after it the analytical stereoplotter—an efficient technique that utilizes computer-driven mathematical models of the terrain. The latest technique (developed in the 1990s) is **soft-copy (digital) stereoplotting**. Each new generation of stereoplotting reflects revolutionary improvements in the mechanical, optical, and computer components of the system. Common features of the first three techniques were the size, complexity, high capital costs, and high operating costs of the equipment— and the degree of skill required by the operator. Soft-copy photogrammetry utilizes (1) high-resolution scanners to digitize the aerial photos and (2) sophisticated algorithms to process the digital images on workstations (e.g., Sun, Unix, etc.) and on personal computers with the Windows NT operating system.

7.7 Airborne and Satellite Imagery

Remote sensing is a term used to describe geodata collection and interpretive analysis for both airborne and satellite imagery. Satellite platforms collect geodata using various digital sensors, for example, multispectral scanners, lidar, and radar. Multispectral scanning techniques—in addition to the traditional aerial photography techniques—are now also available on airborne platforms.

Both airborne digital imagery and satellite imagery have some advantages over airborne film-based photography:

- Large data-capture areas can be processed more quickly using computer-based analyses.
- Satellite imagery can be recaptured every few days or weeks, permitting the tracking of rapidly changing conditions (for example, disaster response for forest fires, flooding, etc.).
- Feature identification can be much more effective when combining multispectral scanners with panchromatic imagery. For example, with the analysis of spectral reflectance variations, even tree foliage differentiation is possible. In near-IR, for instance, coniferous trees are distinctly darker in tone than are deciduous trees.
- Ongoing measurements track slowly changing conditions (for example, crop diseases, etc.).
- With the relatively low cost of imaging, the consumer need purchase only that level of image processing needed for a specific project.
- Data in digital format is ready for computer processing.
- With the use of radar, which has the ability to penetrate cloud cover, remotely sensed data can be collected under an expanded variety of weather conditions, night and day.

- With the use of lidar measurements, DEMs can be created from ground surface measurements relatively inexpensively day or night from aircraft or from satellites.
- Imaging results are a good source for the data needed to build GIS layers.

As with all measurement techniques, however, multispectral scanned imagery is susceptible to errors and other problems requiring analysis, as the following list demonstrates:

- Although individual ground features typically reflect light from unique portions of the spectrum, reflectance can vary for the same type of features.
- Sensors are sensitive only to specified wavelengths, and there is a limit to the number of sensors that can be deployed on a satellite or on an aircraft.
- The sun's energy, the source of reflected and emitted signals, can vary over time and location.
- Atmospheric scattering and absorption of the sun's radiation can vary unpredictably over time and location.
- Depending on the resolution of the images, some detail may be missed or mistakenly identified.
- Although vast amounts of data can be collected very quickly, processing the data takes considerable time. For example, the space shuttle *Endeavor's* ten-day radar topography of the earth mission in 2000 collected about 1 trillion images; it took more than two years to process the data into map form.
- Scale, or resolution, may be a limiting factor on some projects.
- The effect of ground moisture on longer wavelengths, microwaves in particular, makes some analyses complex.

Because aerial imaging is usually at a much larger scale than is satellite imaging, aerial images are presently more suitable for projects requiring maximum detail and precision, i.e., engineering works. However, the functional planning of corridor works such as highways, railways, canals, electrical transmission lines, etc., have been greatly assisted with the advances in satellite imagery.

Remotely sensed data do not necessarily have their absolute geographic positions recorded at the time of data acquisition (lidar positioning can be an exception). Each point (pixel or picture element) within an image is located with respect to other pixels within a single image, but the image must be geocorrected before it can be inserted accurately into a GIS or a design document. In addition to the geocorrections needed to establish spatial location (GPS and inertial measurement techniques are now used for much of this work), the identification of remotely sensed image features must be verified using ground-truth techniques, that is, analyzing aerial photos, on-site reconnaissance with visual confirmations, map analysis (e.g., soils maps), etc. Ground-truth techniques are not only required to verify or classify surface features but also to calibrate sensors. A major advantage of remote sensing information is the ability to provide a basis for repeated and inexpensive updates on GIS information. "A chief use of remote sensing is the classification of the myriad of features in a scene—usually presented as an image—into meaningful categories or classes. The image then becomes a thematic map (the theme is selectable, e.g., land use, geology, vegetation types, rainfall)."[*]

7.7.1 Techniques of Remote Sensing

Two main categories of remote sensing exist: active and passive. Active remote sensing instruments (e.g., radar and lidar) transmit their own electromagnetic waves and then develop images of the earth's surface as the electromagnetic pulses (known as backscatter) are reflected back from the target surface. Passive remote sensing instruments develop images of the ground surface as they detect the natural energy that is either reflected (if the sun is the signal source) or emitted from the observed target area.

7.7.1.1 Lidar Mapping.

Light detection and ranging (**lidar**) is a laser mapping technique that has recently become popular in both topographic and hydrographic surveying. Over land, laser pulses can be transmitted and then returned from ground surfaces. The time required to send and then receive the laser pulses is used to create a digital elevation model (DEM) of the earth's surface. Processing software can separate rooftops from ground surfaces, and treetops and other vegetation from the bare ground surface beneath the trees. Although the laser pulses cannot penetrate very heavily foliaged trees, they can penetrate tree cover and other lower-growth vegetation at a much more efficient rate than does either aerial photography and digital imaging because of the huge number of measurements—thousands of terrain measurements every second. Bare-earth DEMs are particularly useful for design and estimating purposes.

One of the important advantages to using this technique is the rapid processing time. One supplier claims that 1,000 km^2 of hilly, forested terrain can be surveyed by laser in less than twelve hours and that the DEM data are available within twenty-four hours of the flight. That is, data processing doesn't take much longer than does data collection. Because each laser pulse is georeferenced individually, there is no need for the orthorectification steps, which are necessary in aerial photo processing. Additional advantages include the following:

- Laser mapping can be done during the day or at night when there are fewer clouds and calmer air.
- Vertical accuracies of 15 cm (or better) can be achieved.
- No shadow or parallax problems occur, as they do with aerial photos.
- Laser data is digital and is thus ready for digital processing.
- It is less expensive than other techniques of aerial imaging for the creation of DEMs.

Lidar can be mounted in a helicopter or fixed-wing aircraft, and the lower the altitude the better the resulting ground resolution. Typically, lidar can be combined with a digital imaging (panchromatic and multispectral) sensor; an inertial measurement unit (IMU) to provide data to correct pitch, yaw, and roll; and GPS receivers to provide precise positioning. When the data-gathering package also includes a digital camera to collect

---

[*]From NASA's home page: *http://www.nasa.gov/.* Accessed June 2002.

panchromatic imagery and appropriate processing software, it is possible to produce high-quality orthorectified aerial imagery so that each pixel can be assigned x, y, and z values.

Ground lidar operates in the near-IR portion of the spectrum, which produces wavelengths that tend to be absorbed by water; asphalt surfaces, such as roofing and highways; and rain and fog. These surfaces produce "holes" in the coverage, which can be recognized as such and then edited during data processing. Even rolling traffic on a highway at the time of data capture can be removed through editing.

The growing number of applications for lidar include:

- Highway design and redesign.
- Flood plain mapping.
- Forest inventory, including canopy coverage, tree density and heights, and timber output.
- Line-of-sight modeling using lidar-generated three-dimensional building renderings, which are used in telecommunications and airport facilities design.
- Shallow-water hydrographic soundings.

In the past few years, airborne laser bathymetry (ALB) has become operational in the field of shallow-water hydrographic surveying. Most systems employ two spectral bands, one to detect the water surface (1064 nm infrared band) and the other to detect the bottom (532 nm blue/green band). The depth of the water is determined from the time difference of laser returns reflected from the surface and from the waterbed.

7.7.2 Electromagnetic Spectrum

The fundamental unit of electromagnetic radiation is the photon. The photon, which has energy but no mass, moves at the speed of light. The wave theory of light holds that light travels in wavelike patterns, with its energy characterized by its wavelength or frequency. The speed of light is generally considered to be 300,000 km/s, or 186,000 miles/s. From Chapter 2, we have the following equation (Equation 2.4) for the speed of light:

$$c = f \lambda$$

where c = the velocity of light in meters/second

f = the frequency of the light energy in hertz[*]

λ = the wavelength, in meters, as measured between successive wave crests (see Figures 2.25 and 2.26)

Figure 7.24 shows the electromagnetic spectrum, with wavelengths ranging in size from the very small (cosmic, gamma, and x-rays) to the very large (radio and television) waves. In most remote sensing applications, radiation is described by its wavelength, although applications using microwave (radar) sensing have traditionally used frequency instead to describe these much longer wavelength signals.

---

[*]Hertz (Hz) is a frequency of one cycle per second; kilohertz (KHz) is 10^3 Hz; megahertz (MHz) is 10^6 Hz; and gigahertz (GHz) is 10^9 Hz. Frequency is the number of cycles of a wave passing a fixed point in a given time period.

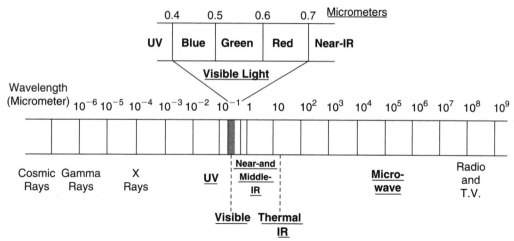

FIGURE 7.24 Electromagnetic spectrum.

Table 7.3 VISIBLE SPECTRUM

Color	Bandwidth
Violet	0.4–0.446 μm
Blue	0.446–0.500 μm
Green	0.500–0.578 μm
Yellow	0.578–0.592 μm
Orange	0.593–0.620 μm
Red	0.620–0.7 μm

Visible light is that very small part of the electromagnetic spectrum ranging from about 0.4 μm to 0.7 μm (μm is the symbol for micrometer or micron, a unit that is one millionth of a meter; see Table 7.3). Infrared radiation (IR) ranges from about 0.7 μm to 100 μm, and that range can be further subdivided into reflected IR (ranging from about 0.7 μm to about 3.0 μm) and the thermal region (ranging from about 3.0 μm to about 100 μm). The other commonly used region of the spectrum is that of microwaves (e.g., radar), which range from about 1 mm to 1 m.

Radiation, having different wavelengths, travels through the atmosphere with varying degrees of success. Radiation is scattered unpredictably by particles in the atmosphere, and atmospheric constituents such as water vapor, carbon dioxide, and ozone absorb radiation. Radiation scattering and absorption rates vary with the radiation wavelengths and the atmospheric particle size. Only radiation in the visible, infrared (IR), and microwave sections of the spectrum travel through the atmosphere relatively free of blockage. These atmospheric transmission anomalies, known as atmospheric windows, are utilized for most common remote sensing activities; see Figure 7.25.

Radiation that is not absorbed or scattered in the atmosphere (incident energy) reaches the surface of the earth, where three types of interaction can occur (see Figure 7.26).

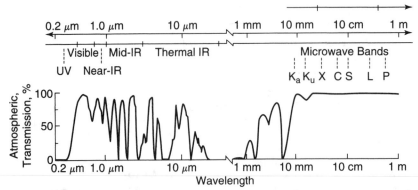

FIGURE 7.25 Electromagnetic spectrum showing atmospheric transmission windows in the visible, near, middle, and thermal infrared and microwave regions. Notice the excellent atmospheric transmission capabilities in the entire range of the microwave. (Courtesy of NASA)

- Transmitted energy passes through an object surface with a change in velocity.
- Absorbed energy is transferred through surface features by way of electron or molecular reactions.
- Reflected energy is reflected back to a sensor, with the incident angle equal to the angle of reflection (those wavelengths that are reflected determine the color of the surface).

Incident energy = transmitted energy + absorbed energy + reflected energy

For example, chlorophyll—a constituent of plants' leaves—absorbs radiation in the red and blue wavelengths but reflects the green wavelengths. Thus, at the height of the growing season, when chlorophyll is strongly present, leaves appear greener. Water absorbs the longer wavelengths in the visible and near-IR portions of the spectrum and reflects the shorter wavelengths in the blue/green part of the spectrum, which results in the often seen blue/green hues of water.

In addition, energy emitted from an object can be detected in the thermal infrared section of the spectrum and recorded. Emitted energy can be the result of the previous absorption of energy—a process that creates heat. Remote sensing instruments can be designed to detect reflected and/or emitted energy (see Figure H.8).

7.7.3 Reflected Energy

Different types of feature surfaces reflect radiation differently. Smooth surfaces act like mirrors and reflect most light in a single direction, a process called specular reflection. At the other extreme, rough surfaces reflect radiation equally in all directions, a process called diffuse reflection. Between these two extremes are an infinite number of reflection possibilities, which are often a characteristic of a specific material or surface condition (see Figure 7.27).

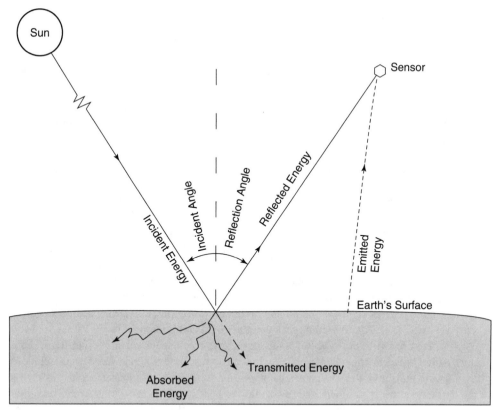

FIGURE 7.26 Interaction of electromagnetic radiation with the earth.

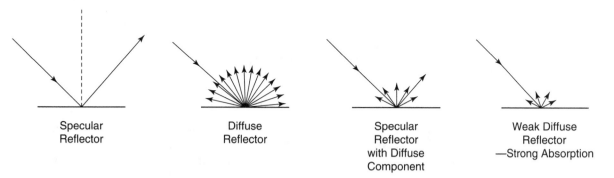

FIGURE 7.27 Specular and diffuse reflection.

7.7.4 Selection of Radiation Sensors

Satellite sensors are designed to meet various conditions, like the following:

- Using those parts of the electromagnetic spectrum that are less affected by scatter and absorption.
- Sensing data from different sources (active or passive sensing).
- Selecting the spectral ranges of energy that react most favorably with specified target surfaces.

Because various types of ground cover, rocks and soils, water, and built features react differently with radiation from different parts of the spectrum, it is common to install sensors (e.g., multispectral scanners) on spacecraft that can detect radiation from several appropriate wavelength ranges.

Most remote-sensing instruments use scanners to evaluate small portions of the earth at instants in time. At each instant, the sensor quantizes the earth's surface into pixels (picture elements) or bands. Pixel resolution can range from 0.6 m (QuickBird satellite) to greater than 1 km, depending on the instrumentation and altitude of the sensing platform. As in aerial photography, spectral scanning utilizes lenses and, like aerial photography, stereo-paired images can be captured and digital elevation models (DEMs) can be created; stereo (three-dimensional) viewing is a great aid in visual imaging.

There are two main types of scanners: across-track scanners and along-track scanners. Across-track scanners, also called whisk-broom scanners, scan lines perpendicular to the forward motion of the spacecraft using rotating mirrors. As the spacecraft moves across the earth, a series of scanned lines are collected, producing a two-dimensional image of the earth. Each on-board sensor in the vehicle collects reflected or emitted energy in a specific range of wavelengths; these electrical signals are converted to digital data and stored until they are transmitted (satellite) or returned (aerial) to earth. Along-track scanners, also called push-broom scanners, utilize the forward motion of the spacecraft itself to record successive scan lines. Instead of rotating mirrors, these scanners are equipped with linear array charge-coupled devices (CCDs) that build the image as the satellite moves forward.

Once the data has been captured and downloaded to earth, much more work has to be done to convert the data into a usable format. These processes, which include image enhancement and image classification, are beyond the scope of this text.

7.7.5 An Introduction to Image Analysis

Analysis of remotely sensed images is a highly specialized field. Because environmental characteristics include most of the physical sciences, the image interpreter should have an understanding of the basic concepts of geography, climatology, geomorphology, geology, soil science, ecology, hydrology, and civil engineering. For this reason, multidisciplinary teams of scientists, geographers, and engineers, all trained in image interpretation, are frequently used to help develop the software utilized in this largely automated process.

Just as photogrammetry/air photo interpretation helps the analyst to identify objects in an aerial photograph, digital image analysis helps the analyst to identify objects depicted in satellite and airborne images. With the blossoming of computerized image analyses that employ automated digital image processing, the relative numbers of trained analysts needed to process data has declined sharply. Satellite imagery has always been processed in digital format, and aerial photos (films) can be scanned easily to convert from analog to digital format. As with aerial photography, satellite images must have ground control and must be corrected for geometric distortion and relief displacement, and also for distortions caused by the atmosphere, by radiometric errors, and by the earth's rotation. Satellite imagery can be purchased with some or many of these distortions removed. The price of the imagery is directly related to the degree of processing requested.

Unlike aerial photography images, different satellite images of the same spatial area—taken by two or more different sensors—can be fused together to produce a new image with distinct characteristics and thus provide even more data. A digital image target may be a point, line, or area (polygon) feature that is distinguishable from other surrounding features. Digital images are comprised of many pixels, each of which has been assigned a digital number (DN). The digital number represents the brightness level for that specific pixel—e.g., in eight-bit (2^8) imagery, the 256 brightness levels range from 0 (black) to 255 (white).

Each satellite has the capability of capturing images and then transmitting the images, in digital format, to ground receiving stations that are located around the globe. Modern remote-sensing satellites are equipped with sensors collecting data from selected bands of the electromagnetic spectrum. For example, Landsat 7's sensors collect data from eight bands (channels)—seven bands of reflected energy and one band of emitted energy. Because radiation from various portions of the electromagnetic spectrum reacts in a unique fashion with different materials found on the surface of the earth, it is possible to develop a signature of the reflected and emitted signal responses and to identify the object or class of objects from those signatures.

When the energy reflections from the ground surface (collected by the various on-board sensors) are plotted, the resultant unique curves, called spectral signatures, are used to help identify various surface materials. Figure 7.28 shows four signatures of different ground surfaces, with the reflection as a function of wavelength used for plotting purposes. The curves were plotted from data received from the eight sensors on board Landsat 7. When the results are plotted as a percentage reflectance for two or more bands

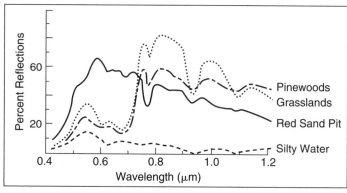

FIGURE 7.28 Spectral reflectance curves of four different targets. (Adapted from NASA's *Landsat 7 Image Assessment System Handbook*, 2000; courtesy NASA)

(see Figure 7.29), in multidimensional space, the ability to identify the surface materials precisely increases markedly. This spectral separation permits the effective analysis of satellite imagery, most of which can be performed through computer applications. Image analysis can be applied productively to a broad range of scientific inquiry (see Table 7.4).

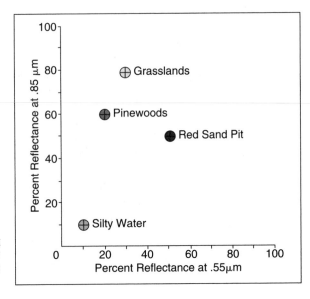

FIGURE 7.29 Spectral separability using just two bands. (Adapted from NASA's *Landsat 7 Image Assessment Handbook*, 2000; courtesy NASA)

Table 7.4 APPLICATIONS OF REMOTELY SENSED IMAGES

Field	Applications
Forestry/agriculture	Inventory of crop and timber acreage, estimating crop yields, crop disease tracking, determination of soil conditions, assessment of fire damage, precision farming, and global food analysis.
Land use/mapping	Classification of land use, mapping and map updating, categorization of land use capability, monitoring urban growth, local and regional planning, functional planning for transportation corridors, mapping land/water boundaries, and flood plain management.
Geology	Mapping of major geologic units, revising geologic maps, recognition of rock types, mapping recent volcanic surface deposits, mapping land forms, mineral and gas/oil exploration, and estimating slope failures.
Natural resources	Natural resources exploration and management.
Water resources	Mapping of floods and flood plains, determining extent of areal snow and ice, measurement of glacial features, measurement of sediment and turbidity patterns, delineation of irrigated fields, inventory of lakes, and estimating snow-melt runoff.
Coastal resources	Mapping shoreline changes; mapping shoals and shallow areas; and tracing oil spills, pollutants, and erosion.
Environment	Monitoring effect of human activity (lake eutrophication, defoliation), measuring the effects of natural disasters, monitoring surface mining and reclamation, assessing drought impact, siting for solid-waste disposal, siting for power plants and other industries, regulatory compliance studies, and development impact analysis.

7.7.6 Classification and Feature Extraction

Satellite imagery contains a wealth of information, all of which may not be relevant for specific project purposes. A decision must be made about which feature classifications are most relevant. Once the types of features to be classified have been determined, a code library giving a unique code for each included feature is constructed. The pixel identification can then be represented by a unique code (e.g., R for residential) rather than a digital number (DN) giving the gray-scale designation. For example, a typical GIS project may require the following features to be classified:

- Pervious/impervious surfaces (needed for rainfall runoff calculations)
- Water
- Wetland
- Pavement
- Bare ground
- Residential areas
- Commercial areas
- Industrial areas
- Grass
- Tree stands
- Transportation corridors

Feature extraction (image interpretation) can be accomplished using manual methods or automated methods. Automated methods employ computer algorithms to classify images in relation to surface features, and four techniques are presently in use:

- Unsupervised: this technique relies on color and tone as well as statistical clustering to identify features.
- Supervised: this technique requires comparative examples of imaging for each ground feature category.
- Hybrid: this technique is a combination of the first two (unsupervised and supervised).
- Classification and regression tree (CART) analysis: CART analysis uses binary partitioning software to analyze and arrive eventually at a best estimate about the ground feature identification.

7.7.7 Ground-Truthing, or Accuracy Assessment

How do we know if the sophisticated image analysis techniques in use are giving us accurate identifications? We have to establish a level of reliability to give credibility to the interpretation process. With aerial photo interpretation, we begin with a somewhat intuitive visual model of the ground feature; with satellite imagery, we begin with pixels identified by their gray-scale digital number (DN)—characteristics that are not intuitive. What we need is a process whereby we can determine the accuracy and reliability of interpretation results. In addition to providing reliability, such a process enables the operator to correct errors and to compare the successes of the various types of feature extraction that may have been used in the project. This process is also invaluable in the calibration of imaging sensors.

Ground-truthing, or accuracy assessment, can be accomplished by comparing the automated feature extraction identifications with feature identifications given by other methods for a given sample size. Alternative interpretation or extraction techniques may include air photo interpretation, analyses of existing thematic maps of the same area, and field work involving same-area ground sampling.

It is not practical to check the accuracy of each pixel identification, so a representative sample is chosen for accuracy assessment. The sample is representative with respect to both the size and locale of the sampled geographic area. Factors to be considered when defining the sample include the following:

- Is the land privately owned and/or restricted to access?
- How will the data be collected?
- How much money can you afford to dedicate to this process?
- Which geographic areas should be sampled: areas that consist of homogeneous features together with areas consisting of a wide mix of features, even overlapping features?
- How can the randomness of the sample points be ensured, given practical constraints (e.g., restricted access to specified ground areas, lack of current air photos and thematic maps, etc.)?
- If verification data is to be taken from existing maps and plans, what impact will their dates of data collection have on the analysis (i.e., the more current the data, the better the correlation)?

The answers to many of the questions above come only with experience. Much of this type of work is based on trial-and-error decisions. For example, when the results of small samples compare favorably with the results of large samples, it may be assumed that, given similar conditions, such small samples may be appropriate in subsequent investigations. On the other hand, if large samples give significantly different (better) results than do small sample sizes, one may assume that such small sample sizes should not be used again under similar circumstances.

Figure 7.30 shows an error matrix consisting of ground point identifications that have been extracted using automated techniques and ground point identifications determined using other techniques (including field sampling). With the example shown here, seventeen of twenty-three sampling sites for water were verified, twenty-two of forty-two sampling sites for agricultural were verified, and twenty-eight of thirty-five sampling points for urban were verified. These verified points, shown along the diagonal, total sixty-seven from a total of 100. Thus, the verified accuracy was 67 percent in this example.

Is 67 percent accuracy acceptable? The answer depends on the intended use of the data and, once again, experience is a determining factor. It may be decided that this method of data collection (e.g., satellite imagery) is inappropriate (i.e., not accurate enough or too expensive) for the intended purposes of the project.

Reference Data (Sampling Data)

Diagonal	Water	Agricultural	Urban	Row Totals
Water	17 Correct	5	1	23
Agricultural	10	22 Correct	10	42
Urban	2	5	28 Correct	35
Column Total	29	32	39	100

(Interpreted Data — row labels at left)

FIGURE 7.30 Error matrix example.

7.8 Remote-Sensing Satellites

7.8.1 Landsat 7 Satellite

Landsat 7 was launched April 15, 1999, from Vandenburg Air Force Base on a Delta-11 launch vehicle. The spacecraft weighs 4,800 lbs, measures about 14 ft long by 9 ft in diameter, and flies at an altitude of 705 km. Landsat 7 employs an extended thematic mapper (ETM+ is a scan mirror spectrometer) and an eight-band multispectral scanner capable of providing high-resolution image information of the earth's surface as it collects seven bands or channels of reflected energy and one band of emitted energy. The multispectral scanner's panchromatic band (0.52–0.90 μm) has a resolution of 15 m (see Table 7.5). The Landsat program was designed to monitor small-scale processes seasonally on a global scale, for example, cycles of vegetation growth, deforestation, agricultural land use, erosion, etc.

Another NASA satellite, the experimental TERRA (EO-1), launched December 18, 1999, employs a push-broom spectrometer/radiometer—the Advanced Land Imager (ALI)—with many more spectral ranges (hyperspectral scanning) and flies the same orbits with the same altitude as does Landsat 7. It is designed to sample similar surface features at roughly the same time (only minutes apart) for comparative analyses. TERRA's panchromatic band (0.48–0.68 μm) has a resolution of 10 m. This experimental satellite is designed to obtain much more data at reduced costs and will influence the design of the present stage (2000–2015) of U.S. exploration satellites.

Whether by spectral analysis, radar imaging, or lidar (light detection and ranging), a huge amount of surface data can be collected and analyzed. Airborne imagery is flown at

Table 7.5 BANDWIDTH CHARACTERISTICS FOR LANDSAT 7

TM and ETM+ Spectral Bandwidths (μm)

Sensor	Band 1	Band 2	Band 3	Band 4	Band 5	Band 6	Band 7	Band 8
TM	.45–.52	.52–.60	.63–.69	.76–.90	1.55–1.75	10.4–12.5	2.08–2.35	N/A
ETM+	.45–.52	.53–.61	.63–.69	.78–.90	1.55–1.75	10.4–12.5	2.09–2.35	.52–.90
	Blue	Green	Red	Near-IR	Shortwave IR	Thermal IR	Shortwave IR	Panchromatic
Resolution (pixel size)	30 m	30 m	30 m	30 m	30 m	60 m	30 m	15 m

Band	Use
1	Soil/vegetation discrimination; bathymetry/coastal mapping; cultural/urban feature identification.
2	Green vegetation mapping; cultural/urban feature identification.
3	Vegetated versus nonvegetated and plant species discrimination.
4	Identification of plant/vegetation types, health and biomass content, water body delineation, soil moisture.
5	Sensitive to moisture in soil and vegetation; discriminating snow and cloud-covered areas.
6	Vegetation stress and soil moisture discrimination related to thermal radiation; thermal mapping (urban, water).
7	Discrimination of mineral and rock types; sensitive to vegetation moisture content.
8	Panchromatic: mapping, planning, design.

Source: Landsat 7 is operated under the direction of NOAA. Data can be obtained from USGS- EROS Data Center, Sioux Falls, SD 57198. Courtesy of NOAA.

relatively low altitudes, giving the potential for high ground resolution. When these techniques are employed on satellite platforms, we receive huge amounts of data that can be collected again at regular intervals, thus adding a temporal (time) dimension to the measurement process. In all cases of satellite imagery, the data output is in digital form and ready for computer processing to elicit the required information. Although satellite imagery cannot give us the same resolutions as airborne imagery, we presently have QuickBird resolution at the level of 0.6 meter, and predictions state that the next generation of satellite imagery will be in the one-third of a meter (or less) range for ground resolution. Satellite imagery has been used for many years to assess crop inventories, flood damages, and other large-scale geographic projects.

7.9 Geographic Information System (GIS)

7.9.1 General Background

GIS is a tool used by engineers, planners, geographers, and other social scientists, so it is not surprising that there seems to be as many definitions for GIS as there are fields in which GIS is employed. My choice for a definition comes from the University of Edinborough's GIS faculty: "GIS is a computerized system for capturing, storing, checking, integrating, manipulating, analyzing, and displaying data related to positions on the earth's surface."

Since the earliest days of mapping, topographic features have routinely been portrayed on scaled maps and plans. These maps and plans provided an inventory of selected or general features that were found in a given geographic area. With the emergence of large databases, which were collected primarily for mapping, attention was given to new techniques for analyzing and querying the computer-stored data. The introduction of topological techniques permitted the data to be connected in a relational sense, in addition to their spatial connections. Thus, it became possible not only to determine where a point (e.g., hydrant), a line (e.g., road), or an area (park, neighborhood, etc.) was located, but also to analyze those features with respect to the adjacency of other spatial features, connectivity (network analysis), and direction of vectors. Adjacency, connectivity, and direction opened the database to a wider variety of analyzing and querying techniques. For example, it is possible in a GIS real-estate application to display, in municipal map form, all the industrially zoned parcels of land with rail-spur possibilities, within 1.5 miles of freeway access, in the range of 1.2 to 3.1 acres in size. Relational characteristics also permit the database to be used for routing of, for example, emergency vehicles and vacationers.

GIS data can be assembled from existing databases; digitized or scanned from existing maps and plans; or collected using conventional surveying techniques, including global positioning system (GPS) surveying techniques. One GPS method that has recently become very popular for GIS data collection is that of differential GPS (see Section 8.6). This technique utilizes a less expensive GPS receiver and radio signal corrections from a base station receiver to provide submeter accuracies that are acceptable for mapping and GIS database inventories. See Figure 7.31 for typical differential GPS equipment.

The ability to store a wide variety of data on feature-unique layers in computer memory permits the simple production of special-feature maps. For example, maps can be produced that show only the registered parcel outlines of an area, or only drainage and contour information, or any other feature-specific data. In fact, any selected layer or combination of selected layers can be depicted on a computer screen or on a hard-copy map (produced on a digital plotter) at any desired scale (see Figure 7.32). In a GIS, spatial entities have two key characteristics: location and attributes. Location can be given by coordinates, street addresses, etc., and attributes describe some characteristic(s) of the feature being analyzed.

GIS has now blossomed into a huge and diverse field of activity. Most activity can be identified as being in one of two broad fields: (1) geographic, feature-specific activities such as mapping, engineering, routing, environment, resources, and agriculture; and (2) cultural/social activities such as marketing research, census, demographics, and socioeconomic studies.

The switch from hard-copy maps to computerized GIS has provided many benefits. For example, now we can:

- Store and easily update large amounts of data.
- Sort and store spatial features, called entities, into thematic layers. Data are stored in layers so that complex spatial data can be manipulated and analyzed efficiently by layer rather than trying to deal with the entire database at the same time.
- Zoom into sections of the displayed data to generate additional graphics that may be hidden at default scales. We can also query items of interest to obtain tables of attribute information that may have been tagged to coordinated points of interest.

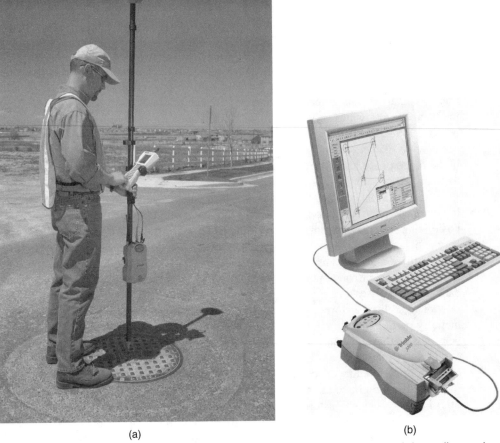

FIGURE 7.31   (a) Trimble's 5700 GPS receiver with the pole-mounted antenna and data collector shown being used to capture municipal asset data location for a GIS database. (b) Trimble's 5700 GPS receiver shown connected to a computer directly downloading to Trimble's Geomatics Office software. (Portions © 2001 Trimble Navigation Limited. All Rights Reserved.)

- Analyze both entities and their attribute data using sophisticated computer programs.
- Prepare maps showing only selected thematic layers of interest (and thus reduce clutter). We can update maps quickly as new data are assembled.
- Use the stored data to prepare maps for different themes and scales, for a wide range of purposes.
- Import stored data (both spatial and nonspatial) electronically from different agencies (via CDs, DVDs, and the Internet) and thus save the costs of collecting the data.

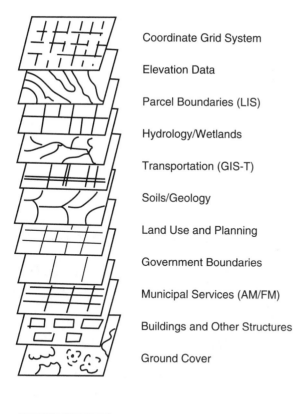

Coordinate Grid System

Elevation Data

Parcel Boundaries (LIS)

Hydrology/Wetlands

Transportation (GIS-T)

Soils/Geology

Land Use and Planning

Government Boundaries

Municipal Services (AM/FM)

Buildings and Other Structures

Ground Cover

Composite, Showing
a Combination of
Selected Layers

FIGURE 7.32 Illustration of thematic layers.

- Build and augment a database by combining digital data from all the data-gathering techniques illustrated in Figure 7.1—field surveying, remote sensing, map digitization, scanning, and Internet transfer/CD-ROM imports.
- Create new maps by modeling or re-interpreting existing data.

Some ask, How do GIS and CAD compare? Both systems are computer-based, but CAD is often associated with higher-precision engineering design and surveying applications, and GIS is more often associated with mapping and planning—activities requiring lower levels of precision. CAD is similar to mapping because it is essentially an inventory of entered data and computed data that can provide answers to the question, Where is it located? GIS has the ability to model data and to provide answers to the spatial and temporal questions, What occurred? What if . . . ? etc. Topology gives GIS the ability to determine spatial relationships such as adjacency and connectivity between physical features or entities. Typical subsets of geographic feature-specific GIS include land information systems (LISs), which deal exclusively with parcel title description and registration; automated mapping and facility management (AM/FM), which models the location of all utility/plant facilities; and transportation design and management (GIS-T).

7.9.2 Components of a GIS

GIS can be divided into four major activities:

- Data collection and input.
- Data storage and retrieval.
- Data analysis.
- Data output and display.

GIS may be further described by listing its typical components:

- The computer, which is the heart of GIS, along with the GIS software, typically uses Windows XT, Windows NT, or Unix operating systems.
- Data collection, which can be divided into the geomatics components shown in Figure 7.1: field surveying, remote sensing, digitization of existing maps and plans, digital data transfer via the Internet or CD/DVD.
- Computer storage: hard drives, optical disks, etc.
- Software designed to download, edit, sort, and analyze data—e.g., database software, relational database software, GIS software, geometric and drawing software (COGO, CAD, etc.), soft-copy photogrammetry, and satellite imagery analysis software.
- Software designed to process and present data in the form of graphics and maps and plans.
- Hardware components, including the computer, surveying and remote-sensing equipment, CD and DVD, digitizers, scanners, interactive graphics terminals, and plotters and printers.

The growth in GIS is assured because it has been estimated that as much as 80 percent of all information used by local governments is geographically referenced. GIS is a tool that encourages planners, designers, and other decision makers to study and analyze spatial data along with an enormous amount of attribute data (cultural, social, geographic, economic, resources, environmental, infrastructure, etc.)—data that can be tagged to specific georeferenced spatial entities.

7.9.3 Sources for GIS Data

The most expensive part of any GIS is data collection. Obviously, if you can obtain suitable data collected from other sources, the efficiency of the process increases. When importing data from other sources, it is imperative that the accuracy level of the collected data be certified as appropriate for its intended use. For example, if you are building a database for use in high-accuracy design applications, it would not be suitable to import data digitized from U.S. Geological Survey (USGS) 1:100,000 topographic maps (as is much of the U.S. Census TIGER data), where accuracy can be restricted to plus or minus 170 feet. However, for small-scale projects (e.g., planning, resources/environmental studies), the nationwide

TIGER file has become one of the prime sources of easily available data in the United States. Government data sources (e.g., USGS, TIGER, etc.) provide free data (except for the cost of reproduction) to the general public. Traditional sources for data collection include:

- Field surveying.
- Remotely sensed images—rectified and digitized aerial photographs (orthophotos) and processed aerial and satellite imagery.
- Existing topographic maps, plans, and photos—via digitizing and/or scanning.
- Census data.
- Electronic transfer of previously digitized data from government agencies or commercial firms.

7.9.4 Georeferencing

Like the map makers of the past, GIS specialists must find some way to relate geospatial data to the surface of the earth. If all or most geographic data users employ the same (or well-recognized) earth-reference techniques, data may be economically shared among agencies using different computer systems. Now, the most widely accepted shape of the earth has been geometrically modeled as an ellipsoid of revolution with a semimajor axis of 6,378,135 m called the GRS-80 ellipsoid (see Section 9.1, and Figure 8.24). Once the shape of the earth has been modeled and a geodetic datum defined, some method must be used to show the earth's curved surface on plane-surface map sheets with minimal distortion. Several map projections have been developed for this purpose. In North America, the projections used most are the Lambert projection, the transverse Mercator projection (used in the United States), and the universal transverse Mercator projection (used primarily in Canada). Along with a specific projection, it is possible to define a coordinate grid that can be used to minimize distortion as we move from a spherical surface to a plane surface. (See Chapter 9 for more on this topic.) Measurements on the grid can be in feet, U.S. feet, or meters (see Table 1.1).

7.9.4.1 Coordinate Grids.
U.S. states have adopted either the Lambert conical projection or the transverse Mercator cylindrical projection and have created coordinate grid systems (false northing and easting) defined for each state. The State Plane Coordinate System 1983 (SPCS83), which superseded the SPCS27, has been applied to all states, and the universal transverse Mercator grid has been applied to Canada. Software programs readily convert coordinates based on one projection to any other commonly used projection. (Coordinate grids are discussed in Chapter 9.)

7.9.4.2 Transformation.
When data is being imported from various sources, it is likely that different earth-reference models may have been used. If a GIS is tied to a specific coordinate grid and specific orientation, new data may have to be transformed to fit the working model. Transformations can be made in grid reference, scale, and orientation, and GIS programs are designed to translate from one grid reference to another, to convert from one scale to another, and to rotate to achieve appropriate orientation (see Figure 7.33).

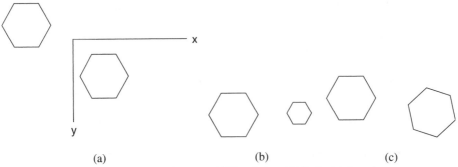

FIGURE 7.33 Transformation techniques. (a) Translation. (b) Scale. (c) Rotation.

7.9.4.3 Geocoding. Spatial data describes spatial location (e.g., geographic coordinates, map coordinates, street addresses, postal codes, etc.) of an entity, whereas attribute data (often shown in related tables) describes the different aspects of the entity being collected. Examples of attribute data include the population of towns, ground cover (including crop types), number of occupants in a dwelling, etc. The linking of entity and attribute data to a specific geographic location is known as **geocoding**. GIS programs should be able to recognize the defining characteristics of imported data in the geocodes from the metadata information supplied and then make the necessary conversions so that the imported data is tied to the same reference datum as are the working data.

7.10 Database Management

A GIS is concerned with large amounts of both spatial data and nonspatial (attribute) data. Data must be organized so that information about entities and their attributes may be accessed by rapid computerized search-and-retrieval techniques. When data concerned with a specific category is organized in such a fashion, that data collection is known as a database. Data collections can range in complexity from simple unstructured lists and tables to ordered lists and tables (e.g., alphabetical, numerical, etc.); to indexes (see the index at the back of this text); and finally to sets of more complex, categorized tables.

The computerized tabular data structure used by many GIS programs is called a relational database. A relational database is a set theory–based data structure comprised of ordered sets of attribute data (entered in rows) known as tuples, and grouped in two-dimensional tables called relations. When searching the columns of one or more databases for specific information, a key (called a primary key) or some other unique identifier must be used. A primary key is an unambiguous descriptor(s) such as a place name, the coordinates of a point, a postal code address, etc. The data in one table is related to similar data in another table by a GIS technique of relational join. Any number of tables can be related if they all share a common column.

For example, with a postal code address or even state plane coordinates, a wide variety of descriptive tables can be accessed that contain information relevant to that specific location. One table may contain parcel data, including current and former ownership, assessment, etc., while other tables may contain municipal services available at that location, rainfall/snowfall statistics, pedestrian and vehicular traffic counts at different periods of the day, etc.

Queries of a relational database are governed by a standard query language (SQL), developed by IBM in the 1970s, to access data in a logically structured manner. Traditional SQL continues to be modified by software designers (with assistance from the Open GIS Consortium—see the next section) for specific applications in GIS. Modifications include spatial operators and tools such as adjacency, overlap, buffering, area, etc. (See Section 7.17 at the end of this chapter for more information on this topic.)

7.11 Metadata

Metadata is "data about data." Metadata describes the content, quality, condition, and other characteristics of data. In 1994, the Federal Geographic Data Committee (FGDC) of the National Spatial Data Infrastructure (NSDI) approved standards for "Coordinating Geographic Data Acquisition and Access" (a pamphlet that was updated in 1998), and in 2000 FGDC published another pamphlet called "Content Standard for Geospatial Metadata Workbook"—available on-line at *http://www.fgdc.gov/publications/documents/ metadata/_workbook_0501_bmk.pdf*.

As we noted earlier in this chapter, the cost of obtaining data is one of the chief expenses in developing a database. Many agencies and commercial firms cannot afford to create all the data that may be needed for various GIS analyses. Thus, the need for standards in the collection, dissemination, and cataloging of data sets (a data set is a collection of related data) is readily apparent.

FGDC lists the different aspects of data described by metadata:

- Identification: What is the name of the data set? Who developed the data set? What geographic area does it cover? What themes of information does it include? How current are the data? Are there restrictions on accessing or using the data?

- Data quality: Are the data reliable? Is information available that allows the user to decide if the data are suitable for the intended purpose? What is the positional and attribute accuracy? Are the data complete? What data were used to create the data set, and what processes were applied to these sources?

- Spatial data organization: What spatial data model was used to code the spatial data? How many spatial objects are there? Are methods other than coordinates, such as street addresses, used to encode locations?

- Spatial reference: Are coordinate locations encoded using longitude and latitude? Is a map projection or grid system such as the State Plane Coordinate System used? What horizontal and vertical datums are used? What parameters should be used to convert the data to another coordinate grid system?

- Entity and attribute information: What geographic information (roads, houses, elevation, temperature, etc.) is included? How is this information encoded? Were codes used? What do the codes mean?

- Distribution: From whom can you obtain the data? What formats are available? What media are available? Are the data available on-line? How much does acquisition of the data cost?

- Metadata reference: When were the metadata compiled? Who compiled the metadata?

7.12 Spatial Entities or Features

Spatial data have two characteristics: location and descriptive attributes. In GIS, entities are modeled as either points, lines, or areas (polygons). When a polygon is given the dimension of height, an additional model, surface, is created (e.g., contoured drawing, rainfall runoff, etc.). Each entity can have various attributes assigned to it that describe something about that entity. Attributes can be kept in relational tables, as in the vector model, or attached to the layer grid cells themselves, as in the raster model. Spatial data describe the geographic location of a feature or entity, and its location with respect to other area features is also given either by rectangular coordinates (vector model) or by grid cells (raster model) described by grid column and row. Other locators, such as postal codes, route mileage posts, etc., can also be used.

Point is the simplest spatial entity because it has zero dimension. In vector representation [see Figure 7.34(a)], a point is described by a set of x/y or east/north coordinates. In raster representation [see Figure 7.34(b)], a point is described by a single grid cell. Points have attributes that describe the entity—e.g., a utility pole entity may have an attribute such as a traffic signal or traffic sign or electrical distribution, etc. In Figure 7.34, for example, the entity being described by the code number 6 could be industrially zoned property.

Line (arc) is a spatial entity having one dimension that joins two points; a string is comprised of a sequence of connected lines or arc segments. The line or arc can be defined by an ordered series of coordinated points (vector representation) or by a series of grid cells (raster representation). Lines or arcs have attributes that describe those particular entities. For example, a line entity may represent a ₵ section of a municipal pipeline, and attributes for that entity could include use, pipe type, slope, diameter, inverts, date of installation, date of last maintenance, etc. In vector models, this attribute data is stored in a table in a relational database.

Polygon is a spatial entity having two dimensions that is described by an enclosed area defined by an arc or a series of arcs closing back to the starting point. Polygons have attributes that describe the whole area, such as tree cover, land zoning, etc. In the case of vector models, the polygon has a single number representing the attribute of the enclosed area. In the case of raster models, the code number representing the attribute for an area is displayed in each grid cell within that enclosed area. See Figure 7.34.

7.13 Typical Data Representation

The representation of data depends on the defined scale or the resolution. Table 7.6 shows typical representations for a point, a line or string, and a polygon. Consider the following when defining how data will be represented:

1. On larger-scale maps, large rivers, expressways, etc., are shown as having width and then classified as being polygons (when section end lines are drawn).
2. Before entities can be modeled as points, lines, polygons, or surfaces, the scale must be taken into consideration. For example, on small-scale maps, towns may show up as dots (points); on larger-scale maps, towns probably show up as polygons or areas.
3. Scale also affects the way entities are shown on a map. On small-scale maps, meandering rivers are shown with fewer and smoother bends, and small crop fields may not show up at all.

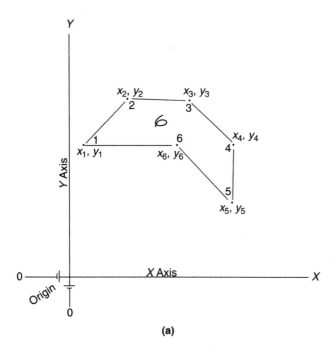

(a)

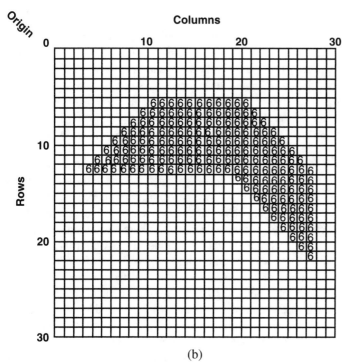

(b)

FIGURE 7.34 Data models used in GIS. (a) Vector model. (b) Raster model. 6 is a numeric attribute code that describes a function or description of the ground area being modeled. For example, 6 can be land-zoned for industrial development.

Table 7.6 DATA TYPES

Point	Line or String	Polygon (Area)
Well	River or shoreline	Pond or lake
Utility pole or tower	Distribution line	Transmission station
Hydrant or valve	Water line	Water line grid
Bus shelter	Sidewalk or road	Parking lot
Storm-water manhole	Storm water pipeline	Catchment/runoff area
Property monument	Lot line/fence line	Land parcel
Land parcel	Neighborhood	Town, city, or political boundary
Deciduous tree	Row of trees	Tree stand
Roads ℄ intersection	Road ℄	Section of road right-of-way (ROW)
Farm house	Irrigation channel	Crop field
Refreshment booth	Path	Park

7.14 Spatial Data Models

In GIS, real-world physical features are called entities and are modeled using one or both of the following techniques: vector model or raster model. We discuss both in the following sections.

7.14.1 Vector Model

Surveying and engineering students are already familiar with the concept of vectors, where collected data are described and defined by their discrete Cartesian coordinates (x, y) or (easting, northing). The origin (0, 0) of a rectangular grid is at the lower left corner of the grid. Points are identified by their coordinates; lines, or a series of lines, are identified by the coordinates of their endpoints. Areas or polygons are described by a series of lines (arcs) that loop back to the coordinated point of beginning. See Figure 7.34(a). Points are described as having zero dimension, lines have one dimension, and areas or polygons have two dimensions.

In addition to treating points, lines, and polygons, vector GIS also recognizes line/polygon intersections and the topological aspects of network analysis (connectivity and adjacency). These capabilities give GIS its unique capabilities of querying and routing.

Attribute data, which describe or classify entities, are located in a database (tables) linked to the coordinated vector model. Attributes can be numbers, characters, images, or even CAD drawings. Data captured using field data techniques usually come with acceptably accurate coordinates. When data is captured by digitizing (using a digitizing table or digitizing tablet) from existing maps and plans, some additional factors must be considered. First, the scale of the map or plan being digitized is a limiting factor in the accuracy of the resulting coordinates. Second, when irregular features are captured by sampling with the digitizing cursor or puck, some errors are introduced, depending on the sampling interval. For example, it is generally agreed that feature sample points should be digitized at each beginning and end point and at each change in direction, but how often do you sample curving lines? If the curved lines are geometrically defined (e.g., circular, spiral), a few sampling points may be sufficient, but if the curve is irregular, the sampling rate depends on the relative importance of the feature and perhaps the scale of any proposed output graphics.

7.14.2 Raster Model

Raster modeling was a logical consequence of the collection of data through remote sensing and scanning techniques. Here, topographic images are represented by pixels (picture elements) that together form the structure of a multicelled grid. A raster is a GIS data structure comprised of a matrix of rectangular (usually square) grid cells. Each cell represents a specific area on the ground. The resolution of the raster is defined by the ground area represented by the raster grid cell. A cell can represent a 10 ft^2 (m^2), 100 ft^2 (m^2), or even 1,000 ft^2 (m^2) ground area—all depending on the resolution or scale of the grid cell. The higher the resolution of the grid, the more cells are required to portray a given area of the ground surface; the larger the ground area represented by a grid cell, the less precise is the cell's ability to define position. Thus, in the raster model, the considerations of scale are vital at the beginning of a project; the resolution of the raster is often a function of the scale of the map from which the spatial data may have been scanned or digitized [see Figure 7.34(b)]. When you are first using GIS programs, it is extremely important to identify the resolution of the raster grid with which you are working because some GIS programs display cursor coordinates to two or three decimal places that change as the cursor is moved over the screen, even when the grid resolution is only 10 to 20 (or more) ft or m.

Raster grid cells are identified as being in rows and columns, with the location of any raster cell given by the column and row numbers. The zero row and column (the origin) is often located at the upper left of the raster grid (the location of the origin is defined differently by various software agencies). Features that are defined by their raster cell are thus not only geographically located but are also located relative to all other features in the raster. Because raster grid cells represent areas (not points) on the earth, they cannot be used for precise measurements. Each cell contains an attribute value (often a number) that describes the entity being represented by that layer-specific cell. In multispectral imagery, the cell is identified by its gray-scale number, e.g., a number in the range of 0 to 255 for eight-bit imagery.

Pixel is a term also employed in the field of remote sensing. Like grid cells in GIS, pixels portray an area subdivided into very small, (usually) square cells. Pixels are the result of capturing data through the digitization of aerial/satellite imagery. The distribution of the colors and tones in an image is established by assigning DN values to each pixel. Much of this can now be accomplished using automated soft-copy (digital) photogrammetry techniques and digital image analysis. Image resolution is stated by defining the ground area represented by one pixel. The concept of dots per inch (dpi) is also used in some applications, where each dot is a pixel. If an image has a dpi of 100, each dot or pixel has a grid cell side length of 1/100 of an inch.

Pixels are identified by unique numerical codes called digital numbers (DNs). Each cell has one DN only; much GIS software now uses eight-bit (2^8) data, which allows for 256 numbers, from 0 to 255. Early in the GIS design, different attributes were assigned to specific feature layers. As the GIS construction proceeds, each cell in each layer can also be given a unique code number or letter identifying the predominant (or average) attribute for that cell. For example, a grid cell code of 16 could be defined as representing areas zoned residential for housing.

7.15 GIS Data Structures

GIS software programs usually support both raster and vector models, and some programs readily convert from one to the other. The type of model selected is usually determined by the source of the data and the intended use of the data. For example, the vector model is used when there is access to coordinated data consistent with surveying field work, and the raster model is used when the data flows from scanned map data or remotely sensed imagery. When the project objectives concern large-scale engineering design or analysis, almost certainly a vector model GIS will be employed. If the objective concerns small-scale land zoning, planning analysis, or some types of cultural or social studies, a raster model GIS may be the best choice. An additional consideration deals with the nature of the data. For example, features that vary continuously, such as elevation or temperature, lend themselves to raster depiction, whereas discrete features, such as roads and pipelines, lend themselves to vector depiction.

Both vector and raster models permit the storing of thematic data on separate layers (see Figure 7.32). Typical thematic layers include:

- Spatial reference system (e.g., coordinate grid system, etc.). The control data on this layer is used to correlate the placement of spatial data on all related layers.
- Elevation data (including contours).
- Parcel (property) boundaries.
- Hydrology (runoff and catchment areas, drainage, streams and rivers).
- Wetlands.
- Transportation.
- Soil types.
- Geology.
- Land use and zoning.
- Government boundaries.
- Municipal services.
- Buildings and other structures.
- Ground cover (including crop types and tree stands).

Individual layers, or any combination of layers, can be overlaid and analyzed or plotted together using digital plotters. Although vector models provide for a higher level of accuracy than do raster models, GIS vector analyses require a much higher level of processing and computer storage.

If large-scale graphics are an important consideration, we repeat that, in raster plots, lines may have a stepped appearance that detracts from the overall presentation. The lower the resolution of the grid cell, the more pronounced is the appearance of stepping. When performing overlay analysis, raster data is more efficient because all the cells from each layer match up; with network analysis, vector depiction is preferred because here we are dealing with attributes of discrete linear features (e.g., a road's speed limit, the number of stop signs, etc.).

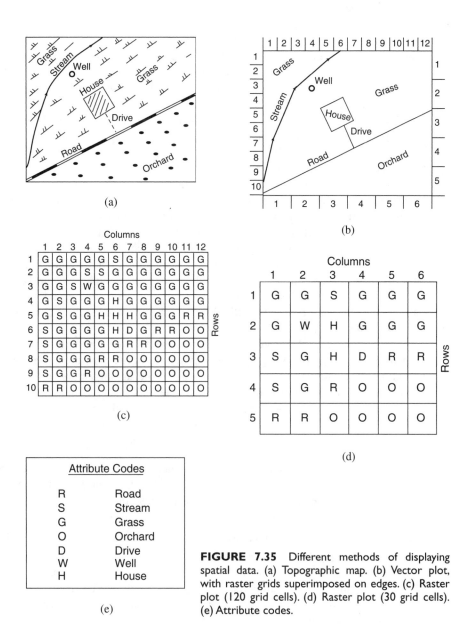

FIGURE 7.35 Different methods of displaying spatial data. (a) Topographic map. (b) Vector plot, with raster grids superimposed on edges. (c) Raster plot (120 grid cells). (d) Raster plot (30 grid cells). (e) Attribute codes.

Attribute Codes	
R	Road
S	Stream
G	Grass
O	Orchard
D	Drive
W	Well
H	House

(e)

7.15.1 Model Data Conversion

At times, data in vector format must be converted to raster format, and vice versa. To understand the working of the various software programs designed to perform those functions, it helps to analyze different data models representing the same planimetric features. Figure 7.35 shows features as depicted on a topographic map, along with both vector and raster models, of a simplified area. Figure 7.35(a) is a mapping representation of collected field data.

The stream looks more natural here because it has been cartographically smoothed for the sake of appearance. In Figure 7.35(b), the data are shown in vectors. Although perhaps more geometrically correct, the stream is less pleasing to the eye. Here, the lines joining coordinated ground points define part of a stream chain, with each segment directionally defined.

To convert the vector model to raster format, a raster of specified resolution is over laid the vector model. The more grid cells, the higher the resolution—and the more realistic the portrayal of the area features. You can see that, in Figure 7.35(c), the length of the sides of each grid cell are half the length of the grid cells in Figure 7.35(d), resulting in four times as many cells. If GIS users were to define very high resolution raster grids in an attempt to improve the model's accuracy, appearance, etc., they may soon run into problems resulting from exponential increases in the quantity of data, with serious implications for computer memory requirements and program analyses runtimes. Another factor that weighs heavily here is the level of the accuracy of the original data and the way the GIS is used.

Vectors represent discrete coordinated locations, and raster grids represent grid cells with areas of some defined size, so precise conversions are not possible. As model data is being converted, some choices will have to be made. For example, each grid cell can have only one attribute, so what happens when more than one feature on a vector model is at least partially captured in a raster grid cell? Which of the possible feature attributes are designated for that specific grid cell?

In Figure 7.35, there are seven identified attributes: grass, stream, drive, road, orchard, well, and house. Many grid cells have at least two possible attributes. Which is more important—stream or grass? If we pick grass, the stream feature will not show up at all. The same is true for the well, the road, and the drive. If we want to portray these features, we have to find some consistent technique for ranking the importance of these features in the raster display. What happens when same-cell features are equally important? Generally, there are four methods of assigning cell attributes:

1. *Presence or absence:* first, we have to determine whether or not a specific feature is present within the area of the cell; a simple Boolean operator can be used in the software to effect this yes/no operation. This technique permits a quick search for specific features.
2. *Centerpoint:* which feature is closest to the cell centerpoint?
3. *Dominant feature:* which feature, if any, is more dominant within the cell (i.e., occupies more than 50 percent of the grid cell)?
4. *Precedence:* the rank of a feature within a cell is considered; that is, which of the captured features are more important? In Figure 7.35, the road, stream, drive, house, and well are more important features than the surrounding grass and orchard. The first five features are more dominant in the cells and closer to many of the cell's centers.

To convert from raster to vector, the cell centerpoints are often identified and then joined, resulting in the stepped appearances displayed in Figure 7.36(c) and (d). As noted earlier, the lower the resolution, the more stepped is the appearance and the less realistic is the portrayal of that specific feature. As line smoothing techniques are used to improve the stepped display appearance in raster models, errors are introduced that may be quite significant. Figure 7.36(a) and (b) shows the stream in topographic and vector modeling—a much smoother presentation. If the stream depictions shown in Figure 7.36(c) and (d) were smoothed, the result would not realistically reflect the poorer resolutions of the data plots.

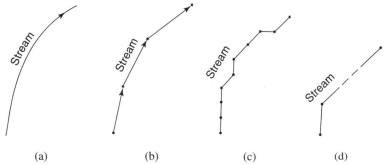

FIGURE 7.36 Stream feature depictions, using four models. (a) Topographic map plot: smoothed plot of data capture. (b) Vector model plot: coordinated tie-points are plotted. (c) Moderate resolution: raster model plot. (d) Coarse resolution: raster model plot. Raster plots are the result of joining the stream feature grid cell's midpoints. Illustrations (a)–(d) are derived from illustrations (a)–(d) of Figure 7.35.

7.16 Topology

Topology describes the relationships among geographic entities (polygons, lines, and points). The relationships may be spatial in nature (proximity, adjacency, connectivity) or the relationships may be based on entity attributes (e.g., tree species planted or harvested within a certain time period). Topology gives GIS its ability to analyze geospatial data. Topological structure is comprised of arcs and nodes. Arcs are one or more line segments that begin and end at a node. Nodes are the points of intersection of arcs, or the terminal points of arcs. Polygons are closed figures consisting of a series of three or more connected chains.

Topology does not define geometric relationships, only relational relationships. The example often used for this distinction is the use of rubber sheeting techniques that join surfaces originally tied to different projections or coordinate grids, where surfaces are stretched to fit with other surfaces. This stretching process may affect distances and angles between entities but will not affect the relational (adjacency, networking, etc.) characteristics of the features.

Topology gives a vector GIS its ability to permit querying of a database and thus differentiates GIS from simple CAD systems, which merely show spatial location of entities and their attributes, not their relationships with each other. Topological relationships include:

- *Connectivity:* used to determine where (e.g., at which node) chains are connected and to give a sense of direction among connected chains by specifying "to nodes" and "from nodes."
- *Adjacency:* used to determine what spatial features (points, lines, and areas) are adjacent to chains and polygons. The descriptions "left" and "right" can be applied once direction has been established by defining to nodes and from nodes.
- *Containment:* used to determine which spatial features (points, lines, and smaller polygons) are enclosed within a specified polygon.

Entities such as polygons, chains, and nodes can have a wide variety of relational characteristics. For example, a polygon defining the location of a stand of spruce trees can overlap or be adjacent to a polygon defining a stand of pine trees (see Figure 7.37). This simple two-attribute data set can be queried to identify six possibilities: those areas

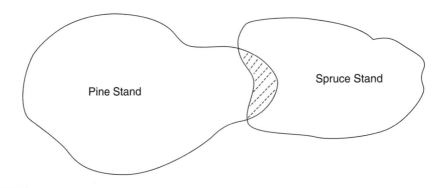

Choices

1. Pine
2. Spruce
3. Pine, Without Spruce
4. Spruce, Without Pine
5. Coniferous (Spruce and Pine)
6. Only Spruce and Pine Together

Boolean Operators

1.	Union		Pine and Spruce
2.	Minus		Pine but Not Spruce
3.	Intersect		Only Spruce and Pine Together
4.	Difference		Only Pine and Only Spruce (No Mixture)

FIGURE 7.37   Polygon analysis.

(polygons) that contain pine trees, spruce trees, only pine trees, only spruce trees, coniferous (spruce and pine) trees, or just that area of overlap containing both pine and spruce trees. Figure 7.37 also shows the Boolean operators that can be used to effect the same type of analysis. Also, if the appropriate data have been collected and input, the database can be further queried to ascertain attribute data, such as the age of various parts of each tree stand, tree diameters, and a potential selected harvest (e.g., board feet of lumber).

7.17   Remote Sensing Internet Web Sites and Further Reading

7.17.1   Satellite Web Sites

ENVISAT   *http://envisat.esa.int/*
Ikonus, IRS, Landsat 7   *www.intecamericas.com/satprices.htm*
IRS   *www.isro.org/programmes.htm*
JERS, Remote Sensing Technology Center of Japan   *www.restec.or.jp/restec_e.html*
Landsat 7   *http://geo.arc.nasa.gov/sge/landsat/landsat.html*

OrbView   *www.orbital.com/*   *www.orbimage.com*
QuickBird   *www.digitalglobe.com/*
Radarsat   *www.rsi.ca/*
Space Imaging   *http://spaceimaging.com*
SPOT   *www.spotimage.fr/html/_167_.php*
Terra (EO-1)   *http://asterweb.jpl.nasa.gov/*

7.17.2   Airborne Imagery Web Sites

Lidar links   *http://www.3dillc.com/rem-lidar.html*
Optech (airborne laser terrain mapper [ALTM])   *www.optech.on.ca*
Scanning Hydrographic Operational Airborne Lidar Survey (**SHOALS**) system,
   U.S. Army Corps of Engineers (USACE)   *http://shoals.sam.usace.army.mil/*

7.17.3   General Reference Web Sites

American Society for Photogrammetry and Remote Sensing (ASPRS)
   www.asprs.org
Australian Surveying and Land Information Group   *http://www.auslig.gov.au/*
   acres/facts.htm
European Space Agency   *www.esa.int/esacp/index.html*
Landsat 7   *http://landsat.usgs.gov*
NASA, EROS Data Center   *http://edcwww.cr.usgs.gov*
Natural Resources, Canada   *www.nrcan_rncan.gc.ca/inter/index_e.php*
Natural Resources Canada Tutorial
   http://www.ccrs.nrcan.gc.ca/ccrs/learn/learn_e.html
Remote Sensing Tutorial (NASA)   *http://rst.gsfc.nasa.gov*

7.17.4   Further Reading

American Society for Photogrammetry and Remote Sensing, *Digital Photogrammetry: An Addendum to the Manual of Photogrammetry* (Bethesda, Maryland: ASPRS, 1996).

Anderson, Floyd M., and Lewis, Anthony J. (eds.), *Principles and Applications of Imaging Radar, Manual of Remote Sensing,* Third Edition, Volume 2 (New York: Wiley, 1998).

Jensen, John R., *Remote Sensing of the Environment,* An Earth Resource Perspective, Prentice Hall Series in Geographic Information Science (Upper Saddle River, New Jersey: Prentice Hall, 2000).

Lillesand, Thomas M., and Kiefer, Ralph W., *Remote Sensing and Image Interpretation,* Fourth Edition (New York: John Wiley and Sons, 2000).

Wolf, Paul R., and Dewitt, Bob A. *Elements of Photogrammetry,* Third Edition (New York: McGraw Hill, 2000).

Review Questions

7.1. What are the chief differences between maps based on satellite imagery and maps based on airborne imagery?

7.2. What are the advantages inherent in the use of hyperspectral sensors over multispectral sensors?

7.3. Describe all the types of data collection that can be used to plan the location of a highway/utility corridor spanning several counties.

7.4. How does the use of lidar imaging enhance digital imaging?

7.5. Why is radar imaging preferred over passive sensors for arctic and antarctic data collection?

7.6. Compare and contrast the two techniques of remote sensing—aerial and satellite imagery—by listing the possible uses for which each technique is appropriate. Use the comparative examples shown in Figures H.5 to H.8 as a basis for your response.

7.7. Why can't aerial photographs be used for scaled measurements?

7.8. What is the chief advantage of using digital cameras over film-based cameras?

7.9. What effects does aircraft altitude have on aerial imagery?

7.10. Why are overlaps and sidelaps designed into aerial photography acquisition?

Problems

A topographic survey was performed on a tract of land, leveling techniques were used to obtain elevations, and total station techniques were used to locate the topographic detail. Figure 7.38 shows the traverse (A to G) used for survey control and the grid baseline (0 + 00 @ A) used to control the leveling survey. Offset distances are at 90° to the baseline. All the necessary information is provided, including:

(a) Bearings and lengths of the traverse sides.

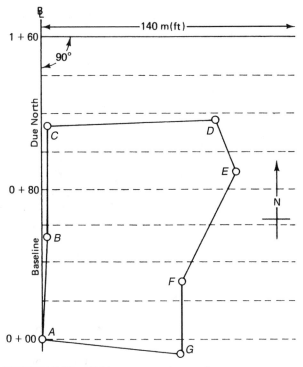

FIGURE 7.38   Grid traverse and control.

Table 7.7 SURVEYING GRID ELEVATIONS

STA.	₵	20 m (ft) E	40 m (ft) E	60 m (ft) E	80 m (ft) E	100 m (ft) E	120 m (ft) E	140 m (ft) E
1 + 60	68.97	69.51	70.05	70.53	70.32			
1 + 40	69.34	69.82	71.12	71.00	71.26	71.99		
1 + 20	69.29	70.75	69.98	71.24	72.07	72.53	72.61	
1 + 00	69.05	71.02	70.51	69.91	72.02	73.85	74.00	75.18
0 + 80	69.09	71.90	74.13	71.81	69.87	71.21	74.37	74.69
0 + 60	69.12	70.82	72.79	72.81	71.33	70.97	72.51	73.40
0 + 40	68.90	69.66	70.75	72.00	72.05	69.80	71.33	72.42
0 + 20	68.02	68.98	69.53	70.09	71.11	70.48	69.93	71.51
0 + 00 @STA.A	67.15	68.11	68.55	69.55	69.92	71.02		

Table 7.8 BALANCED TRAVERSE DATA

Course	Bearing	Distance, m(ft)
AB	N 3°30′ E	56.05
BC	N 0°30′ W	61.92
CD	N 88°40′ E	100.02
DE	S 23°30′ E	31.78
EF	S 28°53′ W	69.11
FG	South	39.73
GA	N 83°37′ W	82.67

Table 7.9 STATION COORDINATES

	Northing	Easting
A	1000.00	1000.00
B	1055.94	1003.42
C	1117.86	1002.88
D	1120.19	1102.87
E	1091.05	1115.54
F	1030.54	1082.16
G	990.81	1082.16

(b) Grid elevations.

(c) Angle and distance ties for the topographic detail.

(d) Northings and eastings of the stations.

Distances used can be in either foot units or metric units. See Figure 7.38 and Tables 7.7 to 7.10.

7.1. Establish the grid, plot the elevations (the decimal point is the plot point), and interpolate the data to establish contours at 1 m (ft) intervals. Scale at 1:500 for metric units or 1 in. = 10 ft for foot units. Use pencil.

7.2. Compute the interior angles of the traverse and check for geometric closure [i.e., $(n - 2)180°$].

7.3. Plot the traverse using the interior angles and the given distances or using the coordinates. Scale as in Problem 7.1.

7.4. Plot the detail using the plotted traverse as control. Scale as in Problem 7.1.

7.5. Determine the area enclosed by the traverse in ft^2 or m^2 using one or more of the following methods.

(a) Use grid paper as an overlay or underlay. Count the squares and partial squares enclosed by the traverse. Determine the area represented by one square at the chosen scale and, from that relationship, determine the area enclosed by the traverse.

(b) Use a planimeter to determine the area.

(c) Divide the traverse into regularly shaped figures (squares, rectangles, trapezoids, triangles) using a scale to determine the figure dimensions. Calculate the areas of

Table 7.10 SURVEY NOTES

STA.	Horizontal Angle	Distance, m(ft)	Description
	π STA. B (SIGHT C, 0°00′)		
1	8°15′	45.5	S. limit of treed area
2	17°00′	57.5	S. limit of treed area
3	33°30′	66.0	S. limit of treed area
4	37°20′	93.5	S. limit of treed area
5	45°35′	93.0	S. limit of treed area
6	49°30′	114.0	S. limit of treed area
	π @ STA. A (SIGHT B, 0°00°)		
7	50°10′	73.5	₵ gravel road (8 m ± width)
8	50°10′	86.0	₵ gravel road (8 m ± width)
9	51°30′	97.5	₵ gravel road (8 m ± width)
10	53°50′	94.5	N. limit of treed area
11	53°50′	109.0	N. limit of treed area
12	55°00′	58.0	₵ gravel road
13	66°15′	32.0	N. limit of treed area
14	86°30′	19.0	N. limit of treed area
	π @ STA. D (SIGHT E, 0°00′)		
15	0°00′	69.5	₵ gravel road
16	7°30′	90.0	N. limit of treed area
17	64°45′	38.8	N.E. corner of building
18	13°30′	75.0	N. limit of treed area
19	88°00′	39.4	N.W. corner of building
20	46°00′	85.0	N. limit of treed area

the individual figures, and sum them to produce the overall traverse area.

(d) Use the given balanced traverse data and the technique of coordinates to compute the traverse area.

A highway is to be constructed so that it passes through points A and E of the traverse. The proposed highway ₵ grade is +2.30 percent rising from A to E (₵ elevation at A = 68.95). The proposed cut-and-fill sections are shown in Figure 7.39(a) and (b).

7.6. Draw profile A–E, showing both the existing ground and the proposed ₵ of the highway. Use the following scales. Metric: horizontal, 1:500; vertical, 1:100. Foot: horizontal, 1 in = 10 ft or 15 ft; vertical, 1 in = 2 ft or 3 ft.

7.7. Plot the highway ₵ and 16 m (ft) width on the plan. Show the limits of cut and fill on the plan. Use the sections shown in Figure 7.39.

7.8. Combine Problems 7.1, 7.3, 7.4, 7.6, and 7.7 on one sheet of drafting paper. Arrange the plan and profile together with the balanced traverse data and a suitable title block. All line work is to be in ink. Use C (Imperial) or A2 (metric) size paper.

7.9. Calculate the flying heights and altitudes, given the following information:
(a) Photographic scale = 1:20,000, lens focal length = 153 mm, elevation of mean

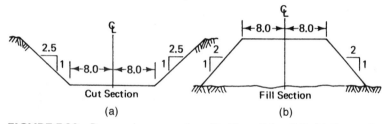

FIGURE 7.39 Proposed cross sections (Problems 7.6 to 7.8). (a) Cut section. (b) Fill section.

datum = 180 m.
    (b) Photographic scale 1 in. = 20,000 ft, lens focal length = 6.022 in., elevation of mean datum = 520 ft.

7.10. Calculate the photo scales, given the following data:
    (a) Distance between points A and B on a topographic map. Scale 1:50,000 = 4.75 cm, distance between same points on air photo = 23.07 cm.
    (b) Distance between points C and D on topographic map. Scale 1:100,000 = 1.85 in., distance between same points on air photo = 6.20 in.

7.11. Calculate the approximate numbers of photographs required for stereoscopic coverage (60 percent forward overlap and 25 percent side overlap) for each of the following conditions:
    (a) Photographic scale = 1:30,000; ground area to be covered is 30 km by 45 km.
    (b) Photographic scale 1:15,000; ground area to be covered is 15 miles by 33 miles.
    (c) Photographic scale is 1 in = 500 ft; ground area to be covered is 10 miles by 47 miles.

7.12. Calculate the dimensions of the area covered on the ground in a single stereo model having a 60 percent forward overlap, if the scale of the photograph (9″ by 9″ format) is:
    (a) 1: 10,000.
    (b) 1 in. = 400 ft.

7.13. Calculate how far the camera would move during the exposure time for each of the following conditions:
    (a) Ground speed of aircraft = 350 km/h; exposure time = 1/100 s.
    (b) Ground speed of aircraft = 350 km/h; exposure time = 1/1,000 s.
    (c) Ground speed of aircraft = 200 miles/h; exposure time = 1/500 s.

Chapter 8

Satellite Positioning

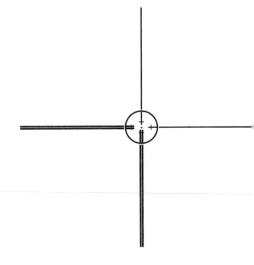

8.1 General Background

Satellite positioning is comprised of three satellite constellations. The oldest, and the only complete system at present (2005), is the U.S. Navigation Satellite Timing and Ranging (NAVSTAR) positioning system, generally referred to as the global positioning system (GPS). The Russian system, GLONASS, is partially completed, and the new European Union (EU) thirty-satellite system, *Galileo,* will be completed by 2008 or earlier. On December 10, 2004, the EU Transport ministers gave final approval to build and deploy the Galileo satellite system. (See *http://europa.eu.int/comm/dgs/energy_transport/galileo/index_en.htm* for current information on Galileo's deployment.)

8.1.1 Existing Systems

By the mid-1980s, the U.S. Department of Defense (DOD) second-generation guidance system—Navigation Satellite Timing and Ranging (NAVSTAR) global positioning system (GPS)—had evolved to many of its present capabilities. In December 1993, the U.S. government officially declared that the system had reached its initial operational capability (IOC) with twenty-six satellites (twenty-three Block II satellites) then potentially available for tracking. Additional Block II satellites continue to be launched (a satellite's life span is thought to be about seven to ten years). Current GPS satellite status and the constellation configuration can be accessed via the Internet at *http://www.navcen.uscg.gov/ftp/ gps/status.txt/* (see Table 8.1). Table 8.2 shows satellite status reports for the Russian GLONASS system (*http://www.glonass-center.ru/nagu.txt*). Some GPS receivers can track both GPS and GLONASS satellite signals, thus providing the potential for improved positioning.

Prior to NAVSTAR, precise positioning was often determined by using low-altitude satellites or inertial guidance systems. The first-generation satellite positioning system,

Table 8.1 GPS STATUS REPORT*

```
GPS OPERATIONAL ADVISORY        038
SUBJ: GPS STATUS          07 FEB 2005
1. SATELLITES, PLANES, AND CLOCKS (CS = CESIUM RB = RUBIDIUM):
A. BLOCK I:     NONE
B. BLOCK II:    PRNS    1,   2,   3,   4,   5,   6,   7,   8,   9,   10,  11,  13,  14,  15
   PLANE  :     SLOT    F6,  D7,  C2,  D4,  B4,  C1,  C4,  A3,  A1,  E3,  D2,  F3,  F1,  D5
   CLOCK  :             CS,  RB,  CS,  RB,  CS,  RB,  RB,  CS,  CS,  CS,  RB,  RB,  RB,  CS
   BLOCK II:    PRNS    16,  17,  18,  19,  20,  21,  22,  23,  24,  25,  26,  27,  28,  29
   PLANE  :     SLOT    B1,  D6,  E4,  C3,  E1,  D3,  E2,  F4,  D1,  A2,  F2,  A4,  B3,  F5
   CLOCK  :             RB,  RB,  RB,  RB,  RB,  RB,  RB,  RB,  CS,  CS,  RB,  RB,  RB,  RB
   BLOCK II:    PRNS    30,  31
   PLANE  :     SLOT    B2,  C5
   CLOCK  :             RB,  RB
```

*Daily updates are available at *www.navcen.uscg.gov/ftp/gps/status.txt.*

Table 8.2 GLONASS STATUS REPORT (AS OF FEBRUARY 8, 2005)*

GLONASS Number	Cosmos Number	Plane/ Slot	Frequency Channel	Launch Date	Introduction Date	Status	Outage Date
796	2413	1/01	02	26.12.2004	06.02.2005	Operating	
794	2402	1/02	04	10.12.2003	02.02.2004	Operating	
789	2381	1/03	12	01.12.2001	04.01.2002	Operating	
795	2403	1/04	06	10.12.2003	30.01.2004	Operating	
711	2382	1/05	02	01.12.2001	15.04.2003	Operating	
701	2404	1/06	01	10.12.2003	09.12.2004	Unusable	21.01.2005
712	2411	1/07		26.12.2004			
797	2412	1/08	06	26.12.2004	06.02.2005	Operating	
787	2375	3/17	05	13.10.2000	04.11.2000	Operating	
783	2374	3/18	10	13.10.2000	05.01.2001	Operating	
792	2395	3/21	05	25.12.2002	31.01.2003	Operating	
791	2394	3/22	10	25.12.2002	10.02.2003	Operating	
793	2396	3/23	11	25.12.2002	31.01.2003	Operating	
788	2376	3/24	03	13.10.2000	21.11.2000	Operating	

*All the dates (DD.MM.YY) are given at Moscow time (UTC + 0300).
*Daily updates are available at *www.glonass-center.ru/nagu.txt.*

called TRANSIT, consisted of six satellites in polar orbit at an altitude of only 1,100 km. Precise surveys, with positioning from 0.2 to 0.3 m, could be accomplished using translocation techniques; that is, one receiver occupied a position of known coordinates while another occupied an unknown position. Data received at the known position were used to model signal transmission and determine orbital errors, thus permitting more precise results. The positioning analysis techniques used in the TRANSIT system utilized a ground receiver capable of noting the change in satellite frequency transmission as the satellite first approached and then receded from the observer. The change in frequency was affected by the velocity of the satellite itself.

The change in velocity of transmissions from the approaching and then receding satellite, known as the Doppler effect, is directly proportional to the shift in frequency of the transmitted signals and thus is proportional to the change of distance between the satellite and the receiver over a given time interval. When the satellite's orbit is known precisely and the position of the satellite in that orbit is also known precisely through ephemeris data and universal time (UT), the position of the receiving station can be computed.

8.1.2 Inertial Measuring Systems

Another, quite different positioning system utilizes inertial measurement units (IMUs) and requires a vehicle (truck, airplane, or helicopter) or a surveyor on foot or in an all-terrain vehicle to occupy a point of known coordinates (northing, easting, and elevation) and remain stationary for a zero-velocity update. As the IMU moves, its location is updated constantly by three computer-controlled accelerometers, each aligned to the north-south, east-west, or vertical axis. The accelerometer platform is oriented north-south, east-west, and plumb by means of three computer-controlled gyroscopes, each of which is aligned to one of three axes; see Figure 8.1(a). Analysis of acceleration data gives rectangular (latitude and longitude) displacement factors for horizontal movement, in addition to vertical displacement.

An IMU used in conjunction with real-time kinematic (RTK) GPS yields several benefits (see Section 8.11.4). First, the roving surveyor, using both IMU and GPS receivers, determines position readily, even if the GPS signals are temporarily blocked by tree canopy or other obstructions. After the obstruction has been bypassed, the rover can continue to receive both IMU and GPS signals, with the IMU readily updating its position from the newly acquired GPS signals. This technique permits positioning to continue under tree canopies (where GPS signals are blocked) and even permits indoor surveying. Second, the combination of IMU and GPS can be used to control data collection for airborne imagery, including **lidar** (**l**ight **d**etection **a**nd **r**anging) images (See Section 7.7.1.1). The presence of IMU data along with the GPS positioning data makes the processing of all collected data possible in a very short period of time. The points within the lidar point cloud are individually referenced with respect to their three-dimensional geospatial positions. Applanix Corporation, a recent Trimble acquisition, is the present leader in this field. The corporation manufactures IMU/GPS positioning devices for use in a backpack, or on motorized vehicles, aircraft, or marine vessels. See Figure 8.1(b) for an illustration of this technology.

8.2 Global Positioning System

Satellite positioning is based on accurate ephemeris data for the real-time location of each satellite and on very precisely kept time. It uses satellite signals, accurate time, and sophisticated algorithms to generate distances to ground receivers and thus provide resectioned positions anywhere on earth. Satellite signals can also provide navigation data, such as the speed and direction of a mobile receiver; directions to an identified (via coordinates) location; and estimated arrival times.

Satellite orbits have been designed so that ground positioning can usually be determined at any location on earth at any time of the day or night. A minimum of four satellites must be tracked to solve the positioning intersection equations dealing with position (x, y, and

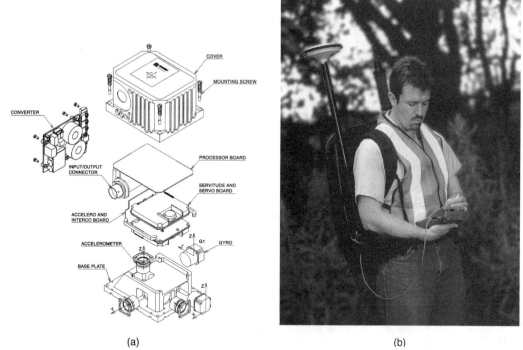

(a)   (b)

FIGURE 8.1 (a) Six degrees of freedom inertial measurement unit (IMU)-SAGEM 33 BM. (Courtesy of SAGEM SA, Paris-La Defense, France) (b) Combined IMU and RTK unit (see Section 8.11.4.) carried in backpack. This unit continues to determine position coordinates, even when GPS signals are temporarily blocked by tree canopy or other obstructions.

z coordinates, which later can be translated to easting, northing, and elevation) and with clock differences between the satellites and ground receivers. In reality, five or more satellites are tracked, if possible, to introduce additional redundancies and to strengthen the geometry of the satellite array. Additional satellites (more than the required minimum) can provide more accurate positioning and can also reduce the receiver occupation time at each survey station.

The U.S. global positioning system (GPS) consists of twenty-four operational satellites (plus spares) deployed in six orbital planes 20,200 km above the earth, with orbit times of twelve hours. Spacing is such that at least six satellites are visible anywhere on earth. In addition to the satellites arrayed in space, the U.S. GPS includes five tracking stations (and three ground antennas) evenly spaced around the earth. Stations are located at Colorado Springs, Colorado (the master control station), and on the islands of Ascension, Diego Garcia, Kwajalein, and Hawaii. All satellites' signals are observed at each station, with all the clock and ephemeris data being transmitted to the control station at Colorado Springs. The system is kept at peak efficiency because corrective data are transmitted back to the satellites from Colorado Springs (and a few other ground stations) every few hours. More recently, satellites have also been given the ability to communicate with each other.

Originally designed for military guidance and positioning, GPS has quickly attracted a wide variety of civilian users in the positioning and navigation applications fields. Additional applications have already been developed in commercial aviation navigation, boating and shipping navigation, trucking and railcar positioning, emergency routing, automobile dashboard electronic charts, and orienteering navigation (see Section 8.12). Also, manufacturers are now installing GPS chips in cell phones to help satisfy the 911-service requirement for precise caller locations.

8.3 Receivers

GPS receivers range in ability (and cost) from survey-level (millimeter) receivers capable of use in surveys requiring high accuracy and costing more than $20,000, to mapping and geographic information system (GIS) receivers (submeter accuracy) costing about $3,000 each, to marine navigation receivers (accuracy 1 to 5 m) costing about $1,000, and finally to orienteering (hiking) and low-precision mapping/GIS receivers costing only a few hundred dollars. See Figures 8.2 to 8.6.

The major differences in the receivers are the number of channels available (the number of satellites that can be tracked at one time) and whether or not the receiver can observe both L1 and L2 frequencies; code phase and carrier phase may also be measured. Generally speaking, the higher-cost dual-frequency receivers require much shorter observation times for positioning measurements than do the less-expensive single-frequency receivers and can be used for real-time positioning. Some low-end general-purpose GPS receivers track only

FIGURE 8.2 Zepher™ geodetic GPS antenna, with a 5700 GPS receiver and radio communications equipment. Zepher antennas are said to have accuracy potentials similar to those of choke-ring antennas—at lower costs. (Courtesy of Trimble)

Chap. 8   Satellite Positioning

FIGURE 8.3 Leica SYS 300. System includes SR9400 GPS receiver (single frequency) with AT201 antenna, CR333 GPS controller (electronic field book) that collects data and provides software for real-time GPS surveying, and a radio modem. It can provide the following accuracies: 10 mm to 20 mm + 2 ppm using differential carrier techniques; and .30 m to .50 m using differential code measurements. (Courtesy of Leica Geosystems Inc.)

one channel at a time (sequencing from satellite to satellite as tracking progresses). An improved low-end, general-purpose receiver tracks on two channels, but it still must sequence the tracking to other satellites to achieve positioning. Some low-end surveying receivers can continuously observe on five channels (sequencing not required), whereas some high-end surveying receivers can observe on twelve channels. Some receivers can control photogrammetric camera operation; some receivers can datalog every second, while others datalog every fifteen seconds. The more expensive receivers can be used in all GPS survey modes with shorter observation times, whereas less expensive receivers can be restricted to a certain type of survey and require longer observation times and perhaps longer processing times as well.

The cost for three precise surveying receivers, together with appropriate software, ranges up from $50,000, but lower costs are expected as production increases and technology improves. In contrast, two lower-order receivers can be used in differential positioning to determine position to within a few meters—a precision that could be acceptable for selected mapping and GIS databases. The total cost (including software) of this system can be as low as $7,000. As noted in Section 8.6, surveys can be performed while using only one receiver if access to radio-transmitted corrections from a permanent receiver/beacon is available, which is the case with the U.S. Coast Guard's differential global positioning system (DGPS) and the national differential global positioning system (NDGPS); see Section 8.6.2.

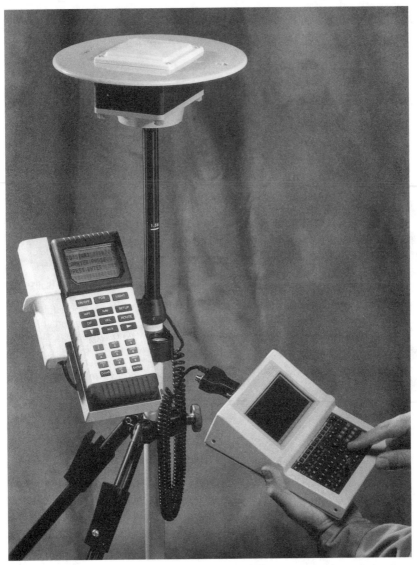

FIGURE 8.4 Geodetic-quality dual-frequency GPS receiver with a choke-ring antenna. (Courtesy of Trimble)

8.4 Satellite Constellations

GPS satellites (Figure 8.7) are manufactured by Rockwell International, weigh about 1,900 pounds, span 17 feet (with solar panels deployed), and orbit the earth at 10,900 nautical miles (20,000 km) in a period of 12 hours (actually 11 hours, 58 minutes). The satellites' constellation (see Figure 8.8) consists of twenty-four satellites (plus spares) placed in six orbital planes. This configuration ensures that at least four satellites (the minimum number needed for precise measurements) are always potentially visible anywhere on earth.

FIGURE 8.5 Trimble navigation total station GPS. Real-time positioning for a wide variety of applications. Features include GPS antenna and electronic field book at the adjustable pole; receiver, radio, and radio antenna in the backpack. The receiver has been equipped with GPS processing software, and the electronic field book has been equipped with the applications software (for example, layout), thus permitting real-time data capture or layout. (Courtesy of Trimble Navigation, Sunnyvale, Calif.)

Russia has also created a satellite constellation, called GLONASS, which was originally planned to consist of twenty-four satellites and is placed in three orbital planes at 19,100 km, with an orbit time of 11 hours, 15 minutes. As with GPS, GLONASS satellites continuously broadcast their own precise positions along with less precise positions for all other same-constellation satellites. Tables 8.1 and 8.2 showed the status for both GPS and GLONASS satellites, respectively. Daily updates of satellite status are available on the Internet at the web sites shown in the tables. Note that in February 2005, there were thirty-one operating satellites in the GPS constellation and twelve operating satellites in the GLONASS constellation. From time to time, operating satellites are set unhealthy (that is, temporarily taken out of service) to make adjustments and recalibrations; knowing the health of the visible satellites on any given day is critical to mission planning.

Some GPS receivers are now capable of tracking both GPS and GLONASS satellite signals. Using both constellations has two additional advantages. The increased number of visible satellites results in the potential for shorter observation times and increased accuracy. The increased number of satellites available to a receiver can also mean that

FIGURE 8.6 Garmin eMap handheld GPS used in navigation and GIS-type data location. It can be used with a DGPS radio beacon receiver to improve accuracy to within a few meters or less.

surveys interrupted by poor satellite geometry or by local obstructions (buildings, tree canopy, and the like) can continue uninterrupted by also tracking satellites from the second constellation.

A new satellite positioning constellation called Galileo is being implemented by the European Union (EU). It will be comprised of twenty-seven operational satellites plus three spares. Satellite construction contracts were awarded beginning in 2004, and the constellation is expected to be operational by 2008 or earlier. Galileo satellites will broadcast on four frequencies:

- E1: 1587 to 1591 MHz; OS, SOL, and CS (see below)
- E2: 1559 to 1563 MHz; PRS
- E5: 1164 to 1188 MHz; OS, SOL, and CS
- E6: 1260 to 1300 MHz; PRS and CS

The services provided by the Galileo system include open services (OS)—free of charge; public regulated service (PRS) (e.g., police, emergency, national security); open and safety

FIGURE 8.7   GPS satellite. (Courtesy of Leica Geosystems Inc.)

FIGURE 8.8   GPS satellites in orbit around the earth.

of life (SOL) services; commercial services (CS), which are available for a fee; and international search and rescue (SAR).

Galileo's satellites will be in circular orbits 23,617 km above the earth, and the system will be under civilian control. Galileo is designed primarily for transportation (rail, auto, and air) and is Europe's contribution to the global navigation satellite system (GNSS); that is, GNSS will be comprised of Galileo + GLONAS + GPS. Galileo will be managed by the Galileo Joint Undertaking, located in Brussels, Belgium. See their web site at *http://europa.eu.int/comm/dgs/energy_transport/galileo/index_en.htm*. In addition to providing positioning data, GNSS provides real-time navigation data for land, marine, and air navigation. Present-day accuracies are now sufficient for most applications, and the Federal Aviation Administration (FAA) is expected to provide certification for precision aircraft flight approaches about the same time as this text is published (2006).

8.5   GPS Satellite Signals

GPS satellites transmit at two L-band frequencies with the following characteristics:

1. The L1 frequency is set at 1575.42 MHz, and because $\lambda = c/f$ (see Equation 2.4), where λ is the wavelength, c is the speed of light, and f is the frequency, the wavelength of L1 is determined as follows:

$$\lambda = \frac{300,000,000 \text{ m/s}}{1,575,420,000 \text{ Hz}} = 0.190 \text{ m}$$

or the wavelength is about 19 cm. The L1 band carries the following codes: C/A code (civil), P code (military), Y code (antispoofing code), and the navigation message, which includes clock corrections and orbital data.

2. The L2 frequency is set at 1227.60 MHz and, using Equation 2.4, the wavelength is determined to 0.244 m, or about 24 cm. The L2 band codes are the same as for L1, except that the CA code is not available.

The civil—or coarse acquisition (C/A)—code is available to the public, whereas the P code was designed for military use exclusively. Only the P code was originally modulated on the L2 band. The DOD originally implemented selective availability (SA), which was designed to degrade signal accuracy (in times of national emergency), resulting in errors in the range of 100 m, and which occurred on both L1 and L2. SA has been permanently deactivated as of May 2000. This policy change improved GPS accuracies in basic point positioning by a fivefold factor (e.g., down from 100 m to about 10 to 20 m). Another method designed to deny accuracy to the user is called antispoofing (AS). Antispoofing occurs when the P code is encrypted to prevent tinkering by hostile forces. At present, AS, along with most natural and other errors associated with the GPS measurements, can be eliminated by using differential global positioning service (DGPS).

The more types of data collected, the faster and more accurate the solutions. For example, some GPS suppliers have computer programs designed to deal with some or all of the following collected data: C/A code pseudorange on L1, both L1 and L2 P code pseudoranges, both L1 and L2 carrier phases, both L1 and L2 Doppler observations, and a measurement of the cross-correlation of the encrypted P code (Y code) on the L1 and L2 frequencies to produce real-time or postprocessing solutions.

The United States announced in early 1999 its intention to modernize the GPS system and to add two additional civilian signals on the next generation of satellites. The second civil signal (L2C) will be added to the L2 signal of future satellites. A third civil signal, L5, will be included on the new Block IIF satellites, scheduled for launch beginning in 2007. The L5 civil signal will transmit on a frequency of 1176.45 MHz. The addition of these two new civil signals will make positioning faster and more accurate under many measuring conditions and will make the GPS system comparable to the proposed Galileo system for civilian applications. More advanced Block III satellites will gradually replace the present Block II satellites in the GPS constellation. Some believe that the L2C and L5 GPS modernizations, as well as the construction of the Block III GPS constellation, are still several years away. Perhaps it will be 2010 before we see a modernized and fully operational GPS constellation complemented by fully operational GLONASS and Galileo constellations.

One key dimension in positioning is the parameter of time. Time is kept on-board the satellites by so-called atomic clocks with a precision of 1 nanosecond (.0000000001 s). The ground receivers are equipped with less precise quartz clocks. Uncertainties caused by these less precise clocks are resolved when observing the signals from four satellites instead of the basic three-satellite configuration required for $x, y,$ and z positioning.

8.6   Position Measurements

Position measurements generally fall into one of two categories: code measurement and carrier phase measurement. Civilian code measurement is presently restricted to the C/A code, which can provide accuracies only in the range of 10 m to 15 m when used in point positioning, and accuracies in the submeter to 5-m range when used in various differential positioning techniques. P code measurements can apparently provide the military with much better accuracies. Point positioning is the technique that employs one GPS receiver to track satellite code signals so that it can directly determine the coordinates of the receiver station.

8.6.1   Relative Positioning

Relative positioning is a technique that employs two GPS receivers to track satellite code signals and/or satellite carrier phases to determine the baseline vector (X, Y, and Z) between the two receiver stations. The two receivers must collect data from the same satellites simultaneously (same epoch), and their observations are then combined to produce results that are superior to those achieved with just a single point positioning. Using this technique, computed (postprocessed) baseline accuracies can be in the submeter range for C/A code measurements and in the millimeter range for carrier phase measurements.

8.6.2   GPS Augmentation Services—Differential GPS

The theory behind differential positioning is based on the use of two or more GPS receivers to track the same satellites simultaneously. Unlike relative positioning, at least one of the GPS receivers must be set up at a station of known coordinates. As the base station receives the satellites' signals, it can compare its computed position to its known position and derive a difference value. Because this difference reflects all the measurement errors (except for multipath) for nearby receivers, this "difference" can be included in the nearby roving GPS receivers' computations to remove the common errors in their measurements. Differential corrections in positioning computations can be incorporated in several ways:

- Postprocessing computations, using data collected manually from the base station and rovers.
- A commercial service (e.g., OmniSTAR, *www.omniSTAR.com*) that collects GPS data at several sites and weights them to produce an optimized set of corrections that are then uplinked to geostationary satellites, which in turn broadcast the corrections

to subscribers, who pay an annual fee for this service. This service provides real-time measurements because the corrections are applied automatically to data collected by the subscribers' GPS roving receivers.

- A radio-equipped GPS receiver at one or more base stations (stations whose coordinates have already been precisely determined) that can compute errors and broadcast corrections to any number of radio-equipped roving receivers in the area, or it can transmit GPS signals to a central server so that the server station can then compute difference corrections (see Section 8.6.5) and transmit them to roving receivers. These techniques are extremely useful for real-time kinematic (RTK) measurements. Cell phones are also used for communications.

- Postprocessing computations using code range and carrier phase data available from the National Geodetic Survey's continuously operating reference system (CORS)— a nationwide network of stations (including links to the Canadian active control system). This data can be downloaded from the Internet at *www.ngs.noaa.gov/CORS* (see also Section 8.8).

- Real-time mapping-level measurements can also be made by accessing DGPS/NDGPS radio corrections that are broadcast from permanent GPS receiver locations established by the U.S Coast Guard and other cooperating agencies (see Sections 8.6.2.1 and 8.6.2.2). The roving receiver must be equipped with an ancillary or built-in radio beacon receiver.

Some GPS receivers are now manufactured with built-in radio receivers for use with each of the above real-time techniques. Roving receivers equipped with appropriate software can then determine the coordinates of the stations in real time.

The types of surveying applications where code pseudorange measurements or carrier measurements are made at a base station and then used to correct measurements made at another survey station are called **differential positioning**. Although the code measurement accuracies (submeter to 10 m) may not be sufficient for traditional control and layout surveys, they are ideal for many navigation needs and for many GIS and mapping surveys.

8.6.2.1 U.S. Coast Guard's DGPS Service.

The U.S. Coast Guard Maritime Differential GPS Service has created a differential global positioning system (DGPS) consisting of more than eighty remote broadcast sites and two control centers (as of 2005). It was originally designed to provide navigation data in coastal areas, the Great Lakes, and major river sites. This system received full operational capability (FOC) in 1999. The system includes continuously operating GPS receivers, at locations of known coordinates, that determine pseudorange values and corrections and then radio-transmit the differential corrections to working navigators and surveyors at distances ranging from 100 to 400 km. The system uses international standards for its broadcasts; the standards are published by the Radio Technical Commission for Maritime (RTCM) services. The surveyor or navigator, using the DGPS broadcasts along with his or her own receiver, has the advantages normally found when using two receivers. The corrections for many of the errors (orbital and atmospheric) are very similar for nearby GPS receivers; once the pseudoranges have been corrected, the accuracies thus become much improved (e.g., vertical accuracies are now in the range of 20 cm).

DGPS SITE STATUS AND OPERATING PARAMETERS
STATUS AS OF 08/02/05

```
Site Name:                   YOUNGSTOWN, NY
Status:                      Operational
RBn Antenna Location:        43° 13.8' N;78° 58.2' W
REFSTA Ant Location (A):     43° 13.8748' N;78° 58.20992' W
REFSTA Ant Location (B):     43° 13.87466' N;78° 58.18778' W
REFSTA RTCM SC-104 ID (A):   118
REFSTA RTCM SC-104 ID (B):   119
REFSTA FIRMWARE VERSION:     RD00-1C19
Broadcast Site ID:           839
Transmission Frequency:      322 KHZ
Transmission Rate:           100 BPS
Signal Strength:             75uV/m at 150 SM
Outages:
No Current Outages.
```

USERS SHOULD NOTIFY THE NIS WATCHSTANDER AT (555)313-5900 OF ANY OBSERVED OUTAGES, PROBLEMS, OR REQUESTS. ALL CURRENT OUTAGE INFORMATION WILL BE LISTED FOLLOWING EACH SITE.

THE COAST GUARD DGPS SERVICE IS AVAILABLE FOR POSITIONING AND NAVIGATION USERS MAY EXPERIENCE SERVICE INTERRUPTIONS WITHOUT ADVANCE NOTICE. COAST GUARD DGPS BROADCASTS SHOULD NOT BE USED UNDER ANY CIRCUMSTANCES WHERE A SUDDEN SYSTEM FAILURE OR INACCURACY COULD CONSTITUTE A SAFETY HAZARD.

NOTE: Differential corrections are based on the NAD 83 position of the reference station (REFSTA) antenna. Positions obtained using DGPS should be referenced to NAD 83 coordinate system only. All sites are broadcasting RTCM Type 9-3 correction messages.

FIGURE 8.9 Printout of status report for a DGPS station. Update status is available at *www. navcen.uscg.gov*. (Courtesy of the U.S. Coast Guard)

Information on DGPS and on individual broadcast sites can be obtained at the U.S. Coast Guard Navigation Center web site: *www.navcen.uscg.gov/*. Figure 8.9 shows typical data available for site 839, Youngstown, New York. See also Section 8.8 on the continuously operating reference station (CORS) network. Effective April 2004, the U.S. Coast Guard, together with the Canadian Coast Guard, implemented a seamless positioning service in the vicinity of their common border.

8.6.2.2 National DGPS (NDGPS).

The success of the U.S. Coast Guard's DGPS prompted the U.S. Department of Transportation (DOT) in 1997 to design a terrestrial expansion over the land surfaces of the conterminous or continental United States (CONUS) and the major transportation routes in Alaska and Hawaii. The U.S. Coast Guard (USCG) acts as the lead agency in this venture, which essentially consisted of expanding the maritime DGPS across all land areas. By 2005, about thirty-six NDGPS station were operating, with plans for complete coverage, consisting of a total of about 140 stations (DGPS and NDGPS), expected in the near future.

8.6.2.3   The Wide Area Augmentation System.

The Wide Area Augmentation System (WAAS) uses twenty-five ground stations for which the spatial coordinates have been determined precisely. The signals from GPS satellites are collected and analyzed by all stations in the network to determine errors (orbit, clock, atmosphere, etc.). The differential corrections are forwarded to one of the two master stations (one located on each U.S. coast). The master stations create differential correction messages that are then relayed to one of two geostationary satellites located near the equator. The WAAS geostationary satellites then rebroadcast the differential correction messages on the same frequency as that used by GPS—i.e., L1 at 1575.42 MHz—to receivers located on the ground, at sea, or in the air. This positional service is free; all you need is a WAAS-equipped receiver.

The WAAS, which is being developed by the Federal Aviation Administration (FAA) and the Department of Transportation (DOT), has not yet (as of 2005) met the criteria for certification in precision aircraft flight approaches. Satellite Pacific Ocean Region (POR) serves the Pacific Ocean region and western North America; satellite Atlantic Ocean Region–West (AOR-W) serves the Atlantic Ocean area and eastern North America, as well as South America. It is planned eventually to cover Europe with satellite Atlantic Ocean Region–East (AOR-E) and Satellite Indian Ocean Region (IOR), as well as the European Space Agency (ESA) satellite, called Aircraft-based Augmentation System (ARTEMIS), which permits GPS and GLONASS augmentation on-board the aircraft. International coverage will be expanded beyond the WAAS area of interest by the creation and development of Europe's Euro Geostationary Navigational Overlay Service (EGNOS) and Japan's Multi-Functional Satellite Augmentation System (MSAS).

8.6.3   Code Measurements

As previously noted, the military can utilize both the P code and the C/A code. The C/A code is used by the military to access the P code quickly. Until the L2C signal becomes operative, the civilian user must be content with using only the C/A code. Both codes are digital codes comprised of zeros and ones [see Figure 8.10(a)], and each satellite transmits codes unique to that satellite. Both codes are carefully structured, but because the codes sound like random electronic noise, they have been given the name pseudo random noise (PRN). The PRN code number indicates which of the thirty-seven seven-day segments of the P code PRN signal is presently being used by each satellite (it takes the P code PRN signal 267 days to transmit). Each one-week segment of the P code is unique to each satellite and is reassigned each week (see Table 8.1). Receivers have replicas of all satellite codes in the on-board memory, and they use these codes to identify the satellite and then to measure the time difference between the signals from the satellite to the receiver. Time is measured as the receiver moves the replica code (retrieved from memory) until a match between the transmitted code and the replica code is achieved [see Figure 8.10(b)]. Errors caused by the slowing effects of the atmosphere on the transmission of satellite radio waves can be corrected by simultaneously performing position measurements utilizing two different wavelengths, such as L1 and L2.

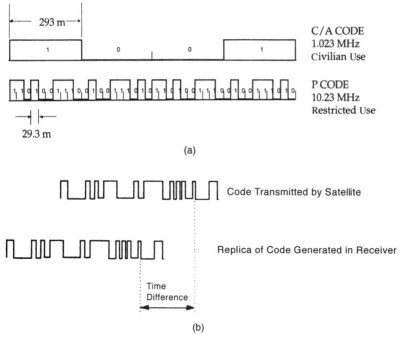

FIGURE 8.10   (a) C/A code and P code. (b) Time determination.

The distance, called pseudorange, is determined by multiplying the time factor by the speed of light:

$$\rho = t(300,000,000) \tag{8.1}$$

where ρ (lowercase Greek letter rho) = the pseudorange

t = travel time of the satellite signal, in seconds

300,000,000 = velocity of light in meters per second

(actually 299,792,458 m/s)

When the pseudorange (ρ) is corrected for clock errors, atmospheric delay errors, multipath errors, and the like, it is then called the range and is abbreviated by the uppercase Greek letter rho, P; that is, P = (ρ + error corrections).

Figure 8.11 shows the geometry involved in point positioning. Computing three pseudoranges (ρ_{AR}, ρ_{BR}, ρ_{CR}) is enough to solve the intersection of three developed spheres—although this computation gives two points of intersection. One point (a superfluous point) will obviously be irrelevant; that is, it probably will not even fall on the surface of the earth. The fourth pseudorange (ρ_{DR}) is required to eliminate the fourth unknown—the receiver clock error.

8.6.4   Carrier Phase Measurement

GPS codes, which are modulations of the carrier frequencies, are comparatively lengthy. Compare the C/A code at 293 m and the P code at 29.3 m (Figure 8.10) with the wavelengths of L1 and L2 at 0.19 m and 0.24 m, respectively. It follows that carrier phase measurements have the potential for much higher accuracies than do code measurements.

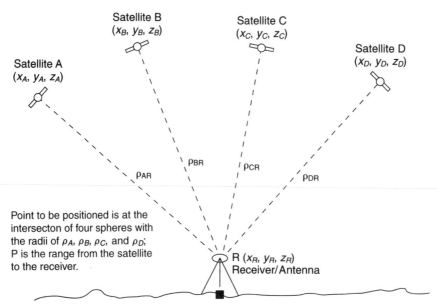

Satellite B
(x_B, y_B, z_B)

Satellite C
(x_C, y_C, z_C)

Satellite D
(x_D, y_D, z_D)

Satellite A
(x_A, y_A, z_A)

ρ_{BR}

ρ_{CR}

ρ_{AR}

ρ_{DR}

Point to be positioned is at the
intersecton of four spheres with
the radii of ρ_A, ρ_B, ρ_C, and ρ_D;
P is the range from the satellite
to the receiver.

R (x_R, y_R, z_R)
Receiver/Antenna

FIGURE 8.11 Geometry of point positioning.

We first encountered phase measurements in Chapter 2, when we observed how EDM equipment measured distances (see Figure 2.26). Essentially, EDM distances are determined by measuring the phase delay required to match the transmitted carrier wave signal with the return signal (two-way signaling), as we saw in Equation 2.5:

$$L = \frac{n\lambda + \varphi}{2}$$

where φ is the partial wavelength determined by measuring the phase delay (through comparison with an on-board reference), n is the number of complete wavelengths [from the electronic distance measurement (EDM) instrument to the prism and back to the EDM instrument], and λ is the wavelength. The integer number of wavelengths is determined as the EDM instrument successively sends out (and receives back) signals at different frequencies.

Because GPS ranging involves only one-way signaling, other techniques must be used to determine the number of full wavelengths. GPS receivers can measure the phase delay (through comparison with on-board carrier replicas) and count the full wavelengths after lock-on to the satellite has occurred, but more complex treatment is required to determine N, the initial cycle ambiguity. That is, N is the number of full wavelengths sent by the satellite prior to lock-on. Because a carrier signal is comprised of a continuous transmission of sinelike waves with no distinguishing features, the wave count cannot be accomplished directly. Equation 8.2 shows the following computation:

$$P = \varphi + N\lambda + \text{errors} \tag{8.2}$$

where P = satellite-receiver range
φ = measured carrier phase

λ = wavelength

N = initial ambiguity (the number of full wavelengths at lock-on)

Once the cycle ambiguity between a receiver and a satellite has been resolved, it does not have to be addressed further unless a loss of lock-on occurs between the receiver and the satellite. When loss of lock-on occurs, the ambiguity must be resolved again. Loss of lock-on results in a loss of the integer number of cycles and is called a cycle slip. The surveyor is alerted to loss of lock-on when the receiver commences a beeping sequence. As an example, loss of lock-on can occur when a roving receiver passes under a bridge, a tree canopy, or any other obstruction that blocks all or some of the satellite signals.

Cycle ambiguity can be determined through the process of differencing. GPS measurements can be differenced between two satellites, between two receivers, and between two epochs. An **epoch** is an event in time—a short observation interval in a longer series of observations. After the initial epoch has been observed, later epochs will reflect the fact that the constellation has moved relative to the ground station and thus presents a new geometrical pattern and consequently new intersection solutions.

8.6.5 Differencing

Relative positioning occurs when two receivers are used simultaneously to observe satellite signals and to compute the vectors (known as a baseline) joining the two receivers. Relative positioning can provide better accuracies because of the correlation possible between measurements made simultaneously over time by two or more different satellite receivers. Differencing is the technique of simultaneous baseline measurements and can be divided into three categories: single difference, double difference, and triple difference:

1. **Single difference:** When two receivers simultaneously observe the same satellite, it is possible to correct for most of the effects of satellite clock errors, orbit errors, and atmospheric delay. See Figure 8.12(a).

2. **Double difference:** When one receiver observes two (or more) satellites, the measurements can be freed of receiver clock error, and atmospheric delay errors can also be eliminated. When using both differences (between the satellites and between the receivers), a double difference occurs. Clock errors, atmospheric delay errors, and orbit errors can all be eliminated. See Figure 8.12(b).

3. **Triple difference:** The difference between two double differences is the triple difference; that is, the double differences are compared over two (or more) successive epochs. This procedure is also effective in detecting and correcting cycle slips and for computing first-step approximate solutions that are used in double-difference techniques. See Figure 8.12(c).

Cycle ambiguities can also be resolved by utilizing algorithms dealing with double differences and triple differences, both code and carrier phase measurements, and the methods of kinematic surveying that utilize start-ups of antenna swap, known location occupation, and on-the-fly (OTF) initialization. OTF initialization is a technique that can be used while the roving receiver is in motion (see Section 8.11.3). See also the references at the end of this chapter for more on the theory of signal observations and ambiguity resolution.

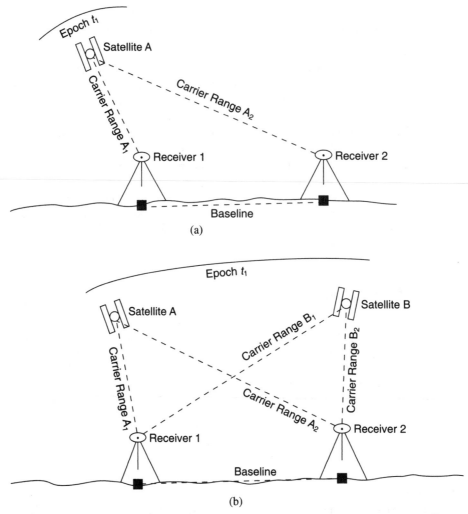

FIGURE 8.12 Differencing. (a) Single difference: two receivers observing the same satellite simultaneously (i.e., difference between receivers). (b) Double difference: two receivers observing the same satellite(s) simultaneously (i.e., difference between receivers and between satellites).

(continued)

8.7   Errors

The chief sources of error in GPS are described below:

1. Clock errors of the receivers.
2. Ionospheric refraction (occurring 50 to 1,000 km above earth) and tropospheric refraction (occurring from the earth's surface to 80 km above earth). Signals are slowed as they travel through these earth-centered layers. The errors worsen as satellite signals are received from directly overhead to down near the horizon.

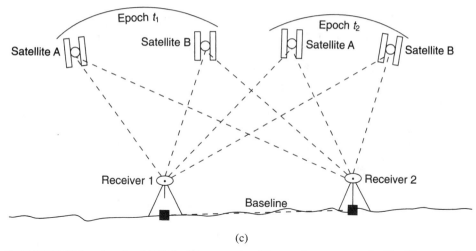

FIGURE 8.12 (*continued*)   (c) Triple difference: the difference between two double differences (i.e., differences among receivers, satellites, and epochs).

These errors can be reduced by scheduling nighttime observations, gathering sufficient redundant data and using reduced baseline lengths (1 to 5 km), or collecting data on both frequencies over long distances (20 km or more). Most surveying agencies do not record observations from satellites below 10° to 15° of elevation above the horizon.

3. Multipath interference, which is similar to the ghosting effect seen on TV. Some signals are received directly, and others are received after they have been reflected from adjacent features such as tall buildings, steel fences, etc. Recent improvements in antenna design have reduced these errors significantly. In data-collection surveys, roving receivers must be positioned according to the presence of topographic features, so the surveyor cannot avoid tall structures, chain link fences, and the like; when considering the placement of base station receivers for RTK surveys, however, the surveyor should make every effort to avoid all features that may contribute to multipath errors.

4. A weak geometric figure that results from poorly located, four-satellite signal intersections. This consideration is called the dilution of precision (DOP). DOP can be optimized if many satellites (beyond the minimum of four) are tracked; the additional data strengthen the position solution. Most survey-level receivers are now capable of tracking five to twelve satellites simultaneously. The geometric effect of satellite vector measurement errors together with receiver clock errors is called the **general dilution of precision (GDOP)**. Elevation solutions require a strong GDOP, and these solutions are strengthened when satellite elevations in excess of 70° are available for the observed satellite orbits. Observations used to be discontinued if the GDOP was above 7; now, some receiver manufacturers suggest that a GDOP of 8 is acceptable.

5. Errors associated with the satellite orbital data.

6. Setup errors. Centering errors can be reduced if the equipment is checked to ensure that the optical plummet is true. Antenna reference height (ARH) measuring errors can be reduced by utilizing equipment that provides a built-in (or accessory) measuring capability to precisely measure the ARH (directly or indirectly) or by using fixed-length tripods and bipods.

7. Selected availability (SA), a denial of accuracy that was turned off in May 2000.

Many of the effects of the above errors, including denial of accuracy by the DOD (if it were to be re-introduced), can be surmounted by using differential positioning surveying techniques. Most of the discussion in this text is oriented to relatively short baselines. For long lines (> 150 km), more sophisticated processing is required to deal with natural and human-made errors.

8.8 Continuously Operating Reference Station (CORS)

The **continuously operating reference station (CORS)** system is a differential measurement system developed by the National Geodetic Survey (NGS) that has now become nationwide (see Figure 8.13). By 2004, the network was comprised of more than 460 stations. The goal is to expand the network until all points within the conterminous

CORS Coverage - June 2005

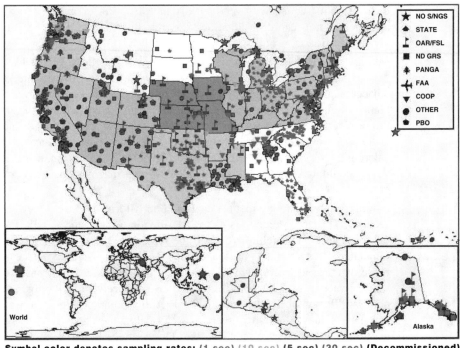

Symbol color denotes sampling rates: (1 sec) (10 sec) (5 sec) (30 sec) (Decommissioned)

Craig 6/21/2005

FIGURE 8.13 CORS coverage as of June 2005. Updated continually at, *http://www.ngs. noaa.gov/CORS/.*

United States (CONUS) will be within 200 km of at least one operational CORS site. The GPS satellite signals observed at each site are used to compute the base station position, which is then compared to the correct position coordinates (which have been previously and precisely determined). The *difference* (thus differential) between the correct position and the computed position is then made available over the Internet for use in the post-processing of the field observations, which are collected by the roving receivers used by a wide variety of government and private surveyors. The CORS network includes stations set up by the National Geodetic Survey (NGS); the U.S. Coast Guard; the U.S. Army Corps of Engineers (USACE), and, more recently, stations set up by other federal and local agencies. CORS stations must be able to determine NAD83 positions for the site that are accurate to 2 cm in horizontal dimensions and 4 cm in vertical dimensions, with 95 percent confidence.

The coordinates at each site are computed from twenty-four-hour data sets collected over a ten- to fifteen-day period. These highly accurate coordinates are then transformed into the NAD83 horizontal datum for use by local surveyors. Each CORS continuously tracks GPS satellite signals and creates files of both carrier measurements and code range measurements, which are available to the public via the Internet, to assist in positioning surveys. NGS converts all receiver data and individual site meteorological data to receiver independent exchange (RINEX) format, version 2. The files can be accessed via the Internet at *http://www.ngs.noaa.gov*. Once in the web site, the new user should select "Products and Services" and then "GPS Continuously Operating Reference Station (CORS)." The user is urged to download and read the "readme" files and the "frequently asked questions" file on the first visit. The NGS stores data from all sites for thirty-one days, and then the data are normally transferred to CD-ROMs, which are available for a fee. The data are often collected at a 30-s epoch rate, although some sites collect data at 1-s, 5-s, 10-s, and 15-s epoch rates. The local surveyor, armed with data sets in his or her locality at the time of an ongoing survey and having entered these data into his or her GPS program, has the equivalent of an additional dual-frequency GPS receiver. A surveyor with just one receiver can proceed as if two receivers were being used in the differential positioning mode. Data from the more than 400 CORS stations form the foundation of the national spatial reference system (NSRS). At the time of this writing, NAD83 positions were referenced to an epoch date of 2002, and published International Terrestrial Reference Frame 2000 (ITRF00) position coordinates were referenced to an epoch date of 1997.

8.8.1 On-line Positioning User Service

As an improvement to their data-download service, in 2001 the National Geodetic Survey (NGS) commenced a positioning processing service for dual-frequency, carrier-phase observations. This on-line positioning user service (OPUS) enables surveyors, working in the static mode with dual-frequency receivers, to send their data files in receiver independent exchange (RINEX) format to NGS via the Internet. Two hours (minimum) of data are required, and processing is based on three close reference stations (CORS)—clients can select reference stations from a scrollable list if they wish. The clients must also provide the antenna reference height (ARH) above the positioning mark and the antenna type, which they select from a list of calibrated antennas. The output coordinates can be selected for international terrestrial reference frame (ITRF), North American datum of 1983 (NAD83), universal transverse Mercator (UTM), and state plane coordinates (SPC). The solution is sent via e-mail within three minutes. See *www.ngs.noaa.gov/OPUS/* for further information.

8.9 Canadian Active Control System

The Geodetic Survey Division (GSD) of Geomatics Canada has combined with the Geological Survey of Canada to establish a network of active control points (ACP) in the Canadian Active Control System. The system includes ten unattended dual-frequency tracking stations (ACPs) that continuously measure and record carrier phase and pseudorange measurements for all satellites in view at a 30-s sampling interval. A master ACP in Ottawa coordinates and controls the system. The data are archived in RINEX format and are available on-line four hours after the end of the day 11:59 P.M. EST. Precise ephemeris data, computed with input from twenty-four globally distributed core GPS tracking stations of the International GPS Service for Geodynamics (IGS), are available on-line within two to five days after the observations; precise clock corrections are also available in the two- to five-day time frame. The Canadian Active Control System is complemented by the Canadian base network, which provides 200-km coverage in Canada's southern latitudes for high accuracy control (centimeter accuracy). These control stations are used to evaluate and to complement a wide variety of control stations established by various government agencies over the years.

These products, which are available for a subscription fee, enable a surveyor to position any point in Canada with a precision ranging from a centimeter to a few meters. Code observation positioning at the meter level, without the use of a base station, is possible using precise satellite corrections. Real-time service at the meter level is also available. The data can be accessed at *http://www.geod.nrcan.gc.ca*.

8.10 Survey Planning

Planning is important for GPS surveys so that almanac data can be analyzed to obtain optimal time sets when a geometrically strong array of satellites is available above 15° of elevation (above the horizon) and to identify topographic obstructions that may hinder signal reception. Planning software (see Figure 8.14) can graphically display geometric dilution of precision (GDOP) at each time of the day (GDOP of 7 or below is usually considered suitable for positioning—a value of 5 or lower is ideal). See Figures 8.15 to 8.18 for various computer screen plots that the surveyor can use to help the mission-planning process. The surveyor can select not only the optimal days for the survey, but also the hours of the day that will result in the best data.

8.10.1 Static Surveys

For static surveys (see Section 8.11.2 for a definition), survey planning includes a visit to the field to inspect existing stations and to place monuments for new stations. A compass and clinometer (Figure 2.7) are handy in sketching the location and elevation of potential obstructions at each station on a visibility (obstruction) diagram (see Figure 8.19). These obstructions are entered into the software for later display. The coordinates (latitude and

FIGURE 8.14 GPS planning software. Graphical depiction of satellite availability, elevation, and GDOP almanac data, which are processed by SKI software for a specific day and location. (Courtesy of Leica Geosystems Inc.)

longitude) of stations should be scaled from a topographic map. Scaled coordinates can help some receivers to lock onto the satellites more quickly.

Computer graphics displays include the number of satellites available (Figure 8.15); satellite orbits and obstructions, showing orbits of satellites as viewed at a specific station on a specific day (Figure 8.16); a visibility plot of all satellites over one day (Figure 8.17); and a polar plot of all visible satellites at a moment in time (Figure 8.18). Figure 8.16 illustrates that, for the location shown (latitude and longitude), most satellite orbits are in the southerly sky; thus, survey stations at that location should be located south of high obstructions, where possible, to minimize topographic interference. Almanac data, used in survey planning, can be updated on a regular basis through satellite observations. Most software suppliers provide almanac updates via their computer bulletin boards.

Static surveys work best with receivers of the same type; that is, they should have the same number of channels, the same type of antennas, and the same and signal-processing techniques. Also, the sampling rate must be set. Faster rates require more storage but can be helpful in detecting cycle slips, particularly on longer lines (> 50 km). A **cycle slip** is a temporary loss of lock on satellite carrier signals causing a miscount in carrier cycles. And the time (hours) of observation must be determined (given the available GDOP and the need for accuracy), as well as the start/stop times for each session. All this information is entered into the software, with the results uploaded into the receivers.

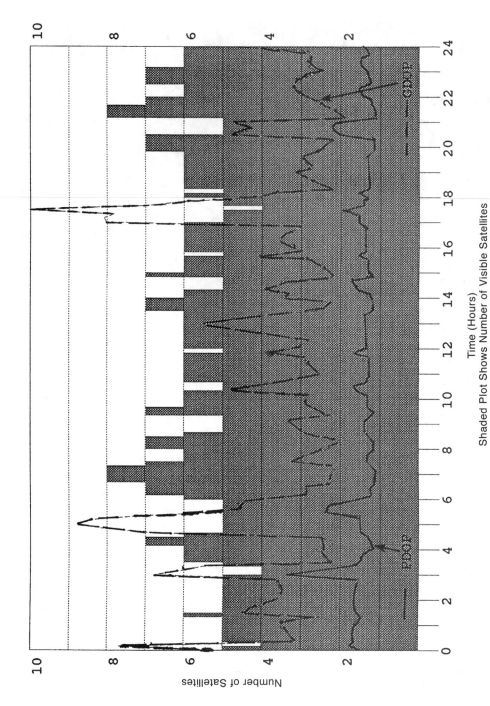

FIGURE 8.15 GPS planning software. GDOP and position dilution of precision (PDOP) during a specific day. (Courtesy of Leica Geosystems Inc.)

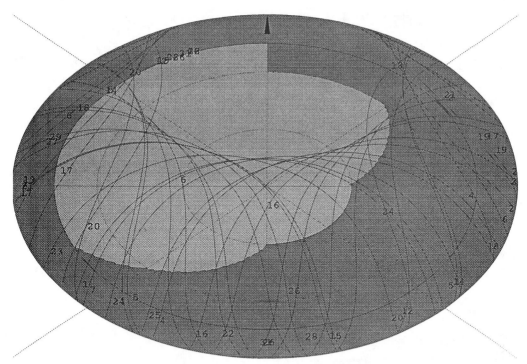

FIGURE 8.16 GPS planning software. Computer "sky plot" display showing satellite orbits and obstructions for a particular station (darker shading). (Courtesy of Leica Geosystems Inc.)

8.10.2 Kinematic Surveys

Much of the discussion about planning static surveys in the previous section also applies to kinematic surveys (see Section 8.11.3 for definitions). For kinematic surveys, the route must also be planned for each roving receiver so that best use is made of control points, crews, and equipment. The type of receiver chosen depends, to some degree, on the accuracy required. For topographic or GIS surveys, low-order (submeter) survey roving receivers (e.g., Figures 8.4 and 8.5) may be good choices. For construction layout in real time, high-end roving receivers capable of centimeter or millimeter accuracy (e.g., Figure 8.6) may be selected. The base station receiver must be compatible with the survey mission and the roving receivers; one base station can support any number of roving receivers. The technique to be used for initial ambiguity resolution must be determined; that is, will it be antenna swap, known station occupation, or on-the-fly (OTF)? And at which stations will this resolution take place?

If the survey is designed to locate topographic and built features, then the codes for the features should be defined or accessed from a symbols library, like they would be for total station surveys (see Chapter 5). If required, additional asset data (attribute data) should be tied to each symbol; the GPS software should permit the entry of several layers of asset data. For example, if the feature to be located is a pole, the pole symbol can be backed up with pole use (illumination, electric wire, telephone, etc.); the type of pole (concrete, wood, metal, etc.) can be booked; and the year of installation can be entered. (Municipal

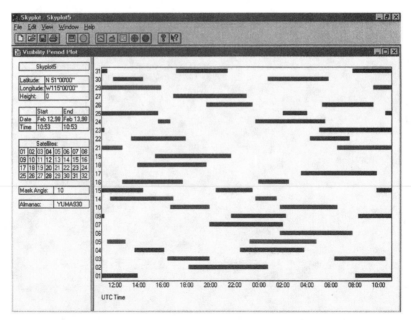

FIGURE 8.17 Visibility plot—from Skyplot series (it is color-coded in its original format). (Courtesy of Position Inc., Calgary, Alberta)

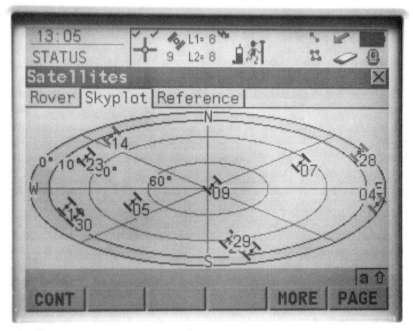

FIGURE 8.18 Skyplot showing the available satellites at one point in time. (Courtesy of Leica Geosystems, Inc.)

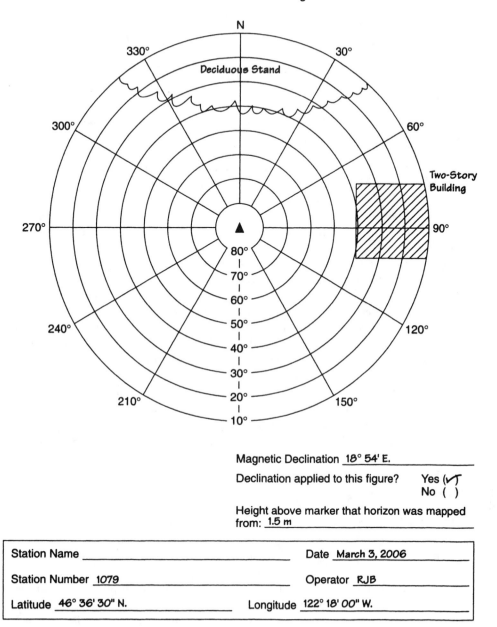

GPS Station Obstruction Diagram

Magnetic Declination _18° 54' E._

Declination applied to this figure?   Yes (✓)
                                        No ( )

Height above marker that horizon was mapped
from: _1.5 m_

Station Name _____	Date _March 3, 2006_
Station Number _1079_	Operator _RJB_
Latitude _46° 36' 30" N._	Longitude _122° 18' 00" W._

FIGURE 8.19 Station visibility diagram.

computers can generate work orders automatically for any facility's maintenance based on the original date of installation.) Planning ensures that when the roving surveyor is at the feature point, all possible prompts will be displayed to capture completely the data for each way point. The success of any kinematic survey depends a great deal on the thoroughness of the presurvey reconnoiter and on the preparation of the job file (using the GPS software).

8.11 GPS Field Procedures

8.11.1 Tripod-Mounted Antenna and Pole-Mounted Antenna Considerations

GPS antennas can be mounted directly on tripods via optical plummet-equipped tribrachs. All static survey occupations and base station occupations for all other types of GPS surveys require the use of a tripod. Care is taken to center the antenna precisely and to measure the antenna reference height (ARH) precisely. All antennas have a direction mark, often an N or an arrow, or a series of numbered notches along the outside perimeter of the antenna. The direction mark enables the surveyor to align the roving antenna in the same direction (usually north) as the base station antenna. This technique helps to eliminate any bias in the antennas. The measured ARH (corrected or uncorrected) is entered into the receiver. When a lock has been established on the satellites, a message is displayed on the receiver display, and observations begin. If the ARH cannot be measured directly, then slant heights are measured. Some agencies measure in two or three slant locations and average the results. The vertical height can be computed using the Pythagorean theorem: $ARH = \sqrt{slant\ height^2 - antenna\ radius^2}$ (see Figure 8.20).

GPS receiver antennas can also be mounted on adjustable-length poles (similar to prism poles) or bipods. These poles and fixed-length bipods may have built-in power supply conduits, a built-in circular bubble, and the ability to display the ARH. When using poles, the GPS receiver program automatically prompts the surveyor to accept the last-used ARH; if there has been no change in the antenna height, the surveyor simply accepts the prompted value. As with most surveys, field notes are important, both as backup and as confirmation of entered data. The ARH at each station is booked, along with equipment numbers, session times, crew, file (job) number, and any other pertinent data (see Figure 8.20).

8.11.2 Static Surveys

8.11.2.1 Traditional Static. In this technique of precise GPS positioning, two or more receivers collect data from the same satellites during the same epochs. Accuracy can be improved by using the differential techniques of relative positioning, whereby one base receiver antenna (single or dual frequency) is placed over a point of known coordinates (X, Y, Z) on a tripod, while other antennas are placed, also on tripods, over permanent stations that are to be positioned. Observation times are 1 hour or more, depending on the receiver, the accuracy requirements, the length of the baseline, the satellites' geometric configuration, and atmospheric conditions. This technique is used for long lines in geodetic control, control densification, and photogrammetric control for aerial surveys and precise engineering surveys; it is also used as a fallback technique when the available geometric array of satellites is not compatible with other GPS techniques. The preplanning of station locations takes into consideration potential obstructions presented by trees and buildings, which must be considered and minimized. Some recommend that, for best

Project Name _____

Project Number _____

Receiver Model/No. _____

Receiver Software Version _____

Data Logger Type/No. _____

Antenna Model/No. _____

Cable Length _____

Ground Plane Extensions Yes ( ) No ( )

Station Name _____

Station Number _____

4-Character ID _____

Date _____

Obs. Session _____

Operator _____

Data Collection

Collection Rate _____

Start Day/Time _____

End Day/Time _____

Receiver Position

Latitude _____

Longitude _____

Height _____

Obstruction or possible interference sources _____

_____

General weather conditions _____

_____

Detailed meteorological observations recorded: Yes ( ) No ( )

Antenna Height Measurement

Show on sketch measurements taken to derive the antenna height. If slant measurements are taken, make measurement on two opposite sides of the antenna. Make measurements before and after observing session.

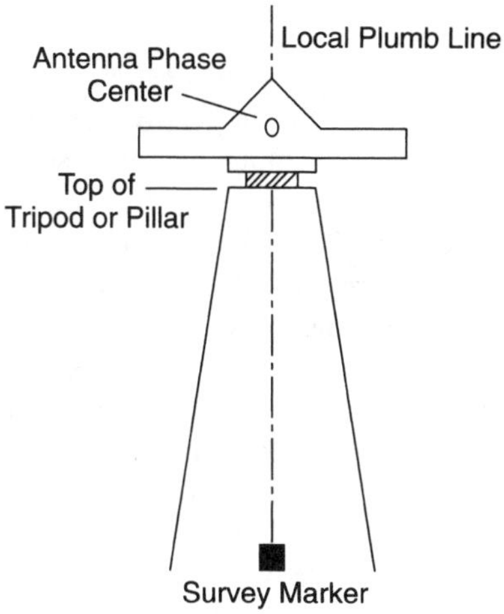

Vertical measurements   ( )

Slant measurements   ( ) : radius _____ m

BEFORE	AFTER
_____ m _____ in.	_____ m _____ in.
_____ m _____ in.	_____ m _____ in.

Mean _____

Corrected to vertical
if slant measurement _____

Vertical offset to
phase center _____

Other offset
(indicate on sketch) _____

TOTAL HEIGHT _____

Verified by: _____

FIGURE 8.20   GPS field log. (Courtesy of Geomatics Canada)

results, dual-frequency receivers be used, the GDOP be less than 8, and a minimum of five satellites be tracked.

8.11.2.2   Rapid Static.

This technique, which was developed in the early 1990s, can be employed where dual-frequency receivers are used over short (up to 15 km) lines. As with static surveys, this technique requires one receiver antenna to be positioned (on a tripod) at a known base station while the roving surveyor moves from station to station with the antenna pole-mounted. With good geometry, initial phase ambiguities can be resolved within a minute (3 to 5 minutes for single-frequency receivers). With this technique, there is no need to maintain lock on the satellites while moving rover receivers—the roving receivers can even be turned off to preserve their batteries. Accuracies of a few millimeters are possible using this technique. Observation times of 5 to 10 minutes are typical.

8.11.2.3   Reoccupation.

This technique is also called pseudo-kinematic and pseudo-static. Reoccupation can be used when fewer than four satellites are available or when GDOP is weak (above a value of 7 or 8). Survey stations are occupied on at least two different occasions at least one hour apart; the solution may be strengthened if the base station receiver has been moved to another control station for the second set of observations. The processing software combines the satellite observations to provide a solution. The data are processed as they are for a static survey; that is, if there are only three satellites available on the first occupation and another three available on the second occupation, the software processes the data as if six satellites are available at once. Observation times of 10 minutes are typical. Roving receivers have their antennas pole-mounted.

8.11.3   Kinematic Surveys

Kinematic surveying is an efficient way to survey detail points for some engineering and topographic surveys. This technique begins with both the base unit and the roving unit occupying a 10-km (or shorter) baseline (two known positions) until ambiguities are resolved. Alternatively, short baselines with one known position can be used where the distance between the stations is short enough to permit antenna swapping. Here, the base station and a nearby (within reach of the antenna cable) undefined station are occupied for a short period of time (say, 2 minutes) in the static mode, after which the antennas are swapped (while still receiving the satellite signals, but now in the rove mode) for an additional few minutes of readings in the static mode (techniques may vary with different manufacturers). Receivers having wireless technology (for example, Bluetooth wireless technology) may not have the baselines restricted by the length of antenna cable. After the antennas have been returned to their original tripods and after an additional short period of observations in the static mode, the roving receiver—in rove mode—then moves (on a pole, backpack, truck, boat, etc.) to position all required detail points while keeping a lock on the satellite signals. If the lock is lost, the receiver is held stationary for a few seconds until the ambiguities are once again resolved so that the survey can continue. The base station stays in the static mode unless it is time to leapfrog the base and rover stations. Another technique, called on-the-fly (OTF) ambiguity resolution, occurs as the base station remains fixed at a position of known coordinates and the rover receiver can determine the integer number of cycles using software designed for that purpose—even as the receiver is on the move.

All kinematic positioning techniques compute a relative differential position at preset time intervals instead of at operator-selected points. Lock must be maintained on a minimum of four satellites, or it must be reestablished (usually by OTF) when lost. This technique is used for road profiling, ship positioning in sounding surveys, and aircraft positioning in aerial surveys. Continuous surveying can be interrupted to take observations (for a few epochs) at any required way points.

8.11.4   Real-Time Kinematic (RTK) Surveying

The real-time combination of GPS receivers, mobile data communications, on-board data processing, and on-board applications software contributes to an exciting new era in surveying. As with the motorized total stations described in Section 5.7, real-time positioning offers the potential of one-person capability in positioning, mapping, and quantity surveys (the base station receiver can be unattended). Layout surveys need two real-time kinematic (RTK) surveyors, one to operate the receiver and one to mark the stations in the field.

RTK surveying requires a base station for receiving the satellite signals and then retransmitting them to the roving surveyor's receiver, which is simultaneously tracking the same satellites. Multibase RTK radio-equipped base receivers, which transmit on different radio channels, require that the roving surveyor switch through the roving receiver's channels to receive all transmissions. The roving surveyor's receiver can then compare the base station signals with the signals received by the roving receiver to process baseline corrections and thus determine accurate positions in real time. The epoch collection rate is usually set to 1 s. Messages to the rover are updated every 0.5 to 2 s, and the baseline processing can be done in 0.5 to 1 s. Figure 8.21 shows a typical radio and amplifier used in code and carrier differential surveys. Radio reception requires a line of sight to the base station.

By 2001, cell phones were also introduced at the base station for field rover communications. The use of cell phones (in areas of cellular coverage) can overcome the interference on radio channels that sometimes frustrates and delays surveyors. Since 2001, rapid improvements have occurred in both cellular coverage and mobile-phone technology. Third-generation (3G) mobile-phone technology has created the prospect of transferring much more than voice communications and small data packets; this evolving technology will enable the surveyor to transfer much larger data files wirelessly from one phone to another and will also enable the surveyor to access computer files wirelessly via the Internet—both with data uploads and downloads. This new wireless technology will greatly affect both the collection of field data and the layout of surveying and engineering works.

RTK positioning can commence without the rover first occupying a known baseline. The base station transmits code and carrier-phase data to the roving receiver, which can use this data to help resolve ambiguities and to solve for coordinate differences between the reference and the roving receivers. The distance and line-of-sight range of the radio transmission (earlier radios) from the base to the rover can be extended by booster radios to a distance of about 10 km. Longer ranges require commercially licensed radios, as do even shorter ranges in some countries. The RTK technique can utilize either single-frequency or dual-frequency receivers. The more expensive dual-frequency receiver has better potential for surveys in areas where satellite visibility may be reduced by topographic obstructions.

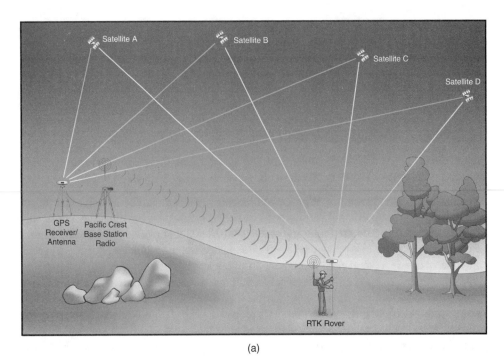

(a)

FIGURE 8.21 A base station GPS receiver and a base station radio, as well as the roving GPS receiver equipped with a radio receiver that allows for real-time positioning. (Courtesy of Pacific Crest Corporation, Santa Clara, Calif.)

(continued)

Loss of signal lock can be regained with single-frequency receivers by reoccupying a point of known position, and with dual-frequency receivers either by remaining stationary for a few minutes or by using OTF resolution, while proceeding to the next survey position. When the second and third civilian frequencies are provided for GPS users in the near future, solutions will be greatly enhanced.

The GPS roving receiver includes the antenna and data collector mounted on an adjustable-length pole, with the receiver, radio, and radio antenna mounted in the backpack. More recent models have all equipment mounted on the pole. The pole kit weighs only 8.5 pounds.

RTK is the positioning technique that has seen the greatest technological advances and greatest surveyor acceptance over the past few years. Several states are now installing (or planning to install) RTK base station networks to service their constituent private and public surveyors; Ohio's Department of Transportation (ODOT) and Michigan's Department of Transportation (MDOT), for example, completed their statewide coverage of base stations by 2004. With the large number of CORS stations now established, the need for additional RTK base stations has been reduced because the existing CORS can also be used in RTK operations. Multibase RTK networks can greatly expand the area in which a wide variety of surveyors can operate. It is reported that multibase RTK base stations, using cellular communications, can cover a much larger area (up to four times) than can radio-equipped RTK base stations. It has been predicted that many local, state/provincial, and

(b)

FIGURE 8.21 (*continued*) (b) Base station radio. (Courtesy of Pacific Crest Corporation, Santa Clara, Calif.)

federal governments will be more inclined to invest in establishing GPS reference station networks rather than maintaining and expanding ground control monuments.

In 2004, Trimble Inc. introduced a variation on the multibase network concept called virtual reference station (VRS) technology. This system requires a minimum of three base stations up to 60 km apart (apparently the maximum spacing for postprocessing in CORS), with communications to rovers and to a processing center via cell phones. This process requires that the rover surveyor dial into the system and provide his or her approximate location. The VRS network processor determines measurements similar to what would have been received had a base station been located at the rover's position and transmited those measurements to the rover receiver. The accuracies and initialization times in this system (using 30-km baselines) are said to be comparable to results obtained in very short (1- to 2-km) baseline situations. Accuracies of 1 cm in northing and easting and 2 cm in elevation are expected.

8.12 GPS Applications

Although GPS was originally devised to assist in military navigation, guidance, and positioning, civil applications continue to evolve at a rapid rate. From its earliest days, GPS was welcomed by the surveying community and recognized as an important tool in precise positioning. Prior to GPS, published surveying accuracy standards, then tied to terrestrial techniques (e.g., triangulation, trilateration, and precise traversing), had an upper limit of 1:100,000. With GPS, accuracy standards have risen as follows: AA (global), 1:100,000,000; A (national—primary network), 1:10,000,000; and B (national—secondary network), 1:1,000,000. In addition to continental control, GPS has now become a widely used technique for establishing and verifying state/provincial and municipal horizontal control, as well as for horizontal control for large-scale engineering and mapping projects.

8.12.1 General Techniques

When in the field, the surveyor first sets the receiver over the point and then measures, records (in the field log), and enters the ARH into the receiver. The session programming is verified, and the mode of operation is selected. Satellite lock and position computation are then verified. As the observations proceed, the process is monitored. When the session is complete, the receiver is turned off (base station) or moved to the next way point (roving receiver).

Newer data collectors (see Figure 8.22), designed for RTK surveys, make the surveyor's job much simpler. Some of these data collectors come with interactive graphics screens (which the surveyor taps with a stylus) and provide features that enable the surveyor to initialize the base station and rovers, set up the radio communications, set up the antenna heights and descriptions, load only that part of the geoid (e.g., GEOID03) that is needed, select a working grid, display how many satellites are visible, display skyplot and PDOP/GDOP, indicate when fixed lock occurs, display horizontal and vertical root mean square error, etc.

The survey controller (data collector) has application programs for topography; radial and linear stakeout; cut and fill; and intersections by bearing-bearing, bearing-distance, and distance-distance. In addition, the controller performs inverse, sea-level, and curvature and refraction calculations; datum transformations [universal transverse Mercator (UTM), state plane, Lambert, etc.]; and geoid corrections. See Figure 8.23 for typical operations capability with data collectors that are used for both total station surveys and GPS surveys.

8.12.2 Topographic Surveys

When GPS is used for topographic surveys, detail is located by short occupation times and described with the input of appropriate coding. Input may be accomplished by keying in, screen-tapping, using bar-code readers, or keying in prepared library codes. Line work may

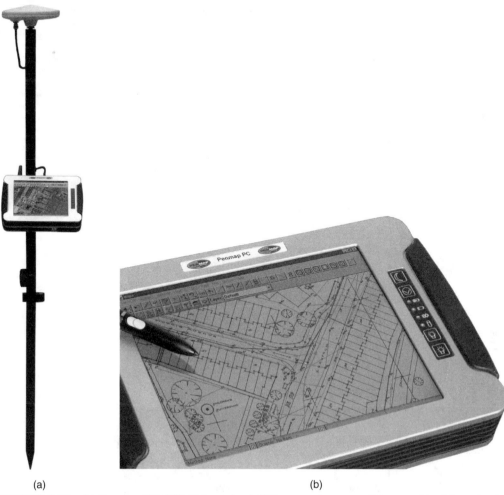

(a) (b)

FIGURE 8.22 (a) Penmap GPS-RTK. This cm-level GPS system features a Pentium tablet PC as the data collector/controller. (b) Its touch-screen capabilities speed up all the data collector/controller operations. (Courtesy of Penmap Strata Software, Bradford, West Yorkshire, U.K.)

require no special coding (see Z codes, Section 5.5) because entities (curbs, fences, etc.), can be joined by their specific codes (curb2, fence3, etc.), or entities may be joined by screen-tapping points on the display. Some software programs display the accuracy of each observation for horizontal and vertical position, giving the surveyor the opportunity to take additional observations if the displayed accuracy does not meet job specifications. In addition to positioning random detail, this GPS technique permits the collection of data on specified profile, cross-section, and boundary locations by utilizing the navigation functions; contours may be readily plotted from the collected data. GPS is also useful when beginning the survey for locating boundary and control markers that may be covered by snow or other ground cover. If the marker's coordinates are in the receiver, the navigation mode can take the surveyor directly to its location. Data captured using these techniques can be added to a mapping or GIS database or directly plotted to scale using a digital plotter (see Chapter 7).

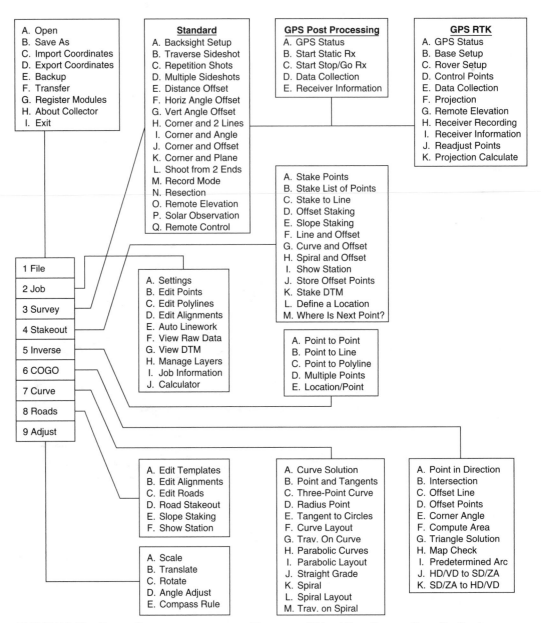

FIGURE 8.23 Data collector menu structure. (Courtesy of Tripod Data System, Corvallis, Ore.)

Although it is widely reported that GPS topographic surveys can be completed more quickly than total station topographic surveys, they do suffer from one major impediment: a GPS receiver must have line of sight to four (preferably five) satellites to determine position. When topographic features are located under tree canopy or hidden by other obstructions, a GPS receiver cannot be used directly to determine position. In this case, offset tools are required, such as handheld and portable total stations (see Figures 8.24 and 8.25).

FIGURE 8.24 MapStar Compass Module II, which has reflectorless EDM capabilities and azimuth determination (accuracy of ±0.6°). (Courtesy of Laser Technology Inc., Centennial, Colorado)

8.12.3 Layout Surveys

For layout work, the coordinates of all relevant control points and layout points are uploaded from computer files before going out to the field. After the base station receiver has been set up over a control point and the roving receiver is appropriately referenced, the layout can begin. As each layout point number is keyed into the collector, the azimuth and distance to the required position are displayed on the screen. Guided by these directions, the surveyor eventually moves to the desired point, which is then staked. One base receiver can support any number of rover receivers, permitting the instantaneous layout of large-project boundaries, pipelines, roads, and building locations by several surveyors, each working only on a specific type of facility, or by all roving surveyors working on all proposed facilities but only on selected geographic sections of the project.

As with topographic applications, the precision of the proposed location is displayed on the receiver as the antenna pole is held on the grade stake or other marker to confirm that layout specifications have been met. If the displayed precision is below specifications, the surveyor simply waits at the location until the processing of data from additional epochs provides him or her with the necessary precision. On road layouts, both line and grade can be given directly to the builder (by marking grade stakes), and progress in cut and fill can be monitored. Slope stakes can be located without any need for intervisibility. The next logical step is to mount GPS antennas directly on various construction excavating equipment to provide line and grade control directly. See Chapter 10 and Figure H.3.

(a)

FIGURE 8.25 (a) Criterion handheld survey laser #300, with data collector; angles by flux-gate compass; and distances without a prism to 1,500 ft (to 40,000 ft with a prism). This instrument is used to collect data with an accuracy of ±0.3° and ±0.3 ft for mapping and GIS databases. It provides a good extension for canopy-obstructed GPS survey points. (Courtesy of Laser Technology Inc., Englewood, Colorado)

(continued)

As with the accuracy/precision display previously mentioned, cut and fill, grades, and the like, are displayed at each step along the way, and a permanent record is kept of these data in case a review is required. Accuracy can also be confirmed by reoccupying selected layout stations and noting and recording the displayed measurements—an inexpensive yet effective method of quality control.

During material inventory measurements, GPS techniques are particularly useful in open-pit mining, where original, in-progress, and final surveys can easily be performed for quantity and payment purposes. Material stockpiles can also be surveyed quickly, and volumes can be computed using appropriate on-board software.

For both GPS topographic and layout work, there is no way to get around the fact that existing and proposed stations must be occupied by the antenna. If some of these specific locations are such that satellite visibility is impossible, perhaps because obstructions are blocking the satellites' signals (even when using receivers capable of tracking both GPS and GLONASS constellations), then ancillary surveying techniques must be used. Some examples of such equipment include backpacked inertial measurement unit (IMU)–equipped GPS receivers, total stations equipped with built-in GPS receivers (see Section 5.9 and Figure 5.29), and handheld and portable total stations (Figures 8.24 and 8.25). In cases where millimeter accuracy is required in vertical dimensions (e.g., structural

(b)

FIGURE 8.25 (*continued*)   (b) Prosurvey 1000 handheld survey laser: used (without prisms) to record distances and angles to survey stations. Shown here being used to determine river width in a hydrographic survey. (Courtesy of Laser Atlanta, Norcross, Georgia)

layouts), one GPS manufacturer has incorporated a precise rotating laser (generating a laser fan) into the GPS receiver-controller instrumentation package. An additional receiver, mounted on the prism pole, is used to detect and interpret the received portion of the laser fan, transmitted by the stand-along rotating laser, and thus generate precise elevations—even when the elevations between the laser and receiver differ by as much as 30 ft.

8.12.4   Additional Applications

GPS is ideal for the precise type of measurements needed in deformation studies—whether they are for geological events (e.g., plate slippage) or for structure stability studies such as the monitoring of bridges and dams. In both cases, measurements from permanently established remote sites can be transmitted to more central control offices for immediate analysis.

In addition to the static survey control described earlier, GPS can also be utilized in dynamic applications of aerial surveying and hydrographic surveying where on-board GPS receivers can be used to supplement existing ground or shore control or where they can now be used in conjunction with inertial guidance equipment (Inertial Navigation System [INS]) for control purposes, without the need for external (shore or ground) GPS receivers. Navigation has always been one of the chief uses of GPS. Civilian use in this area has really taken off. Commercial shipping and pleasure boating now have an accurate and relatively inexpensive navigation device. With the cessation of SA, the precision of low-cost receivers has improved to the < 10 m range, and that range can be further improved to the submeter level using differential (for example, DGPS radio beacon) techniques. Using low-cost GPS receivers, we can now navigate to the correct harbor, and we can navigate to the correct mooring within that harbor. GPS and on-board inertial systems (INS) are rapidly becoming the norm for aircraft navigation during airborne remote-sensing missions.

GPS navigation has now become a familiar tool for backpacking, where the inexpensive GPS receiver (often less than $300) has become a superior adjunct to the compass (see Figure 8.6). Using GPS, a backpacker can determine geographic position at selected points (way points) such as trail intersections, river crossings, campsites, and other points of interest. Inexpensive software (less than $100) can be used to transfer collected way points to the computer, make corrections for DGPS input, and display data on previously loaded maps and plans. The software can also be utilized to identify way points on a displayed map, which can be coordinated and then downloaded to a GPS receiver so that the backpacker can go to the field and navigate to the selected way points. Many backpackers continue to use a magnetic compass while navigating from way point to way point and to maneuver under tree canopy, for example, where GPS signals are blocked, and thus conserve GPS receiver battery life.

8.13 Vertical Positioning

Until recently, most surveyors have been able to ignore the implications of geodesy for normal engineering plane surveys. The distances encountered are so relatively short that global implications are negligible. However, the elevation coordinate (h) given by GPS solutions refers to the height from the surface of the reference ellipsoid (GRS80; see Figure 8.26) to the ground station, whereas the surveyor needs the orthometric height (H). The ellipsoid is referenced to a spatial Cartesian coordinate system (Figure 8.27) called the International Terrestrial Reference Framework (ITRF; ITRF00 [also written as ITRF 2000] is the latest model), in which the center (0, 0, 0) is the center of the mass of the earth, and the X axis is a line drawn from the origin through the equatorial plane to the Greenwich meridian. The Y axis is in the equatorial plane perpendicular to the X axis, and the Z axis is drawn from the origin perpendicular to the equatorial plane, as shown in Figure 8.28.

Essentially, GPS observations permit the computation of X, Y, and Z Cartesian coordinates of a geocentric ellipsoid. These Cartesian coordinates can then be transformed to

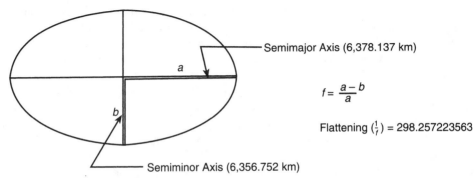

FIGURE 8.26 Ellipse parameters of the GRS80 ellipsoid (flattening is exaggerated).

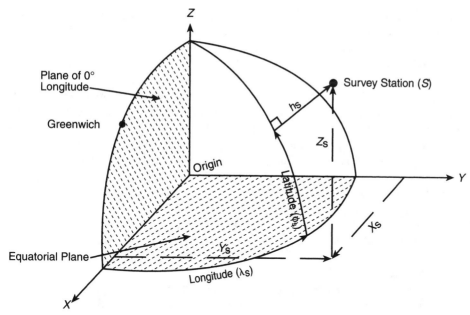

FIGURE 8.27 Cartesian (X, Y, Z) and geodetic $(\varphi_s, \lambda_s, h_s)$ coordinates.

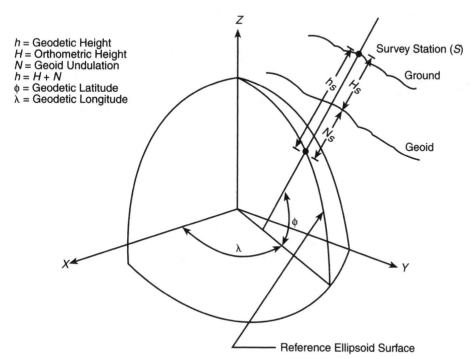

h = Geodetic Height
H = Orthometric Height
N = Geoid Undulation
$h = H + N$
ϕ = Geodetic Latitude
λ = Geodetic Longitude

FIGURE 8.28 Relationship of geodetic height and orthometric height.

geodetic coordinates: latitude (φ), longitude (λ), and ellipsoidal height (h). The geodetic coordinates, together with geoid corrections, can be transformed to UTM, state plane, or other grids (see Figure 8.28 and Section 8.13.1) to provide working coordinates (northing, easting, and elevation) for the field surveyor.

The ellipsoid presently used most often to portray the earth is the WGS84 (as described by World Geodetic System), which is generally agreed to represent the earth more accurately than previous versions. (Ongoing satellite observations permit scientists to improve their estimates about the size and mass of the earth.) An earlier reference ellipsoid GRS80—the Geodetic Reference System of the International Union of Geodesy and Geophysics (IUGG)—was adopted in 1979 by that group as the model then best representing the earth. It is the ellipsoid on which the horizontal datum, the North American Datum of 1983 (NAD83), is based (see Figure 8.26). In this system, the geographic coordinates are given by the ellipsoidal latitude (φ), longitude (λ), and height (h) above the ellipsoidal surface to the ground station. The GEOID03 model (see Section 8.13.1) is based on known relationships between NAD83 and the ITRF spatial reference frames, together with GPS height measurements on the North American Vertical Datum of 1988 (NAVD88) benchmarks.

Surveyors are used to working with spirit levels and reference orthometric heights (H) to the average surface of the earth, as depicted by mean sea level (MSL). The surface of mean sea level can be approximated by the equipotential surface of the earth's gravity field (this surface is everywhere normal to the direction of gravity) and is called the **geoid**. The density of adjacent landmasses at any particular survey station influences the geoid, which has an irregular surface. Thus, its surface does not follow the surface of the ellipsoid; sometimes it is below the ellipsoid, and other times, above it. Wherever the mass of the earth's crust changes, the geoid's gravitational potential also changes, resulting in a nonuniform and unpredictable geoid surface. Because the geoid does not lend itself to mathematical expression, as does the ellipsoid, **geoid undulation** (the difference between the geoid surface and the ellipsoid surface) must be measured or interpolated at specific sites to determine the local geoid undulation value (see Figures 8.29 and 8.30).

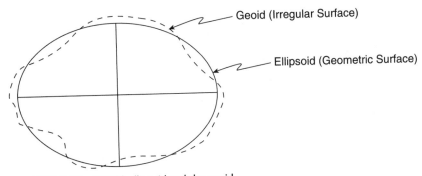

FIGURE 8.29 GRS80 ellipsoid and the geoid.

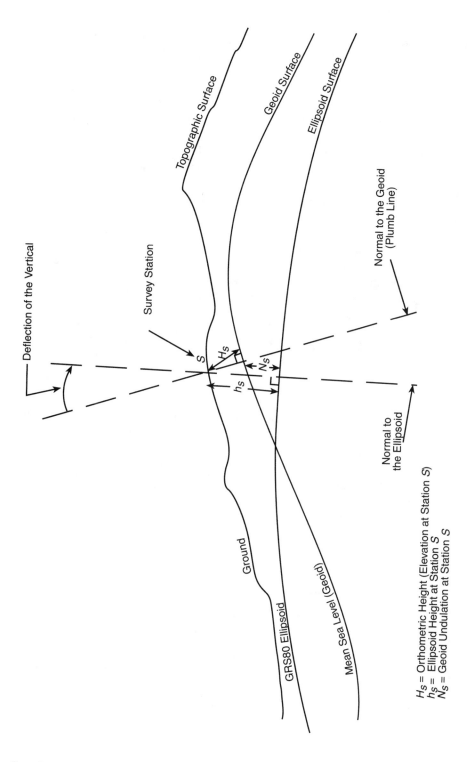

H_S = Orthometric Height (Elevation at Station S)
h_S = Ellipsoid Height at Station S
N_S = Geoid Undulation at Station S

(N Is Positive When Geoid Is Above the Ellipsoid)

$$\boxed{h_S = H_S + N_S}$$

FIGURE 8.30   The three surfaces of geodesy (undulations greatly exaggerated).

8.13.1 Geoid Modeling

Geoid undulations can be determined both by gravimetric surveys and by the inclusion of points of known elevation in GPS surveys. When the average undulation of an area has been determined, the residual undulations over the surveyed area must still be determined. While residual undulations are usually less than 0.020 m over areas of 50 km^2, the earth's undulation itself ranges from $+75$ m at New Guinea to -104 m at the south tip of India.

After all the known geoid separations have been plotted, the geoid undulations (N) at any given survey station can be interpolated; the orthometric height (H) can be determined from the relationship $H = h - N$, where h is the ellipsoid height (N is positive when the geoid is above the ellipsoid and negative when the geoid is below the ellipsoid)—see Figures 8.29 and 8.30. Geoid modeling data can be obtained from government agencies and, in many cases, GPS receiver suppliers provide this data as part of their on-board software.

Because of the uncertainties still inherent in geoid modeling, it is generally thought that accuracies in elevation are only about half the accuracies achievable in horizontal positioning. That is, if a horizontal accuracy is defined to be $\pm$ (5 mm + 1 ppm), the vertical accuracy is probably close to $\pm$ (10 mm + 2 ppm). However, the accuracy of geoid models is improving with each new version. As more and more GPS observations on NAVD88 benchmarks (GPSBMs) are included in the net, we move ever closer to the goal of a geoid accurate to 1 cm. GPS observations can directly deliver ellipsoidal heights, but beginning in 1996, GPS manufacturers made it possible for field surveyors to determine orthometric heights quickly from GPS observations by incorporating geoid models directly into their receivers.

8.13.1.1 CGG2000 GEOID (Canada).

Natural Resources, Canada (NRCan) has developed an improved geoid model, the Canadian Gravimetric Geoid model (CGG2000), which is a refinement of previous models (GSD95 and GSD91). It takes into account about 700,000 surface gravity observations in Canada, with the addition of about 1,477,000 observations taken in the United States, and 117,100 observations taken in Denmark. Although the model covers most of North America, it was designed for use in Canada. Geoid data (the GPS.H package includes the new geoid model and GPS.Hv2.1, which is the latest software) are available at the ministry's web site at *www.geod.nrcan.gc.ca* (go to "Products and Services").

8.13.1.2 GEOID03 (United States).

GEOID03 (a refinement of GEOID99, GEOID96, GEOID93, and GEOID90) is a geoid-elevation estimation model of the conterminous United States (CONUS), referenced to the GRS80 ellipsoid. Thus, it enables mainland surveyors to convert ellipsoidal heights (NAD83/GRS80) reliably to the more useful orthometric heights (NAVD88) at the centimeter level. A more inclusive geoid model—expected in 2006 or 2007—will include all U.S. states and territories. GEOID03 was created from 14,185 points—including 579 in Canada—and used updated GPSBMs that were not available for the previous geoid model (GEOID99). GEOID03 files are among the many services to be found at the National Geodetic Survey's Geodetic toolkit web site at *www.ngs.noaa.gov/TOOLS/*.

8.14 Conclusion

Table 8.3 summarizes the GPS positioning techniques described in this chapter. GPS techniques hold so much promise that most future horizontal and vertical control could be coordinated using these techniques. It is also likely that many engineering, mapping, and GIS surveying applications will be developed using emerging advances in real-time GPS data collection. In the future, the collection of GPS data will be enhanced as more and more North American local and federal agencies install continuously operating receivers and transmitters, which provide positioning solutions for a wide variety of private and government agencies involved in surveying, mapping, planning, GIS-related surveying, and navigation.

8.15 GPS Glossary

Absolute Positioning   The direct determination of a station's coordinates by receiving positioning signals from a minimum of four GPS satellites. Also known as *point positioning*.

Active Control Station (ACS)   *See* continuously operating reference station (CORS).

Ambiguity   The integer number of carrier cycles between the GPS receiver and a satellite.

Continuously Operating Reference Station (CORS)   CORS transmitted data can be used by single-receiver surveyors or navigators to permit higher precision differential positioning through post-processing computations.

Cycle Slip   A temporary loss of lock on satellite carrier signals causing a miscount in carrier cycles; lock-on must be reestablished to continue positioning solutions.

Differential Positioning   Obtaining satellite measurements at a known base station to correct simultaneous same-satellite measurements made at rover receiving stations. Corrections can be postprocessed, or corrections can be real-time kinematic (RTK), for example, when they are broadcast directly to the roving receiver.

Epoch   An observational event in time that forms part of a series of GPS observations.

General Dilution of Precision (GDOP)   A value that indicates the relative uncertainty in position, using GPS observations, caused by errors in time (GPS receivers) and satellite vector measurements. A minimum of four widely spaced satellites at high elevations usually produces accurate results (i.e., lower GDOP values).

Geodetic Height (h)   The distance from the ellipsoid surface to the ground surface.

Geoid   A surface that is approximately represented by mean sea level (MSL). D is, in fact, the equipotential surface of the earth's gravity field. This surface is everywhere normal to the direction of gravity.

Geoid Undulation (N)   The difference in elevation between the geoid surface and the ellipsoid surface. N (the mathematical expression for geoid undulation) is negative if the geoid surface is below the ellipsoid surface. Also known as *geoid height* and as *geoid separation*.

Table 8.3 GPS MEASUREMENTS SUMMARY*

Two Basic Modes:
Code-based measurements: Satellite-to-receiver pseudorange is measured and then corrected to provide the range; four satellite ranges are required to determine position—by removing the uncertainties in X, Y, Z, and receiver clocks. The military can access both P code and C/A code; civilians can access only the C/A code.
Carrier-based measurements: The carrier waves themselves are used to compute the satellite(s)-to-receiver range, similar to EDM. Most carrier receivers utilize both code measurements and carrier measurements to compute positions.

Two Basic Techniques:
Point positioning: Code measurements are used to compute the position of the receiver directly. Only one receiver is required.
Relative positioning: Code and/or carrier measurements are used to compute the baseline vector (ΔX, ΔY, and ΔZ) from a point of known position to a point of unknown position, thus enabling the computation of the coordinates of the new position.

Relative positioning: Two receivers required—one occupying a point of known position.
Static: Accuracy—5 mm + 1 ppm. Observation times—1 hour to many hours. Use—control surveys; standard method when lines are longer than 20 km. Uses dual- or single-frequency receivers.
Rapid static: Accuracy—5 to 10 mm +1 ppm. Observation times—5 to 10 minutes. Initialization time of 1 minute for dual-frequency and about 3 to 5 minutes for single-frequency receivers. Receiver must have specialized, rapid static observation capability. The roving receiver does not have to maintain lock on satellites (useful feature in areas with many obstructions). Used for control surveys, including photogrammetric control for lines 10 km or less. The receiver program determines the total length of the sessions, and the receiver screen displays "time remaining" at each station session.
Reoccupation (also known as pseudostatic and pseudo-kinematic): Accuracy—5 to 10 mm + 1 ppm. Observation times are about 10 minutes, but each point must be reoccupied after at least 1 hour, for another 10 minutes. No initialization time required. Useful when GDOP is poor. No need to maintain satellite lock. Same rover receivers must reoccupy the points they initially occupied. Moving the base station for the second occupation sessions may improve accuracy. Voice communications are needed to ensure simultaneous observations between base receiver and rover(s).
Kinematic: Accuracy—10 mm + 2 ppm. Observation times—1 to 4 epochs (1 to 2 minutes on control points); the faster the rover speed, the quicker must be the observation (shorter epochs). Sampling rate is usually between 0.5 and 5 seconds. Initialization by occupying two known points—2 to 5 minutes, or by antenna swap—5 to 15 minutes. Lock must be maintained on four satellites (five satellites are better in case one of them moves close to the horizon). Good technique for open areas (especially hydrographic surveys) and where large amounts of data are required quickly.
Stop and go: Accuracy—10 to 20 mm + 2 ppm. Observation times—a few seconds to a minute. Lock must be maintained to four satellites and if loss of lock occurs, it must be reinitialized [that is, occupy known point, rapid static techniques, or on-the-fly (OTF) resolution—OTF requires dual-frequency receivers]. This technique is one of the more effective ways of locating topographic and built features, as for engineering surveys.
DGPS: The U.S. Coast Guard's system of providing differential code measurement surveys. Accuracy—submeter to 10 m. Roving receivers are equipped with radio receivers capable of receiving base station broadcasts of pseudorange corrections, using RTCM standards. For use by individual surveyors working within range of the transmitters (100 km to 400 km). Positions can be determined in real time. Surveyors using just one receiver have the equivalent of two receivers.
Real-time differential surveys: Also known as real-time kinematic (RTK). Accuracies—1 to 2 cm. Requires a base receiver occupying a known station, which then radio-transmits error corrections to any number of roving receivers, thus permitting them to perform data gathering and layout surveys in real time. All required software is on board the roving receivers. Dual-frequency receivers permit OTF reinitialization after loss of lock. Baselines are restricted to about 10 km. Five satellites are required. This, or similar techniques, is without doubt the future for many engineering surveys.
CORS: Nationwide differential positioning system—using code and/or carrier observations. When fully implemented this system of continuously operating reference stations will enable surveyors working with one GPS receiver to obtain the same results as if working with two. The CORS receiver is a highly accurate dual-frequency receiver. Surveyors can access station data for the appropriate location, date, and time via the Internet—data is then input to the software to combine with the surveyor's own data to produce accurate (postprocessed) positioning. Canada's nationwide system, active control system (ACS), provides base station data for a fee.

*Observation times and accuracies are affected by the quality and capability of the GPS receivers, by signal errors, and by the geometric strength of the visible satellite array (GDOP). Vertical accuracies are about half the horizontal accuracies.

Global Positioning System (GPS)   A ground positioning (Y, X, and Z) technique based on the reception and analysis of NAVSTAR satellite signals.

Ionosphere   The section of the earth's atmosphere that is about 50 km to 1,000 km above the earth's surface.

Ionospheric Refraction   The impedance in the velocity of signals (GPS) as they pass through the ionosphere.

Lidar   Light detection and ranging. This laser technique, used in airborne and satellite imagery, utilizes laser pulses that are reflected from ground features to obtain topographic and DTM mapping detail.

NAVSTAR   A set of orbiting satellites used in navigation and positioning.

Orthometric Height (H)   The distance, (measured along a plumb line) from the geoid surface to the ground surface. Also known as *elevation*.

Pseudorange   The uncorrected distance from a GPS satellite to a GPS ground receiver determined by comparing the code transmitted from the satellite to the replica code residing in the GPS receiver. When corrections are made for clock and other errors, the pseudorange becomes the range.

Real-Time Positioning   Surveying technique that requires a base station to measure the satellites' signals, process the baseline corrections, and then broadcast the corrections (differences) to any number of roving receivers that are simultaneously tracking the same satellites. Also known as *real-time kinematic (RTK) surveying*.

Relative Positioning   The determination of position through the combined computations of two or more receivers simultaneously tracking the same satellites, resulting in the determination of the baseline vector (X, Y, Z) joining two receivers.

Troposphere   The part of the earth's atmosphere that stretches from the surface to about 80 km upward (includes the stratosphere as its upper portion).

8.16   Recommended Readings

8.16.1   Books and Articles

Geomatics Canada, *GPS Positioning Guide* (Ottowa, Canada: Natural Resources Canada, 1993).

Hofman-Wellenhof et al., *GPS Theory and Practice*, Fourth Edition. (New York: Springer-Verlag Wien, 1997).

Leick, Alfred, *GPS Satellite Surveying*, Second Edition (New York: John Wiley & Sons, 1995).

Spofford, Paul, and Neil Weston, "CORS—The National Geodetic Surveys Continuously Operating Reference Station Project," *ACSM Bulletin*, March/April 1998.

Van Sickle, Jan, *GPS for Land Surveyors*. (Chelsea, Mich.: Ann Arbor Press Inc., 1996).

Wells, David, et al., *Guide to GPS Positioning* (Fredericton, New Brunswick: Canadian GPS Associates, 1986).

8.16.2   Periodicals

Consult these periodicals for general information, including archived articles.

ACSM Bulletin, American Congress on Surveying and Mapping,

http://www.acsm.net

GPS World, *http://www.gpsworld.com*

Point of Beginning (POB), *http://www.pobonline.com*

Professional Surveyor, *http://www.profsurv.com*

8.16.3   Web Sites

Go to these web sites for general information, reference and web links, and GPS receiver manufacturers. See also the list of additional Internet references in Appendix B.

Canada-wide real-time DGPS Service, *www.cdgps.com*

DGPS, *http://www.navcen.uscg.gov/dgps* (U.S. Coast Guard Navigation Center)

Galileo, *http://europa.eu.int/comm/dgs/energy_transport/galileo/index_en.htm*

GLONASS, *http://www.glonass-center.ru/nagu.txt*

Land Surveyors' Reference Page, Stan Thompson, PLS, Huntington Technology Group, *http://www.lsrp.com/*

Leica Geosystems, *http://www.leica.com/*

National Geodetic Survey (NGS), *http://www.ngs.noaa.gov/*

Natural Resources Canada, *http://www.nrcan.gc.ca/*

On-line Positioning Service (OPUS) *www.ngs.noaa.gov/OPUS/*

Sokkia, *http://www.sokkia.com*

Topcon, *www.topconpositioning.com/home.html*

Topcon Equipment, *http://topcongps.com*

Trimble, *http://www.trimble.com/*

Review Questions

8.1. Why is it necessary to observe a minimum of four GPS satellites to solve for position?

8.2. How does the GPS constellation compare with the GLONASS constellation?

8.3. How does differential positioning work?

8.4. What is the difference between range and pseudorange?

8.5. What are the chief sources of error in GPS measurements? How can you minimize or eliminate each of these errors?

8.6. What are the factors that must be analyzed in GPS survey planning?

8.7. Describe RTK techniques used for a layout survey.

8.8. What effect has GPS had on national control surveys?

8.9. Explain why station visibility diagrams are used in survey planning. Why is it better to locate a survey station on the south side of a tall building (rather than on the north side)?

8.10. Explain the difference between orthometric heights and ellipsoid heights.

8.11. Explain the CORS system.

Chapter 9

Horizontal Control Surveys

9.1 General Background

The highest order of control surveys was once thought to be national or continental in scope. With the advent of the global positioning system (GPS; see Chapter 8), control surveys are now based on frameworks that cover the entire surface of the earth, and they must take into account the ellipsoidal shape of the earth. Surveys that take into account the ellipsoidal shape of the earth are called **geodetic surveys.**

The early control net of the United States was tied into the control nets of both Canada and Mexico, giving a consistent continental net. The first major adjustment in control data was made in 1927, which resulted in the North American Datum (NAD27). Since that time, a great deal more has been learned about the shape and mass of the earth; these new and expanded data come to us from releveling surveys, precise traverses, very long baseline interferometry (VLBI), satellite laser ranging (SLR) surveys, earth movement studies, GPS observations, gravity surveys, etc. The data thus accumulated have been utilized to update and expand existing control data. The new geodetic data have also provided scientists with the means to define more precisely the actual geometric shape of the earth.

The reference solid previously used for this purpose (the Clarke spheroid of 1866) was modified to reflect the knowledge of the earth's dimensions at that time. Thus, a world geodetic system, first proposed in 1972 (WGS '72) and later endorsed in 1979 by the International Association of Geodesy (IAG), included proposals for an earth-mass-centered ellipsoid (GRS80 ellipsoid) that would represent more closely the planet on which we live. The ellipsoid was chosen over the spheroid as a representative solid model because of the slight bulging of the earth near the equator. The bulge is caused by the earth spinning on its polar axis. See Table 9.1 for the parameters of four models. See also Figure 8.26.

GRS80 was used to define the North American Datum of 1983 (NAD83), which covers the North American continent, including Greenland and parts of Central America. All individual control nets were included in a weighted simultaneous computation. A good tie

Table 9.1 TYPICAL REFERENCE SYSTEMS

Reference System	a (Semimajor) (m)	b (Semiminor) (m)	1/f (Flattening)
NAD83 (GRS80)	6,378,137.0	6,356,752.3	298.257222101
WGS84	6,378,137.0	6,356,752.3	298.257223563
ITRS	6,378,136.49	6,356,751.75	298.25645
NAD27 (Clarke, 1866)	6,378,206.4	6,356,583.8	294.978698214

to the global system was given by satellite positioning. The geographic coordinates of points in this system are latitude (ϕ) and longitude (λ). Although this system is widely used in geodetic surveying and mapping, it is too cumbersome for use in everyday surveying. For example, the latitude and longitude angles must be expressed to three or four decimals of a second (01.0000″) to give positions to the closest 0.01 ft. At latitude 44°, one second of latitude equals 101 ft and one second of longitude equals 73 ft. Conventional field surveying (as opposed to control surveying) is usually referenced to a plane grid (see Section 9.2).

In most cases, the accuracies between NAD83 first-order stations were better than 1:200,000, which would have been unquestioned in the pre-GPS era. However, the increased use of very precise GPS surveys and the tremendous potential for new applications for this technology created a demand for high-precision upgrades, using GPS techniques, to the control net.

9.1.1 Modern Considerations

A cooperative network upgrading program under the guidance of the National Geodetic Survey (NGS), including both federal and state agencies, began in 1986 in Tennessee and was completed in Indiana in 1997. This high accuracy reference network (HARN)—sometimes called the high precision geodetic network (HPGN)—resulted in about 16,000 horizontal control survey stations being upgraded to either AA-order, A-order, or B-order status. Horizontal AA-order stations have a relative accuracy of 3 mm ± 1:100,000,000 relative to other AA-order stations; horizontal A-order stations have a relative accuracy of 5 mm ± 1:10,000,000 relative to other A-order and AA-order stations; horizontal B-order stations have a relative accuracy of 8 mm ± 1:1,000,000 relative to other AA-order, A-order, and B-order stations. Of the 16,000 survey stations, the National Geodetic Survey has committed to the maintenance of about 1,400 AA- and A-order stations, which form the federal base network (FBN); individual states maintain the remainder of the survey stations, the B-order stations), which form the cooperative base network (CBN).

See Table 9.2 for accuracy standards for the new high accuracy reference network, (HARN) using GPS techniques and for traditional (pre-GPS) terrestrial techniques. The Federal Geodetic Control Subcommittee (FGCS) has published guidelines for the GPS field techniques (see Chapter 8) needed to achieve the various orders of surveys shown in Table 9.2. For example, AA-, A-, and B-order surveys require the use of receivers having both L1 and L2 frequencies, whereas the C-order results can be achieved using only a single-frequency (L1) receiver. Orders AA and A require five receivers observing simultaneously, order B requires four receivers observing simultaneously, and order C requires three receivers observing simultaneously.

Table 9.2 POSITIONING ACCURACY STANDARDS (95 PERCENT CONFIDENT LEVEL; MINIMUM GEOMETRIC ACCURACY STANDARD)

Survey Categories	Order	Base Error e (cm)	Base Error p (ppm)	Line-Length Dependent Error a (1:a)
HARN				
Federal base network (FBN)				
Global-regional geodynamics	**AA**	0.3	0.01	1:100,000,000
National Geodetic Reference System, primary network	**A**	0.5	0.1	1:10,000,000
Cooperative base network (CBN)				
National Geodetic Reference System, secondary networks	**B**	0.8	1	1:1,000,000
Terrestrial-based	**C**			
National Geodetic Reference System				
	1	1.0	10	1:100,000
	2-I	2.0	20	1:50,000
	2-II	3.0	50	1:20,000
	3	5.0	100	1:10,000

Source: From Geometric Geodetic Accuracy Standards Using GPS Relative Positioning Techniques. [Federal Geodetic Control Subcommittee (FGCS) 1988]. Publications are available through the National Geodetic Survey (NGS), (301) 443-8631.

Work was also completed on an improved vertical control net with revised values for about 600,000 benchmarks in the United States and Canada. This work was largely completed in 1988, resulting in a new North American Datum (NAVD88). The original adjustment of continental vertical values was performed in 1927.

The surface of the earth has been approximated by the surface of an ellipsoid, that is, the surface developed by rotating an ellipse on its minor axis. An ellipse (defined by its major axis and flattening), that most closely conformed to the geoid of the area of interest, which was usually continental in scope, was originally chosen. The reference ellipsoid chosen by the United States of America and Canada was one recommended by the International Association of Geodesy called the Geodetic Reference System 1980 (GRS80). See Table 9.1 for reference ellipsoid parameters; see Figures 8.26 through 8.30 for more on this topic.

The origin of a three-dimensional coordinate system was defined to be the center of the mass of the earth (geocentric), which is located in the equatorial plane. The z axis of the ellipsoid was defined as running from the origin through the mean location of the north pole, more precisely, the international reference pole as defined by the International Earth Rotation Service (IERS); z coordinates are measured upward (positive) or downward (negative) from the equatorial plane (see Figures 9.1 and 8.27). The x axis runs from the origin to a point of 0° longitude (Greenwich meridian) in the equatorial plane. The x coordinates are measured from the y-z plane, parallel to the x axis. They are positive from the 0 meridian 90° east and west; for the remaining 180°, they are negative. The y axis forms a right-handed coordinate frame (easterly from the x axis) 90° to the x axis and z axis. The y

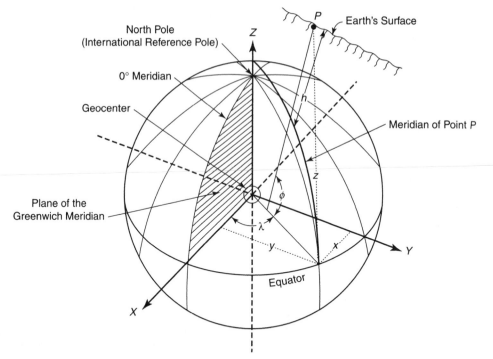

FIGURE 9.1 Ellipsoidal and geographic reference systems.

coordinates are measured perpendicular to the plane through the 0 meridian and are positive in the eastern hemisphere and negative in the western hemisphere. The semimajor axis (a) runs from the origin to the equator, and the semiminor axis (b) runs from the origin to the earth's north pole. Another defining parameter used to define ellipsoids is the flattening (f), which is defined to be $f = (a - b)/a$, or $f = 1 - b/a$ (see Figure 8.26 and Table 9.1).

Figure 9.1 shows the relationships between the ellipsoidal coordinates x, y, and z and the geodetic (or geographic) coordinates of latitude (φ), longitude (λ), and ellipsoidal height (h). Note that the x, y, and z dimensions are measured parallel to the X, Y, and Z axes, respectively, and that h is measured vertically up from the ellipsoidal surface to a point on the surface of the earth. For use in surveying, the ellipsoidal coordinates/geodetic coordinates are transformed into plane grid coordinates such as those used for the state plane grid or the universal transverse Mercator (UTM) grid. The ellipsoidal height (h) is also transformed into an orthometric height (elevation) by determining the geoid separation at a specified geographic location, as described in Section 8.13.

The axes of this coordinate system have not remained static for several reasons; for example, the earth's rotation varies, and the vectors between the positions of points on the surface of the earth do not remain constant because of plate tectonics. It is now customary to publish the x, y, and z coordinates along with velocities of change (plus or minus), in meters/year, for all three directions (vx, vy, and vz) for each station. For this reason, the axes are defined with respect to positions on the earth's surface at a particular epoch. The North American Datum of 1983 (NAD83), adopted in 1986, was first determined through measurements using very long baseline interferometry (VLBI) and satellite ranging.

These ongoing geodetic measurements together with continuous GPS observations (e.g., CORS) have discovered discrepancies, resulting in several upgrades to the parameters of NAD83. The IERS continues to monitor the positioning of the coordinates of their global network of geodetic observation stations, which now include GPS observations. This network is known as the International Terrestrial Reference Frame, with the latest reference epoch, at the time of this writing, set at the year 2000 (ITRF2000 or ITRF00).

For most purposes, the latest versions of NAD83 and WGS84 are considered identical. Because of the increases in accuracy occasioned by improvements in measurement technology, the National Geodetic Survey (NGS) is commencing an adjustment to their National Spatial Reference System (NSRS) of all GPS HARN stations (CORS stations' coordinates will be held fixed). This adjustment, when combined with their newest geoid model, expected by 2006 or 2007, should result in horizontal and vertical coordinate accuracies in the 1- to 2-cm range, including orthometric heights. The adjustments will affect all HARN stations in the federal base network (FBN), specifically its AA- and A-order stations, and all B-order stations in the cooperative base network (CBN). This adjustment (begun in 2005 and expected to be finished by 2007 or earlier) and both NAD83 (NSRS) and ITRF00 [or the latest ITRF (e.g., ITRF200x)] positional coordinates will be produced and published. The ITRF reference ellipsoid is very similar to GRS80 and WGS84, with slight changes in the *a* and *b* parameters and more significant changes in the flattening values (see Table 9.1). NGS reports that NAD83 is not being abandoned because many states have legislation specifying that datum. See the latest NGS news on this topic at *www.ngs.noaa.gov/NationalReadjustment/*.

9.1.2 Traditional Considerations

First-order horizontal control accuracy using terrestrial (preelectronics) techniques were originally established using **triangulation**. This technique involved (1) a precisely measured baseline as a starting side for a series of triangles or chains of triangles; (2) the determination of each angle in the triangle using a precise theodolite, which permitted the computation of the lengths of each side; and (3) a check on the work made possible by precisely measuring a side of a subsequent triangle (the spacing of check lines depended on the desired accuracy level). See Figure 9.2.

Triangulation was originally favored because the basic measurement of angles (and only a few sides) could be taken more quickly and precisely than could the measurement of all the distances (the surveying solution technique of measuring only the sides of a triangle is called **trilateration**). The advent of EDM instruments in the 1960s changed the approach to terrestrial control surveys. It became possible to measure the length of a triangle side precisely in about the same length of time as was required for angle determination.

Figure 9.2 shows two control survey configurations. Figure 9.2(a) depicts a simple chain of single triangles. In triangulation (angles only), this configuration suffers from the weakness that essentially only one route can be followed to solve for side *KL*. Figure 9.2(b) shows a chain of double triangles, or quadrilaterals. This configuration is preferred for triangulation because side *KL* can be solved using different routes (many more redundant measurements). Modern terrestrial control survey practice favors a combination of triangulation and trilateration (i.e., measure both the angles and the distances), which ensures many redundant measurements even for the simple chain of triangles shown in Figure 9.2(a).

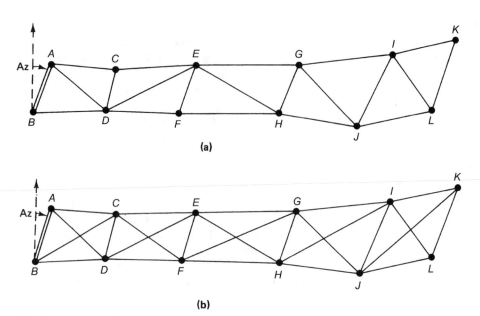

FIGURE 9.2 Control survey configurations. AB is the measured baseline, with known (or measured) azimuth. (a) Chain of single triangles. (b) Chain of double triangles (quadrilaterals).

Table 9.3 TRAVERSE SPECIFICATIONS—UNITED STATES

Classification	First Order	Second Order		Third Order	
		Class I	Class II	Class I	Class II
Recommended spacing of principal stations	Network stations 10–15 km; other surveys seldom less than 3 km	Principal stations seldom less than 4 km except in metropolitan area surveys, where the limitation is 0.3 km	Principal stations seldom less than 2 km except in metropolitan area surveys, where the limitation is 0.2 km	Seldom less than 0.1 km in tertiary surveys in metropolitan area surveys; as required for other surveys	
Position closure After azimuth adjustment	0.04 m $\sqrt{K}$ or 1:100,000	0.08 m $\sqrt{K}$ or 1:50,000	0.2 m $\sqrt{K}$ or 1:20,000	0.4 m $\sqrt{K}$ or 1:10,000	0.8 m $\sqrt{K}$ or 1:5000

Source: Federal Geodetic Control Committee, United States, 1974.

Whereas triangulation control surveys were originally used for basic state or provincial controls, precise traverses and GPS surveys are now used to densify the basic control net. The advent of reliable and precise EDM instruments has elevated the traverse to a valuable role, both in strengthening a triangulation net and in providing a stand-alone control figure itself. To provide reliability, traverses must close on themselves or on previously coordinated points. Table 9.3 summarizes characteristics and specifications for traverses. Tables 3.1 and 3.2 summarized characteristics and specifications for vertical control for the United States and Canada.

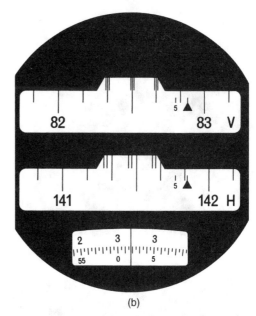

(a)   (b)

FIGURE 9.3 (a) Kern DKM 3 precise theodolite; angles read directly to 0.5 seconds; used in first-order surveys. (b) Kern DKM 3 scale reading (vertical angle = 82°53′01.8″). (Courtesy of Leica Geosystems Inc.)

More recently, with the introduction of the programmed total station, the process called resection is used much more often. **Resection** permits the surveyor to set up the total station at any convenient location and then, by sighting (measuring just angles or both angles and distances) to two or more coordinated control stations, the coordinates of the setup station can then be computed. See Section 5.3.3.

In traditional (pre-GPS) surveying, the surveyor had to use high-precision techniques to obtain high accuracy for conventional field control surveys. Several types of high-precision equipment, used to measure angles and vertical and horizontal slope distances, are illustrated in Figures 9.3, 9.4, and 9.5. Specifications for horizontal high-precision techniques stipulate the least angular count of the theodolite or total station, the number of observations, the rejection of observations exceeding specified limits from the mean, the spacing of major stations, and the angular and positional closures.

Higher-order specifications are seldom required for engineering or mapping surveys. An extensive interstate highway control survey could be one example where higher-order specifications are used in engineering work. Control for large-scale projects (for example, interchanges, large housing projects) that are to be laid out using polar ties (angle/distance) by total stations may require accuracies in the range of 1/10,000 to 1/20,000, depending on the project, and would fall between second- and third-order accuracy specifications (Table 9.2). Control stations established using GPS techniques have the inherent potential for higher orders of accuracy. The lowest requirements are reserved for small engineering or mapping projects that are limited in scope—for example, traffic studies, drainage studies, borrow pit volume surveys, etc.

FIGURE 9.4 Precise level. Precise optical levels have accuracies in the range of 1.5 mm to 1.0 mm for 1-km two-way leveling—depending on the instrument model and the type of leveling rod used. See also Figure 3.7. (Courtesy of Trimble)

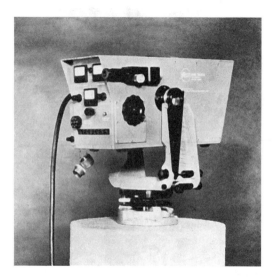

FIGURE 9.5 Kern Mekometer ME 3000, a high-precision EDM [SE = ± (0.2 mm ± 1 ppm)] with a triple-prism distance range of 2.5 km. Used wherever first-order results are required, for example, deformation studies, network surveys, plant engineering, and baseline calibration. (Courtesy of Leica Geosystems Inc.)

The American Congress on Surveying and Mapping (ACSM) and the American Land Title Association (ALTA) collaborated to produce new classifications for cadastral surveys based on present and proposed land use. These 1992 classifications (subject to state regulations) are shown in Table 9.4. Recognizing the impact of GPS techniques on all branches of surveying, in 1997 ACSM and ALTA published positional tolerances for different classes of surveys (see Table 9.5).

To enable the surveyor to perform reasonably precise surveys and still use plane geometry and trigonometry for related computations, several forms of plane coordinate grids have been introduced. These grids will be described in the next section.

Table 9.4 AMERICAN CONGRESS ON SURVEYING AND MAPPING MINIMUM ANGLE, DISTANCE, AND CLOSURE REQUIREMENTS FOR SURVEY MEASUREMENTS THAT CONTROL LAND BOUNDARIES FOR ALTA-ACSM LAND TITLE SURVEYS (1)

Direct Reading of Instrument (2)	Instrument Reading, Estimated (3)	Number of Observations per Station (4)	Spread from Mean of D&R Not to Exceed (5)	Angle Closure Where N = No. of Stations Not to Exceed . . .	Linear Closure (6)	Distance Measurement (7)	Minimum Length of Measurements (8), (9), (10)
20″<1′>$\boxed{10''}$	5″<0.1′>N.A.	2 D&R	5″<0.1′>$\boxed{5''}$	$10''\sqrt{N}$	1:15,000	EDM or double-tape with steel tape	(8) 81m, (9) 153 m, (10) 20 m

Note (1): All requirements of each class must be satisfied to qualify for that particular class of survey. The use of a more precise instrument does not change the other requirements, such as number of angles turned, etc.

Note (2): Instrument must have a direct reading of at least the amount specified (not an estimated reading), i.e.: 20″ = micrometer reading theodolite, < 1′ > = scale reading theodolite, $\boxed{10''}$ = electronic reading theodolite.

Note (3): Instrument must have the capability of allowing an estimated reading below the direct reading to the specified reading.

Note (4): D&R means the direct and reverse positions of the instrument telescope; i.e., urban surveys require that two angles in the direct and two angles in the reverse position be measured and meaned.

Note (5): Any angle measured that exceeds the specified amount from the mean must be rejected, and the set of angles must be remeasured.

Note (6): Ratio of closure after angles are balanced and closure is calculated.

Note (7): All distance measurements must be made with a property calibrated EDM or steel tape, applying atmospheric, temperature, sag, tension, slope, scale factor, and sea-level corrections as necessary.

Note (8): EDM having an error of 5 mm, independent of distance measured (manufacturer's specifications).

Note (9): EDM having an error of 10 mm, independent of distance measured (manufacturer's specifications).

Note (10): Calibrated steel tape.

Table 9.5 STANDARDS FOR LAND TITLE SURVEYS: POSITIONAL TOLERANCES FOR CLASSES OF SURVEY

Urban surveys	0.07 ft (or 20 mm) + 50 ppm
Suburban surveys	0.13 ft (or 40 mm) + 100 ppm
Rural surveys	0.26 ft (or 80 mm) + 200 ppm
Mountain/marshland surveys	0.66 ft (or 200 mm) + 200 ppm

Source: From Classifications of ALTA-ACSM Land Title Surveys, as adopted by American Land Title Association and ACSM, 1997.

9.2 Plane Coordinate Grids

9.2.1 General Background

The earth is ellipsoidal in shape, and if you try to portray a section of the earth on a flat map or plan, a certain amount of distortion is unavoidable. Also, some allowances must be made when you wish to create plane grids and use plane geometry and trigonometry to define the earth's curved surface. Over the years, various grids and projections have been

Table 9.6 STATE PLANE COORDINATE SYSTEMS

Transverse Mercator System		Lambert System		Both Systems
Alabama	Mississippi	Arkansas	North Dakota	Alaska
Arizona	Missouri	California	Ohio	Florida
Delaware	Nevada	Colorado	Oklahoma	New York
Georgia	New Hampshire	Connecticut	Oregon	
Hawaii	New Jersey	Iowa	Pennsylvania	
Idaho	New Mexico	Kansas	Puerto Rico	
Illinois	Rhode Island	Kentucky	South Carolina	
Indiana	Vermont	Louisiana	South Dakota	
Maine	Wyoming	Maryland	Tennessee	
		Massachusetts	Texas	
		Michigan	Utah	
		Minnesota	Virginia	
		Montana	Virgin Islands	
		Nebraska	Washington	
		North Carolina	West Virginia	
			Wisconsin	

employed. The United States uses the state plane coordinate grid system (SPCS), which utilizes both a transverse Mercator cylindrical projection and the Lambert conformal conic projection.

As already noted, geodetic control surveys are based on the best estimates of the actual shape of the earth. For many years, geodesists used the Clarke 1866 spheroid as a base for their work, including the development of the first North American Datum in 1927 (NAD27). The National Geodetic Survey (NGS) created the state plane coordinate system (SPCS27) based on the NAD27 datum. In this system, map projections are used that best suit the geographic needs of individual states (see Table 9.6): the Lambert conformal conical projection is used in states with larger east/west dimensions (see Figures 9.6 and 9.7), and the transverse Mercator cylindrical projection (see Figure 9.8) is used in states with larger north/south dimensions. To minimize the distortion that always occurs when a spherical surface is converted to a plane surface, the Lambert projection grid is limited to a relatively narrow strip of about 158 miles in a north/south direction, and the transverse Mercator projection grid is limited to about 158 miles in an east/west direction. At the maximum distance of 158 miles, or 254 km, a maximum scale factor of 1:10,000 exists at the zone boundaries. Also, as already noted, modernization in both instrumentation and technology permitted the establishment of a more representative datum based on the Geodetic Reference System 1980 (GRS80), which was used to define the new NAD83 datum. A new state plane coordinate system of 1983 (SPCS83) was developed based on the NAD83 datum.

Surveyors using both the old and new versions of SPCS can compute positions using tables and computer programs made available from NGS. SPCS83, which enables the surveyor to work in a more precisely defined datum then did SPCS27, uses similar mathematical approaches with some new nomenclature. For example, in SPCS27, the Lambert coordinates were expressed as X and Y, with values given in feet (U.S. survey foot, see Table 1.1), and the convergence angle (mapping angle) was displayed as θ; alternately, in the transverse Mercator grid, the convergence angle (in seconds) was designated by $\Delta\lambda''$. SPCS83 uses metric values for coordinates (designated as eastings and northings) as well as foot units

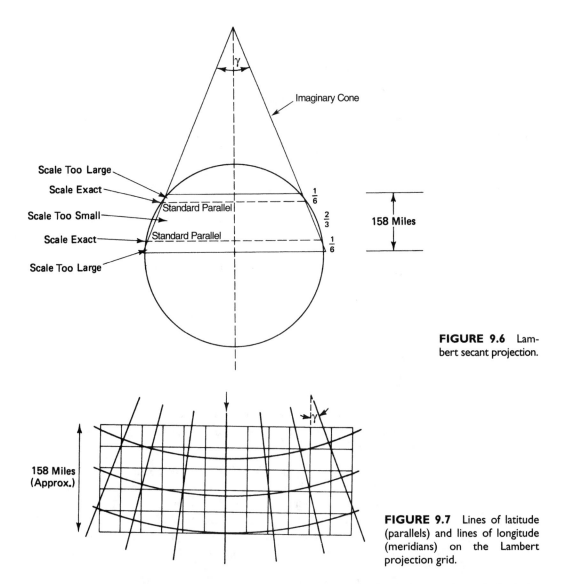

FIGURE 9.6 Lambert secant projection.

FIGURE 9.7 Lines of latitude (parallels) and lines of longitude (meridians) on the Lambert projection grid.

(U.S. survey foot or international foot). The convergence angle is now shown in both the Lambert and transverse Mercator projections as γ.

In North America, the grids used most often are the state plane coordinate grids. These grids are used in each U.S. state; the universal transverse Mercator grid is used in much of Canada. The Federal Communications Commission recently mandated that all cell phones be able to provide the spatial location of all 911 callers. By 2002, half the telephone carriers had opted for network-assisted GPS (NA-GPS) for 911 caller-location. The other carriers proposed to implement caller-location using the enhanced observed time difference of arrival (E-OTD); this technique utilizes the cellular network itself to pinpoint the caller-location.

With such major initiatives in the use of GPS to help provide caller-location, some believe that to provide seamless service, proprietary map databases should be referenced to

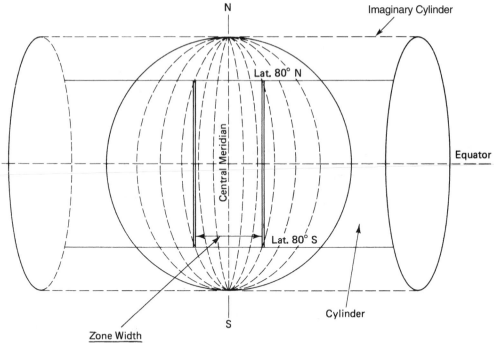

Imaginary Cylinder

Lat. 80° N

Central Meridian

Equator

Lat. 80° S

Cylinder

S

Zone Width

1. 6° for UTM

2. 3° for MTM (Some Regions in Canada)

3. About 158 Miles for State Plane Coordinate Grid (United States)

FIGURE 9.8 Transverse Mercator projection cylinder *tangent* to the earth's surface at the central meridian (CM) (see Figure 9.12 for zone number).

a common grid, that is, a U.S. national grid (USNG) for spatial addressing (see Section 9.5.3). One such grid being considered is the military grid reference system (MGRS), which is based on the universal transverse Mercator (UTM) grid. In addition to the need for a national grid for emergency (911) purposes, the Federal Geographic Data Committee (FGDC) recognizes the benefits of such a national grid for the many applications now developed for the GIS field. The ability to share data from one proprietary software program to another depends on a common grid, such as that proposed in the U.S. National Grid (USNG). See *http://www.fgdc.gov/standards/documents/proposals/usngprop.html*.

In addition to supplying tables for computations in SPCS83, the NGS provides both interactive computations on the Internet, and PC software available for downloading at *www.ngs.noaa.gov* [See also Section 8.8.1, which describes the NGS on-line user positioning system (OPUS)]. Many surveyors prefer computer-based computations to working with cumbersome tables. A manual that describes SPCS83 in detail, NOAA Manual NOS NGS 5: *State Plane Coordinate System of 1983*, is available from NGS.[*] This manual contains an introduction to SPCS, a map index showing all state plane coordinate zone numbers (zones are tied to state counties); which are required for converting state plane coordinates to geodetic

---

[*]To obtain NGS publications, contact NOAA, National Geodetic Survey, N/NGS12, 1315 East-West Highway, Station 9202, Silver Springs, MD 20910-3282. Publications can also be ordered by phoning (301) 713-3242.

positions; a table showing the SPCS legislative status of all states (1988); a discussion of the (t-T) convergence second term correction, which is needed for precise surveys of considerable extent (see Figure 9.18); and the methodology required to convert NAD83 latitude/longitude to SPCS83 northing/easting, plus the reverse process. This manual also contains the four equations needed to convert from latitude/longitude to northing/easting (that is, for northing, easting, convergence, and the grid scale factor) and the four equations to convert from northing/easting to latitude/longitude (latitude, longitude, convergence factor, and grid scale factor). Refer to the NGS manual for these conversion equation techniques. NGS uses the term *conversion* to describe this process and reserves the term *transformation* to describe the process of converting coordinates from one datum or grid to another, for example, from NAD27 to NAD83, or from SPCS27 to SPCS83 to UTM, etc.

The National Geodetic Survey (NGS) also has a range of software programs designed to assist the surveyor in several areas of geodetic inquiry. You can find the NGS toolkit at *www.ngs.noaa.gov/TOOLS/*. This site has on-line calculations capability for many of the geodetic activities listed below (to download PC software programs, go to *www.ngs.noaa.gov/* and click on the PC software icon):

- DEFLEC99—computes deflections of the vertical at the surface of the earth for the continental United States, Alaska, Puerto Rico, Virgin Islands, and Hawaii.
- G99SSS—computes the gravimetric height values for the continental United States.
- GEOID99—computes geoid height values for the continental United States.
- HTDP—time-dependent horizontal positioning software that allows users to predict horizontal displacements and/or velocities at locations throughout the United States.
- NADCON—transforms geographic coordinates between the NAD27, Old Hawaiian, Puerto Rico, or Alaska Island data and NAD83 values.
- State plane coordinate GPPCGP—converts NAD27 state plane coordinates to NAD27 geographic coordinates (latitudes and longitudes), and vice versa.
- SPCS83—converts NAD83 state plane coordinates to NAD83 geographic positions, and vice versa.
- Surface gravity prediction—predicts surface gravity at a specified geographic position and topographic height.
- Tidal information and orthometric elevations of a specific survey control mark—can be viewed graphically; this data can be referenced to NAVD88, NGVD29, and mean lower low water (MLLW) data.
- VERTCON—computes the modeled difference in orthometric height between the North American Vertical Datum of 1988 (NAVD88) and the National Geodetic Vertical Datum of 1929 (NGVD29) for any given location specified by latitude and longitude.

In Canada, software programs designed to assist the surveyor in various geodetic applications are available on the Internet from the Canadian Geodetic Survey at *www.geod.nrcan.gc.ca*. The following list is a selection of available services, including on-line applications and programs that can be downloaded:

- Precise GPS satellite ephemerides.
- GPS satellite clock corrections.

- GPS constellation information.
- GPS calendar.
- National gravity program.
- Universal transverse Mercator (UTM) to and from geographic coordinate conversion (UTM is in 6° zones with a scale factor of 0.9996).
- Transverse Mercator (TM) to and from geographic coordinate conversion (TM is in 3° zones with a scale factor of 0.9999, similar to U.S. state plane grids).
- GPS height transformation (based on GSD99; see Section 8.13).

■ **EXAMPLE 9.1** *Use of the NGS Toolkit to Convert Coordinates*

(a) Convert geodetic positions to state plane coordinates.

(b) Convert state plane coordinates to geodetic positions.

Solution

(a) When user selects *http://www.ngs.noaa.gov//TOOLS/spc.html* and selects latitude/longitude > SPC, he or she is asked to choose NAD83 or NAD27, and to enter the geodetic coordinates and the zone number (the zone number is not really required here because the program automatically generates the zone number directly from the geodetic coordinates of latitude and longitude). The longitude degree entry must always be three digits, 079 in this example:

O NAD 83

O NAD 27

Latitude **N421423.0000**

Longitude **W0792035.0000**

Zone [   ] (This can be left blank.)

The program response is:

INPUT =	Latitude	Longitude	Datum	Zone
	N421423.0000	W0792035.0000	NAD 83	3103
North (Y)	**East (X)**	**Area**	**Convergence**	**Scale**
Meters	**Meters**		**DD MM SS.ss**	
248999.059	**287296.971**	**NY W**	**−0 30 38.62**	**0.99998586**

(b) When the user selects *http://www.ngs.noaa.gov/TOOLS/spc.html* and then selects SPC > latitude/longitude, he or she must select either NAD83 or NAD27, and must enter the state plane coordinates and the SPCS zone number:

O NAD 83

O NAD 27

Northing = **248999.059**

Easting = **287296.971**

Zone = **3103**

The program response is:

INPUT =	North (Meters)	East (Meters)		Datum	Zone
	248999.059	287296.971		NAD 83	3103

Latitude DD MM SS.sssss	Longitude DD MM SS.sssss	Area	Convergence	Scale Factor
42 14 23.00000	079 20 35.00001	NY W	$-0°30\ 38.62$	0.9999859

■ **EXAMPLE 9.2** *Use of the Canadian Geodetic Survey On-line Sample Programs* The programs can be downloaded free. Use the same geographic position as in Section 9.2.2:

(a) Convert geographic position to universal transverse Mercator.

(b) Convert universal transverse Mercator to geographic position.

(a) Go to: *http://www.geod.nrcan.gc.ca* and select *English—Online Applications— GSRUG* and then select *Geographic to UTM*. Enter the geographic coordinates of the point you want to compute. For this example, enter the following:

Latitude: **42** degrees **14** minutes **23.0000** seconds **north**

Longitude: **079** degrees **20** minutes **35.0000** seconds **west**

Ellipsoid: **GRS80 (NAD 83, WGS84)**

Zone width: **6° UTM**

The desired ellipsoid and zone width, 6° or 3°, are selected by highlighting the appropriate entry while scrolling through the list. The program response is:

Input Geographic Coordinates

Latitude: 42 degrees 14 minutes 23.0000 seconds North

Longitude: 079 degrees 20 minutes 35.0000 seconds West

Ellipsoid: NAD 83 (WGS84)

Zone Width: 6 UTM

Output: UTM Coordinates:

UTM Zone: 17

Northing: 4677721.911 meters North

Easting: 636709.822 meters

(b) Go to: *http://www.geod.nrcan.gc.ca* and select *English—Online Applications— GSRUG* and then select *UTM to Geographic*. Enter the UTM coordinates of the point you want to compute. For this example, enter the following:

Zone: **17**

Northing: **4677721.911** meters **North**

Easting: **636709.822** meters

Ellipsoid: **GRS80 (NAD83, WGS84)** Zone Width **6 UTM**

The program's response is:

Input Geographic Coordinates
UTM Zone 17
Northing: 4677721.911 meters North
Easting: 636709.822 meters
Ellipsoid: NAD83 (WGS84)
Zone Width: 6 UTM

Output Geographic Coordinates
Latitude: 42 degrees 14 minutes 23.000015 seconds North
Longitude: 79 degrees 20 minutes 34.999989 seconds West

9.3 Lambert Projection

The Lambert projection is a conical conformal projection. The imaginary cone is placed around the earth so that the apex of the cone is on the earth's axis of rotation above the north pole for northern hemisphere projections, and below the south pole for southern hemisphere projections. The location of the apex depends on the area of the ellipsoid that is being projected. Figures 9.6 and 9.7 confirm that, although the east-west direction is relatively distortion-free, the north-south coverage must be restrained (e.g., to 158 miles) to maintain the integrity of the projection; therefore, the Lambert projection is used for states having a greater east-west dimension, such as Pennsylvania and Tennessee. Table 9.6 lists all the states and the type of projection each uses; New York, Florida, and Alaska utilize both the transverse Mercator and the Lambert projections. The NGS publication *State Plane Coordinate Grid System of 1983* gives a more detailed listing of each state's projection data.

9.4 Transverse Mercator Grid System

The transverse Mercator projection is created by placing an imaginary cylinder around the earth, with the cylinder's circumference tangent to the earth along a meridian (central meridian; see Figure 9.8). When the cylinder is flattened, a plane is developed that can be used for grid purposes. At the central meridian, the scale is exact [Figures 9.8 and 9.9(a)], and the scale becomes progressively more distorted as the distance east and west of the central meridian increases. This projection is used in states with a more predominant north/south dimension, such as Illinois and New Hampshire. The distortion (which is always present when a spherical surface is projected onto a plane) can be minimized in two ways. First, the distortion can be minimized by keeping the zone width relatively narrow (158 miles in SPCS); second, the distortion can be lessened by reducing the radius of the projection cylinder (secant projection) so that, instead of being tangent to the earth's surface, the cylinder cuts through the earth's surface at an optimal distance on either side of the central meridian [see Figures 9.9(b) and 9.10]. Thus, the scale factor at the central meridian is less than unity (0.9999); it is unity at the line of intersection at the earth's surface and greater than unity between the lines of intersection and the zone limit meridians.

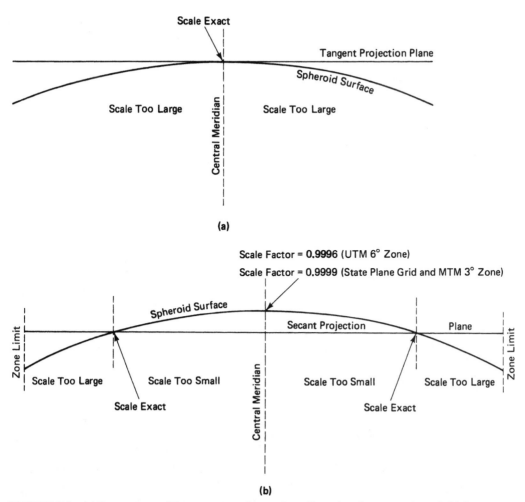

FIGURE 9.9 (a) Section view of the projection plane and earth's surface (*tangent projection*). (b) Section view of the projection plane and the earth's surface (*secant projection*).

Figure 9.11 shows a cross section of an SPCS transverse Mercator zone. For both the Lambert and transverse Mercator grids, the scale factor of 0.9999 (this value is much improved for some states in the SPCS83) at the central meridian gives surveyors the ability to work within a specification of 1:10,000 while neglecting the impact of scale distortion.

9.5 Universal Transverse Mercator (UTM) Grid System

9.5.1 General Background

The UTM grid is much as described above except that the zones are wider—set at a width 6° of longitude. This grid is used worldwide for both military and mapping purposes. UTM coordinates are now published (in addition to SPCS and geodetic coordinates) for all

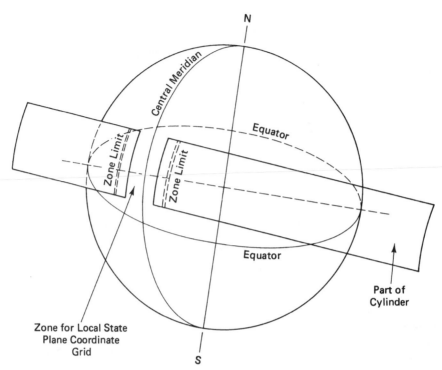

FIGURE 9.10 Transverse Mercator projection. *Secant* cylinder for state plane coordinate grids.

NAD83 control stations. With a wider zone width than the SPCS zones, the UTM has a scale factor at the central meridian of only 0.9996. Surveyors working at specifications better than 1:2,500 must apply scale factors in their computations.

UTM zones are numbered beginning at longitude 180°W from 1 to 60. Figure 9.12(a) shows that U.S. territories range from zone 1 to zone 20 and that Canada's territory ranges from zone 7 to zone 22. The central meridian of each zone is assigned a false easting of 500,000 m, and the northerly is based on a value of 0 at the equator.

CHARACTERISTICS OF THE UNIVERSAL TRANSVERSE MERCATOR (UTM) GRID SYSTEM

1. Zone is 6° wide. Zone overlap of 0°30′ (see Table 9.7)
2. Latitude of the origin is the equator, 0°.
3. Easting value of each central meridian = 500,000.000 m.
4. Northing value of the equator = 0.000 m (10,000,000.000 m in the southern hemisphere).
5. Scale factor at the central meridian is 0.9996 (i.e., 1/2,500).
6. Zone numbering commences with 1 in the zone 180°W to 174°W and increases eastward to zone 60 at the zone 174°E to 180°E [see Figure 9.12(a)].
7. Projection limits of latitude 80°S to 80°N.

See Figure 9.13 for a cross section of a 6° zone (UTM).

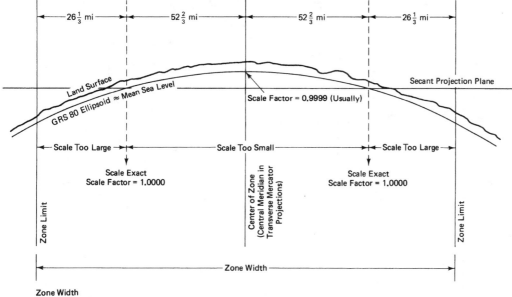

Zone Width

About 158 Miles for State Plane Coordinate Systems:
— East-West Orientation for Transverse Mercator Projections.
— North-South Orientation for Lambert Projections.

FIGURE 9.11 Section of the projection plane and the earth's surface for state plane grids (*secant projection*).

9.5.2 Modified Transverse Mercator (MTM) Grid System

Some regions and agencies outside the United States have adopted a modified transverse Mercator system. The modified projection is based on 3° wide zones instead of 6° wide zones. By narrowing the zone width, the scale factor at the central meridian is improved from 0.9996 (1/2,500) to 0.9999 (1/10,000), the same as for the SPCS grids. The improved scale factor permits surveyors to work at moderate levels of accuracy without having to account for projection corrections. The zone width of 3° (about 152 miles wide at latitude 43°) compares very closely with the 158-mile-wide zones used in the United States for transverse Mercator and Lambert projections in the state plane coordinate system.

CHARACTERISTICS OF THE 3° ZONE

1. Zone is 3° wide.
2. Latitude of origin is the equator, 0°.
3. Easting value of the central meridian, for example, 1,000,000.000 ft or 304,800.000 m, is set by the agency.
4. Northing value of the equator is 0.000 ft or m.
5. Scale factor at the central meridian is 0.9999 (i.e., 1/10,000).

Keep in mind that narrow grid zones (1:10,000) permit the surveyor to ignore only corrections for scale and that other corrections to field measurements, such as for elevation,

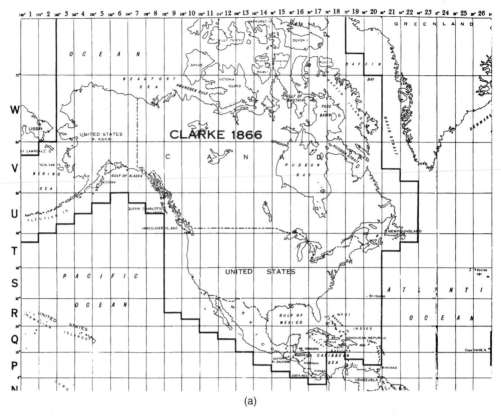

FIGURE 9.12 (a) Universal transverse Mercator grid zone numbering system.

(*continued*)

temperature, sag, etc., and the balancing of errors are still routinely required. See Figures 9.13 and 9.14 for cross sections of the UTM and modified transverse Mercator projection planes.

9.5.3 The United States National Grid (USNG)

As noted in previous sections, points on the surface of the earth can be identified by several types of coordinate systems, for example, the geographic coordinates of latitude and longitude, state plane coordinates, UTM coordinates, and MTM coordinates. Points on the earth's surface in the United States can be identified (georeferenced) by using the United States national grid (USNG); the USNG is an expansion of the long-established military grid reference system (MGRS), which is used in many countries. The need for a simpler georeferencing system became apparent with the advent of 911 emergency procedures. A GPS-equipped cell phone (or handheld GPS receiver) can display location information in different coordinate systems such as latitude and longitude, UTM coordinates,

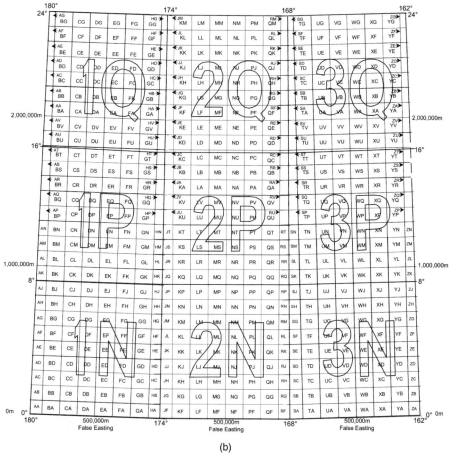

(b)

FIGURE 9.12 (*continued*) (b) Basic plan of the 100,000-meter square identification of the United States national grid (USNG).

(*continued*)

USNG coordinates, etc. Many think that the USNG is easier for the general population to comprehend and that it could be a good choice when a standard grid is selected for nationwide use.

Based on the framework of the worldwide UTM coordinate system, USNG is an alphanumeric reference grid that includes three levels of identity. It covers the earth's surface from 80° south to 84° north of the equator. The first level of location precision is denoted by the UTM zone number and the latitude band letter, and is known as the *grid zone designation* (GZD). It usually covers an area of 6° in an east/west direction and 8° in a north/south direction. UTM zones, each covering 6° of longitude, are numbered eastward from the 180° meridian [see Figure 9.12(a)]. For example, the state of Florida is in zone 17. In North America, the zones in the conterminous United States run from zone 10 (meridian

ZONES	SET 1 1, 7, 13, 19, 25, 31, 37, 43, 49, 55								SET 2 2, 8, 14, 20, 26, 32, 38, 44, 50, 56								SET 3 3, 9, 15, 21, 27, 33, 39, 45, 51, 57								SET 4 4, 10, 16, 22, 28, 34, 40, 46, 52, 58								SET 5 5, 11, 17, 23, 29, 35, 41, 47, 53, 59								SET 6 6, 12, 18, 24, 30, 36, 42, 48, 54, 60							
2,000,000 m	AV	BV	CV	DV	EV	FV	GV	HV	JE	KE	LE	ME	NE	PE	QE	RE	SV	TV	UV	VV	WV	XV	YV	ZV	AE	BE	CE	DE	EE	FE	GE	HE	JV	KV	LV	MV	NV	PV	QV	RV	SE	TE	UE	VE	WE	XE	YE	ZE
	AU	BU	CU	DU	EU	FU	GU	HU	JD	KD	LD	MD	ND	PD	QD	RD	SU	TU	UU	VU	WU	XU	YU	ZU	AD	BD	CD	DD	ED	FD	GD	HD	JU	KU	LU	MU	NU	PU	QU	RU	SD	TD	UD	VD	WD	XD	YD	ZD
	AT	BT	CT	DT	ET	FT	GT	HT	JC	KC	LC	MC	NC	PC	QC	RC	ST	TT	UT	VT	WT	XT	YT	ZT	AC	BC	CC	DC	EC	FC	GC	HC	JT	KT	LT	MT	NT	PT	QT	RT	SC	TC	UC	VC	WC	XC	YC	ZC
	AS	BS	CS	DS	ES	FS	GS	HS	JB	KB	LB	MB	NB	PB	QB	RB	SS	TS	US	VS	WS	XS	YS	ZS	AB	BB	CB	DB	EB	FB	GB	HB	JS	KS	LS	MS	NS	PS	QS	RS	SB	TB	UB	VB	WB	XB	YB	ZB
1,500,000 m	AR	BR	CR	DR	ER	FR	GR	HR	JA	KA	LA	MA	NA	PA	QA	RA	SR	TR	UR	VR	WR	XR	YR	ZR	AA	BA	CA	DA	EA	FA	GA	HA	JR	KR	LR	MR	NR	PR	QR	RR	SA	TA	UA	VA	WA	XA	YA	ZA
	AQ	BQ	CQ	DQ	EQ	FQ	GQ	HQ	JV	KV	LV	MV	NV	PV	QV	RV	SQ	TQ	UQ	VQ	WQ	XQ	YQ	ZQ	AV	BV	CV	DV	EV	FV	GV	HV	JQ	KQ	LQ	MQ	NQ	PQ	QQ	RQ	SV	TV	UV	VV	WV	XV	YV	ZV
	AP	BP	CP	DP	EP	FP	GP	HP	JU	KU	LU	MU	NU	PU	QU	RU	SP	TP	UP	VP	WP	XP	YP	ZP	AU	BU	CU	DU	EU	FU	GU	HU	JP	KP	LP	MP	NP	PP	QP	RP	SU	TU	UU	VU	WU	XU	YU	ZU
	AN	BN	CN	DN	EN	FN	GN	HN	JT	KT	LT	MT	NT	PT	QT	RT	SN	TN	UN	VN	WN	XN	YN	ZN	AT	BT	CT	DT	ET	FT	GT	HT	JN	KN	LN	MN	NN	PN	QN	RN	ST	TT	UT	VT	WT	XT	YT	ZT
	AM	BM	CM	DM	EM	FM	GM	HM	JS	KS	LS	MS	NS	PS	QS	RS	SM	TM	UM	VM	WM	XM	YM	ZM	AS	BS	CS	DS	ES	FS	GS	HS	JM	KM	LM	MM	NM	PM	QM	RM	SS	TS	US	VS	WS	XS	YS	ZS
1,000,000 m	AL	BL	CL	DL	EL	FL	GL	HL	JR	KR	LR	MR	NR	PR	QR	RR	SL	TL	UL	VL	WL	XL	YL	ZL	AR	BR	CR	DR	ER	FR	GR	HR	JL	KL	LL	ML	NL	PL	QL	RL	SR	TR	UR	VR	WR	XR	YR	ZR
	AK	BK	CK	DK	EK	FK	GK	HK	JQ	KQ	LQ	MQ	NQ	PQ	QQ	RQ	SK	TK	UK	VK	WK	XK	YK	ZK	AQ	BQ	CQ	DQ	EQ	FQ	GQ	HQ	JK	KK	LK	MK	NK	PK	QK	RK	SQ	TQ	UQ	VQ	WQ	XQ	YQ	ZQ
	AJ	BJ	CJ	DJ	EJ	FJ	GJ	HJ	JP	KP	LP	MP	NP	PP	QP	RP	SJ	TJ	UJ	VJ	WJ	XJ	YJ	ZJ	AP	BP	CP	DP	EP	FP	GP	HP	JJ	KJ	LJ	MJ	NJ	PJ	QJ	RJ	SP	TP	UP	VP	WP	XP	YP	ZP
	AH	BH	CH	DH	EH	FH	GH	HH	JN	KN	LN	MN	NN	PN	QN	RN	SH	TH	UH	VH	WH	XH	YH	ZH	AN	BN	CN	DN	EN	FN	GN	HN	JH	KH	LH	MH	NH	PH	QH	RH	SN	TN	UN	VN	WN	XN	YN	ZN
	AG	BG	CG	DG	EG	FG	GG	HG	JM	KM	LM	MM	NM	PM	QM	RM	SG	TG	UG	VG	WG	XG	YG	ZG	AM	BM	CM	DM	EM	FM	GM	HM	JG	KG	LG	MG	NG	PG	QG	RG	SM	TM	UM	VM	WM	XM	YM	ZM
500,000 m	AF	BF	CF	DF	EF	FF	GF	HF	JL	KL	LL	ML	NL	PL	QL	RL	SF	TF	UF	VF	WF	XF	YF	ZF	AL	BL	CL	DL	EL	FL	GL	HL	JF	KF	LF	MF	NF	PF	QF	RF	SL	TL	UL	VL	WL	XL	YL	ZL
	AE	BE	CE	DE	EE	FE	GE	HE	JK	KK	LK	MK	NK	PK	QK	RK	SE	TE	UE	VE	WE	XE	YE	ZE	AK	BK	CK	DK	EK	FK	GK	HK	JE	KE	LE	ME	NE	PE	QE	RE	SK	TK	UK	VK	WK	XK	YK	ZK
	AD	BD	CD	DD	ED	FD	GD	HD	JJ	KJ	LJ	MJ	NJ	PJ	QJ	RJ	SD	TD	UD	VD	WD	XD	YD	ZD	AJ	BJ	CJ	DJ	EJ	FJ	GJ	HJ	JD	KD	LD	MD	ND	PD	QD	RD	SJ	TJ	UJ	VJ	WJ	XJ	YJ	ZJ
	AC	BC	CC	DC	EC	FC	GC	HC	JH	KH	LH	MH	NH	PH	QH	RH	SC	TC	UC	VC	WC	XC	YC	ZC	AH	BH	CH	DH	EH	FH	GH	HH	JC	KC	LC	MC	NC	PC	QC	RC	SH	TH	UH	VH	WH	XH	YH	ZH
	AB	BB	CB	DB	EB	FB	GB	HB	JG	KG	LG	MG	NG	PG	QG	RG	SB	TB	UB	VB	WB	XB	YB	ZB	AG	BG	CG	DG	EG	FG	GG	HG	JB	KB	LB	MB	NB	PB	QB	RB	SG	TG	UG	VG	WG	XG	YG	ZG
0 m	AA	BA	CA	DA	EA	FA	GA	HA	JF	KF	LF	MF	NF	PF	QF	RF	SA	TA	UA	VA	WA	XA	YA	ZA	AF	BF	CF	DF	EF	FF	GF	HF	JA	KA	LA	MA	NA	PA	QA	RA	SF	TF	UF	VF	WF	XF	YF	ZF
	200,000 m 300,000 400,000 500,000 600,000 700,000 800,000 m								200,000 m 300,000 400,000 500,000 600,000 700,000 800,000 m								200,000 m 300,000 400,000 500,000 600,000 700,000 800,000 m								200,000 m 300,000 400,000 500,000 600,000 700,000 800,000 m								200,000 m 300,000 400,000 500,000 600,000 700,000 800,000 m								200,000 m 300,000 400,000 500,000 600,000 700,000 800,000 m							

FIGURE 9.12 (*continued*)   (c) Organization of the U.S. national grid (USNG) 100,000-m grid squares.

Table 9.7   UTM ZONE WIDTH

North Latitude	Width (km)
42°00′	497.11827
43°00′	489.25961
44°00′	481.25105
45°00′	473.09497
46°00′	464.79382
47°00′	456.35005
48°00′	447.76621
49°00′	439.04485
50°00′	430.18862

Source: Ontario Geographical Referencing Grid,
Ministry of Natural Resources, Ontario, Canada.

126°W) eastward to zone 20 (meridian 60°W), and the zones in Canada run from zone 7 (meridian 144°W) eastward to zone 22 (meridian 48°W).

Latitude bands, for the most part, cover 8° of latitude (the exception is band X, which covers from 72° north to 84° north) and are identified by letters. In the northern hemisphere, latitude band coverage is as follows:

LATITUDE BAND	COVERAGE
N	0° to 8°N
P	8°N to 16°N
Q	16°N to 24°N
R	24°N to 32°N
S	32°N to 40°N
T	40°N to 48°N
U	48°N to 56°N
V	56°N to 64°N
W	64°N to 72°N
X	72°N to 84°N

Note that the letters O, Y, and Z are not used for designating a latitude band. In the conterminous United States, latitude band letters run northerly from letter R (latitude 24°N) in the southern United States to letter U (latitude 56°N) for the northern states. The latitude bands covering Canada run from the letter T (latitude 40°N) in southern Ontario and Quebec up to letter W (latitude 72°N) for the rest of the country. Refer to Figure 9.12(a) to see that the GZD for the state of Florida is 17R.

The second level of location precision is a 100,000-m^2 designation. This designation is given by two unique letters that repeat every three UTM zones (east/west) and every 2,000,000 m north of the equator; thus, there is little opportunity for mistaken identification. These 100,000-m^2 identifiers are defined in the document *United States National Grid,* Standards Working Group, Federal Geographic Data Committee, December 2001, available at www.fgdc.gov/standards/status/usng.html. See Figure 9.12(b) and 9.12(c). Note that the northerly progression of letters in the first column (180° meridian) begins at AA at the equator

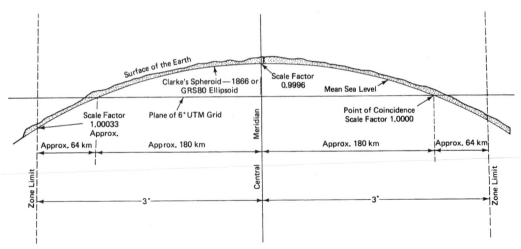

FIGURE 9.13 Cross section of a 6° zone (UTM).

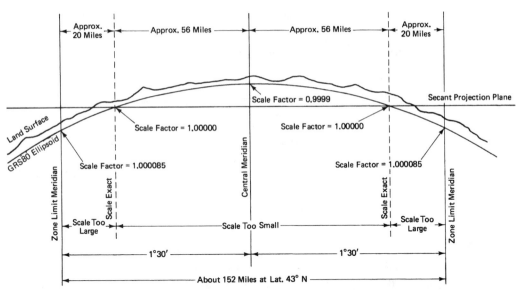

FIGURE 9.14 Section of the projection plane for the modified transverse Mercator (3°) grid and the earth's surface (secant projection).

and progresses northerly for twenty bands (2,000,000 m) to AV (letters I, O, X, W, Y, and Z are not used). They commence again at AA and continue on northerly. Also note that in the easterly progression of letters, at the equator, the letters progress easterly from AA (at 180° meridian) through twenty-four squares to AZ (letters I and O are not used). In each zone, the most westerly and easterly columns of "squares" become progressively narrower than 100,000 m as the meridians converge northerly. Just as with the UTM, the false northing is 0.00 m at the equator, and the false eastings are set at 500,000 m at the central meridian of each zone.

The third level of location precision is given by coordinates unique to a specific 100,000-m^2 grid. The coordinates are an even number of digits ranging from 2 to 10. The coordinates are written in a string, with no space between easting and northing values (easting is always listed first). These grid coordinates follow the grid zone designation and the 100,000-m^2 identification letters in the string identification. For example, the *United States National Grid* (page 8) shows the following coordinates:

- 18SUJ20 locates a point with 10-km precision (18S is the grid zone designation and UJ is the 100,000-m^2 identification, 18 is the UTM zone extending from longitude 78°W to 72°W, and band S extends from 32°N to 40°N).
- 18SUJ2306 locates a point with 1-km precision. (The first half of the grid numbers is the grid easting and the second half is the grid northing—i.e., 23 km east, 06 km north.)
- 18SUJ234064 locates a point with 100-m precision.
- 18SUJ23480647 locates a point with 10-m precision.
- 18SUJ2348306479 locates a point with 1-m precision (23,483 m east and 6,479 m north, measured from the southwest corner of square UJ, in the S band and in UTM zone 18).

The NGS toolkit contains interactive software to convert UTM and latitude/longitude to USNG, and vice versa; see www.ngs.noaa.gov/TOOLS/usng.html.

9.6 Use of Grid Coordinates

9.6.1 Grid/Ground Distance Relationships: Elevation and Scale Factors

When local surveys (traverse or trilateration) are tied into coordinate grids, corrections must be provided so that:

1. Grid and ground distances can be reconciled by applying elevation and scale factors.
2. Grid and geodetic directions can be reconciled by applying convergence corrections.

9.6.1.1 Elevation Factor. Figures 9.14 and 9.15 show the relationship among ground distances, sea-level distances, and grid distances. A distance measured on the earth's surface must first be reduced for equivalency at sea level, and then it must be further reduced (in Figure 9.14) for equivalency on the projection plane. The first reduction involves multiplication by an **elevation factor** (sea-level factor); the second reduction (adjustment) involves multiplication by the **scale factor**.

The elevation (sea-level) factor can be determined by establishing a ratio, as is illustrated in Figures 9.15(a) and 9.16:

$$\text{Elevation factor} = \frac{\text{sea-level distance}}{\text{ground distance}} = \frac{R}{R + H} \qquad (9.1)$$

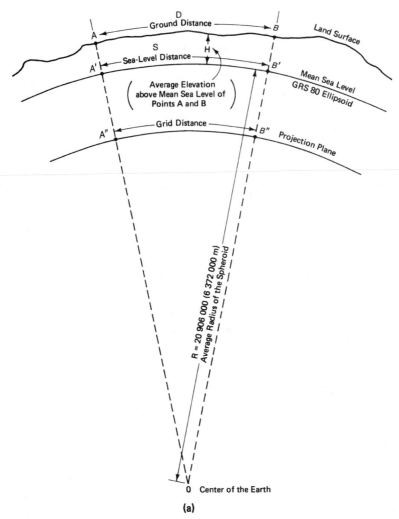

FIGURE 9.15 (a) General case: relationship of ground distances to sea-level distances and grid distances.
(*continued*)

where R is the average radius of the earth (average radius of sea-level surface = 20,906,000 ft or 6,372,000 m), and H is the elevation above mean sea level. For example, at 500 ft, the elevation factor would be:

$$\frac{20,906,000}{20,906,500} = 0.999976$$

and a ground distance of 800.00 ft at an average elevation of 500 ft would become 800 × 0.999976 = 799.98 at sea level.

Elevation Factor

$$\frac{S}{D} = \frac{R}{R + h}$$

$$S = D\left(\frac{R}{R + h}\right)$$

$$h = N + H$$

$$S = D\left(\frac{R}{R + N + H}\right)$$

Where
- S = Geodetic Distance
- D = Horizontal Distance
- H = Mean Elevation
- N = Mean Geoid Height
- R = Mean Radius of Earth
  (6,372,000 m or
  20,906,000 ft)

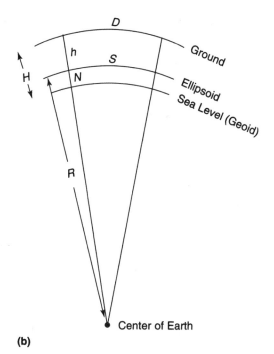

(b)

FIGURE 9.15 *(continued)* (b) Impact of geoid separation on elevation factor determination. Used only on very precise surveys.

Note: Figure 9.15(b) shows the case (encountered in very precise surveys) where the geoid separation (N) must also be considered. Geoid separation (also known as geoid height and geoid undulation) is the height difference between the sea-level surface and the ellipsoid surface. NOAA Manual NOS NGS5, *State Plane Coordinate System of 1983*, notes, for example, that a geoid height of -30 m (in the conterminous United States, the ellipsoid is above the geoid) systematically affects reduced distances by -4.8 ppm (1:208,000), which is certainly not a factor in any but the most precise surveys. See also Section 8.13.

9.6.1.2 Scale Factor.

For state plane projections, the computer solution gives scale factors for positions of latitude difference (Lambert projection) or for distances east or west of the central meridian (transverse Mercator projections). By way of illustration, scale factors for the transverse Mercator projections may also be computed by the following equation:

$$M_p = M_o(1 + x^2/2R^2) \tag{9.2}$$

where M_p = the scale factor at the survey station
M_o = the scale factor at the central meridian (CM)
x = the east/west distance of the survey station from the central meridian
R = the average radius of the spheroid.
$x^2/2R^2$ can be expressed in feet, meters, miles, or kilometers.

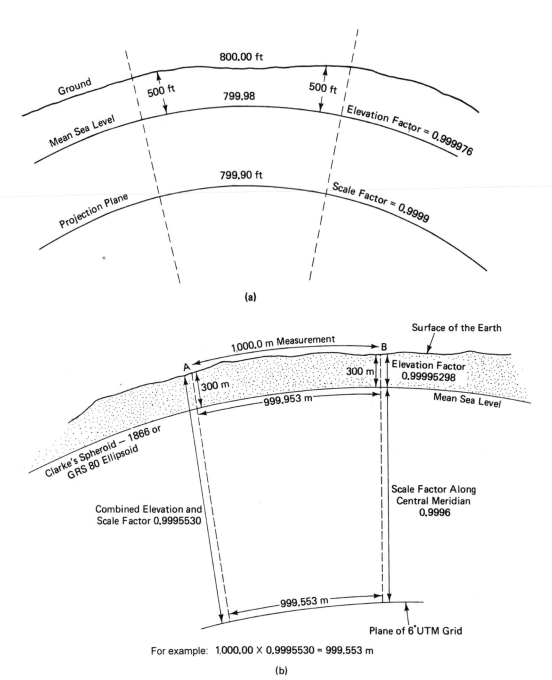

FIGURE 9.16 Conversion of a ground distance to a grid distance using the elevation factor and the scale factor. (a) SPCS83 grid. (b) Universal transverse Mercator (UTM) grid, 6° zone.

For example, survey stations 12,000 ft from a central meridian, and having a scale factor of 0.9999, have the scale factor determined as follows:

$$M_p = 0.9999 \left(1 + \frac{12,000^2}{2 \times 20,906,000^2} \right) = 0.9999002$$

9.6.1.3 Combined Factor.
When the elevation factor is multiplied by the scale factor, the result is known as the combined factor. This relationship is expressed in the following equation:

$$\text{Ground distance} \times \text{combined factor} = \text{grid distance} \tag{9.3}$$

Equation 9.3 can also be expressed as follows:

$$\frac{\text{grid distance}}{\text{combined factor}} = \text{ground distance}$$

In practice, it is seldom necessary to use Equations 9.1 and 9.2 because computer programs are now routinely used for computations in all state plane grids and the universal transverse Mercator grid. Computations were previously based on data from tables and graphs (see Figure 9.17 and Tables 9.8 and 9.9).

■ EXAMPLE 9.3
Using Table 9.8 (MTM projection), determine the combined scale and elevation factor (combined factor) of a point 125,000 ft from the central meridian (scale factor = 0.9999) and at an elevation of 600 ft above mean sea level.

Solution

Elevation	Distance from Central Meridian (Thousands of Feet)		
	100	150	125 (Interpolated)
500	.999888	.999902	.999895
750	.999876	.999890	.999883
			.000012

To determine the combined factor for a point 125,000 ft from the central meridian and at an elevation of 600 ft, a solution involving double interpolation must be used. We must interpolate between the combined values for 500 ft and for 750 ft elevation, and between combined values for 100,000 ft and for 150,000 ft from the central meridian. First, the combined value for 125,000 ft from the central meridian can be interpolated simply by averaging the values for 100,000

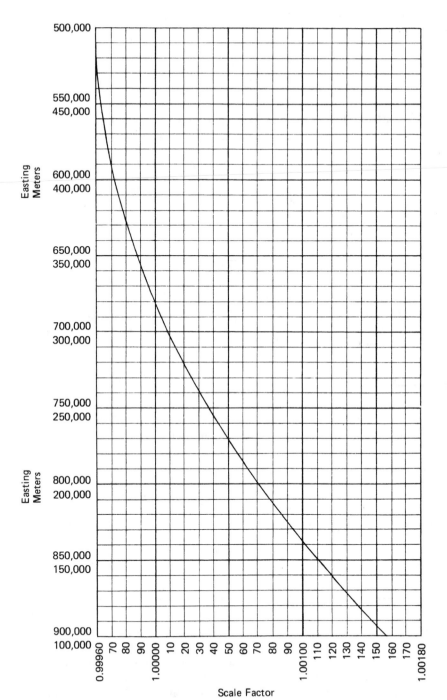

FIGURE 9.17 Universal transverse Mercator scale factors. (Courtesy of U.S. Department of the Army, TM5-241-4/1)

Table 9.8 3° MTM COMBINED GRID FACTOR BASED ON CENTRAL SCALE FACTOR OF 0.9999 FOR THE MODIFIED TRANSVERSE MERCATOR PROJECTION*

Elevation (ft)	Distance from Central Meridian (Thousands of Feet)												
	0	50	100	150	200	250	300	350	400	450	500	550	600
0	0.999900	0.999903	0.999911	0.999926	0.999946	0.999971	1.000003	1.000040	1.000083	1.000131	1.000186	1.000245	1.000311
250	0.999888	0.999891	0.999899	0.999914	0.999934	0.999959	0.999991	1.000028	1.000071	1.000119	1.000174	1.000234	1.000299
500	0.999876	0.999879	0.999888	0.999902	0.999922	0.999947	0.999979	1.000016	1.000059	1.000107	1.000162	1.000222	1.000287
750	0.999864	0.999867	0.999876	0.999890	0.999910	0.999936	0.999967	1.000004	1.000047	1.000095	1.000150	1.000210	1.000275
1,000	0.999852	0.999855	0.999864	0.999878	0.999898	0.999924	0.999955	0.999992	1.000035	1.000083	1.000138	1.000198	1.000263
1,250	0.999840	0.999843	0.999852	0.999864	0.999886	0.999912	0.999943	0.999980	1.000023	1.000072	1.000126	1.000186	1.000251
1,500	0.999828	0.999831	0.999840	0.999854	0.999874	0.999900	0.999931	0.999968	1.000011	1.000060	1.000114	1.000174	1.000239
1,750	0.999816	0.999819	0.999828	0.999842	0.999862	0.999888	0.999919	0.999956	0.999999	1.000048	1.000102	1.000162	1.000227
2,000	0.999804	0.999807	0.999816	0.999830	0.999850	0.999876	0.999907	0.999944	0.999987	1.000036	1.000090	1.000150	1.000216
2,250	0.999792	0.999795	0.999804	0.999818	0.999838	0.999864	0.999895	0.999932	0.999975	1.000024	1.000078	1.000138	1.000204
2,500	0.999781	0.999784	0.999792	0.999806	0.999826	0.999852	0.999883	0.999920	0.999963	1.000012	1.000066	1.000126	1.000192
2,750	0.999769	0.999771	0.999780	0.999794	0.999814	0.999840	0.999871	0.999908	0.999951	1.000000	1.000054	1.000114	1.000180

*Ground distance × combined factor = grid distance.

Source: Adapted from the "Horizontal Control Survey Precis," Ministry of Transportation and Communications, Ottawa, Canada, 1974.

Table 9.9 COMBINED SCALE AND ELEVATION FACTORS: UTM

Distance from CM (km)	Elevation Above Mean Sea Level (m)										
	0	100	200	300	400	500	600	700	800	900	1,000
0	0.999600	0.999584	0.999569	0.999553	0.999537	0.999522	0.999506	0.999490	0.999475	0.999459	0.999443
10	0.999601	0.999586	0.999570	0.999554	0.999539	0.999523	0.999507	0.999492	0.999476	0.999460	0.999445
20	0.999605	0.999589	0.999574	0.999558	0.999542	0.999527	0.999511	0.999495	0.999480	0.999464	0.999448
30	0.999611	0.999595	0.999580	0.999564	0.999548	0.999553	0.999517	0.999501	0.999486	0.999470	0.999454
40	0.999620	0.999604	0.999588	0.999573	0.999557	0.999541	0.999526	0.999510	0.999494	0.999479	0.999463
50	0.999631	0.999615	0.999599	0.999584	0.999568	0.999552	0.999537	0.999521	0.999505	0.999490	0.999474
60	0.999644	0.999629	0.999613	0.999597	0.999582	0.999566	0.999550	0.999535	0.999519	0.999503	0.999488
70	0.999660	0.999645	0.999629	0.999613	0.999598	0.999582	0.999566	0.999551	0.999535	0.999519	0.999504
80	0.999679	0.999663	0.999647	0.999632	0.999616	0.999600	0.999585	0.999569	0.999553	0.999538	0.999522
90	0.999700	0.999684	0.999668	0.999653	0.999637	0.999621	0.999606	0.999590	0.999574	0.999559	0.999543
100	0.999723	0.999707	0.999692	0.999676	0.999660	0.999645	0.999629	0.999613	0.999598	0.999582	0.999566
110	0.999749	0.999733	0.999717	0.999702	0.999686	0.999670	0.999655	0.999639	0.999623	0.999608	0.999592
120	0.999777	0.999761	0.999746	0.999730	0.999714	0.999699	0.999683	0.999667	0.999652	0.999636	0.999620
130	0.999808	0.999792	0.999777	0.999761	0.999745	0.999730	0.999714	0.999698	0.999683	0.999667	0.999651
140	0.999841	0.999825	0.999810	0.999794	0.999778	0.999763	0.999747	0.999731	0.999716	0.999700	0.999684
150	0.999877	0.999861	0.999845	0.999830	0.999814	0.999798	0.999783	0.999767	0.999751	0.999736	0.999720
160	0.999915	0.999899	0.999884	0.999868	0.999852	0.999837	0.999821	0.999805	0.999790	0.999774	0.999758
170	0.999955	0.999940	0.999924	0.999908	0.999893	0.999877	0.999861	0.999846	0.999830	0.999814	0.999799
180	0.999999	0.999983	0.999967	0.999952	0.999936	0.999920	0.999905	0.999889	0.999873	0.999858	0.999842
190	1.000044	1.000028	1.000013	0.999997	0.999981	0.999966	0.999950	0.999934	0.999919	0.999903	0.999887
200	1.000092	1.000076	1.000061	1.000045	1.000029	1.000014	0.999998	0.999982	0.999967	0.999951	0.999935
210	1.000142	1.000127	1.000111	1.000095	1.000080	1.000064	1.000048	1.000033	1.000017	1.000001	0.999986
220	1.000195	1.000180	1.000164	1.000148	1.000133	1.000117	1.000101	1.000086	1.000070	1.000054	1.000039
230	1.000251	1.000235	1.000219	1.000204	1.000188	1.000172	1.000157	1.000141	1.000125	1.000110	1.000094
240	1.000309	1.000293	1.000277	1.000262	1.000246	1.000230	1.000215	1.000199	1.000183	1.000168	1.000152
250	1.000369	1.000353	1.000338	1.000322	1.000306	1.000290	1.000275	1.000259	1.000243	1.000228	1.000212

and 150,000. Second, the value for 600 ft can be determined as follows (from Table 9.8, for 600 ft):

$$100/250 \times 12 = 5$$

Combined factor = 0.999895
$$\underline{-0.000005}$$
$$= 0.999890$$

For important survey lines, grid factors can be determined for both ends and then averaged. For lines longer than five miles, intermediate computations are required to maintain high precision.

9.6.2 Grid/Geodetic Azimuth Relationships

9.6.2.1 Convergence. In plane grids, the difference between grid north and geodetic north is called convergence (also called the mapping angle). In the SPCS27 transverse Mercator grid, convergence was denoted by $\Delta\alpha''$; in the Lambert grid, it was denoted as θ. In SPCS83, convergence (in both projections) is denoted by γ (gamma). On a plane grid, grid north and geodetic north coincide only at the central meridian. As you work farther east or west of the central meridian, convergence becomes more pronounced.

Using SPCS83 symbols, approximate methods can be determined as follows:

$$\gamma'' = \Delta\lambda'' \sin\varphi\, P \tag{9.4a}$$

where $\Delta\lambda''$ = the difference in longitude, in seconds, between the central meridian and point P
φP = the latitude of point P

When long sights (>5 miles) are taken, a second term (present in the interactive Internet programs used in Examples 9.1 and 9.2 and in all precise computations) is required to maintain directional accuracy. See also Section 9.6.2.2.

When the direction of a line from P_1 to P_2 is being considered, the expression becomes:

$$\gamma'' = \Delta\lambda'' \left(\frac{\sin(\varphi P_1 + \varphi P_2)}{2} \right) \tag{9.4b}$$

If the distance from the central meridian is known, the expression can be written as follows:

$$\gamma'' = 32.370\, dk \tan\varphi \tag{9.5}$$

or

$$\gamma'' = 52.09d \tan\varphi \tag{9.6}$$

where γ = convergence angle, in seconds
d = departure distance from the central meridian, in miles (dk is the same distance, in kilometers)
φ = average latitude of the line

9.6.2.2 Corrections to Convergence. When high precision and/or long distances are involved, there is a second-term correction required for convergence. This term

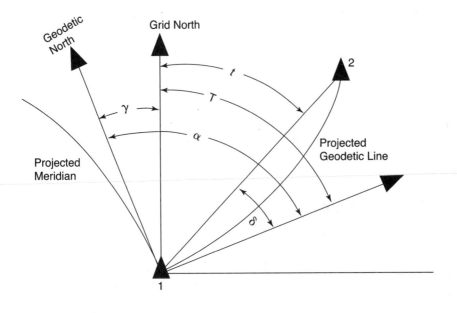

$$t = \alpha - \gamma + \delta$$

α = Geodetic Azimuth Reckoned from North

T = Projected Geodetic Azimuth

t = Grid Azimuth Reckoned from North

γ = Convergence Angle (Mapping Angle)

$\delta = t - T$ = Second Term Correction = Arc-to-Chord Correction

FIGURE 9.18   Relationships among geodetic and grid azimuths. (From the NOAA Manual NOS NGS 5, *State Plane Coordinate System of 1983*, Section 2.5)

δ refers to $t - T$ (the grid azimuth − the projected geodetic azimuth) and results from the fact that the projection of the geodetic azimuth δ onto the grid is not the grid azimuth but the projected geodetic azimuth, symbolized as T. Figure 9.18 (which comes from Section 2.5 of the *State Plane Coordinate System of 1983*) shows the relationships between geodetic and grid azimuths.

9.7   Illustrative Examples

These approximate methods are included only to broaden your comprehension in this area. For current precise solution techniques, see NOAA Manual NOS NGS 5, *State Plane Coordinate System of 1983,* or the geodetic toolkit discussed in Section 9.2.

■ EXAMPLE 9.4

You are given the transverse Mercator coordinate grid of two horizontal control monuments and their elevations (see Figure 9.19 for additional given data). Compute the ground distance and geodetic direction between them.

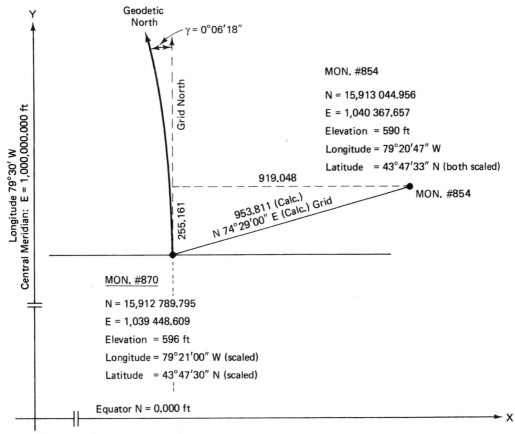

FIGURE 9.19 Illustration for Example 9.4.

Station	Elevation	Northing	Easting
Monument 870	595 ft	15,912,789.795 ft	1,039,448.609 ft
Monument 854	590 ft	15,913,044.956 ft	1,040,367.657 ft

Scale factor at CM = 0.9999.

Solution

By subtraction of coordinate distances, $\Delta N = 255.161$ ft, $\Delta E = 919.048$ ft. The solution is obtained as follows:

1. Grid distance 870 to 854:

$$\text{Distance} = \sqrt{255.161^2 + 919.048^2} = 953.811 \text{ ft}$$

2. Grid bearing:

$$\tan \text{bearing} = \frac{\Delta E}{\Delta N} = \frac{919.048}{255.161} = 3.6018357$$

$$\text{grid bearing} = 74.48\ 342°$$
$$= \text{N } 74°29'00'' \text{ E}$$

3. Convergence: use method (a) or (b). Average latitude = 43°47'31''; average longitude = 79°20'54'' (see Figure 9.19).

   (a) We can use Equation 9.6:

$$\gamma'' = 52.09d \tan \varphi$$
$$= 52.09 \times \frac{(40,367.657 + 39,448.609)}{2 \times 5280} \tan 43°47'31''$$
$$= 377.45''$$
$$\gamma = 0°06'\ 17.5''$$

   (b) We can use Equation 9.4:

$$\gamma'' = \Delta\lambda'' \sin \varphi \ P$$
$$= (79°30'' - 79°20'54'') \sin 43°47'31''$$
$$= 546'' \times \sin 43°47'31''$$
$$= 377.85''$$
$$\gamma = 0°06'\ 17.9''$$

The slight difference here (0.4'') between methods (a) and (b) reflect this approximate approach. Use a convergence of 6°18'.

Convergence, in this case, neglects the second-term correction and is computed only to the closest second of arc. (*Note:* The average latitude need not have been computed because the latitude range, 03'', was, in this case insignificant.)

Refer to Figure 9.20:

$$\text{Grid bearing} = \text{N } 74°29'00'' \text{ E}$$
$$+ \text{ convergence} = 0°06'18''$$
$$\text{geodetic bearing} = \text{N } 74°35'18'' \text{ E}$$

4. Scale factor:

$$\text{Scale factor at CM, } M_o = 0.9999$$
$$\text{Distance } (x) \text{ from CM} = \frac{40,368 + 39,449}{2} = 39,908 \text{ ft}$$

From Equation 9.2, we can calculate the scale factor at midpoint 870 to 854:

$$M_p = M_o \left(1 + \frac{x^2}{2R^2} \right)$$
$$= 0.9999 \left(1 + \frac{39,908^2}{2 \times 20,906,000^2} \right)$$
$$= 9999018$$

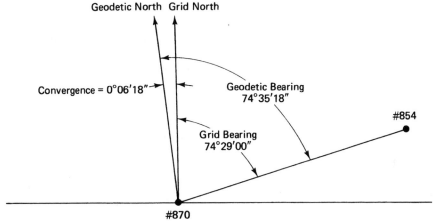

FIGURE 9.20 Illustration for Example 9.4.

5. We use Equation 9.1 for the elevation factor:

$$\text{Elevation factor} = \frac{\text{sea-level distance}}{\text{ground distance}} = \frac{R}{R + H}$$

$$= \frac{20,906,000}{20,906,000 + 593} = 0.999716$$

593 ft is the midpoint (average) elevation.

6. Combined factor: Use method (a), which is a computation involving Equation 9.5, or method (b).

   (a)   Combined factor = scale factor × elevation factor
   $$= 0.9999018 \times 0.9999716$$
   $$= 0.9998734$$

   (b) The combined factor can also be determined through double interpolation of Table 9.8. The following values are taken from Table 9.8. The required elevation is 593 ft at 39,900 ft from the CM.

	Distance from Central Meridian (Thousands of Feet)		
Elevation, ft	0	50	39.9 (Interpolated)
500	0.999876	0.999879	0.999878
593			
750	0.999864	0.999867	0.999866

After first interpolating the values at 0 and 50 for 39,900 ft from the CM, it is a simple matter to interpolate for the elevation of 593 ft:

$$0.999878 - (0.000012) \times \frac{93}{250} = 0.999873$$

Thus, the combined factor at 39,900 ft from the CM at an elevation of 593 ft is 0.999873.

7. For ground distance, we can use the relationship given previously in Equation 9.3:

$$\text{Ground distance} \times \text{combined factor} = \text{grid distance}$$

or

$$\text{Ground distance} = \frac{\text{grid distance}}{\text{combined factor}}$$

In this example:

$$\text{Ground distance (870 to 854)} = \frac{953.811}{0.9998734} = 953.93 \text{ ft}$$

■ **EXAMPLE 9.5**

You are given the coordinates (on the UTM 6° coordinate grid, based on the Clarke 1866 ellipsoid) of two horizontal control monuments (Mon. 113 and Mon. 115) and their elevations. Compute the ground distance and geodetic direction between them [see Figure 9.21(a)].

Station	Elevation	Northing	Easting
113	181.926 m	4,849,872.066 m	632,885.760
115	178.444 m	4,849,988.216 m	632,971.593

Zone 17 UTM; CM at 81° longitude west [see Figure 9.12(a)]
Scale factor at CM = 0.9996
ϕ (lat.) = 43°47′33″ [scaled from topographic map for midpoint of line joining Mon. 113 and Mon. 115, or as shown in Example 9.2(b).]
λ (long.) = 79°20′52″

Solution

Analysis of the coordinates of Stations 113 and 115 gives the coordinate distances of ΔN = 116.150 m, ΔE = 85.833 m.

1. Grid distance from Mon. 113 to Mon. 115:

$$\text{Distance} = \sqrt{116.150^2 + 85.833^2} = 144.423 \text{ m}$$

2. Grid bearing:

$$\text{tan bearing} = \frac{\Delta E}{\Delta N} = \frac{85.833}{116.150}$$

$$\text{Bearing} = 36.463811°$$

$$\text{Grid bearing} = \text{N. } 36° \ 27''50'' \text{ E}$$

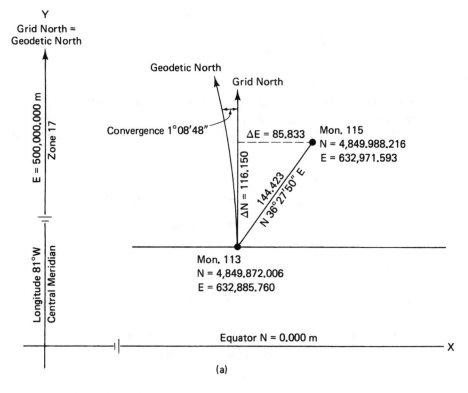

(a)

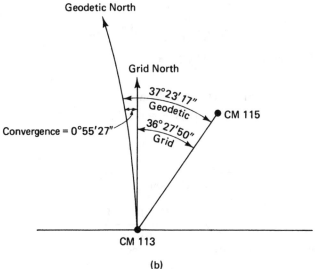

(b)

FIGURE 9.21 Illustration for Example 9.3. (a) Grid. (b) Application of convergence.

3. From Equation 9.4, we can calculate the convergence:

$$\gamma'' = \Delta\lambda'' \sin\varphi\,P$$
$$= (81° - 79°\ 20'52'')\sin 43°47'33''$$
$$= 4116''$$
$$\gamma = 1°08'36''$$

See Figure 9.21(b) for application of convergence. As we saw in Example 9.4, this technique yields convergence only to the closest second of arc. For more precise techniques, including second-term correction for convergence, see the NGS toolkit.

4. Scale factor:

$$\text{Scale factor at CM} = 0.9996$$
$$\text{Distance from CM} = \frac{132,885.760 + 132,971.593}{2}$$
$$= 132,928.677 \text{ m, or } 132.929 \text{ km}$$

The scale factor at midpoint line from Mon. 113 to Mon. 115 is computed from Equation 9.2:

$$M_p = M_o\left(\frac{1 + x^2}{2R^2}\right) \qquad M_p = 0.9996\left(1 + \frac{132.929^2}{2 \times 6372^2}\right) = 0.999818$$

The average radius of the sea-level surface = 20,906,000 ft or 6,372,000 m.

5. The elevation factor is calculated using Equation 9.1:

$$\text{Elevation factor} = \frac{\text{sea-level distance}}{\text{ground distance}} = \frac{R}{R + H}$$
$$= \frac{6372}{6372.00 + 0.180} = 0.999972$$

0.180 is the midpoint elevation divided by 1,000.

6. Combined factor:

(a) Combined factor = elevation factor × scale factor
$$= 0.999972 \times 0.999818 = 0.999790$$

or

(b) Use Table 9.9:

	Distance from CM (km)		
Elevation	130	140	132.9 (Interpolated)
100	0.999792	0.999825	0.999802
200	0.999777	0.999810	0.999787

The last computation is the interpolation for the elevation value of 180 m; that is:

$$0.999802 - \left(\frac{80}{100} \times 0.000015\right) = 0.999790$$

7. Ground distance:

$$\text{Ground distance} = \frac{\text{grid distance}}{\text{combined factor}}$$

In this example:

$$\text{Ground distance from Mon. 113 to Mon. 115} = \frac{144{,}423}{0.999790} = 144.453 \text{ m}$$

9.8 Horizontal Control Techniques

Typically, the highest-order control is established by federal agencies, the secondary control is established by state or provincial agencies, and the lower-order control is established by municipal agencies or large-scale engineering works' surveyors. Sometimes the federal agency establishes all three orders of control when requested to do so by the state, province, or municipality.

In triangulation surveys, a great deal of attention was paid to the geometric strength of figure of each control configuration. Generally, an equilateral triangle is considered strong, whereas triangles with small (less than 10°) angles are considered relatively weak. Trigonometric functions vary in precision as the angle varies in magnitude. The sines of small angles (near 0°), the cosines of large angles (near 90°), and the tangents of both small (0°) and large (90°) angles are all relatively imprecise. That is, there are relatively large changes in the values of the trigonometric functions that result from relatively small changes in angular values. For example, the angular error of 5 seconds in the sine of 10° is 1/7,300, whereas the angular error of 5 seconds in the sine of 20° is 1/15,000, and the angular error of 5 seconds in the sine of 80° is 1/234,000 (see Example 9.6).

You can see that if sine or cosine functions are used in triangulation to calculate the triangle side distances, care must be exercised to ensure that the trigonometric function itself is *not* contributing to the solution errors more significantly than the specified surveying error limits. When all angles and distances are measured for each triangle, the redundant measurements ensure an accurate solution, and the configuration strength of figure becomes somewhat less important. Given the opportunity, however, most surveyors still prefer to use well-balanced triangles and to avoid using the sine and tangent of small angles and the cosine and tangent of large angles to compute control distances. This concept of strength of figure helps to explain why GPS measurements are more precise when the observed satellites are spread across the visible sky instead of being bunched together in one portion of the sky.

■ **EXAMPLE 9.6** *Effect of the Angle Magnitude on the Accuracy of Computed Distances*

(a) Consider the right-angle triangle in Figure 9.22, with a hypotenuse 1,000.00 ft long. Use various values for θ to investigate the effect of 05″ angular errors.

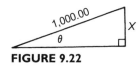

FIGURE 9.22

1. $\theta = 10°$ $X = 173.64818$ ft
   $\theta = 10°00'05''$ $X = 173.67205$ ft
   Difference = 0.02387 in 173.65 ft, an accuracy of 1/7,300.
2. $\theta = 20°$ $X = 342.02014$ ft
   $\theta = 20°00'05''$ $X = 342.04292$ ft
   Difference = 0.022782 in 342.02 ft, an accuracy of 1/15,000.
3. $\theta = 80°$ $X = 984.80775$ ft
   $\theta = 80°00'05''$ $X = 984.81196$ ft
   Difference = 0.00421 in 984.81 ft, an accuracy of 1/234,000.

(b) Consider the right triangle in Figure 9.23, with the adjacent side 1,000.00 ft long. Use various values for θ to investigate the effect of 05'' angular errors.

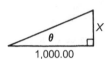

1,000.00

FIGURE 9.23

1. $\theta = 10°$ $X = 176.32698$ ft
   $\theta = 10°00'05''$ $X = 176.35198$ ft
   Difference = 0.025 accuracy of 1/7,100
2. $\theta = 45°$ $X = 1,000.00$ ft
   $\theta = 45°00'05''$ $X = 1,000.0485$ ft
   Difference = 0.0485, an accuracy of 1/20,600.
3. $\theta = 80°$ $X = 5,671.2818$ ft
   $\theta = 80°00'05''$ $X = 5,672.0858$ ft
   Difference = 0.804, an accuracy of 1/7,100.
4. In part (b)3, if the angle can be determined to the closest second, the accuracy would be as follows:
   $\theta = 80°$ $X = 5,671.2818$ ft
   $\theta = 80°00'01''$ $X = 5,671.4426$ ft
   Difference = 0.1608, an accuracy of 1/35,270.

Example 9.6 illustrates that the surveyor should avoid using weak (relatively small) angles in distance computations. If weak angles must be used, they should be measured more precisely than would normally be required. Also illustrated in the example is the need for the surveyor to analyze the proposed control survey configuration beforehand to determine optimal field techniques and attendant precisions.

9.9 Project Control

9.9.1 General Background

Project control begins with either a boundary survey (e.g., for large housing projects) or an all-inclusive peripheral survey (e.g., for construction sites). If possible, the boundary or site peripheral survey is tied into state or provincial grid control monuments or is located precisely using appropriate GPS techniques so that references can be made to the state or provincial coordinate grid system. The peripheral survey is densified with judiciously placed control stations over the entire site. The survey data for all control points are entered into the computer for accuracy verification, error adjustment, and finally for coordinate determination of all control points. All key layout points (e.g., lot corners, radius points, ℄ stations, curve points, construction points) are also coordinated using coordinate geometry computer programs. Printout sheets are used by the surveyor to lay out (using total stations) the proposed facility from coordinated control stations. The computer results give the surveyor the azimuth and distance from one, two, or perhaps three different control points to one layout point. Positioning a layout point from more than one control station provides the opportunity for an exceptional check on the accuracy of the work. When GPS is used, the surveyor first uploads the relevant stations' coordinates into the receiver-controller before going to the field so that the GPS receiver can lead the surveyor directly to the required point.

To ensure that the layout points have been accurately located (e.g., with an accuracy level of between 1/5,000 and 1/10,000), the control points themselves must be located to an even higher level of accuracy (i.e., typically better than 1/15,000). These accuracies can be achieved using GPS techniques for positions, and total stations for distances and angles. As we noted earlier, the surveyor must use "quality" geometrics, in addition to quality instrumentation, in designing the shape of the control net. A series of interconnected equilateral triangles provides the strongest control net.

When positioning control points, keep in mind the following:

1. Good visibility to other control points and an optimal number of layout points is important.
2. The visibility factor is considered not only for existing ground conditions but also for potential visibility lines during all stages of construction.
3. At least two reference ties (three is preferred) are required for each control point so that it can be reestablished if it is destroyed. Consideration must be given to the availability of features suitable for referencing (i.e., features into which nails can be driven or cut-crosses chiseled, etc.). Ideally, the three ties are each 120° apart.
4. Control points should be placed in locations that will not be affected by primary or secondary construction activity. In addition to keeping clear of the actual construction site positions, the surveyor must anticipate temporary disruptions to the terrain resulting from access roads, materials stockpiling, and so on. If possible, control points are safely located adjacent to features that will *not* be moved (e.g., electrical or communications towers; concrete walls; large, valuable trees).
5. Control points must be established on solid ground (or rock). Swampy areas or loose fill areas must be avoided.

Once the control point locations have been tentatively chosen, they are plotted so that the quality of the control net geometrics can be considered. At this stage, it may be necessary to return the field and locate additional control points to strengthen weak geometric figures. When the locations have been finalized on paper, each station is given a unique identification code number, and then the control points are set in the field. Field notes, showing reference ties to each point, are carefully taken and then filed. Now the actual measurements of the distances and angles of the control net are taken. When all the field data have been collected, the closures and adjustments are computed. The coordinates of

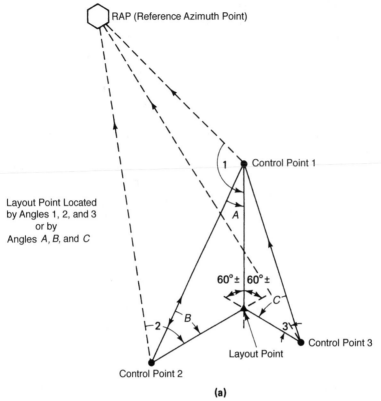

FIGURE 9.24 Examples of coordinate control for polar layout. (a) Single point layout, located by three angles.
(*continued*)

any layout points are then computed, with polar ties being generated for each layout point, from two or possibly three control stations.

Figure 9.24(a) shows a single layout point being positioned by angle only from three control sights. The three control sights can simply be referenced to the farthest of the control points themselves (e.g., angles *A, B,* and *C*). If a reference azimuth point (RAP) has been identified and coordinated in the locality, it would be preferred because it is no doubt farther away and thus capable of providing more precise sightings (e.g., angles 1, 2, and 3). RAPs are typically communications towers, church spires, or other identifiable points that can be seen from widely scattered control stations. Coordinates of RAPs are computed by turning angles to the RAP from project control monuments or preferably from state or provincial control grid monuments. Figure 9.24(b) shows a bridge layout involving azimuth and distance ties for abutment and pier locations. Note that, although the perfect case of equilateral triangles is not always present, the figures are quite strong, with redundant measurements providing accuracy verification.

Figure 9.25 illustrates a method of recording angle directions and distances to control stations with a list of derived azimuths. Station 17 can be found quickly by the surveyor from the distance and alignment ties to the hydrant, the cut-cross on the curb, and the nail

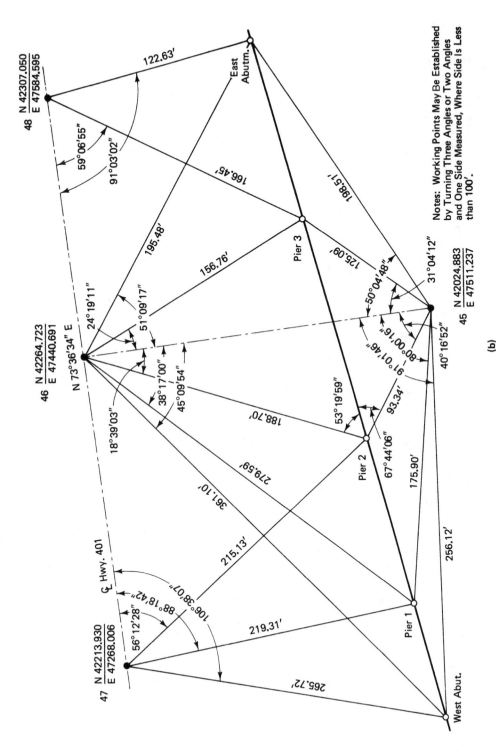

FIGURE 9.24 (continued) (b) Bridge layout, located by angle and distance. (Adapted from the *Construction Manual*, Ministry of Transportation, Ontario)

Notes: Working Points May Be Established by Turning Three Angles or Two Angles and One Side Measured, Where Side Is Less than 100'.

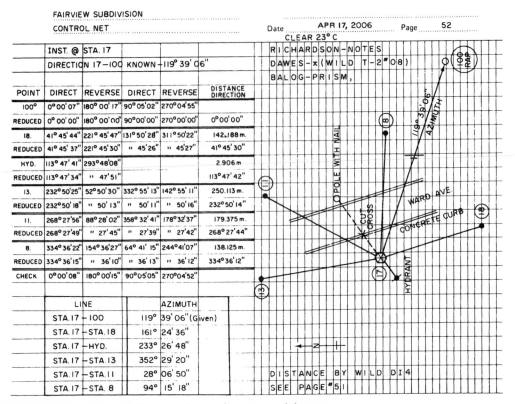

FAIRVIEW SUBDIVISION
CONTROL NET

Date APR 17, 2006   Page 52
CLEAR 23° C

INST. @ STA. 17
DIRECTION 17–100 KNOWN +119° 39' 06"

RICHARDSON-NOTES
DAWES-× (WILD T-2 #08)
BALOG-PRISM,

POINT	DIRECT	REVERSE	DIRECT	REVERSE	DISTANCE DIRECTION
100°	0°00'07"	180°00'17"	90°05'02"	270°04'55"	
REDUCED	0°00'00"	180°00'00"	90°00'00"	270°00'00"	0°00'00"
18.	41°45'44"	221°45'47"	131°50'28"	311°50'22"	142.188 m.
REDUCED	41°45'37"	221°45'30"	" 45'26"	" 45'27"	41°45'30"
HYD.	113°47'41"	293°48'08"			2.906 m
REDUCED	113°47'34"	" 47'51"			113°47'42"
13.	232°50'25"	52°50'30"	332°55'13"	142°55'11"	250.113 m.
REDUCED	232°50'18"	" 50'13"	" 50'11"	" 50'16"	232°50'14"
11.	268°27'56"	88°28'02"	358°32'41"	178°32'37"	179.375 m.
REDUCED	268°27'49"	" 27'45"	" 27'39"	" 27'42"	268°27'44"
8.	334°36'22"	154°36'27"	64°41'15"	244°41'07"	138.125 m.
REDUCED	334°36'15"	" 36'10"	" 36'13"	" 36'12"	334°36'12"
CHECK	0°00'08"	180°00'15"	90°05'05"	270°04'52"	

LINE	AZIMUTH
STA.17 – 100	119° 39' 06" (Given)
STA.17 – STA.18	161° 24' 36"
STA.17 – HYD.	233° 26' 48"
STA.17 – STA.13	352° 29' 20"
STA.17 – STA.11	28° 06' 50"
STA.17 – STA.8	94° 15' 18"

DISTANCE BY WILD DI4
SEE PAGE #51

FIGURE 9.25 Field notes for control point directions and distances.

in the pole. (Had station 17 been destroyed, it could have been reestablished from these and other reference ties.) The row marked "check" in Figure 9.25 indicates that the surveyor has "closed the horizon" by continuing to revolve the theodolite back to the initial target point (100 in this example) and then reading the horizontal circle. An angle difference of more than 5″ between the initial reading and the check reading usually means that the series of angles in that column must be repeated.

After the design of a facility has been coordinated, polar layout coordinates can be generated for points to be laid out from selected stations. The surveyor can copy the computer data directly into the field book (see Figure 9.26) for use later in the field. On large projects (expressways, dams, etc.), it is common practice to print bound volumes that include polar coordinate data for all control stations and all layout points. Modern total station and GPS practices also permit the direct uploading of the coordinates of control points and layout points to be used in layout surveys (see Chapter 5).

Figure 9.27 shows a primary control net established to provide control for a construction site. The primary control stations are tied in to a national, state, or provincial coordinate grid by a series of precise traverses or triangular networks. Points on baselines (secondary points) can be tied in to the primary control net by polar ties, intersection, or resection. The actual layout points of the structure (columns, walls, footings, etc.) are established from these secondary points. International standard ISO 4463 (from the International

FAIRVIEW SUBDIVISION
LAYOUT DATA
OAK ROAD

Job

Date APR 24, 2006 Page 16

POLAR CO-ORDINATES ABOUT STATION 12.

POINT	AZIMUTH	DISTANCE	*
100	122° 31' 17"		
11	202° 46' 20"		*DISTANCES SHOWN ARE ~~GROUND~~ DISTANCES
365	358° 47' 38"	31.727 m.	
369	12° 26' 22"	50.813	
1405	49° 00' 09"	44.391	
368	91° 00' 03"	49.521	GRID FACTOR
367	104° 09' 30"	30.105	= 0.999903
534	117° 40' 41"	36.412	
536	127° 44' 04"	37.027	
482	190° 50' 07"	64.301	
483	185° 01' 58"	60.399	
484	199° 41' 51"	45.360	
485	202° 29' 29"	47.616	
486	171° 59' 20"	31.739	
487	158° 29' 00"	30.571	
488	159° 59' 04"	50.661	
489	163° 34' 24"	50.969	
537	139° 44' 43"	54.800	
539	137° 24' 57"	60.823 m.	

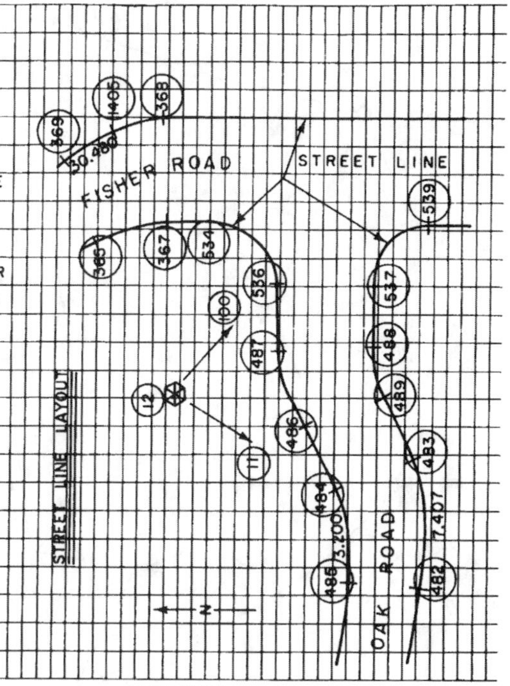

Party Chief ____________

Weather ____________

Party Chief RICHARDSON.

Weather ____________

FIGURE 9.26 Prepared polar coordinate layout notes.

Organization for Standardization) points out that the accuracy of key building or structural layout points should not be influenced by possible discrepancies in the state or provincial coordinate grid. For that reason, the primary project control net is analyzed and adjusted independently of the state or provincial coordinate grid. This "free net" is tied to the state or provincial coordinate grid without becoming an integrated adjusted component of that grid. The relative positional accuracy of project layout points to each other is more important than the positional accuracy of these layout points relative to a state or provincial coordinate grid.

9.9.2 Positional Accuracies (ISO 4463)

9.9.2.1 Primary System Control Stations.

1. Permissible deviations of the distances and angles obtained when measuring the positions of primary points, and those calculated from the adjusted coordinates of these points, shall not exceed:

$$\text{Distances: } \pm 0.75\sqrt{L} \text{ mm}$$

$$\text{Angles: } \pm 0.045 \text{ degrees}/\sqrt{L}$$

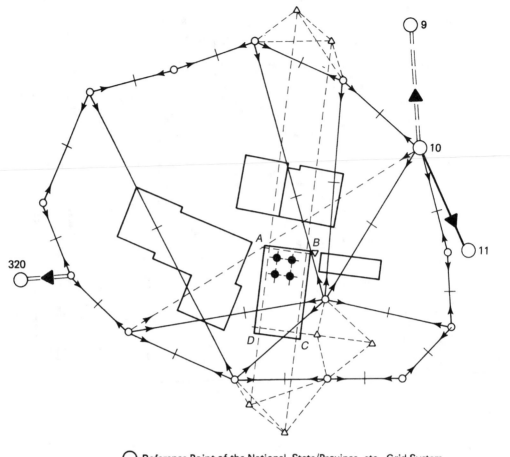

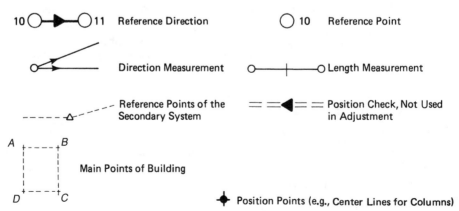

○ Reference Point of the National, State/Province, etc., Grid System

○ Reference Point of the Primary Project Control System

10 ○━━▶○ 11 Reference Direction   ○ 10   Reference Point

○━▶━━ Direction Measurement   ○━━┼━━○ Length Measurement

─ ─ ─△─ ─ Reference Points of the   ═ ═◀═ ═ Position Check, Not Used
           Secondary System   in Adjustment

A +┄┄┄+ B
 ┊   ┊   Main Points of Building
 ┊   ┊
D +┄┄┄+ C

✦ Position Points (e.g., Center Lines for Columns)

FIGURE 9.27 Project control net. (Adapted from International Organization for Standardization [ISO], Standard 4463)

or

$$\pm\frac{0.05}{\sqrt{L}} \text{ gon}$$

where L is the distance in meters between primary stations; in the case of angles, L is the shorter side of the angle. One revolution $= 360° = 400$ gon (also grad—a European angular unit); 1 gon $= 0.9$ degrees (exactly).

2. Permissible deviations of the distances and angles obtained when checking the positions of primary points shall not exceed:

$$\text{Distances: } \pm 2\sqrt{L} \text{ mm}$$

$$\text{Angles: } \pm\frac{0.135}{\sqrt{L}} \text{ degrees}$$

or

$$\pm\frac{0.15}{\sqrt{L}} \text{ gon}$$

where L is the distance in meters between primary stations; in the case of angles, L is the shorter side of the angle.

Angles are measured with a 1-second theodolite, with the measurements made in two sets. (Each set is formed by two observations, one on each face of the instrument; see Figure 9.26). Distances can be measured with steel tapes or EDMs, and they are measured at least twice by either method. Steel tape measurements are corrected for temperature, sag, slope, and tension. A tension device is used while taping. EDM instruments should be checked regularly against a range of known distances.

9.9.2.2 Secondary System Control Stations.

1. Secondary control stations and main layout points (e.g., *ABCD*, Figure 9.27) constitute the secondary system. The permissible deviations for a checked distance from a given or calculated distance between a primary control station and a secondary point shall not exceed:

$$\text{Distances: } \pm 2\sqrt{L} \text{ mm}$$

2. The permissible deviations for a checked distance from the given or calculated distance between two secondary points in the same system shall not exceed:

$$\text{Distances: } \pm 2\sqrt{L} \text{ mm}$$

where L is the distance in meters. For L less than 10 m, permissible deviations are ± 6 mm:

$$\text{Angles: } \pm\frac{0.135}{\sqrt{L}} \text{ degrees}$$

or

$$\pm\frac{0.15}{\sqrt{L}}\text{ gon}$$

where L is the length in meters of the shorter side of the angle.

Angles are measured with a theodolite or total station reading to at least one minute. The measurement shall be made in at least one set (i.e., two observations, one on each face of the instrument). Distances can be measured using steel tapes or EDMs and are measured at least twice by either method. Taped distances are corrected for temperature, sag, slope, and tension. A tension device is to be used with the tape. EDM instruments should be checked regularly against a range of known distances.

9.9.2.3 Layout Points.
The permissible deviations of a checked distance between a secondary point and a layout point, or between two layout points, are $\pm K\sqrt{L}$ mm, where L is the specified distance in meters, and K is a constant taken from Table 9.10. For L less than 5 m, the permissible deviation is $\pm 2K$ mm.

The permissible deviations for a checked angle between two lines, dependent on each other, through adjacent layout points are:

$$\pm\frac{0.0675}{\sqrt{L}}\text{ K degrees}$$

or

$$\pm\frac{0.075}{\sqrt{L}}\text{ K gon}$$

where L is the length in meters of the shorter side of the angle and K is a constant from Table 9.10.

Figure 9.28 illustrates the foregoing specifications for the case involving a stakeout for a curved concrete curb. The layout point on the curve has a permissible area of uncertainty generated by $\pm 0.015/m$ due to angle uncertainties and by $\pm 0.013/m$ due to distance uncertainties.

Table 9.10 ACCURACY REQUIREMENT CONSTANTS FOR LAYOUT SURVEYS

K	Application
10	Earthwork without any particular accuracy requirement (e.g., rough excavation, embankments)
5	Earthwork subject to accuracy requirements (e.g., roads, pipelines, structures)
2	Poured concrete structures (e.g., curbs, abutments)
1	Precast concrete structures, steel structures (e.g., bridges, buildings)

Source: Adapted from Table 8-1, ISO 4463.

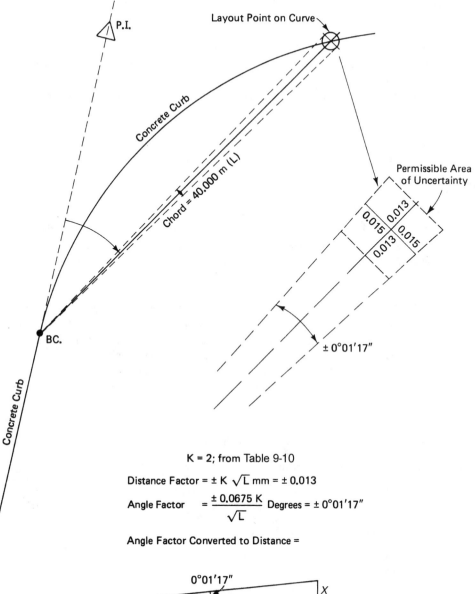

K = 2; from Table 9-10

Distance Factor = $\pm$ K $\sqrt{L}$ mm = $\pm$ 0.013

Angle Factor $\quad = \dfrac{\pm 0.0675 \text{ K}}{\sqrt{L}}$ Degrees = $\pm$ 0°01'17"

Angle Factor Converted to Distance =

X = 40 tan 0°01'17" = 0.015 m

FIGURE 9.28 Accuracy analysis for a concrete curb layout point. (See ISO Standard 4463.)

Review Questions

9.1. What are the advantages of referencing a survey to a recognized plane grid?

9.2. Describe why the use of a plane grid distorts spatial relationships.

9.3. How can the distortions in spatial relationships that you described in Review Question 9.2 be minimized?

9.4. Describe the factors you would consider in establishing a net of control survey monuments to facilitate the design and construction of a large engineering works, for example, an airport.

9.5. Some have said that, with the advent of the global positioning system (GPS), the need for extensive ground control survey monumentation has been much reduced. Explain.

Problems

Problems 9.1 through 9.5 use the control point data shown in Table 9.11.

9.1. Draw a representative sketch of the four control points and then determine the grid distances and grid bearings of sides *AB*, *BC*, *CD*, and *DA*.

9.2. From the grid bearings computed in Problem 9.1, compute the interior angles (and their sum) of the traverse *A*, *B*, *C*, *D*, *A*, thus verifying the correctness of the grid bearing computations.

9.3. Determine the ground distances for the four traverse sides by applying the scale and elevation factors (i.e., grid factors). Use average latitude and longitude.

9.4. Determine the convergence correction for each traverse side and determine the geodetic bearings for each traverse side. Use average latitude and longitude.

9.5. From the geodetic bearings computed in Problem 9.4, compute the interior angles (and their sum) of the traverse *A*, *B*, *C*, *D*, *A*, thus verifying the correctness of the geodetic bearing computations.

Table 9.11 CONTROL POINT DATA FOR PROBLEMS 9.1 THROUGH 9.5

Monument	Elevation	Northing	Easting	Latitude	Longitude
A	179.832	4,850,296.103	317,104.062	43°47′33″ N	079°20′50″ W
B	181.356	4,480,218.330	316,823.936	43°47′30″ N	079°21′02″W
C	188.976	4,850,182.348	316,600.889	43°47′29″ N	079°21′12″ W
D	187.452	4,850,184.986	316,806.910	43°47′29″ N	079°21′03″ W

Average longitude   = 079°21′02″ (Mon. B)
Average latitude   = 43°47′30″ N (Mon. B)
Central meridian
(CM) at longitude   = 079°30″ W
Easting at CM   = 304,800.000 m
Northing at equator = 0.000 m
Scale factor at CM = 0.9999

Data are consistent with the 38° transverse Mercator projection, related to NAD83.

PART II
Construction Applications: General Background

II.1 General Background

Building on the skills developed over the first nine chapters, the topics in Part II describe how these basic skills are applied to engineering and construction surveying. Chapter 10 introduces construction surveying in general and machine guidance/control in particular. Also described in Chapter 10 is the topic of digital data files; these files contain the coordinated horizontal and vertical positions of existing ground (EG) points and surfaces, as well as design points and surfaces. Digital data files are at the heart of modern project design, and they form the cornerstone for all modern electronic surveying layout practices. They are also vital for the successful implementation of machine guidance and control practices.

Chapter 11 describes the circular, parabolic, and spiral curves used in highway and road design and layout. The remaining chapters (Chapters 12 to 17) describe survey practices used in highway construction, municipal street construction, pipeline and tunnel construction, culvert and bridge construction, building construction, and quantity and final surveys.

Chapter 10

Modern Construction Surveying Practices

10.1 General Background

Construction surveys provide for the horizontal and vertical layout for every key component of a construction project. Only surveyors experienced in both project design and appropriate construction techniques can accomplish the provision of line and grade. A knowledge of related civil design is essential to interpret the design drawings effectively for layout purposes, and knowledge of construction techniques is required to ensure that the survey layout is optimal for line-and-grade transfer, construction scheduling, and ongoing payment measurements.

We have seen that data can be gathered for engineering projects (preengineering surveys) and other works in various ways. Modern practice favors total station surveys for high-density areas of limited size, and aerial imaging techniques for high-density areas covering large tracts. Global positioning system (GPS) techniques are now also being implemented successfully in areas of moderate density. The survey method chosen by a survey manager is usually influenced by the costs (per point) and the reliability of the point positioning.

In theory, GPS techniques seem to be ideal because roving receiver-equipped surveyors move quickly to establish precise locations for layout points, in which both line and grade are promptly determined and marked; however, when GPS signals are blocked by tree canopies or other obstructions, additional surveying methods must be used. Attained accuracies can be observed simply by remaining at the layout point as the position location is continuously updated.

Surveyors have found that, to utilize real-time kinematic (RTK) surveying successfully in construction surveying, much work had to be done to establish sufficient horizontal and vertical control on the project site; however, the recent increase in coverage of GPS/RTK base stations has lessened the need for as many on-site ground control stations. An RTK layout has many components that may be cause for extra vigilance. Some examples are short observation times, problems associated with radio or cellular transmissions, problems with satellite signal reception due to canopy obstructions and possibly weak constellation geometry (multiconstellation receivers can help here), instrument calibration, multipath errors, and other errors (some of which will not be evident in error displays). For these reasons, like all other layout procedures, layouts using GPS techniques must be verified.

If possible, verification can be accomplished through independent surveys; for example, check GPS surveys—based on different control stations, tape measurements, or total station sightings taken from point to point where feasible, and even by pacing. In the real world of construction works, the problems surrounding layout verification are compounded by the fact that the surveyor often doesn't have unlimited time to perform measurement checks. The contractor may actually be waiting on site to commence construction. A high level of planning and a rigid and systematic method (proven successful in past projects) of performing the layout survey are recommended.

Unlike other forms of surveying, construction surveying is often associated with the speed of the operation. Once contracts have been awarded, contractors may wish to commence construction immediately because they likely have commitments for their employees and equipment. They do not want to accept delay. A hurried surveyor is more likely to make mistakes in measurements and calculations, and thus even more vigilance than normal is required. Construction surveying is not an occupation for the faint of heart; the responsibilities are great and the working conditions are often less than ideal. However, the sense of achievement when viewing the completed facility can be very rewarding.

10.2 Grade

The word *grade* has several different meanings. In construction work alone, it is often used in three distinctly different ways:

1. To refer to a proposed elevation.
2. To refer to the slope of profile line (i.e., gradient).
3. To refer to cuts and fills—vertical distances measured below or above grade stakes.

The surveyor should be aware of these different meanings and should always note the context in which the word *grade* is being used.

10.3 Machine Guidance and Control

10.3.1 General Background

The role of the construction surveyor is changing. Before machine guidance and control techniques were introduced—and even today on many smaller construction sites—surveyors manually provide grade stakes indicating design line and grade to the contractor. On construction sites where machine guidance and control techniques are not used, the surveyor is responsible for the following:

- The control survey (horizontal and vertical) over the construction site on which the preliminary survey and layout survey are to be based (this includes the placement and tie-ins of control monuments).

- A preengineering topographic survey, which is used as a basis for the project design and is performed over the site.

- Once the design is completed and the project commenced, staking art the facility on centerline or on offset—to provide alignment and grade control to the contractor. The stakeouts are repeated until the contractor finally brings the facility to the required design elevations and alignments.

- At periodic intervals (often monthly), measuring construction quantities (item count, lengths, areas, and volumes) reflecting the progress made by the contractor during that time period. Typical quantities include earth volumes computed from end areas (for cut and fill), length of pipe, lengths of curb, and fence; tons of asphalt and granular material delivered on site; areas of sod laid; areas of the work surface receiving dust control; volumes of concrete placed; etc. Interim payments to the contractor are based on these interim measurements taken by the project surveyor or works inspector. Contractors normally employ their own surveyors to confirm the owner's survey measurements.

- After the completion of construction, performing a **final** or **as-built survey** to confirm that the facility was built according to the design criteria and that any in-progress design changes were suitably recorded.

Regardless of the layout technique, traditional (preelectronic) layout activity often takes up much of construction surveyors' time and attention. For example, in highway construction, where cuts and fills can be large, surveyors often have to restake the project continuously (sometimes daily in large cut/fill situations) because the grade and alignment stakes are knocked out or buried during the cut-and-fill process. As the facility nears design grade, restaking becomes less frequent.

Recent advances in machine guidance and machine control have resulted in layout techniques that significantly improve the efficiency of the stakeout (by as much as 30 percent according to some reports). Also, by reducing the need for as many stakeouts and thus the need for as many layout surveyors and grade checkers working near the moving equipment, these techniques provide a significant improvement in personnel safety.

With advancements in machine guidance and control, the surveyor's job has been greatly simplified in some aspects and become somewhat more complicated in others. The major impetus for the development and acceptance of guidance and control techniques in construction surveying has been the increase in efficiency of the operation, with a resulting reduction in costs. By reducing continual and repetitive human operations and computations, these techniques also reduce the chances for costly errors and mistakes. The tedious job of staking and restaking has been greatly reduced, and the ongoing measurements needed to record the contractor's progress can now be automated using the guidance/control software. Similarly, the measurements needed for a final or as-built survey already exist in the database at the conclusion of the project.

Simple machine guidance techniques have been with us for a long time. Laser beams, rotating in a fixed plane (horizontal or sloped), at a known vertical offset to finished grade, have long been used to guide earth-moving equipment. Ultrasonic detectors guide pavers by analyzing the timed sonic signal returns from string lines or other vertical references— see Figure H.1. Large tracts (e.g., airports, shopping centers, parking lots, etc.) are brought to grade through the use of rotating lasers and machine-mounted laser detectors, which convey to the operator of the machine (bulldozer, grader, or scraper) the up/down operations

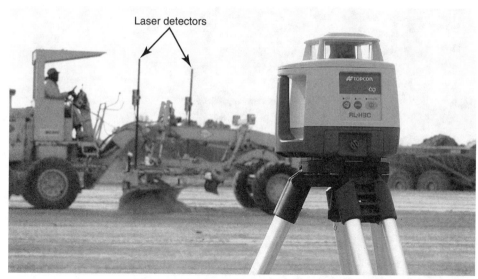

FIGURE 10.1 Visual grading control is provided by a rotating laser referenced to a laser receiver/display that is attached to the blade. (Courtesy of Topcon Positioning Systems, Inc., Pleasanton, Calif.)

required to bring that part of the project to the designed grade elevation. Laser grade displays are mounted outside the cab, where flashing colored lights guide the operator to move the blade up or down to be at design grade, or the grade display can be brought inside the cab, where the operator is guided by viewing the display monitor, which shows the position of the equipment's blade (or bucket teeth) in relation to design grade (see Figures 10.1 and 10.2). Accuracies are said to be as reliable as those used for most earthwork techniques.

The difference between machine guidance and machine control is the level of machine automation. More sophisticated systems can send signals to machine receptors that automatically open or close the valves needed to direct the machine to the proper alignment and grade. Presumably, the day may arrive when there will be no need for an operator in the machine—the machine's capabilities and limitations will be programmed into the project data file so that the machine does not attempt to perform large cuts and fills all at once (beyond the capability of the machine).

Sophisticated guidance and control techniques presently utilize either motorized total station techniques [local positioning system (LPS)] or real-time (RTK) GPS techniques. GPS techniques have been used where only a moderate level of accuracy is needed, such as in the earth-grading applications found in the construction of highways, shopping centers, airports, etc. Total station techniques are used where a higher level of accuracy is required, such as in the final grading of large projects, slip-form concrete pavers, etc. Presently, accuracies that can be achieved using GPS grade control are in the ± 0.10 ft (± 3 cm) range, whereas the accuracies using LPS grade control are in the ± 0.02 ft to 0.04 ft (± 5- to 10-mm) range.

Machine guidance includes the capability of informing the machine operator about cut/fill and left/right movements. Light bars located inside or outside the cab within the operator's field of view indicate the required cut/fill and left/right movements; some systems also provide audible tones that increase (decrease) as the operation approaches (retreats

FIGURE 10.2 BucketPro Excavator cab display system used in machine-controlled excavation. The screen display shows the bucket teeth in relation to the proposed gradient. (Courtesy of Trimble Geomatics and Engineering Division)

from) design grade. Machine control includes the capability of signaling the machine directly so that valves open and close automatically, thus driving the various components of the machine to perform the needed functions. Cross-slope grading is accomplished on bladed machines by mounting GPS receivers (or laser sensors) at either end of the blade and letting the GPS signals (or laser beam) trigger the software to adjust the slope angle. Slope sensors can also be mounted directly on the blade to have the blade adjust automatically and thus have it conform with the preselected design cross-slope. The in-cab touch-screen monitor shows the location and orientation of the machine with respect to the plan, profile, and cross-section views, each may of which be toggled onto the screen as needed.

The surveyor or perhaps a new-breed of project designer must create three-dimensional (3D) data files portraying all the original features and the existing ground (EG) digital terrain model (DTM). All design data information must also be converted to a site-defined DTM. These 3D files are the foundation for the use of machine guidance and control, and for much of the automated interim and final measurements needed for payment purposes. See Section 10.3.4.

In summary, automated layout techniques provide the following benefits: reduced costs, increased safety of construction personnel, and reduced number of mistakes (however, mistakes that are not noticed can quickly become quite large). Additions and revisions to the design DTM can also be updated electronically at the machine's computer. Data can be transferred to the machine electronically (connected or wireless) or via data cards.

10.3.2 Motorized Total Station Control and Guidance

Some manufacturers produce layout software that can integrate motorized total stations and appropriately programmed personal computers (PCs). These radio- or optical-controlled systems can monitor work progress and give real-time direction for line-and-grade operations of various types of construction equipment in a wide selection of engineering works (e.g., tunnels, road and railway construction, site development, drilling, etc.). The motorized total station tracks one machine as it moves (up to 28 mph) while keeping a lock on the machine-mounted reflecting prism. The coordinates of the machine are continuously updated (up to eight times a second), and the machine's cutting-edge coordinates are compared to the design DTM coordinates (located in the in-cab computer) to determine cut/fill and left/right operations. Directions are sent via radio or optical communications to and from the machine control-center computer. Manufacturers claim measurement standard deviations of 2 mm in height and 5 mm in position. See Figures 10.3 and H.2. Also see Section 10.5.

Motorized total stations can be set up in convenient locations so that an optimal area can be controlled (the instrument station's coordinates can be determined using resection techniques if necessary—see Section 5.3.3). The drawbacks to this system are that the motorized total station and its computer can direct only one machine at a time, and the unimpeded line-of-sight range is restricted to 15 to 700 m. Much time can be lost when the instrument has to be relocated to another previously established control station to overcome blocked sight lines. Having said that, the recent introduction in early 2005 of the first integrated total station/GPS receiver (see Figure 5.29) signals the beginning of a new era when this guidance/control technique will be all the more effective because it allows the surveyor to establish control immediately anywhere on the construction site. The surveyor simply sets up this integrated instrument in a new convenient location and takes satellite signals as well as the differencing signals sent by a GPS base station to determine precisely, in real time, the new setup point's horizontal coordinates, and elevation—as long as the setup point is within 50 km of an RTK GPS base station.

FIGURE 10.3 Paving-machine operation controlled by a motorized total station/radio control modem/PC computer instrumentation package. (Courtesy of Leica Geosystems Inc.)

10.3.3 GPS Guidance and Control

In addition to laser- and computer-controlled total station techniques, machine guidance and control is also available, in real time, with the use of layout programs featuring GPS receivers. As with total station techniques, receptors are mounted on the various types of construction equipment (the height of the GPS receivers above the ground is measured and entered into the controller as part of the site calibration measurements), and the readings are transmitted to the in-cab GPS controller or integrated PC computers (see Figure H.3). In-cab displays show the operator how much up/down and left/right movement is needed on the cutting edge of the blade (grader or bulldozer) to achieve or maintain design alignment. In addition to providing guidance to machine operators, these systems can also be configured to provide control of the machine—that is, to send signals directly to the machine to operate valves that control the operation of various cutting-edge components such as grader or bulldozer blades, backhoe buckets, etc.

At the beginning of a project, the GPS calibration must be performed and verified. Several on-site control monuments are needed for site calibration. Calibration is performed to equate the GPS horizontal coordinates (WGS84) with the coordinates used for the project design—usually state plane coordinates or UTM coordinates. The GPS heights must also be reconciled to the orthometric heights used in the design. A local geoid model and several benchmarks are used in this process. GPS observations taken on existing and nearby horizontal control monuments (whose local coordinates are known) and on nearby benchmarks (whose local elevations are known), are necessary for this project startup calibration. Machine-accessible control monuments are established in secure locations so machines can revisit the control monuments at regular intervals to check their positional calibration settings.

A base station equipped with a dual-frequency GPS receiver and a radio transmitter (10- to 20-km range for local base stations) can be established close to or on the construction site and can help in the guidance or control of any number of GPS-equipped construction machines. If radio transmission is used, the range can be extended by adding repeater stations; some systems use digital cell phones for all communications and data transmission. The base station broadcasts the differential GPS signals to all machine-mounted dual-frequency GPS receivers within range of the base station so that differential corrections can be made. The accuracies of the resulting RTK GPS positions are usually in the range of 2 to 4 cm and are ideal for all rough grading practices. Machine calibration includes measuring the location of the GPS antennas with respect to the existing ground, and then these vertical and horizontal offsets are recorded in the in-cab machine computer via the touch-screen monitor. The machine computers, loaded with appropriate layout software, receive design DTM coordinates each day via data cards or via connected or wireless electronics. In highway construction, the computer monitor displays plan, profile, and cross-section views of the proposed grade and existing grade with an indication of where (horizontally and vertically) the machine and its blades are with respect to the required grade. The program computes and stores cuts and fills for interim and final surveys and payments. One typical program keeps the operator up to date by displaying the cut areas in red, the fill areas in blue, and the at-grade areas without color. These progress displays are usually governed by some predefined tolerance (for example, ±5 cm) that has been entered into the program.

GPS vertical accuracies can now be improved to that of motorized total stations (see Section 10.3.2) by incorporating laser guidance. In 2005, Topcon Corporation introduced a rotating fan laser for use in GPS stakeouts. The rotating laser, set up on a control station, can help guide or control any number of laser zone sensor–equipped construction machines or roving GPS surveyors. A laser zone sensor can be mounted on the machine or on a rover pole just below the GPS antenna/receiver. These sensors can decode the laser signal and calculate the height difference to within 1/3,600 of a degree, giving an elevation difference to within a few (6 to 12) millimeters. The laser fan sends out laser signals in a vertical working zone 33 ft high (10 m high at a distance of 150 m), in a working diameter of 2,000 ft (600 m). Multiple lasers can be used to provide guidance for larger differentials in height and larger working areas.

We noted in Section 8.11.4 that in 2004 Trimble introduced software that permitted accurate surveying based on a multibase network concept called virtual reference station (VRS) technology. This system permitted base stations to be placed farther apart while still producing accurate results. One example of this technology was instituted in 2004, when the Ohio Department of Transportation (ODOT) completed statewide CORS coverage. This coverage, together with the new VRS software, enables Ohio surveyors to work across the state at cm-level accuracy—with no need for their own base stations or repeaters; digital cell phones are used for communications. Operational capability occurred in January 2005. ODOT's CORS network consists of fifty-two stations; they will be added to the NGS national CORS network. This development signals the beginning of widely available base stations for general surveying use. It has been predicted that the local and state/province government agencies may be more inclined to spend money on densifying networks of GPS base stations instead of spending more money on replacing (and referencing) destroyed ground control monuments and creating ground control monuments in new locations.

As with all construction processes, accuracy verification is essential. Random GPS observations can be used. Some contractors install GPS receivers, together with all construction layout software, in superintendents' pickup trucks or all-terrain vehicles (ATVs) so they can monitor the quality and progress of large-scale construction. When the construction is nearing completion, the grading status can be transferred (electronically or via data cards) to the office computer so that the operation can be analyzed for authorization of the commencement of the next stage of operation—for example, to move from subgrade to granular base course in the roadbed, or to move from final earth grading to seeding or sod placement. Another feature of this technique is that the machine operator can update the database by identifying and storing the location of discovered features that were not part of the original data (for example, buried pipes that may become apparent only during the excavation process).

This RTK GPS system, which can measure the ground surface ten times a second, cannot be used with less accurate GPS systems such as DGPS (which normally are accurate only to within ± 3 ft). Normally, five satellites in view are required for initialization, and thereafter four satellites will suffice. If loss of lock occurs on some or all of the four satellites, the receiver display should indicate the problem and signal when the system is functioning properly again. When more satellites are launched in the GLONASS system and when the Galileo system becomes functional, multiconstellation receivers will diminish loss-of-lock problems.

10.4 Three-Dimensional Data Files

10.4.1 Surface Models

As noted earlier, to take advantage of machine guidance and control capabilities, both the existing ground (EG) surface model (DTM) and the design surface model (a second DTM) must be created. Surface models are usually created as triangulated networks (TINs). DTM files can be created using a wide variety of software programs, or they can be uploaded from standard design packages, such as AutoCad (ACAD). The EG surface model data in the computer can be collected from field work using total stations, GPS, or remote sensing techniques, or the surface model data can be digitized from contoured topographic plans prepared at a suitable scale, thus creating three-dimensional data files.

Each type of construction has attracted software developers who create programs designed specifically for those applications. For example, in highway work, the program is designed to create finished cross sections at regular station intervals based solely on proposed elevations along the centerline (derived from design TINs) and on the proposed cross section of the facility (pavement widths and crowns, shoulder widths and cross falls, ditch depths and side slopes, curb cross sections, etc.). These are called roading templates (see Figures 11.5 and 11.12). The program can also determine the elevation and distance from centerline where the design boulevard slope or ditch back slope rises or falls and thus intersects the existing ground (EG)—also known as original ground (OG). Existing or original ground is defined as the ground surface at the time of the preliminary survey.

By analyzing the EG model surface and the design model surface, cross-section end areas and volumes of cut and fill can be generated automatically. Volumes can also be computed using software based on the prismoidal formula; see Chapter 17 for an explanation of these techniques. Volumes can also be generated using software based on grid techniques, where grid cells are defined by size (1 ft on a side is common). The size of the grid chosen usually reflects the material-handling unit costs; for example, it is less expensive, per cubic yard (or cubic meter), to cut/fill using scrapers then it is using loaders or backhoes, and more approximate computation methods can be used when unit costs are lower. The elevation difference between the EG and design surface models can be interpolated at each corner of each grid cell, with the average height difference multiplied by the grid area to give the volume of cut or fill at each grid area. Finally, some software developers compute volumes by directly analyzing existing ground (EG) TINs and design TINs.

While there is only one EG surface elevation model, there can be several design surface elevation models, each of which can be stored on separate CAD layers in the digital file. For example, in highway work or roadwork, design surface models can be generated for the surfaces representing the following: the surface after the topsoil has been stripped, the subgrade surface, the top of granular surface, and the finished asphalt surface. Additional surfaces can be generated for off-the-roadway sites such as borrow/fill areas and sod/seeding areas. In municipal design, typical design surfaces to be modeled depict some or all of the following: storm-water detention basins, front and rear yard surfaces, boulevard surfaces, building pad surfaces, excavated pipeline trenches, back slopes, etc.

One of the objectives of engineering works designers is to minimize project costs. In highway work and in large site developments (for example, airports, large commercial developments), one of the major costs is the excavating and filling of material. Allowances can be made for the inclusion of shrinkage and swell factors that reflect closely the end result of the compaction of fill material and the placement of shattered rock in fill areas. By adjusting the design elevations or design gradients up and down (for example, building pad or first-floor elevations, road centerline profiles, pipeline profiles), the designer can use the software to determine quickly the effects that such adjustments have on the overall cut/fill quantities. The ideal (seldom realized) for cost effectiveness would be to balance cut and fill completely. Additionally, design software can have unit costs (estimated or bid) tagged to all construction quantities—for example cuts/fills per cubic yard or meter (tied to the use of various machines such as scrapers, loaders, backhoes, trenchers, etc.); unit lengths of curb, sidewalk, and pipelines (and other buried services); unit areas of asphalt or concrete; sod, seeding, dust control applications; unit weights of materials trucked on site (for example, granular material, asphalt), etc. The designer can thus keep abreast of cost factors when design changes are proposed.

Once the final design elevations and slopes have been chosen, the data stored on CAD layers in the digital file are used to generate quantities and costs for various stages of the project. The quantity estimates are used first to generate cost estimates, which are useful in obtaining approval to let a project out for bids. Then the cost estimates are useful to contractors as they prepare their final bids. Once the contract has been awarded and construction has commenced, the in-progress surface models are upgraded (sometimes monthly) to reflect construction progress. Progress payments to the contractor are based on measurements and computations based on those interim surface models. Both the owner and the contractor employ their own surveyors to resurvey the in-progress surface using LPS, GPS or other field techniques. By the completion of the project, the digital files already show the final (as-built) survey data, and no additional is work is required.

10.4.2 Horizontal Design Alignments

In addition to the EG surface elevation model and the design surface elevation models, the digital files contain the horizontal location of all existing and proposed features. In machine guidance and control situations, the complete digital file is available on the in-cab computer. When design revisions are made, the in-cab computer files can be upgraded using data cards or, more recently, even through wireless communications directly from the design office. When machine operators discover features during construction (for example, buried pipes) that were not included in the original project digital data file or were located incorrectly, the file can be updated right in the machine cab and the updates can be transmitted to the design office. When stakeouts are required, LPS, GPS, or other field techniques can be used. Modern software, working with the data in the project digital file and the stored coordinates of the project control monuments, can generate the horizontal and vertical alignment measurements needed to locate a facility centerline, or a feature location directly or on some predefined offset.

10.5   Summary of the 3D Design Process

10.5.1   Data File Construction

Data files are a combination of existing ground elevations and proposed design elevations. They are generated by commercially available software programs (see Section 10.6). The following list shows some typical steps in the process:

1. Access the design software and create a working file.
2. Select the source of the data (e.g., CAD data file, digitizer, etc.).
3. When importing CAD files, first identify the layers containing existing grade, proposed grade, subgrade, etc.
4. Import soils bore-hole data (if applicable).
5. Import the existing grade files (from CAD) first, and assign elevations to contours (use the pull-down menu) to convert from 2D to 3D. The area of specific interest may first need to be identified (by cursor-boxing) if the CAD file extends beyond the area of interest.
6. If digitized data is to be added to CAD file data, identify the digitized point on the CAD drawing (in at least two locations) to place data from the two source on the same datum.
7. Repeat the process for proposed grade files, which will include general stripping limits and depths, subgrade elevations, etc.
8. Identify the structure (e.g., road, building, pipeline, etc.) to be built and identify it (using the cursor) on the plan display. Each type of structure has its own design routines. In the case of a road, first identify the centerline and grade-point elevations, and then select (from pull-down menus) the lane widths, cross fall, depth of roadbed materials, side slopes, offsets, and staking intervals. The software will compute and plot the intermediate centerline elevations and pavement edge elevations, and the top and bottom of slope locations. All this computed data can be shown on the plan, or some (or all) such data can be turned off. The plotted road can now be shown in 3D for visual inspection.
9. Trace the road area stripping limits and select for import. Select the stripping depth for that area (refer to the imported bore-hole data if relevant), and then the volumes of topsoil stripping can be computed.
10. Existing and proposed cross sections and profiles for any defined line (including subgrade and final grades) can now be generated. Thus, volumes of general cut and fill can be determined for the entire road structure.
11. If the bore holes had identified rock strata in the general area, the software can determine and graphically display the location and elevation of any rock quantities that need to be computed separately (at much higher costs of excavation).

Most commercial software treats pipeline construction similarly (see Chapter 14), where invert elevations define the grade line and define the depth of cut so that cut volumes can be computed. The type of pipe bedding (granular or concrete) can be selected from pull-down menus, and the bedding volume can be determined by factoring in the trench lengths and widths as well as the bedding depth (see Figure 14.1). Other backfill material volumes, such as volumes for placed and compacted excavated trench material can also determined using the trench width and depth dimensions.

Building construction (see Chapter 16) can also be dealt with by locating the building footprint on the plan display and then entering first floor (or other) elevations as directed by a pull-down menu. Basement or foundation elevations are entered (as directed by the subgrade menu) to provide cut computations. Offsets (also selected from a pull-down menu) are defined and then displayed on the plan graphics. Parking areas and driveways are defined with cursor clicks, and selected locations have their proposed elevations entered; subgrade elevations are determined after material depths are determined. Thus, cut/fill volumes can be determined from the existing surface model (which was imported into the working file) and the newly determined subgrade surface model.

10.5.2 Layout

Layout can be accomplished using the design data from the 3D working file. The layout can be performed using conventional theodolite or tape surveys, total station surveys, robotic total station surveys, or GPS surveys, or by machine guidance and control techniques.

When using the first two techniques mentioned above (conventional theodolite or tape surveys), the northing, easting, and elevation of each layout point can be downloaded from the data file for use in the field. Some software will generate the angle and distance to be measured from preselected control stations to each of the layout points. The occupied and backsight points are identified first so that the relevant coordinates can be used (see Figure 5.20)

When using total stations and GPS receivers, the 3D data software (including the existing ground surface model and the design surface model) can reside on the field instrument controller. Once the base station (occupied point) and roving station (backsight point) have been set over their control points and tied into each other, the rover can, in real time, continually display its 3D position ground coordinates as it is moved, and its can be directed to selected layout points. (Selection can be made by tapping on a touch-screen display or by selecting a layout point by entering its number.) Once the layout point has been occupied by the rover, the layout data can then be saved in a layout file for documentation purposes, and the layout position accuracy can be noted (GPS surveys). This type of layout work is facilitated by using a large touch-screen tablet controller (see Figure 8.22), which can display a large section of the proposed layout works.

The 3D data files are also useful for layout by machine guidance and/or control using either RTK GPS techniques or local positioning system (LPS) (robotic total station) techniques. In the case of LPS, the 3D data files are transferred to the computer controlling the robotic total station. The robotic total station can thus control or guide the piece of construction equipment (e.g., bulldozer, grader, loader, etc.) by sighting the position-calibrated prisms attached to the equipment. In the case of RTK GPS layouts, the 3D files are transferred directly to the on-board controller (computer) located in the construction equipment cab. The interfaced GPS receivers are calibrated for position (relative to the ground), and the construction site itself is calibrated by taking GPS reading on all available horizontal and vertical control monuments.

10.6 Web Site References for Data Collection, DTM, and Civil Design

AGTEK Development   *www.AGTEK.com*
Autodesk   *www.autodesk.com*
Bentley   *www.bentley.com*
CAiCE   *www.CAiCe.com*
Carlson   *www.carlsonsw.com*
Eagle Point   *www.eaglepoint.com*
InSite Software   *www.insitesoftware.com*
Leica   *www.leica-geosystems.com*
Trimble   *www.trimble.com*
Tripod Data System   *www.tdsway.com*

Review Questions

10.1. What is the purpose of a GPS base station?

10.2. What is the meaning of the term *line and grade*?

10.3. How is the word *grade* used in construction work?

10.4. What is the difference between machine guidance and machine control?

10.5. What data would you typically find in a project 3D data file?

10.6. Compared to GPS techniques, what are the advantages of using motorized total stations to guide and control construction machines? What are the disadvantages?

10.7. Compared to motorized total station techniques, what are the advantages of using GPS techniques to guide and control construction machines? What are the disadvantages?

10.8. What recent development has now enabled RTK GPS techniques to rival motorized total station techniques in the vertical accuracies of layouts?

10.9. What recent development has enabled motorized total stations to become much more effective in machine guidance and control?

10.10. What is involved in project calibration when using guidance and control methods?

10.11. How can 3D data files be used to compute volumes of cut and fill?

Chapter 11

Highway Curves

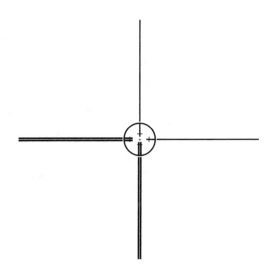

11.1 Route Surveys

Highway and railroad routes are chosen only after a complete and detailed study of all possible locations has been completed. Functional planning and route selection usually involves the use of aerial imagery, satellite imagery, and ground surveys, as well as the analysis of existing plans and maps. The selected route is chosen because it satisfies all design requirements with minimal social, environmental, and financial impact.

The proposed centerline (℄) is laid out in a series of straight lines (tangents) beginning at 0 + 00 (0 + 000, metric) and continuing to the route terminal point. Each time the route changes direction, the deflection angle between the back tangent and forward tangent is measured and recorded. Existing detail that might have an effect on the highway design is tied in either by conventional ground surveys, by aerial surveys, or by a combination of the two methods. Typical detail includes lakes, streams, trees, structures, existing roads and railroads, and so on. In addition to the detail location, the surveyor runs levels along the proposed route, with rod shots being taken across the route width at right angles to the ℄ at regular intervals (full stations, half stations, etc.) and at locations dictated by changes in the topography. The elevations thus determined are used to aid in the design of horizontal and vertical alignments; in addition, these elevations form the basis for the calculation of construction cut-and-fill quantities (see Chapter 17).

Advances in aerial imaging, including lidar and radar imaging (see Chapter 7), have resulted in ground-surface measuring techniques that can eliminate much of the time-consuming field surveying techniques described above. The location of detail and the determination of elevations are normally confined to that relatively narrow strip of land representing the highway right-of-way (ROW). Exceptions include potential river, highway, and railroad crossings, where approach profiles and sight lines (railroads) may have to be established.

11.2 Circular Curves: General Background

We noted in the previous section that a highway route survey is initially laid out as a series of straight lines (tangents). Once the ℄ location alignment has been confirmed, the tangents are joined by circular curves that allow for smooth vehicle operation at the speeds for which the highway was designed. Figure 11.1 illustrates how two tangents are joined by a circular curve and shows some related circular curve terminology. The point at which the alignment changes from straight to circular is known as the BC (beginning of curve). The BC is located distance T (subtangent) from the PI (point of tangent intersection). The length of the circular curve (L) depends on the central angle and the value of the radius (R). The point at which the alignment changes from circular back to tangent is known as the EC (end of curve). Because the curve is symmetrical about the PI, the EC is also located distance T from the PI. Recall from geometry that the radius of a circle is perpendicular to the tangent at the point of tangency. Therefore, the radius is perpendicular to the back tangent at the BC and to the forward tangent at the EC. The terms BC and EC are also referred to by some agencies as PC (point of curve) and PT (point of tangency), respectively, and by others as TC (tangent to curve) and CT (curve to tangent), respectively.

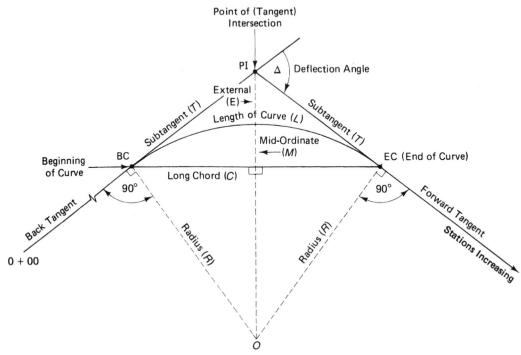

FIGURE 11.1 Circular curve terminology.

11.3 Circular Curve Geometry

Most curve problems are calculated from field measurements (Δ and the chainage or stationing of the PI) and from design parameters (R). Given R (which depends on the design speed) and Δ, all other curve components can be computed.

Analysis of Figure 11.2 shows that the curve deflection angle at the BC (PI-BC-EC) is $\Delta/2$, and that the central angle at O is equal to Δ, the tangent deflection angle. The line (O–PI), joining the center of the curve to the PI, effectively bisects all related lines and angles.

Tangent: In triangle BC-O-PI,

$$\frac{T}{R} = \tan\frac{\Delta}{2}$$

$$T = R\tan\frac{\Delta}{2} \tag{11.1}$$

Chord: In triangle BC-O-B,

$$\frac{1/2C}{R} = \sin\frac{\Delta}{2}$$

$$C = 2R\sin\frac{\Delta}{2} \tag{11.2}$$

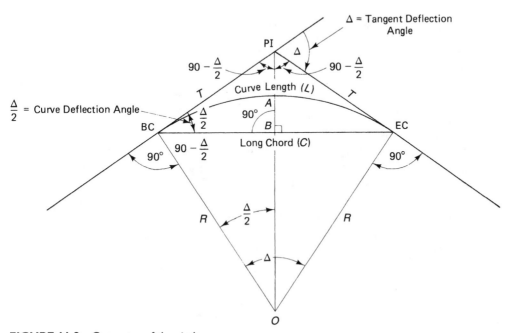

FIGURE 11.2 Geometry of the circle.

Midordinate:

$$\frac{OB}{R} = \cos\frac{\Delta}{2}$$

$$OB = R\cos\frac{\Delta}{2}$$

But:

$$OB = R - M$$

$$R - M = R\cos\frac{\Delta}{2}$$

$$M = R\left(1 - \cos\frac{\Delta}{2}\right) \tag{11.3}$$

External: In triangle BC-O-PI, O to PI $= R + E$:

$$\frac{R}{(R+E)} = \cos\frac{\Delta}{2}$$

$$E = R\left(\frac{1}{\cos\dfrac{\Delta}{2}} - 1\right) \tag{11.4}$$

$$= R\left(\sec\frac{\Delta}{2} - 1\right) \qquad \text{(alternate)}$$

Arc: From Figure 11.3, we can determine the following relationship:

$$\frac{L}{2\pi R} = \frac{\Delta}{360}$$

$$L = 2\pi R\left(\frac{\Delta}{360}\right) \tag{11.5}$$

where Δ is expressed in degrees and decimals of a degree.

The sharpness of the curve is determined by the choice of the radius (R); large radius curves are relatively flat, whereas small radius curves are relatively sharp. Many highway agencies use the concept of degree of curve (D) to define the sharpness of the curve. Degree of curve (D) is defined to be that central angle subtended by 100 ft of arc. (In railway design, D is defined to be the central angle subtended by 100 ft of chord.) From Figure 11.3, we can determine the following:

D and R:

$$D/360 = \frac{100}{2\pi R}$$

$$D = \frac{5729.58}{R} \tag{11.6}$$

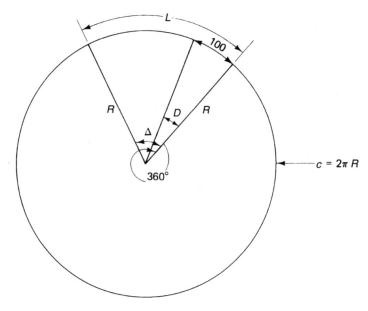

FIGURE 11.3 Relationship between the degree of curve (*D*) and the circle.

Arc:

$$\frac{L}{100} = \frac{\Delta}{D}$$

$$L = 100 \left(\frac{\Delta}{D}\right) \qquad (11.7)$$

■ EXAMPLE 11.1

Refer to Figure 11.4. You are given the following information:

$$\Delta = 16°38'$$
$$R = 1,000 \text{ ft}$$
$$\text{Pl at } 6 + 26.57$$

Calculate the station of the BC and EC; also calculate lengths *C*, *M*, and *E*.

Solution

We can use Equation 11.1 to determine the tangent distance (*T*) and Equation 11.5 to calculate the length of arc (*L*):

$$T = R \tan \frac{\Delta}{2}$$
$$= 1,000 \tan 8°19'$$
$$= 146.18 \text{ ft}$$

$$L = 2R \frac{\Delta}{360}$$
$$= 2\pi \times 1,000 \times \frac{16.6333}{360}$$
$$= 290.31 \text{ ft}$$

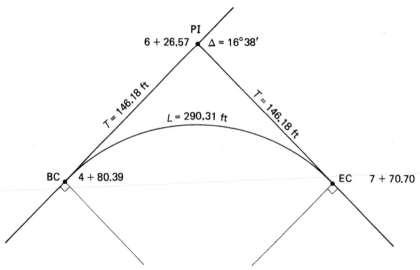

PI
6 + 26.57   Δ = 16°38′

T = 146.18 ft

T = 146.18 ft

L = 290.31 ft

BC   4 + 80.39

EC   7 + 70.70

FIGURE 11.4 Sketch for Example 11.1. *Note:* To aid in comprehension, the magnitude of the Δ angle has been exaggerated in this section.

Determine the BC and EC stations as follows:

$$
\begin{array}{rl}
\text{PI} & \text{at } 6 + 26.57 \\
-T & 1\quad 46.18 \\
\text{BC} = & \overline{4 + 80.39} \\
+L & 2\quad 90.31 \\
\text{EC} = & \overline{7 + 70.70}
\end{array}
$$

Use Equation 11.2 to calculate the length of C (see Figure 11.1):

$$
\begin{aligned}
C &= 2R \sin \frac{\Delta}{2} \\
&= 2 \times 1{,}000 \times \sin 8°19′ \\
&= 289.29 \text{ ft}
\end{aligned}
$$

Use Equation 11.3 to calculate the length of M (see Figure 11.1):

$$
\begin{aligned}
M &= R \left(1 - \cos \frac{\Delta}{2} \right) \\
&= 1{,}000 \left(1 - \cos 8°19′ \right) \\
&= 10.52 \text{ ft}
\end{aligned}
$$

Use Equation 11.4 to calculate the length of E (see Figure 11.1):

$$
\begin{aligned}
E &= R \left(\sec \frac{\Delta}{2} - 1 \right) \\
&= 1{,}000 \left(\sec 8°19′ - 1 \right) \\
&= 10.63 \text{ ft}
\end{aligned}
$$

Note: A common mistake made by students when they first study circular curves is to determine the station of the EC by adding the T distance to the PI. Although the EC is physically a distance of T from the PI, the stationing (chainage) must reflect the fact that the centerline (℄) no longer goes through the PI. The ℄ now takes the shorter distance (L) from the BC to the EC.

■ EXAMPLE 11.2

Refer to Figure 11.5. You are given the following information:

$$\Delta = 12°51'$$
$$R = 400 \text{ m}$$
$$\text{PI at } 0 + 241.782$$

Calculate the station of the BC and EC.

Solution

Use Equation 11.1 and Equation 11.5 to determine the T and L distances:

$$\begin{aligned} T &= R \tan \frac{\Delta}{2} \\ &= 400 \tan 6°25'30'' \\ &= 45.044 \text{ m} \end{aligned}$$

$$\begin{aligned} L &= 2\pi R \frac{\Delta}{360} \\ &= 2\pi \times 400 \times \frac{12.850}{360} \\ &= 89.710 \text{ m} \end{aligned}$$

Determine the BC and EC stations as follows:

PI	at 0 + 241.782
$-T$	45.044
BC	= 0 + 196.738
$+L$	89.710
EC	= 0 + 286.448

■ EXAMPLE 11.3

Refer to Figure 11.6. You are given the following information:

$$\Delta = 11°21'35''$$
$$\text{PI at } 14 + 87.33$$
$$D = 6°$$

Calculate the station of the BC and EC.

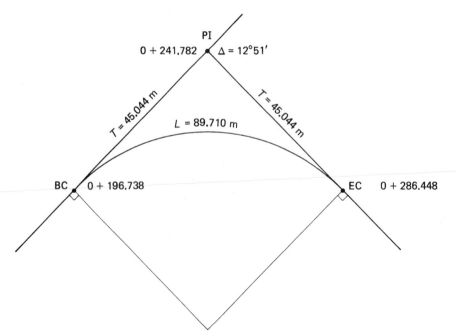

FIGURE 11.5 Sketch for Example 11.2.

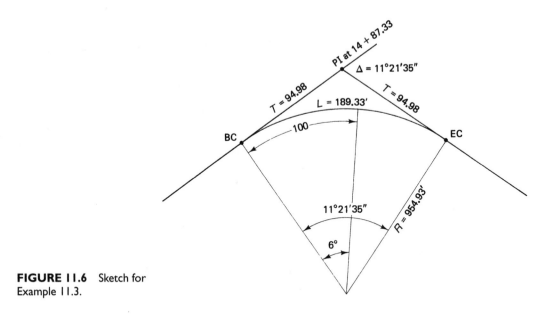

FIGURE 11.6 Sketch for
Example 11.3.

Solution

Use Equations 11.6, 11.1, and 11.7:

$$R = \frac{5,729.58}{D} = 954.93 \text{ ft}$$

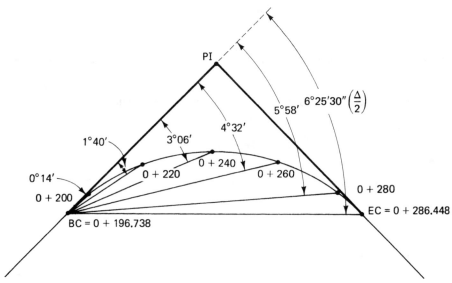

FIGURE 11.7 Field location for deflection angles. See Example 11.2.

$$T = R \tan \frac{\Delta}{2} = 954.93 \tan 5.679861° = 94.98 \text{ ft}$$

$$L = 100 \frac{\Delta}{D} = 100 \times \frac{11.359722}{6} = 189.33 \text{ ft}$$

Another alternative is to use Equation 11.5:

$$L = 2\pi R \frac{\Delta}{360} = 2\pi \times 954.93 \times \frac{11.359722}{360} = 189.33 \text{ ft}$$

$$
\begin{array}{rll}
\text{PI} & \text{at} & 14 + 87.33 \\
-T & & \underline{\quad 94.98} \\
\text{BC} & = & 13 + 92.35 \\
+L & \underline{1 \quad 89.33} \\
\text{EC} & = & 15 + 81.68
\end{array}
$$

11.4 Circular Curve Deflections

A common method of locating a curve in the field is by deflection angles. Typically, the theodolite is set up at the BC, and the deflection angles are turned from the tangent line (see Figure 11.7). If we use the following data from Example 11.2:

$$\text{BC at } 0 + 196.738$$

$$\text{EC at } 0 + 286.448$$

$$\frac{\Delta}{2} = 6°25'30'' = 6.4250°$$

$$L = 89.710 \text{ m}$$
$$T = 45.044 \text{ m}$$

and if the layout is to proceed at 20-m intervals, the procedure would be as described below:

1. Compute the deflection angles for the three required arc distances.

$$\text{Deflection angle} = \frac{\text{arc}}{L} \times \frac{\Delta}{2}$$

(a) BC to first even station (0 + 200): $(0 + 200) - (0 + 196.738) = 3.262$

$$\frac{3.262}{89.710} \times 6.4250 = 0.2336° = 0°14'01''$$

(b) Even station interval:

$$\frac{20}{89.710} \times 6.4250 = 1.4324° = 1°25'57''$$

(c) Last even station (0 + 280) to EC:

$$\frac{6.448}{89.710} \times 6.4250 = 0.4618° = 0°27'42''$$

2. Prepare a list of appropriate stations and *cumulative* deflection angles.

Stations	Deflection Angles
BC 0 + 196.738	0°00'00''
0 + 200	0°14'01'' + 1°25'57''
0 + 220	1°39'58'' + 1°25'57''
0 + 240	3°05'55'' + 1°25'57''
0 + 260	4°31'52'' + 1°25'57''
0 + 280	5°57'49'' + 0°27'42''
EC 0 + 286.448	6°25'31'' ≈ 6°25'30'' = $\frac{\Delta}{2}$

For many engineering layouts, the deflection angles are rounded to the closest minute or half-minute.

Another common method of locating a curve in the field is by using the "setting out" feature of total stations (see Chapter 5). The coordinates of each station on the curve are first uploaded into the total station, permitting its processor to compute the angle and distance from the instrument.

11.5　Chord Calculations

In Section 11.4, we determined that the deflection angle for station $0 + 200$ was $0°14'01''$. It follows that $0 + 200$ can be located by placing a stake on the theodolite line at $0°14'$ and at a distance of 3.262 m ($200 - 196.738$) from the BC. Furthermore, station $0 + 220$ can be located by placing a stake on the theodolite line at $1°40'$ (rounded) and at a distance of 20 m along the arc from the stake that locates $0 + 200$. The remaining stations can be located in a similar manner. Note, however, that this technique contains some error because the distances measured with a steel tape are not arc distances; they are straight lines known as chords or subchords (see Figure 11.8).

Equation 11.8 can be used to calculate the subchord. This equation, derived from Figure 11.2, is the special case of the long chord and the total deflection angle, as given by Equation 11.2. The general case can be stated as follows:

$$C = 2R \, (\sin \text{deflection angle}) \tag{11.8}$$

Any subchord can be calculated if its deflection angle is known.

Relevant chords for Section 11.4 can be calculated as follows (see Figure 11.8):

First chord: $C = 2 \times 400 \, (\sin 0°14'01'') = 3.2618$ m $= 3.262$ m (at three decimals, chord = arc)

Even station chord: $C = 2 \times 400 \, (\sin 1°25'57'') = 19.998$ m

Last chord: $C = 2 \times 400 \, (\sin 0°27'42'') = 6.448$ m

If these chord distances are used, the curve layout can proceed without error.

Note: Although the calculation of the first and last subchord shows these chords and arcs to be equal (i.e., 3.262 m and 6.448 m), the chords are always marginally shorter than the arcs. In the cases of short distances (above) and in the case of flat (large radius) curves, the arcs and chords can often appear to be equal. If more decimal places are introduced into the calculation, the marginal difference between arc and chord become evident.

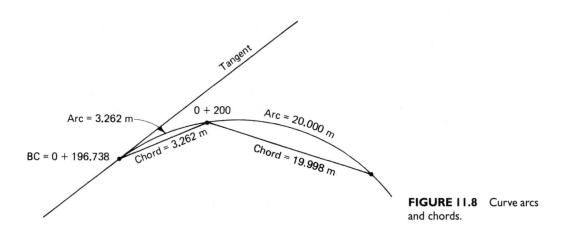

FIGURE 11.8 Curve arcs and chords.

11.6 Metric Considerations

Countries that use metric (SI) units have adopted for highway use a reference station of 1 km (e.g., 1 + 000); cross sections at 50-m, 20-m, and 10-m intervals; and a curvature design parameter based on a rational (even meter) value radius, as opposed to a rational value degree (even degree) of curve (D). The degree of curve originally found favor with most highway agencies because of the somewhat simpler calculations associated with its use. This factor was significant in the preelectronic age, when most calculations were performed by using logarithms. A comparison of techniques involving both D and R (radius) shows that the only computation in which the rational aspect of D is carried through is that for the arc length—that is, $L = 100\Delta/D$ (from Equation 11.7)—and even in that one case, the ease of calculation depends on delta (Δ) also being a rational number. In all other formulas, the inclusion of trigonometric functions or pi (π) ensures a more complex computation requiring the use of a calculator.

The widespread use of handheld calculators and computers has greatly reduced the importance of techniques that permit only marginal reductions in computations. Surveyors are now routinely solving their problems with calculators and computers rather than with the seemingly endless array of tables that once characterized the back section of survey texts. An additional reason for the lessening importance of D in computing deflection angles is that many curves (particularly at interchanges) are now being laid out by control point–based polar or intersection techniques (i.e., angle/distance or angle/angle) instead of deflection angles (see Chapter 5). Those countries using the metric system, almost without exception, use a rational value for the radius (R) as a design parameter.

11.7 Field Procedure

With the PI location and Δ angle measured in the field, and with the radius or degree of curve (D) chosen consistent with the design speed, all curve computations can be completed. The surveyor then returns to the field and measures off the tangent (T) distance from the PI to locate the BC and EC on the appropriate tangent lines. The theodolite or total station is then set up at the BC and zeroed and sighted in on the PI. The $\Delta/2$ angle ($6°25'30''$ in Example 11.2) is then turned off in the direction of the EC mark (wood stake, nail, etc.). If the computations for T and the field measurements of T have been performed correctly, the line of sight of the Δ angle will fall over the EC mark. If this does not occur, the T computations and then the field measurements are repeated.

Note: The $\Delta/2$ line of sight over the EC mark will, of necessity, contain some error. In each case, the surveyor will have to decide if the resultant alignment error is acceptable for the type of survey in question. For example, if the $\Delta/2$ line of sight misses the EC mark by 0.10 ft (30 mm) in a ditched highway ℄ survey, the surveyor would probably find the error acceptable and then proceed with the deflections. However, a similar error in the $\Delta/2$ line of sight in a survey to lay out an elevated portion of urban freeway would not be acceptable; in that case, an acceptable error would be roughly one-third of the preceding error (0.03 ft or 10 mm).

After the $\Delta/2$ check has been satisfactorily completed, the curve stakes are set by turning off the deflection angle and measuring the chord distance for the appropriate stations. If possible, the theodolite is left at the BC (see next section) for the entire curve stakeout, whereas the distance measuring moves continually forward from station to station. The rear taping surveyor keeps his or her body to the outside of the curve to avoid blocking the line of sight from the instrument. If total stations or EDMs are used for the deflections, there is no rear surveyor to obstruct the line of sight.

A final verification of the work is available after the last even station has been set, when the chord distance from the last even station to the EC stake is measured and compared to the theoretical value. If the check indicates an unacceptable discrepancy, the work is checked and the discrepancy is solved. Finally, after the curve has been deflected in, the party chief usually walks the curve, looking for any abnormalities. If a mistake has been made (e.g., putting in two stations at the same deflection angle is a common mistake), it will probably be evident. The circular curve's symmetry is such that even minor mistakes are obvious in a visual check.

Note here that many highway agencies use polar layout for interchanges and other complex features. If the coordinates of centerline alignment stations are determined, they can be used to locate the facility. In this application, the total station is placed at a known (or resection) station and aligned with another known station so that the instrument's processor can compute and display the angle and distance needed for layout (see also Chapters 5 and 12).

11.8 Moving up on the Curve

The curve deflections shown in Section 11.4 are presented in a form suitable for deflecting in while set up at the BC, with a zero setting at the PI. Often, however, the entire curve cannot be deflected in from the BC, and two or more instrument setups may be required before the entire curve has been located. The reasons for this situation include a loss of line of sight due to intervening obstacles (i.e., detail or elevation rises).

In Figure 11.9, the data of Example 11.2 are used to illustrate the geometric considerations in moving up on the curve. In this case, station 0 + 260 cannot be established with the theodolite at the BC (as were the previous stations) because a large tree obscures the line of sight from the BC to 0 + 260. To establish station 0 + 260, the instrument is moved forward to the last station (0 + 240) established from the BC. The horizontal circle is zeroed, and the BC is then sighted with the telescope in its inverted position. When the telescope is transited, the theodolite is once again oriented to the curve; that is, to set off the next (0 + 260) deflection, the surveyor refers to the previously prepared list of deflections and sets the appropriate deflection (4°32′—rounded) for the desired station location and then for all subsequent stations.

Figure 11.9 shows the geometry involved in this technique. A tangent to the curve is shown by a dashed line through station 0 + 240 (the proposed setup location). The angle from that tangent line to a line joining 0 + 240 to the BC is the deflection angle 3°06′. When the line from the BC is produced through station 0 + 240, the same angle (3°06′) occurs between that line and the tangent line through 0 + 240 (opposite angles). We have already determined that the deflection angle for 20 m was 1°26′ (see Section 11.4). When 1°26′ is added to 3°06′, the angle of 4°32′ for station 0 + 260 results, the same angle that was previously calculated for that station.

This discussion has limited the move up to one station. In fact, the move up can be repeated as often as necessary to complete the curve layout. The technique can generally be stated as follows: *When the instrument is moved up on the curve and the instrument is back-sighted with the telescope inverted at any other station, the theodolite will be oriented to the curve if the horizontal circle is first set to the value of the deflection angle for the sighted*

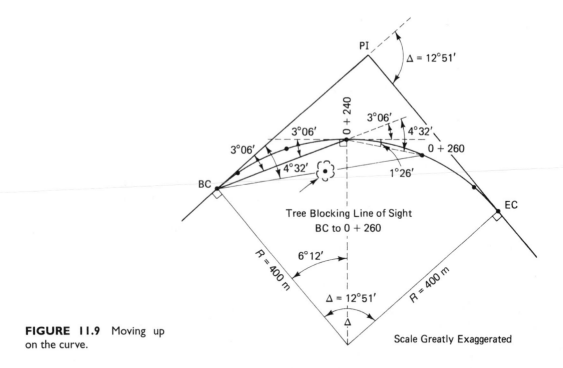

FIGURE 11.9 Moving up on the curve.

Scale Greatly Exaggerated

station. That is, in the case of a BC sight, the deflection angle to be set is obviously zero; if the instrument were set on $0 + 260$ and sighting $0 + 240$, a deflection angle of $3°06'$ would first be set on the scale.

When the inverted telescope is transited to its normal position, all subsequent stations can then be sighted by using the original list of deflections. This is the meaning of the phrase *theodolite oriented to the curve.* It is also why the list of deflections can be made first, before the instrument setup stations have been determined and (as we shall see in the next section) even before it has been decided whether to run in the curve centerline (₵) or whether it would be more appropriate to run in the curve on some offset line.

11.9 Offset Curves

Curves laid out for construction purposes must be established on offsets so that the survey stakes are not disturbed by construction activities. Many highway agencies prefer to lay out the curve on ₵ (centerline) and then offset each ₵ stake a set distance left and right. (Left and right are oriented by facing to a forward station.)

The stakes can be offset to one side by using the arm-swing technique described in Section 7.4.1.3, with the hands pointing to the two adjacent stations. If this method is done with care, the offsets on that one side can be established on radial lines without too much error. After one side has been offset in this manner, the other side is then offset by lining up the established offset stake with the ₵ stake and measuring the offset distance, while ensuring that all three stakes are visually in a straight line. Keeping the three stakes in a straight line ensures that any alignment error existing at the offset stakes steadily diminishes as one moves toward the ₵ and the construction works.

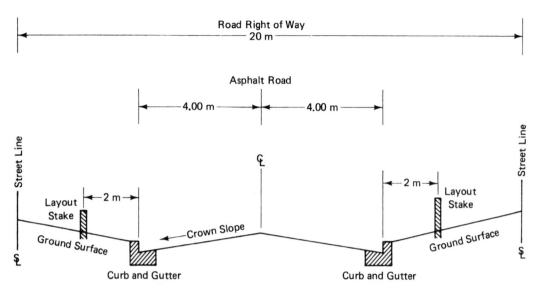

FIGURE 11.10 Municipal road cross section.

In the construction of most municipal roads, particularly curbed roads, the centerline may not be established. Instead, the road alignment is established directly on offset lines located a safe distance from the construction works. To illustrate, consider the curve in Example 11.2 used to construct a curbed road, as shown in Figure 11.10. The face of the curb is to be 4.00 m left and right of the centerline. Assume that the curb layout can be offset 2 m (each side) without interfering with construction. (Generally, the less cut or fill required, the smaller can be the offset distance.)

Figure 11.11 shows that if the layout is to be kept on radial lines through the ₵ stations, the station arc distances on the left-side (outside) curve will be longer than the corresponding ₵ arc distances, whereas the station arc distances on the right-side (inside) curve will be shorter than the corresponding ₵ arc distances. The figure also shows clearly that the ratio of the outside arc to the ₵ arc is identical to the ratio of the ₵ arc to the inside arc. (See the arc computations in Example 11.4.) *By keeping the offset stations on radial lines, the surveyor can use the ₵ deflections computed previously.*

When using setting-out programs in total stations to locate offset stations in the field, the surveyor can simply identify the offset value (when prompted by the program). The processor can then compute the coordinates of the offset stations and then inverse to determine (and display) the required angle and distance from the instrument station. Civil COGO-type software can also be used to compute coordinates of all offset stations, and the layout angles and distances from selected proposed instrument stations.

■ **EXAMPLE 11.4** *Illustrative Problem for Offset Curves (Metric Units)*
Consider the problem of a construction offset layout using the data of Example 11.2, the deflections developed in Section 11.4, and the offset of 2 m introduced already in this section.

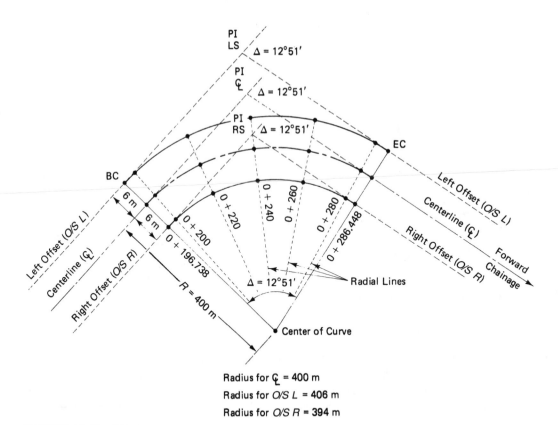

Radius for C_L = 400 m
Radius for O/S L = 406 m
Radius for O/S R = 394 m

FIGURE 11.11 Offset curves.

Given Data:

$$\Delta = 12°51'$$
$$R = 400 \text{ m}$$
$$\text{PI at } 0 + 241.782$$

Calculated Data:

$$T = 45.044 \text{ m}$$
$$L = 89.710 \text{ m}$$
$$\text{BC at } 0 + 196.738$$
$$\text{EC at } 0 + 286.448$$

Required: Curbs to be laid out on 2-m offsets at 20-m stations.

Solution

The following chart shows the calculated field deflections:

Station	Computed Deflection	Field Deflection
BC 0 + 196.738	0°00'00"	0°00'
0 + 200	0°14'01"	0°14'

0 + 220	1°39′58″	1°40′
0 + 240	3°05′55″	3°06′
0 + 260	4°31′52″	4°32′
0 + 280	5°57′49″	5°58′
EC 0 + 286.448	6°25′31″	$6°25′30″ = \dfrac{\Delta}{2}$; Check

Figures 11.10 and 11.11 show that the left-side (outside) curb face has a radius of 404 m. A 2-m offset for that curb results in an offset radius of 406 m. Similarly, the offset radius for the right-side (inside) curb is $400 - 6 = 394$ m.

Because we will use the deflections already computed, we must calculate only the corresponding left-side arc or chord distances and the corresponding right-side arc or chord distances. Although layout procedure (angle and distance) indicates that chord distances are required, for illustrative purposes, we will compute both the arc and chord distances on offset.

Arc Distance Computations: Figure 11.12 shows that the offset (o/s) arcs can be computed by direct ratio:

$$\frac{\text{o/s arc}}{\text{\textcentoldstyle arc}} = \frac{\text{o/s radius}}{\text{\textcentoldstyle radius}}$$

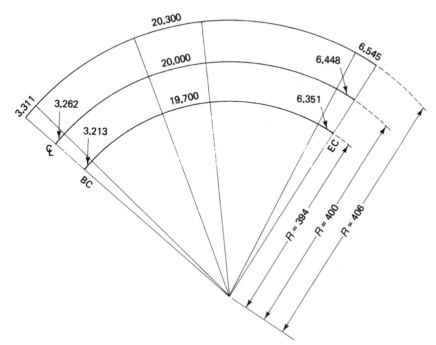

FIGURE 11.12 Offset arc lengths calculated by ratios.

For the first arc (BC to 0 + 200):

$$\text{Left side: o/s arc} = 3.262 \times \frac{406}{400} = 3.311 \text{ m}$$

$$\text{Right side: o/s arc} = 3.262 \times \frac{394}{400} = 3.213 \text{ m}$$

For the even station arcs:

$$\text{Left side: o/s arc} = 20 \times \frac{406}{400} = 20.300 \text{ m}$$

$$\text{Right side: o/s arc} = 20 \times \frac{394}{400} = 19.700 \text{ m}$$

For the last arc (0 + 280 to EC):

$$\text{Left side: o/s arc} = 6.448 \times \frac{406}{400} = 6.545 \text{ m}$$

$$\text{Right side: o/s arc} = 6.448 \times \frac{394}{400} = 6.351 \text{ m}$$

Arithmetic check:

$$\text{LS} - ₵ = ₵ - \text{RS}$$

Chord Distance Computations: Refer to Figure 11.13. For any deflection angle, the equation for chord length (see Equation 11.8 in Section 11.5) is:

$$C = 2R \,(\sin \text{ deflection angle})$$

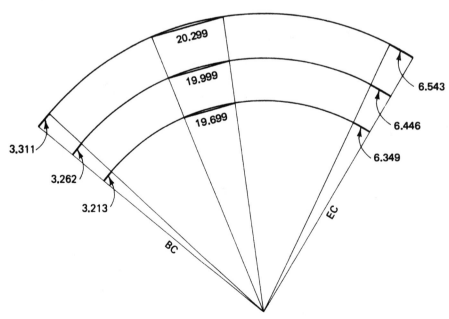

FIGURE 11.13 Offset chords calculated from deflection angles and offset radii.

In this problem, the deflection angles have been calculated previously, and the radius (R) is the variable. For the first chord (BC to 0 + 200):

$$\text{Left side: } C = 2 \times 406 \times \sin 0°14'01'' = 3.311 \text{ m}$$
$$\text{Right side: } C = 2 \times 394 \times \sin 0°14'01'' = 3.213 \text{ m}$$

For the even station chords (see Section 11.4):

$$\text{Left side: } C = 2 \times 406 \times \sin 1°25'57'' = 20.299 \text{ m}$$
$$\text{Right side: } C = 2 \times 394 \times \sin 1°25'57'' = 19.699 \text{ m}$$

For the last chord:

$$\text{Left side: } C = 2 \times 406 \times \sin 0°27'42'' = 6.543 \text{ m}$$
$$\text{Right side: } C = 2 \times 394 \times \sin 0°27'42'' = 6.349 \text{ m}$$

Arithmetic check:

$$\text{LS chord} - \text{₵ chord} = \text{₵ chord} - \text{RS chord}$$

■ **EXAMPLE 11.5** *Curve Problem (Foot Units)*
You are given the following ₵ data:

$$D = 5°$$
$$\Delta = 16°28'30''$$
$$\text{PI at } 31 + 30.62$$

You must furnish stakeout information for the curve on 50-ft offsets left and right of ₵ at 50-ft stations.

Solution
From Equation 11.6:

$$R = \frac{5,729.58}{D} = 1,145.92 \text{ ft}$$

From Equation 11.1:

$$T = R \tan \frac{\Delta}{2} = 1,145.92 \tan 8°14'15'' = 165.90 \text{ ft}$$

And from Equation 11.7:

$$L = 100 \left(\frac{\Delta}{D} \right) = 100 \times \left(\frac{16.475}{5} \right) = 329.50 \text{ ft}$$

we can also use Equation 11.5 to find the arc length:

$$L = 2\pi R \left(\frac{\Delta}{360} \right) = 329.50 \text{ ft}$$

$$
\begin{array}{rll}
\text{PI} & \text{at} & 31 + 30.62 \\
T & & \underline{1 \quad 65.90} \\
\text{BC} & = & 29 + 64.72 \\
+L & & \underline{3 \quad 29.50} \\
\text{EC} & = & 32 + 94.22
\end{array}
$$

Computation of Deflections:

$$\text{Total deflection for curve} = \frac{\Delta}{2} = 8°14'15'' = 494.25'$$

$$\text{Deflection per foot} = \frac{494.25}{329.50} = 1.5' \text{ per ft}$$

Also, because $D = 5°$, the deflection for 100 ft is $D/2$ or $2°30' = 150'$. Therefore, the deflection for 1 ft is 150/100 or 1.5' per ft.

Deflection for the first station:

$$35.28 \times 1.5 = 52.92' = 0°52.9'$$

Deflection for the even 50-ft stations:

$$50 \times 1.5 = 75' = 1°15'$$

Deflection for the last station:

$$44.22 \times 1.5 = 66.33' = 1°06.3'$$

The following chart shows the calculated deflections:

Stations	Deflections (Cumulative)	
	Office	Field (Closest Minute)
BC 29 + 64.72	0°00.0'	0°00'
30 + 00	0°52.9'	0°53'
30 + 50	2°07.9'	2°08'
31 + 00	3°22.9'	3°23'
31 + 50	4°37.9'	4°38'
32 + 00	5°52.9'	5°53'
32 + 50	7°07.9'	7°08'
EC 32 + 94.22	8°14.2'	8°14'
	$\approx 8°14.25'$	
	$= \dfrac{\Delta}{2}$, Check	

Chord calculations for left- and right-side curves on 50-ft (from ₵) offsets (see the chord calculation table and Figure 11.14):

$$\text{Radius for } ₵ = 1,145.92 \text{ ft}$$
$$\text{Radius for LS} = 1,195.92 \text{ ft}$$
$$\text{Radius for RS} = 1,095.92 \text{ ft}$$

Chap. 11   Highway Curves

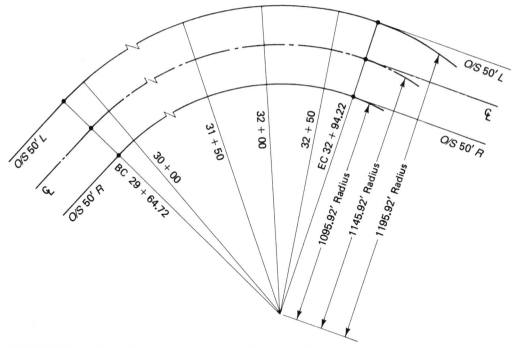

FIGURE 11.14 Sketch for the curve problem of Example 11.5.

CHORD CALCULATIONS FOR EXAMPLE 11.5

Interval	Left Side	℄	Right Side
BC to 30 + 00	$C = 2 \times 1195.92$ $\times \sin 0°52.9'$ $= 36.80$ ft	$C = 2 \times 1145.92$ $\times \sin 0°52.9'$ $= 35.27$ ft	$C = 2 \times 1095.92$ $\times \sin 0°52.9'$ $= 33.73$ ft
	Diff. = 1.53	Diff. = 1.54	
50-ft stations	$C = 2 \times 1195.92$ $\times \sin 1°15'$ $= 52.18$ ft	$C = 2 \times 1145.92$ $\times \sin 1°15'$ $= 50.00$ (to 2 decimals)	$C = 2 \times 1095.92$ $\times \sin 1°15'$ $= 47.81$ ft
	Diff. = 2.18	Diff. = 2.19	
32 + 50 to EC	$C = 2 \times 1195.92$ $\times \sin 1°06.3'$ $= 46.13$ ft	$C = 2 \times 1145.92$ $\times \sin 1°06.3'$ $= 44.20$ ft	$C = 2 \times 1095.92$ $\times \sin 1°06.3'$ $= 42.27$ ft
	Diff. = 1.93	Diff. = 1.93	

11.10 Compound Circular Curves

A compound curve consists of two (usually) or more circular arcs between two main tangents turning in the same direction and joining at common tangent points. Figure 11.15 shows a compound curve consisting of two circular arcs joined at a point of compound curve (PCC). The lower station (chainage) curve is number 1, whereas the higher station

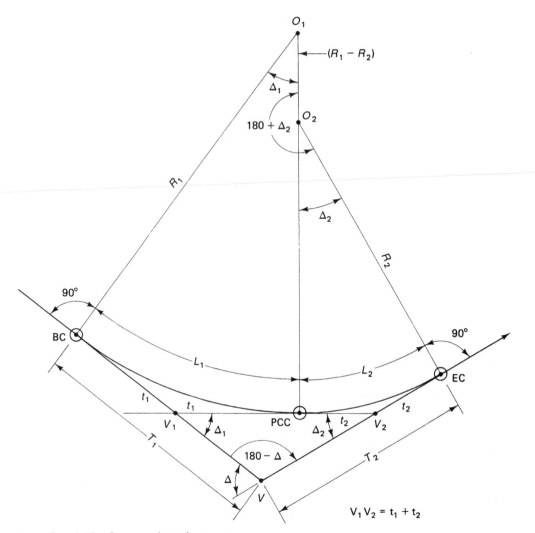

FIGURE 11.15 Compound circular curve.

curve is number 2. The parameters are R_1, R_2, Δ_1, Δ_2 ($\Delta_1 + \Delta_2 = \Delta$), T_1, and T_2. If four of these six or seven parameters are known, the others can be solved. Under normal circumstances, Δ_1 and Δ_2 (or Δ) are measured in the field, and R_1 and R_2 are given by design considerations, with minimum values governed by design speed.

Although compound curves can be manipulated to provide practically any vehicle path desired by the designer, they are not employed where simple curves or spiral curves can be used to achieve the same desired effect. Compound curves are reserved for those applications where design constraints (topographic or cost of land) preclude the use of simple or spiral curves, and they are now usually found chiefly in the design of interchange loops and ramps. Smooth driving characteristics require that the larger radius be no more than 1 1/3 times larger than the smaller radius. (This ratio increases to 1 1/2 when dealing with interchange curves.)

Solutions to compound curve problems vary because several possibilities exist according to which of the data are known in any one given problem. All problems can be solved by use of the sine law or cosine law or by the omitted measurement traverse technique illustrated in Example 6.2. If the omitted measurement traverse technique is used, the problem becomes a five-sided traverse (see Figure 11.15), with sides R_1, T_1, T_2, R_2, and $(R_1 - R_2)$, and with angles $90°$, $180 - \Delta°$, $90°$, $180 + \Delta°_2$, and $\Delta°_1$. An assumed azimuth can be chosen to simplify the computations (i.e., set the direction of R_1 to $0°00'00''$).

11.11 Reverse Curves

Reverse curves [see Figure 11.16(a) and (b)] are seldom used in highway or railway alignment. The instantaneous change in direction occurring at the point of reverse curve (PRC) would cause discomfort and safety problems for all but the slowest of speeds. Also, because the change in curvature is instantaneous, there is no room to provide superelevation transition from cross slope right to cross slope left. Reverse curves can be used to advantage, however, when the instantaneous change in direction poses no threat to safety or comfort.

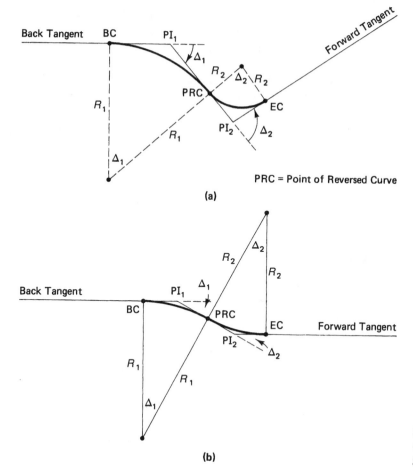

FIGURE 11.16 Reverse curves. (a) Nonparallel tangents. (b) Parallel tangents.

The reverse curve is particularly pleasing to the eye and is used with great success for park roads, formal paths, waterway channels, and the like. This curve is illustrated in Figure 11.16(a) and (b); the parallel tangent application is particularly common (R_1 is often equal to R_2). As with compound curves, reverse curves have six independent parameters (R_1, Δ_1, T_1, R_2, Δ_2, T_2). The solution technique depends on which parameters are unknown, and the techniques noted for compound curves also provide the solution to reverse curve problems.

11.12 Vertical Curves: General Background

Vertical curves are used in highway and street vertical alignment to provide a gradual change between two adjacent gradelines. Some highway and municipal agencies introduce vertical curves at every change in gradeline slope, whereas other agencies introduce vertical curves into the alignment only when the net change in slope direction exceeds a specific value (e.g., 1.5 percent or 2 percent).

In Figure 11.17, vertical curve terminology is introduced: g_1 is the slope (percentage) of the lower chainage grade line, g_2 is the slope of the higher chainage grade line, BVC is the beginning of the vertical curve, EVC is the end of the vertical curve, and PVI is the point of intersection of the two adjacent gradelines. The length of vertical curve (L) is the projection of the curve onto a horizontal surface and, as such, corresponds to plan distance. The algebraic change in slope direction is A, where $A = g_2 - g_1$. For example, if $g_1 = +1.5$ percent and $g_2 = -3.2$ percent, A would be equal to $(-3.2 - 1.5) = -4.7$.

The geometric curve used in vertical alignment design is the vertical axis parabola. The parabola has the desirable characteristics of (1) a constant rate of change of slope, which contributes to a smooth alignment transition, and (2) ease of computation of vertical offsets, which permits easily computed curve elevations. The general equation of the parabola is:

$$y = ax^2 + bx + c \tag{11.9}$$

The slope of this curve at any point is given by the first derivative:

$$\frac{dy}{dx} = 2ax + b \tag{11.10}$$

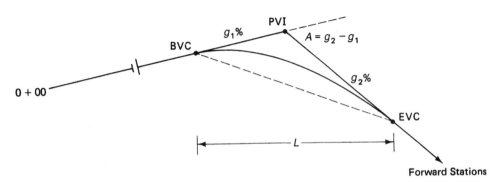

FIGURE 11.17 Vertical curve terminology (profile view shown).

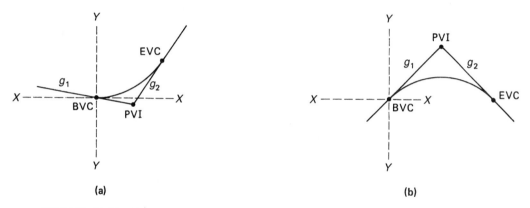

FIGURE 11.18 Types of vertical curves. (a) Sag curve. (b) Crest curve.

and the rate of change of slope is given by the second derivative:

$$\frac{d^2 y}{dx^2} = 2a \qquad (11.11)$$

which is a constant, as we noted previously. The rate of change of slope ($2a$) can also be written as A/L.

If, for convenience, the origin of the axes is placed at the BVC (see Figure 11.18), the general equation becomes:

$$y = ax^2 + bx$$

and because the slope at the origin is g_1, the expression for the slope of the curve at any point becomes:

$$\frac{dy}{dx} = \text{slope} = 2ax + g_1 \qquad (11.12)$$

The final general equation can be written as:

$$y = ax^2 + g_1 x \qquad (11.13)$$

11.13 Geometric Properties of the Parabola

Figure 11.19 illustrates the following relationships:

- The difference in elevation between the BVC and a point on the g_1 gradeline at a distance x units (feet or meters) is $g_1 x$ (g_1 is expressed as a decimal).
- The tangent offset between the gradeline and the curve is given by ax^2, where x is the horizontal distance from the BVC; that is, tangent offsets are proportional to the squares of the horizontal distances.
- The elevation of a crest curve at distance x from the BVC is given by BVC + $g_1 x$ − ax^2 = curve elevation. (The sign is reversed in a sag curve.)

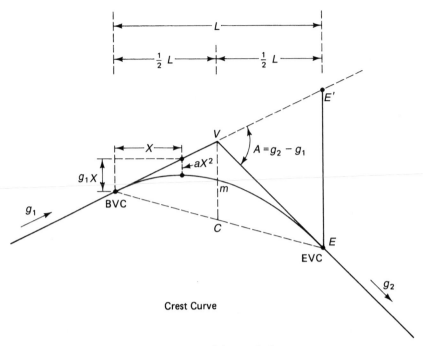

FIGURE 11.19 Geometric properties of the parabola.

- The gradelines (g_1 and g_2) intersect midway between the BVC and the EVC; that is, BVC to $V = \frac{1}{2} L = V$ to EVC.
- Offsets from the two gradelines are symmetrical with respect to the PVI (V).
- The curve lies midway between the PVI and the midpoint of the chord; that is, CM = MV.

11.14 Computation of the High or the Low Point on a Vertical Curve

The locations of curve high and low points (if applicable) are important for drainage considerations; for example, on curbed streets, catch basins must be installed precisely at the drainage low point. We noted in Equation 11.12 that the slope was given by:

$$\text{Slope} = 2ax + g_1$$

Figure 11.20 shows a sag vertical curve with a tangent drawn through the low point. It is obvious that the tangent line is horizontal with a slope of zero; that is:

$$2ax + g_1 = 0 \tag{11.14}$$

Had a crest curve been drawn, the tangent through the high point would have exhibited the same characteristics. Because $2a = A/L$, Equation 11.14 can be rewritten as

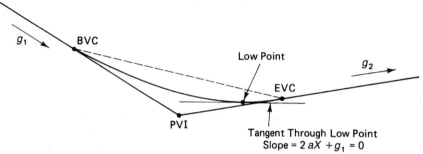

FIGURE 11.20 Tangent at curve low point.

$$x = -g_1 \left(\frac{L}{A} \right) \tag{11.15}$$

where x is the distance from the BVC to the high or low point.

11.15 Computing a Vertical Curve

Use the following procedure for computing a vertical curve:

1. Compute the algebraic difference in grades: $A = g_2 - g_1$.
2. Compute the chainage of the BVC and EVC. If the chainage of the PVI is known, $\frac{1}{2}L$ is simply subtracted and added to the PVI chainage.
3. Compute the distance from the BVC to the high or low point (if applicable) using Equation 11.15:

$$x = -g_1 \left(\frac{L}{A} \right)$$

and determine the station of the high/low point.

4. Compute the tangent gradeline elevation of the BVC and the EVC.
5. Compute the tangent gradeline elevation for each required station.
6. Compute the midpoint of the chord elevation:

$$\frac{\text{elevation of BVC} + \text{elevation of EVC}}{2}$$

7. Compute the tangent offset (d) at the PVI (i.e., distance VM in Figure 11.19):

$$d = \frac{\text{difference in elevation of PVI and midpoint of chord}}{2}$$

8. Compute the tangent offset for each individual station (see line ax^2 in Figure 11.19):

$$\text{Tangent offset} = \frac{d(x)^2}{(L/2)^2}, \text{ or } \frac{(4d)\, x^2}{L^2} \tag{11.16}$$

where x is the distance from the BVC or EVC (whichever is closer) to the required station.

9. Compute the elevation on the curve at each required station by combining the tangent offsets with the appropriate tangent gradeline elevations. Add for sag curves and subtract for crest curves.

■ EXAMPLE 11.6

The techniques used in vertical curve computations are illustrated in this example. You are given the following information: $L = 300$ ft, $g_1 = -3.2\%$, $g_2 = +1.8\%$, PVI at 30 + 30 with elevation = 465.92. Determine the location of the low point and the elevations on the curve at even stations, as well as at the low point.

Solution

Refer to Figure 11.21.

1. $A = 1.8 - (-3.2) = 5.0$

2. $\text{PVI} - \frac{1}{2}L = \text{BVC}$; BVC at $(30 + 30) - 150 = 28 + 80.00$

   $\text{PVI} + \frac{1}{2}L = \text{EVC}$; EVC at $(30 + 30) + 150 = 31 + 80.00$

   $\text{EVC} - \text{BVC} = L$; $(31 + 80) - (28 + 80) = 300$; Check

3. Elevation of PVI = 465.92

   150 ft at 3.2% = 4.80 (see Figure 11.21)

   Elevation BVC = 470.72

   Elevation PVI = 465.92

   150 ft at 1.8% = 2.70

   Elevation EVC = 468.62

4. Location of low point is calculated using Equation 11.15:

$$x = \frac{-g_1 L}{A}$$

$$= \frac{3.2 \times 300}{5} = 192.00 \text{ ft (from the BVC)}$$

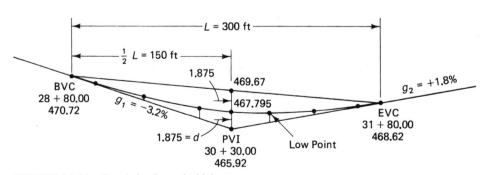

FIGURE 11.21 Sketch for Example 11.6.

Table 11.1 PARABOLIC CURVE ELEVATIONS BY TANGENT OFFSETS

Station	Tangent Elevation	+ Tangent Offset $\left(\dfrac{x}{\frac{1}{2}L}\right)^2 d^*$	= Curve Elevation
BVC 28 + 80	470.72	$(0/150)^2 \times 1.875 = 0$	470.72
29 + 00	470.08	$(20/150)^2 \times 1.875 = .03$	470.11
30 + 00	466.88	$(120/150)^2 \times 1.875 = 1.20$	468.08
PVI 30 + 30	465.92	$(150/150)^2 \times 1.875 = 1.875$	467.80
Low 30 + 72	466.68	$(108/150)^2 \times 1.875 = .97$	467.65
31 + 00	467.18	$(80/150)^2 \times 1.875 = .53$	467.71
EVC 31 + 80	468.62	$(0/150)^2 \times 1.875 = 0$	468.62
See Section 11.16			
$\begin{cases} 30+62 \\ 30+72 \\ 30+82 \end{cases}$	466.50 466.68 466.86	$(118/150)^2 \times 1.875 = 1.16$ $(108/150)^2 \times 1.875 = 0.97$ $(98/150)^2 \times 1.875 = 0.80$	$\left.\begin{matrix}467.66 \\ 467.65 \\ 467.66\end{matrix}\right\}$

*Where x is distance from BVC or EVC, whichever is closer.

5. Tangent gradeline computations are entered in Table 11.1. Example: Elevation at $29 + 00 = 470.72 - (0.032 \times 20) = 470.72 - 0.64 = 470.08$
6. Midchord elevation:

$$\frac{470.72 \,(\text{BVC}) \; + \; 468.62 \,(\text{EVC})}{2} = 469.67 \text{ ft}$$

7. Tangent offset at PVI (d):

$$d = \frac{\text{difference in elevation of PVI and midchord}}{2}$$
$$= \frac{469.67 - 465.92}{2} = 1.875 \text{ ft}$$

8. Tangent offsets are computed by multiplying the distance ratio squared

$$\left(\frac{x}{L/2}\right)^2$$

by the maximum tangent offset (d). See Table 11.1.
9. The computed tangent offsets are added (in this example) to the tangent elevation to determine the curve elevation.

11.15.1 Parabolic Curve Elevations Computed Directly from the Equation

In addition to the tangent offset method shown earlier, vertical curve elevations can also be computed directly from the general equation.

$$y = ax^2 + bx + c, \text{ where } a = (g_2 - g_1)/2L$$
$$L = \text{horizontal length of vertical curve}$$

Table 11.2 PARABOLIC CURVE ELEVATIONS COMPUTED FROM THE EQUATION
$Y = ax^2 + bx + c$

Station	Distance from BVC	ax^2	bx	c	y (Elevation on the Curve)
BVC 28 + 80	0				470.72
29 + 00	20	0.03	−0.64	470.72	470.11
30 + 00	120	1.20	−3.84	470.72	468.08
PVI 30 + 30	150	1.88	−4.80	470.72	467.80
Low 30 + 72	192	3.07	−6.14	470.72	467.65
31 + 00	220	4.03	−7.04	470.72	467.71
EVC 31 + 80	300	7.50	−9.60	470.72	468.62

$b = g_1$

c = elevation at BVC

x = horizontal distance from BVC

y = elevation on the curve at distance x from the BVC

This technique is illustrated in Table 11.2 using the data from Example 11.6.

11.16 Design Considerations

From Section 11.14, $2a = A/L$ is an expression giving the constant rate of change of slope for the parabola. Another useful relationship is the inverse, or:

$$K = \frac{L}{A} \tag{11.17}$$

where K is the horizontal distance required to effect a 1 percent change in slope on the vertical curve. Substituting for L/A in Equation 11.15 yields:

$$x = -g_1 K \tag{11.18}$$

(the result is always positive), where x is the distance to the low point from the BVC, or:

$$x = +g_2 K \tag{11.19}$$

where x is the distance to the low point from the EVC.

You can see in Figure 11.19 that EE' is the distance generated by the divergence of g_1 and g_2 over the distance $L/2$:

$$EE' = \frac{(g_2 - g_1)}{100} \times \frac{L}{2}$$

Figure 11.19 also shows that VC = $\frac{1}{2}$ EE' (similar triangles) and that VM = $d = \frac{1}{4}$ EE'; thus:

$$d = \left(\frac{1}{4}\right)\frac{(g_2 - g_1)}{100} \times \frac{L}{2} = \frac{AL}{800} \tag{11.20}$$

Table 11.3 TYPICAL DESIGN CONTROLS FOR CREST AND SAG VERTICAL CURVES BASED ON MINIMUM STOPPING SIGHT DISTANCES FOR VARIOUS DESIGN SPEEDS

Design Speed, V (km/h)	Min. Stopping Sight Distance, S (m)	K Factor	
		Crest (m)	Sag (m)
40	45	4	8
50	65	8	12
60	85	15	18
70	110	25	25
80	135	35	30
90	160	50	40
100	185	70	45
110	215	90	50
120	245	120	60
130	275	150	70
140	300	180	80

$$K_{crest} = \frac{S^2}{200\, h_1\, (1 + \sqrt{h_2/h_1})^2}$$

$h_1 = 1.05$ m, height of driver's eye

$h_2 = 0.38$ m, height of object

$L = KA$, from Equation 11.15, where L(m) cannot be less than V(km/h). When $KA < V$, use $L = V$.

$$K_{sag} = \frac{S^2}{200\, (h + S\, tan\alpha)}$$

$h = 0.60$ m, height of headlights

$\alpha = 1°$, angular spread of light beam

Source: Vertical Curve Tables, Ministry of Transportation, Ontario.

or from Equation 11.17:

$$d = \frac{KA^2}{800} \qquad (11.21)$$

Equations 11.18, 11.19, and 11.21 are useful when design criteria are defined in terms of K.

Table 11.3 shows values of K for minimum stopping sight distances. On crest curves, it is assumed that the driver's eye height is at 1.05 m and the object that the driver must see is at least 0.38 m. The defining conditions for sag curves would be nighttime restrictions that relate to the field of view given by headlights with an angular beam divergence of 1°.

In practice, the length of a vertical curve is rounded to the nearest even meter, and if possible the PVI is located at an even station so that the symmetrical characteristics of the curve can be fully used. To avoid the aesthetically unpleasing appearance of very short vertical curves, some agencies insist that the length of the vertical curve (L) be at least as long in meters as the design velocity is in kilometers per hour.

Closer analysis of the data shown in Table 11.1 indicates a possible concern. The vertical curve at the low or high point has a relatively small change of slope. This is not a problem for crest curves or for sag curves for ditched roads and highways (ditches can have gradelines independent of the ℄ gradeline). However, when sag curves are used for curbed municipal street design, a drainage problem is introduced. The curve elevations in brackets in Table 11.1 cover 10 ft on either side of the low point. These data illustrate that, for a distance of 20 ft, there is almost no change in elevation (i.e., only 0.01 ft). If the road were built according to the design, chances are that the low-point catch basin would not completely drain the extended low-point area. The solution to this problem is for the surveyor to lower the catch-basin grate arbitrarily (1 in. is often used) to ensure proper drainage, or to install additional catch basins in the extended low area.

11.17 Spiral Curves: General Background

A spiral is a curve with a uniformly changing radius. Spirals are used in highway and railroad alignment to overcome the abrupt change in direction that occurs when the alignment changes from a tangent to a circular curve, and vice versa. The length of the spiral curve is also used for the transition from normally crowned pavement to fully superelevated (banked) pavement.

Figure 11.22 illustrates how the spiral curve is inserted between tangent and circular curve alignment. You can see that, at the beginning of the spiral (T.S. = tangent to spiral), the radius of the spiral is the radius of the tangent line (infinitely large), and that the radius of the spiral curve decreases at a uniform rate until, at the point where the circular curve begins (S.C. = spiral to curve), the radius of the spiral equals the radius of the circular curve. In the previous section, we noted that the parabola, which is used in vertical alignment, had the important property of a uniform rate of change of slope. Here, we find that the spiral, used in horizontal alignment, has a uniform rate of change of radius (curvature). This property permits the driver to leave a tangent section of highway at relatively high rates of speed without experiencing problems with safety or comfort.

Figure 11.23 illustrates how the circular curve is moved inward (toward the center of the curve), leaving room for the insertion of a spiral at either end of the shortened circular curve. The amount that the circular curve is shifted in from the main tangent line is known

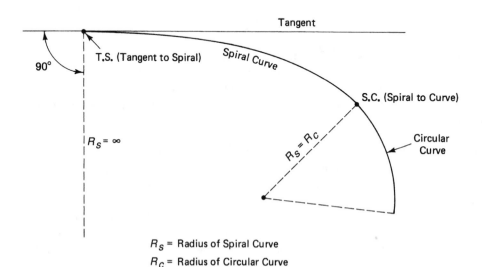

R_S = Radius of Spiral Curve
R_C = Radius of Circular Curve

FIGURE 11.22 Spiral curves.

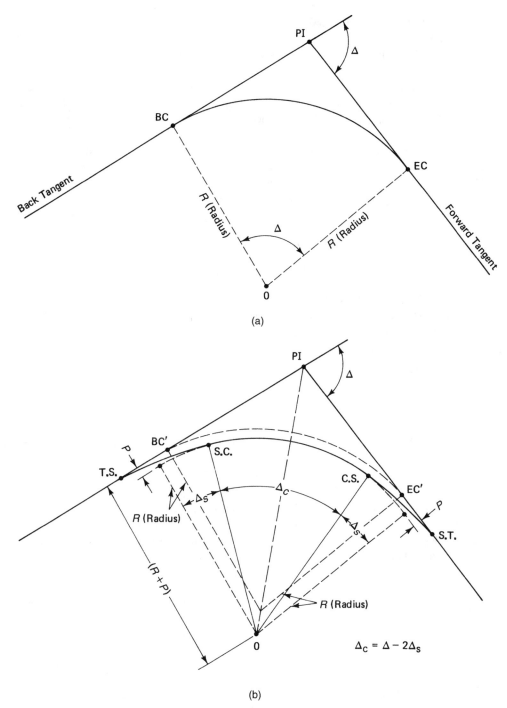

(a)

(b)

FIGURE 11.23 Shifting the circular curve to make room for the insertion of spirals. (a) Circular curve joining two tangents. (b) Circular curve shifted inward (toward curve center) to make room for the insertion of spiral curves at either end of the circular curve.

Sec. 11.17   Spiral Curves: General Background   **415**

as P. This shift results in the curve center (O) being at the distance $(R + P)$ from the main tangent lines.

The spirals illustrated in this text reflect the common practice of using equal spirals to join the ends of a circular or compound curve to the main tangents. For more complex spiral applications, such as unequal spirals and spirals joining circular arcs, refer to a text on route surveying. This text shows excerpts from spiral tables (see Tables 11.4 through 11.7). Each U.S. state and Canadian province prepares and publishes similar tables for use by their personnel. A wide variety of spirals, both geometric and empirical, have been used to develop spiral tables. Geometric spirals include the cubic parabola and the clothoid curve, and empirical spirals include the A.R.E.A. 10-chord spiral, used by many railroads. Generally, the use of tables is giving way to computer programs for spiral solutions. All spirals give essentially the same appearance when staked out in the field.

11.18 Spiral Curve Computations

Usually, data for a spiral computation are obtained as follows (refer to Figure 11.24):

1. Δ is determined in the field.
2. R or D (degree of curve) is given by design considerations (usually defined by design speed, but sometimes by property constraints).
3. Stationing (chainage) of PI is determined in the field.
4. L_s is chosen with respect to design speed and the number of traffic lanes.

All other spiral parameters can be determined by computation and/or by use of spiral tables.

$$\text{Tangent to spiral (see Figure 11.24): } T_s = (R + P) \tan \frac{\Delta}{2} + q \qquad (11.22)$$

$$\text{Spiral tangent deflection: } \Delta_s = \frac{L_s D}{200} \qquad (11.23)$$

In circular curves, $\Delta = LD/100$ (see Equation 11.7). Because the spiral has a uniformly changing D, the spiral angle (Δ_s) = the length of the spiral (L_s) in stations times the average degree of curve ($D/2$). The total length of the curve system is comprised of the circular curve and the two spiral curves, that is:

$$\text{Total length: } L = L_c + 2L_s \qquad (11.24)$$

See Figure 11.24, where L is the *total length of the curve system*.

$$\text{Total deflection: } \Delta = \Delta_c + 2\Delta_s \qquad (11.25)$$

See Figure 11.24.

$$\text{Spiral deflection: } \theta_s = \frac{\Delta_s}{3} \quad \text{(approximate)} \qquad (11.26)$$

Table 11.4 SPIRAL TABLES FOR L_s = 150 FEET

D	Δ_s	R	p	$R+p$	q	LT	ST	Δ_s	X_c	Y_c	$\dfrac{D}{10\,L_s}$
7°30'	5°37'30"	763.9437	1.2268	765.1705	74.9759	100.0305	50.0459	5.62500°	149.86	4.91	0.00500
8°00'	6°00'00"	716.1972	1.3085	717.5057	74.9726	100.0575	50.0523	6.00000°	149.84	5.23	0.00533
30'	6°22'30"	674.0680	1.3902	675.4582	74.9691	100.0649	50.0590	6.37500°	149.81	5.56	0.00567
9°00'	6°45'00"	636.6198	1.4719	638.0917	74.9653	100.0728	50.0662	6.75000°	149.79	5.88	0.00600
30'	7°07'30"	603.1135	1.5536	604.6671	74.9614	100.0811	50.0738	7.12500°	149.77	6.21	0.00633
10°00'	7°30'00"	572.9578	1.6352	574.5930	74.9572	100.0899	50.0817	7.50000°	149.74	6.54	0.00667
30'	7°52'30"	545.6741	1.7169	547.3910	74.9528	100.0991	50.0901	7.87500°	149.72	6.86	0.00700
11°00'	8°15'00"	520.8707	1.7985	522.6692	74.9482	100.1088	50.0989	8.25000°	149.69	7.19	0.00733
30'	8°37'30"	498.2242	1.8802	500.1044	74.9434	100.1190	50.1082	8.62500°	149.66	7.51	0.00767
12°00'	9°00'00"	477.4648	1.9618	479.4266	74.9384	100.1295	50.1178	9.00000°	149.63	7.84	0.00800
13°00'	9°45'00"	440.7368	2.1249	442.8617	74.9277	100.1521	50.1383	9.75000°	149.57	8.49	0.00867
14°00'	10°30'00"	409.2556	2.2880	411.5436	74.9161	100.1765	50.1605	10.50000°	149.50	9.14	0.00933
15°00'	11°15'00"	381.9719	2.4510	384.4229	74.9037	100.2027	50.1843	11.25000°	149.42	9.79	0.01000
16°00'	12°00'00"	358.0986	2.6139	360.7125	74.8905	100.2307	50.2098	12.00000°	149.34	10.44	0.01067
17°00'	12°45'00"	337.0340	2.7767	339.8107	74.8764	100.2606	50.2370	12.75000°	149.26	11.09	0.01133
18°00'	13°30'00"	318.3099	2.9394	321.2493	74.8614	100.2924	50.2659	13.50000°	149.17	11.73	0.01200
19°00'	14°15'00"	301.5567	3.1020	304.6587	74.8456	100.3259	50.2964	14.25000°	149.07	12.38	0.01267
20°00'	15°00'00"	286.4789	3.2645	289.7434	74.8290	100.3614	50.3287	15.00000°	148.98	13.03	0.01333
21°00'	15°45'00"	272.8370	3.4269	276.2639	74.8115	100.3987	50.3627	15.75000°	148.87	13.67	0.01400
22°00'	16°30'00"	260.4354	3.5891	264.0245	74.7932	100.4379	50.3983	16.50000°	148.76	14.31	0.01467
23°00'	17°15'00"	249.1121	3.7512	252.8633	74.7740	100.4790	50.4357	17.25000°	148.65	14.96	0.01533
24°00'	18°00'00"	238.7324	3.9132	242.6456	74.7539	100.5219	50.4748	18.00000°	148.53	15.60	0.01600
25°00'	18°45'00"	229.1831	4.0750	233.2581	74.7331	100.5668	50.5157	18.75000°	148.40	16.26	0.01667
26°00'	19°30'00"	220.3684	4.2367	224.6051	74.7114	100.6135	50.5582	19.50000°	148.27	16.88	0.01733

Source: Spiral Tables (foot units), courtesy Ministry of Transportation, Ontario.

Table 11.5 SPIRAL CURVE LENGTHS AND SUPERELEVATION RATES: SUPERELEVATION (e) MAXIMUM OF 0.06, TYPICAL FOR NORTHERN CLIMATES*

D	V = 30 e	V = 30 L, ft 2 Lane	V = 30 L, ft 4 Lane	V = 40 e	V = 40 L, ft 2 Lane	V = 40 L, ft 4 Lane	V = 50 e	V = 50 L, ft 2 Lane	V = 50 L, ft 4 Lane	V = 60 e	V = 60 L, ft 2 Lane	V = 60 L, ft 4 Lane	V = 70 e	V = 70 L, ft 2 Lane	V = 70 L, ft 4 Lane	V = 80 e	V = 80 L, ft 2 Lane	V = 80 L, ft 4 Lane
0°15'	NC	0	0	NC	0	0	NC	0	0	NC	0	0	NC	0	0	RC	250	250
0°30'	NC	0	0	NC	0	0	NC	0	0	RC	200	200	RC	200	200	.023	250	250
0°45'	NC	0	0	NC	0	0	RC	150	150	.021	200	200	.026	200	200	.033	250	250
1°00'	NC	0	0	RC	150	150	.020	150	150	.027	200	200	.033	200	200	.041	250	250
1°30'	RC	100	100	.020	150	150	.028	150	150	.036	200	200	.044	200	200	.053	250	300
2°00'	RC	100	100	.026	150	150	.035	150	150	.044	200	200	.052	200	250	.059	250	300
2°30'	.020	100	100	.031	150	150	.040	150	150	.050	200	200	.057	200	300	.060	250	300
3°00'	.023	100	100	.035	150	150	.044	150	150	.054	200	250	.060	200	300	$D_{max} = 2°30'$		
3°30'	.026	100	100	.038	150	150	.048	150	200	.057	200	250	$D_{max} = 3°00'$					
4°00'	.029	100	100	.041	150	150	.051	150	200	.059	200	250						
5°00'	.034	100	100	.046	150	150	.056	150	200	.060	200	250						
6°00'	.038	100	100	.050	150	200	.059	150	250	$D_{max} = 4°30'$								
7°00'	.041	100	150	.054	150	200	.060	150	250									
8°00'	.043	100	150	.056	150	200	$D_{max} = 7°00'$											
9°00'	.046	100	150	.058	150	200												
10°00'	.048	100	150	.059	150	200												
11°00'	.050	100	150	.060	150	200												
12°00'	.052	100	150	$D_{max} = 11°00'$														
13°00'	.053	100	150															
14°00'	.055	100	150															
16°00'	.058	100	200															
18°00'	.059	150	200															
20°00'	.060	150	200															
21°00'	.060	150	200															
	$D_{max} = 21°00'$																	

Source: Ministry of Transportation, Ontario.

Legend: V, design speed, mph

e, rate of superelevation, feet per foot of pavement width

L, length of superelevation runoff or spiral curve

NC, normal crown section

RC, remove adverse crown, superelevate at normal crown slope

D, degree of circular curve

*Above the heavy line, spirals are not required, but superelevation is to be run off in distances shown.

Table 11.6 SPIRAL CURVE LENGTHS (FEET) AND SUPERELEVATION RATES: SUPERELEVATION (e) MAXIMUM OF 0.100, TYPICAL FOR SOUTHERN CLIMATES*

| | | V = 30 | | | V = 40 | | | V = 50 | | | V = 60 | | | V = 70 | | |
| | | | L | | | L | | | L | | | L | | | L | |
D	R	e	2 Lane	4 Lane	e	2 Lane	4 Lane	e	2 Lane	4 Lane	e	2 Lane	4 Lane	e	2 Lane	4 Lane
0°15'	22918'	NC	0	0	NC	0	0	NC	0	0	NC	0	0	RC	200	200
0°30'	11459'	NC	0	0	NC	0	0	RC	150	150	RC	175	175	RC	200	200
0°45'	7639'	NC	0	0	RC	125	125	RC	150	150	0.018	175	175	0.020	200	200
1°00'	5730'	NC	0	0	RC	125	125	0.018	150	150	0.022	175	175	0.028	200	200
1°30'	3820'	RC	100	100	0.020	125	125	0.027	150	150	0.034	175	175	0.042	200	200
2°00'	2865'	RC	100	100	0.027	125	125	0.036	150	150	0.046	175	190	0.055	200	250
2°30'	2292'	0.020	100	100	0.033	125	125	0.045	150	160	0.059	175	240	0.069	210	310
3°00'	1910'	0.024	100	100	0.038	125	125	0.054	150	190	0.070	190	280	0.083	250	370
3°30'	1637'	0.027	100	100	0.045	125	140	0.063	150	230	0.081	220	330	0.096	290	430
4°00'	1432'	0.030	100	100	0.050	125	160	0.070	170	250	0.090	240	360	0.100	300	450
5°00'	1146'	0.038	100	100	0.060	130	190	0.083	200	300	0.099	270	400	$D_{max} = 3°9'$		
6°00'	955'	0.044	100	120	0.068	140	210	0.093	220	330	0.100	270	400			
7°00'	819'	0.050	100	140	0.076	160	240	0.097	230	350	$D_{max} = 5°5'$					
8°00'	716'	0.055	100	150	0.084	180	260	0.100	240	360						
9°00'	637'	0.061	110	160	0.089	190	280	0.100	240	360						
10°00'	573'	0.065	120	180	0.093	200	290	$D_{max} = 8.3°$								
11°00'	521'	0.070	130	190	0.096	200	300									
12°00'	477'	0.074	130	200	0.098	210	310									
13°00'	441'	0.078	140	210	0.099	210	310									
14°00'	409'	0.082	150	220	0.100	210	320									
16°00'	358'	0.087	160	240	$D_{max} = 13.4°$											
18°00'	318'	0.093	170	250												
20°00'	286'	0.096	170	260												
22°00'	260'	0.099	180	270												
24.8°	231'	0.100	180	270												
$D_{max} = 24.8°$																

*NC = normal crown section. RC = remove adverse crown, superelevate at normal crown slope. Spirals desirable but not as essential above the heavy line. Lengths rounded in multiples of 25 or 50 ft permit simpler calculations. The higher maximum e value (0.100) in this table permits a sharper maximum curvature.

Table 11.7 SPIRAL CURVE LENGTHS AND SUPERELEVATION RATES (METRIC)*

Design Speed km/h	40			50			60			70			80		
			A			A			A			A			A
Radius m	e	2 Lane	3 & 4 Lane	e	2 Lane	3 & 4 Lane	e	2 Lane	3 & 4 Lane	e	2 Lane	3 & 4 Lane	e	2 Lane	3 & 4 Lane
7000	NC			NC			NC			NC			NC		
5000	NC			NC			NC			NC			NC		
4000	NC			NC			NC			NC			NC		
3000	NC			NC			NC			NC			NC		
2000	NC			NC			NC			RC	275	275	RC	300	300
1500	NC			NC			RC	225	225	RC	250	250	0.024	250	250
1200	NC			NC			RC	200	200	0.023	225	225	0.028	225	225
1000	NC			RC	170	170	0.021	175	175	0.027	200	200	0.032	200	200
900	NC			RC	150	150	0.023	175	175	0.029	180	180	0.034	200	200
800	NC			RC	150	150	0.025	160	160	0.031	175	175	0.036	175	175
700	NC			0.021	140	140	0.027	150	150	0.034	175	175	0.039	175	175
600	NC	120	120	0.024	125	125	0.030	140	140	0.037	150	150	0.042	175	175
500	RC	100	100	0.027	120	120	0.034	125	125	0.041	140	150	0.046	150	160
400	0.023	90	90	0.031	100	100	0.038	115	120	0.045	125	135	0.051	135	150
350	0.025	90	90	0.034	100	100	0.041	110	115	0.048	120	125	0.054	125	140
300	0.028	80	80	0.037	90	100	0.044	100	110	0.051	120	125	0.057	125	135
250	0.031	75	80	0.040	85	90	0.048	90	100	0.055	110	120	0.060	125	125
220	0.034	70	80	0.043	80	90	0.050	90	100	0.057	110	110	0.060	125	125
200	0.036	70	75	0.045	75	90	0.052	85	100	0.059	110	110	Minimum R = 250		
180	0.038	60	75	0.047	70	90	0.054	85	95	0.060	110	110			
160	0.040	60	75	0.049	70	85	0.056	85	90	Minimum R = 190					
140	0.043	60	70	0.052	65	80	0.059	85	90						
120	0.048	60	65	0.055	65	75	0.060	85	90						
100	0.049	50	65	0.058	65	70	Minimum R = 130								
90	0.051	50	60	0.060	65	70									
80	0.054	50	60	0.060	65	70									
70	0.058	50	60	Minimum R = 90											
60	0.059	50	60												
	0.059	50	60												
Minimum R = 55															

Source: Roads and Transportation Association of Canada (RTAC).

$*D_{max} = 0.06$

e is superelevation.

A is spiral parameter in meters.

NC is normal cross section.

RC is remove adverse crown and superelevate at normal rate.

Spiral length, $L = A^2 \div$ radius.

Spiral parameters are minimum and higher values should be used where possible.

Spirals are desirable but not essential above the heavy line.

For 6-lane pavement: above the dashed line use 4-lane values; below the dashed line use 4-lane values $\times$ 1.15.

A divided road having a median less than 7 m may be treated as a single pavement.

Table 11.7 (*continued*)

| 90 | | | 100 | | | 110 | | | 120 | | | 130 | | | 140 | | |
| | A | | | A | | | A | | | A | | | A | | | A | |
e	2 Lane	3 & 4 Lane	e	2 Lane	3 & 4 Lane	e	2 Lane	3 & 4 Lane	e	2 Lane	3 & 4 Lane	e	2 Lane	3 & 4 Lane	e	2 Lane	3 & 4 Lane
NC			NC			NC			NC			NC	700	700	RC	700	700
NC			NC			NC	500	500	RC	600	600	0.021	600	600	0.024	625	625
NC			RC	480	480	RC	500	500	0.022	500	500	0.025	500	500	0.028	560	560
RC	390	400	0.025	400	400	0.023	450	450	0.027	450	450	0.030	450	450	0.034	495	495
0.023	300	350	0.027	340	340	0.031	350	350	0.035	350	350	0.039	400	400	0.043	400	400
0.029	270	275	0.033	300	300	0.037	300	300	0.041	300	300	0.046	330	330	0.050	345	340
0.033	240	240	0.038	250	250	0.042	275	275	0.047	285	285	0.051	300	300	0.056	330	330
0.037	225	225	0.042	240	240	0.046	250	260	0.051	250	275	0.055	300	300	0.060	325	325
0.039	200	200	0.044	225	225	0.049	230	250	0.053	250	270	0.057	300	300	0.060	325	325
0.042	200	200	0.047	200	225	0.051	225	250	0.056	250	260	0.060	275	275	Minimum R = 1000		
0.045	185	185	0.049	200	220	0.054	225	235	0.059	250	250	0.060	275	275			
0.048	175	185	0.053	200	200	0.057	220	220	0.060	250	250	Minimum R = 800					
0.052	160	175	0.057	200	200	0.060	220	220	Minimum R = 650								
0.057	160	165	0.060	200	200	Minimum R = 525											
0.059	160	160	Minimum R = 420														
0.060	160	160															
Minimum R = 340																	

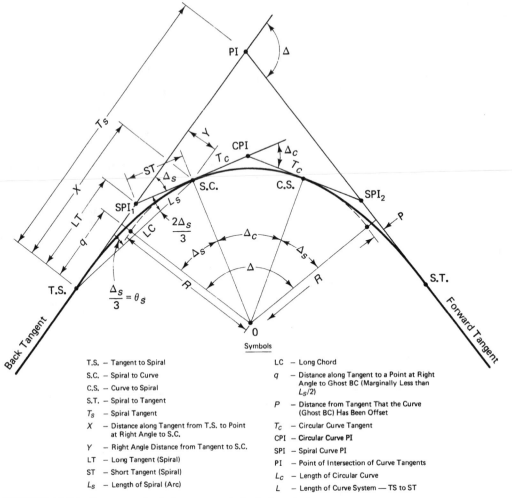

FIGURE 11.24 Summary of spiral geometry and spiral symbols.

Symbols

T.S.	– Tangent to Spiral	LC	– Long Chord
S.C.	– Spiral to Curve	q	– Distance along Tangent to a Point at Right Angle to Ghost BC (Marginally Less than $L_S/2$)
C.S.	– Curve to Spiral		
S.T.	– Spiral to Tangent	P	– Distance from Tangent That the Curve (Ghost BC) Has Been Offset
T_S	– Spiral Tangent		
X	– Distance along Tangent from T.S. to Point at Right Angle to S.C.	T_C	– Circular Curve Tangent
		CPI	– **Circular Curve PI**
Y	– Right Angle Distance from Tangent to S.C.	SPI	– Spiral Curve PI
LT	– Long Tangent (Spiral)	PI	– Point of Intersection of Curve Tangents
ST	– Short Tangent (Spiral)	L_C	– Length of Circular Curve
L_S	– Length of Spiral (Arc)	L	– Length of Curve System — TS to ST

θ_s is the total spiral deflection angle; compare to circular curves, where the deflection angle is $\Delta/2$. The approximate formula in Equation 11.26 gives realistic results for the vast majority of spiral problems. For example, when Δ_s is as large as 21° (which is seldom the case), the correction to Δ_s is approximately +30″.

$$L_c = \frac{2\pi R \Delta_c}{360} \text{ (foot or meter units)} \tag{11.27}$$

$$L_c = \frac{100\Delta_c}{D} \text{ (foot units)} \tag{11.28}$$

$$\Delta_s = \frac{90}{\pi} \times \frac{L_s}{R} \tag{11.29}$$

$$\phi = \left(\frac{l}{L_s}\right)^2 \theta_s \tag{11.30}$$

Equation 11.30 comes from the spiral definition, where ϕ is the deflection angle for any distance l, and ϕ and θ_s are in the same units. Practically:

$$\phi' = l^2 \frac{(\theta_s \times 60)}{L_s^2}$$

where ϕ' is the deflection angle in minutes for any distance measured from the T.S. or the S.T.

Other values, such as x, y, P, q, ST, and LT, are found routinely in spiral tables issued by state and provincial highway agencies and can also be found in route surveying texts. For the past few years, solutions to these problems have been achieved almost exclusively by the use of computers and appropriate coordinate geometry computer software.

With the switch to metric that took place in the 1970s in Canada, a decision was reached to work exclusively with the radius in defining horizontal curves. It was decided that new spiral tables, based solely on the definition of R, were appropriate. The Roads and Transportation Association of Canada (RTAC) prepared spiral tables based on the spiral property defined as follows:

> The product of any instantaneous radius r and the corresponding spiral length λ (i.e., l) from the beginning of the spiral to that point is equal to the product of the spiral end radius R and the entire length (L_s) of that spiral, which means it is a constant.

Thus, we have the following equation:

$$r\lambda = RL_s = A^2 \tag{11.31}$$

The constant is denoted as A^2 to retain dimensional consistency because it represents a product of two lengths. The following relationship can be derived from Equation 11.31:

$$\frac{A}{R} = \frac{L_s}{A} \tag{11.32}$$

The RTAC tables are based on design speed and the number of traffic lanes, together with a design radius. The constant value A is taken from one of the tables and is used in conjunction with R to find all the spiral table curve parameters noted earlier in this section.

Note that, concurrent with the changeover to metric units, a major study in Ontario of highway geometrics resulted in spiral curve lengths that reflected design speed, attainment of superelevation, driver comfort, and aesthetics. Thus, the spiral lengths vary somewhat between the foot unit system and the metric system.

11.19 Spiral Layout Procedure Summary

Refer to Figure 11.24. The following list summarizes the spiral layout procedure using traditional techniques:

1. Select L_s (foot units) or A (metric units) in conjunction with the design speed, number of traffic lanes, and sharpness of the circular curve (radius or D).
2. From the spiral tables, determine P, q, x, y, and so on.

3. Compute the spiral tangent (T_s) using Equation 11.22 and the circular tangent (T_c) using Equation 11.1.

4. Compute the spiral angle (Δ_s). Use Equation 11.33 for foot units or Equation 11.29 for metric units (see Tables 11.3 and 11.7).

5. Prepare a list of relevant layout stations. This list should include all horizontal alignment key points (e.g., T.S., S.C., C.S., and S.T.), as well as all vertical alignment key points, such as grade points, BVC, low point, and EVC.

6. Calculate the deflection angles. See Equation 11.30.

7. From the established PI, measure out the T_s distance to locate the T.S. and S.T.

8. (a) From the T.S., turn off the spiral deflection ($\theta_s = \frac{1}{3}\Delta_s$ approximately), measure out the long chord (LC), and thus locate the S.C.

   or

   (b) From the T.S., measure out the LT distance along the main tangent and locate the spiral PI (SPI); the spiral angle (Δ_s) can now be turned, and the ST distance can be measured out to locate the S.C.

9. From the S.C., measure out the circular tangent (T_c) along the line SPI$_1$–SC to establish the CPI.

10. Steps 8 and 9 are repeated, starting at the S.T. instead of at the T.S.

11. The key points are verified by checking all angles and redundant distances. Some surveyors prefer to locate the CPI by intersecting the two tangent lines (i.e., lines through SPI$_1$ and S.C. and through SPI$_2$ and C.S.). The locations can be verified by checking the angle Δ_c and by checking the two tangents (T_c).

12. Only after all key control points have been verified can the deflection angle stakeout commence. The lower chainage spiral is run in from the T.S., whereas the higher chainage spiral is run in from the S.T. The circular curve can be run in from the S.C., although it is common practice to run half the curve from the S.C. and the other half from the C.S. so that any acceptable errors that accumulate can be isolated in the middle of the circular arc where they will be less troublesome, relatively speaking.

When employing layout procedures using polar layouts by total station, or positioning layouts using GPS techniques, the surveyor must first compute the position coordinates of all layout points and then upload these coordinates into the instrument's controller.

■ **EXAMPLE 11.7** *Illustrative Spiral Problem (Foot Units)*

You are given the following data:

$$\Delta = 25°45'RT$$
$$V = 40 \text{ mph}$$
$$D = 9°$$
$$\text{PI at } 36 + 17.42$$
$$\text{Two-lane highway, 24 ft wide}$$

Compute the stations for all key curve points (in foot units) and for all relevant spiral and circular curve deflections.

Solution

1. From Table 11.5:

$$L_s = 150 \text{ ft}$$
$$e = 0.058 \text{ (superelevation rate)}$$

2. From Table 11.4:

$$
\begin{aligned}
\Delta_s &= 6.75000° \\
R &= 636.6198 \text{ ft} \\
P &= 1.4719 \text{ ft} \\
R + P &= 638.0917 \text{ ft} \\
q &= 74.9653 \text{ ft} \\
LT &= 100.0728 \text{ ft} \\
ST &= 50.0662 \text{ ft} \\
X &= 149.79 \text{ ft} \\
Y &= 5.88 \text{ ft}
\end{aligned}
$$

You can also use Equation 11.23 to determine:

$$
\begin{aligned}
\Delta_s &= \frac{L_s D}{200} \\
&= \frac{150}{200} \times 9 = 6.75000°
\end{aligned}
$$

3. From Equation 11.22, we have:

$$
\begin{aligned}
T_s &= (R + p) \tan \frac{\Delta}{2} + q \\
&= 638.0917 \tan 12°52.5' + 74.9653 = 220.82 \text{ ft}
\end{aligned}
$$

4.
$$
\begin{aligned}
\Delta_c &= \Delta - 2\Delta_s \\
&= 25°45' - 2\,(6°45') = 12°15'
\end{aligned}
$$

5. From Equation 11.7, we have:

$$
\begin{aligned}
L_c &= 100\Delta_c \\
&= \frac{(100 \times 12.25)}{9} = 136.11 \text{ ft}
\end{aligned}
$$

6. Key station computation:

$$
\begin{array}{lll}
\text{PI} & \text{at} & 36 + 17.42 \\
T_s & 2 & 20.82 \\
\hline
\text{T.S.} & = 33 & + 96.60 \\
+L_s & 1 & 50.00 \\
\hline
\text{S.C.} & = 35 & + 46.60 \\
+L_c & 1 & 36.11 \\
\hline
\end{array}
$$

$$C.S. = 36 + 82.71$$
$$+L_s \quad 1 \quad 50.00$$
$$S.T. = 38 + 32.71$$

7.
$$\theta_s = \frac{\Delta_s}{3}$$
$$= \frac{6.753000°}{3} = 2.25° = 2°15'00''$$

8. Circular curve deflections. See Table 11.8:

$$\Delta_c = 12°15'$$
$$\frac{\Delta_c}{2} = 6°07.5' = 367.5'$$

From Section 11.4, the deflection angle for one unit of distance is $(\Delta/2)/L$. Here, the deflection angle for one foot of arc (minutes) is:

$$\frac{\Delta_c/2}{L_c} = \frac{367.5}{136.11} = 2.700'$$

Table 11.8 CURVE SYSTEM DEFLECTION ANGLES

Station	Distance* from T.S. (or S.T.) l (ft)	l^2	$\dfrac{\theta_s° \times 60}{L_s^2}$	$\dfrac{l^2(\theta_s \times 60)}{L_s^2}$ Deflection Angle (Minutes)	Deflection
T.S. 33 + 96.60	0	0	0.006	0	0°00'00''
34 + 00	3.4	11.6	0.006	0.070	0°00'04''
34 + 50	53.4	2851.4	0.006	17.108	0°17'06''
35 + 00	103.4	10,691.6	0.006	64.149	1°04'09''
S.C. 35 + 45.60	150	22,500	0.006	135	$\theta_s = 2°15'00''$

	Circular Curve Data			Deflection Angle (Cumulative)	
S.C. 35 + 46.60	$\Delta_c = 12°15', \dfrac{\Delta_c}{2} = 6°07'30''$			0°00.00'	0°00'00''
35 + 50	Deflection for 3.40' = 9.18'			0°09.18'	0°09'11''
36 + 00	Deflection for 50' = 135'			2°24.18'	2°24'11''
36 + 50	Deflection for 32.71' = 88.32'			4°39.18'	4°39'11''
C.S. 36 + 82.71				6°07.50'	6°07'30''

C.S. 36 + 82.71	150	22,500	0.006	135	$\theta_s = 2°15'00''$
37 + 00	132.71	17,611.9	0.006	105.672	1°45'40''
37 + 50	82.71	6840.9	0.006	41.046	0°41'03''
38 + 00	32.71	1069.9	0.006	6.420	0°06'25''
S.T. 38 + 32.71	0	0	0.006	0	0°00'00''

*Note that l is measured from the S.T.

Alternatively, because $D = 9°$, then the deflection for 100 ft is $D/2$ or $4°30'$, which is $270'$. The deflection angle for 1 ft = $270/100 = 2.700'$ (as previously noted). The required distances (from Table 11.8) are:

$(35 + 50) - (35 + 46.60) = 3.4'$; deflection angle $= 3.4 \times 2.7 = 9.18'$

Even interval $= 50'$; deflection angle $= 50 \times 2.7 = 135'$

$(36 + 82.71) - (36 + 50) = 32.71$; deflection angle $= 32.71 \times 2.7 = 88.32'$

These values are now entered cumulatively in Table 11.8.

■ **EXAMPLE 11.8** *Illustrative Spiral Problem (Metric Units)*
You are given the following data:

$$\text{PI at } 1 + 086.271$$
$$V = 80 \text{ kmh}$$
$$R = 300 \text{ m}$$
$$\Delta = 16°00' RT$$
$$\text{Two-lane road (7.5 m wide)}$$

From Table 11.7, $A = 125$ and $e = 0.057$. (See Section 11.21 for a discussion of superelevation.) From Table 11.9, we obtain the following:

Solution
Compute the stations for all key curve points (in metric units) and all relevant spiral and circular curve deflections.

Steps	For $A = 125$	and $R = 300$
1 and 2	$L_s = 52.083$ m	$LT = 34.736$ m
	$P = 0.377$ m	$ST = 17.374$ m
	$X = 52.044$ m	$\Delta_s = 4°58'24.9''$
	$Y = 1.506$ m	$\theta_s = \frac{1}{3}\Delta_s = 1°39'27.9''$
	$q = 26.035$ m	$LC = 52.066$ m long chord

You can also use Equation 11.29:

$$\Delta_s = \frac{90}{\pi} \times \frac{L_s}{R}$$
$$= \frac{90}{\pi} \times \frac{52.083}{300} = 4.9735601° = 4°58'24.8''$$

Step 3. From Equation 11.22, we have:

$$T_s = (R + p) \tan\frac{\Delta}{2} + q$$
$$= 300.377 \tan 8° + 26.035 = 68.250 \text{ m}$$

Table 11.9 FUNCTIONS OF THE STANDARD SPIRAL* FOR A = 125 m

R (m)	A/R	L_s	X	Y	q	P	LT	ST	L_c	Δ_s Deg.	Min.	Sec.	θ_s Deg.	Min.	Sec.
				Meters											
115	1.0870	135.870	131.204	26.095	67.152	6.606	92.293	46.851	133.774	33	50	48.3	11	14	55.1
120	1.0417	130.208	126.428	23.057	64.471	5.825	88.183	44.658	128.513	31	05	05.8	10	20	08.3
125	1.0000	125.000	121.911	20.464	61.983	5.162	84.451	42.685	123.617	28	38	52.4	9	31	44.3
130	0.9615	120.192	117.649	18.240	59.671	4.595	81.044	40.898	119.055	26	29	11.7	8	48	46.1
140	0.8929	111.607	109.847	14.661	55.509	3.686	75.034	37.755	110.821	22	50	16.5	7	36	08.5
150	0.8333	104.167	102.918	11.953	51.875	3.001	69.888	35.126	103.610	19	53	39.7	6	37	28.8
160	0.7813	97.656	96.751	9.868	48.677	2.475	65.425	32.844	97.253	17	29	07.0	5	49	25.8
170	0.7353	91.912	91.242	8.239	45.844	2.065	61.511	30.852	91.614	15	29	19.3	5	09	34.9
180	0.6944	86.086	86.302	6.948	43.319	1.741	58.048	29.096	86.581	13	48	55.9	4	36	10.5
190	0.6579	82.237	81.853	5.913	41.054	1.481	54.960	27.535	82.066	12	23	58.3	4	07	53.5
200	0.6250	78.125	77.828	5.072	39.013	1.270	52.188	26.187	77.993	11	11	26.1	3	43	44.4
210	0.5952	74.405	74.172	4.384	37.163	1.097	49.685	24.876	74.301	10	09	00.7	3	22	57.0
220	0.5682	71.023	70.838	3.814	35.481	0.954	47.413	23.733	70.941	9	14	54.3	3	04	55.6
230	0.5435	67.935	67.787	3.339	33.943	0.835	45.342	22.692	67.869	8	27	42.1	2	49	12.1
240	0.5208	65.104	64.984	2.940	32.532	0.735	43.455	21.739	65.051	7	46	16.5	2	35	24.0
250	0.5000	62.500	62.402	2.601	31.234	0.651	41.701	20.864	62.457	7	09	43.1	2	23	13.2
280	0.4464	55.804	55.748	1.852	27.983	0.463	37.222	18.619	55.779	5	24	34.1	1	54	10.8
300	0.4167	52.083	52.044	1.506	26.035	0.377	34.736	17.374	52.066	4	58	24.9	1	39	27.9
320	0.3906	48.828	48.800	1.241	24.409	0.310	32.562	16.285	48.815	4	22	16.8	1	27	25.3
340	0.3676	45.956	45.935	1.035	22.974	0.259	30.645	15.325	45.947	3	52	19.8	1	17	26.4
350	0.3571	44.643	44.625	0.949	22.318	0.237	29.768	14.887	44.635	3	39	14.6	1	13	04.7
380	0.3289	41.118	41.106	0.741	20.557	0.185	27.416	13.710	41.113	3	05	59.6	1	01	59.8
400	0.3125	39.063	39.063	0.636	19.530	0.159	26.045	13.024	39.058	2	47	51.5	0	55	57.1
420	0.2976	37.202	37.195	0.549	18.600	0.137	24.804	12.403	37.195	2	32	15.2	0	50	45.0
450	0.2778	34.722	34.717	0.446	17.360	0.112	23.150	11.576	34.720	2	12	37.7	0	44	12.5
475	0.2632	32.895	32.891	0.380	16.447	0.095	21.931	10.966	32.893	1	59	02.1	0	39	40.7
500	0.2500	31.250	31.247	0.325	15.624	0.081	20.834	10.418	31.249	1	47	25.8	0	35	48.6
525	0.2381	29.762	29.760	0.281	14.881	0.070	19.842	9.921	29.761	1	37	26.5	0	32	20.8
550	0.2273	28.409	28.407	0.245	14.204	0.061	18.940	9.470	28.408	1	28	47.1	0	29	35.7
575	0.2174	27.174	27.172	0.214	13.587	0.054	18.116	9.058	27.173	1	21	13.9	0	27	04.6

Source: Metric Curve Tables, Table IV, Roads and Transportation Association of Canada (RTAC).
*Short radius (i.e., < 150 m) may require a correction to $\Delta_s/3$ to determine the precise value of θ_s.

Step 4.
$$\Delta_c = \Delta - 2\Delta_s$$
$$= 16° - 2(4° \ 58'24.8'') \ = 6°03'10.2''$$
$$\frac{\Delta_c}{2} = 3°01'35.1''$$

Step 5. From Equation 11.5, we have:

$$L_c = \frac{2\pi R\Delta_c}{360}$$
$$= \frac{(2\pi \times 300 \times 6.052833)}{360} = 31.693 \text{ m}$$

Step 6. Key station computation:

$$
\begin{array}{lr}
\text{PI} \quad \text{at } 1 + 086.271 \\
T_s \qquad \underline{68.250} \\
\text{T.S.} \ = 1 + 018.021 \\
+L_s \qquad \underline{52.083} \\
\text{S.C.} \ = 1 + 070.104 \\
+L_c \qquad 31.693 \\
\text{C.S.} \ = 1 + 101.797 \\
+L_s \qquad 52.083 \\
\text{S.T.} \ = 1 + 153.880
\end{array}
$$

Step 7. Circular curve deflections. See Table 11.10:

$$\Delta_c = 6°03'10''$$
$$\frac{\Delta_c}{2} = 3°01'35'' = 181.58'$$

From Section 11.4, the deflection angle for one unit of distance is $(\Delta/2)/L$. Here, the deflection angle for 1 m of arc is:

$$\frac{\Delta_c/2}{L_c} = \frac{181.58}{31.693} = 5.7293'$$

The required distances (deduced from Table 11.10) are:

$(1 + 080) - (1 + 070.104) = 9.896$; deflection angle $= 5.7293 \times 9.896 = 56.70'$
Even interval $= 20.000$; deflection angle $= 5.7293 \times 20 = 114.59'$
$(1 + 101.797) - (1 + 100) = 1.797$; deflection angle $= 5.7293 \times 1.787 = 10.30'$

These values are now entered cumulatively in Table 11.10.

Table 11.10 CURVE SYSTEM DEFLECTION ANGLES

Station	Distance* from T.S. (or S.T.) l (in meters)	l^2	$\dfrac{\theta_s \times 60}{L_s^2}$	$\dfrac{l^2 (\theta_s \times 60)}{L_s^2}$ Deflection Angle (Minutes)	Deflection
T.S. 1 + 018.021	0	0			0°00′00″
1 + 020	1.979	3.9	0.036667	0.1436	0°00′09″
1 + 040	21.979	483.1	0.036667	17.71	0°17′43″
1 + 060	41.979	1762.2	0.036667	64.62	1°04′37″
S.C. 1 + 070.104	52.083	2712.6	0.036667	99.464	1°39′28″

	Circular Curve Data	Deflection Angle (Cumulative)	Deflection
S.C. 1 + 070.104	$\Delta_c = 6°03′10″,\ \dfrac{\Delta_c}{2} = 3°01′35″$	0°00.00′	0°00′00″
1 + 080	Deflection for 9.896 m = 56.70′	0°56.70′	0°56′42″
1 + 100	Deflection for 20 m = 114.59′	2°51.29′	2°51′17″
C.S. 1 + 101.797	Deflection for 1.797 m = 10.30′	3°01.59′	3°01′35″

Station	l (in meters)	l^2	$\dfrac{\theta_s \times 60}{L_s^2}$	Deflection Angle (Minutes)	Deflection
C.S. 1 + 101.797	52.083	2712.6	0.036667	99.464	1°39′28″
1 + 120	33.880	1147.9	0.036667	42.09	0°42′05″
1 + 140	13.880	192.7	0.036667	7.06	0°07′04″
S.T. 1 + 153.880	0	0	0.036667	0	0°00′00″

*Note that l is measured from the S.T.

11.20 Approximate Solution for Spiral Problems

It is possible to lay out spirals by using the approximate relationships illustrated in Figure 11.25. Because $L_s \approx$ LC (long chord), the following can be assumed:

$$\frac{Y}{L_s} = \sin \theta_s \qquad Y = L_s \sin \theta_s \tag{11.33}$$

$$X^2 = L_s^2 - Y^2$$
$$X = \sqrt{L_s^2 - Y^2} \tag{11.34}$$

$$q = \frac{1}{2} X \tag{11.35}$$

$$p = \frac{1}{4} Y \tag{11.36}$$

Using the sine law, we obtain the following:

$$LT = \sin \frac{2}{3}\Delta_s \times \frac{\sin L_s}{\sin \Delta_s} \tag{11.37}$$

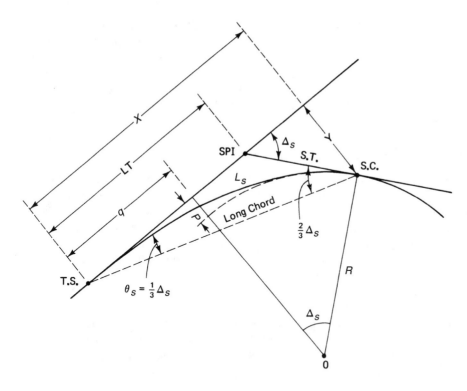

Basic Assumption: $L_s \approx$ Long Chord

FIGURE 11.25 Sketch for approximate formulas.

Using the sine law yields:

$$ST = \sin \frac{1}{3}\Delta_s \times \frac{L_s}{\sin \Delta_s} \tag{11.38}$$

For comparison, the values from Examples 11.7 and 11.8 are compared to the values obtained by the approximate methods:

	Precise Methods		Approximate Methods	
Parameter	Example 11.7 (ft)	Example 11.8 (m)	Example 11.7 (ft)	Example 11.8 (m)
Y	5.88	1.506	5.89	1.507
X	149.79	52.044	149.88	52.061
q	74.97	26.035	74.94	26.031
P	1.47	0.377	1.47	0.377
LT	100.07	34.736	100.13	34.744
ST	50.07	17.374	50.10	17.379

You can see from the data summary that the precise and approximate values for Y, X, q, P, LT, and ST are quite similar. The largest discrepancy shows up in the X value, which is not required for spiral layout. The larger the Δ_s values, the larger is the discrepancy between the precise and approximate values. For the normal range of spirals in use, the approximate method is adequate for the layout of an asphalt-surfaced ditched highway. For curbed highways or elevated highways, precise methods should be employed.

11.21 Superelevation: General Background

If a vehicle travels too fast on a horizontal curve, the vehicle may either skid off the road or overturn. The factors that cause this phenomenon are based on the radius of curvature and the velocity of the vehicle: the sharper the curve and the higher the velocity, the larger will be the centrifugal force requirement. Two factors can be called on to help stabilize the radius and velocity factors: (1) side friction, which is always present to some degree between the vehicle tires and the pavement, and (2) superelevation (e), which is a banking of the pavement toward the center of the curve.

The side friction factor (f) has been found to vary linearly with velocity. Design values for f range from 0.16 at 30 mph (50 km/h) to 0.11 at 80 mph (130 km/h). Superelevation must satisfy normal driving practices and climatic conditions. In practice, values for superelevation range from 0.125 (i.e., 0.125 ft/ft or 12.5 percent cross slope) in warmer southern U.S. states to 0.06 in the northern U.S. states and Canadian provinces. Typical values for superelevation can be found in Tables 11.5 and 11.6.

11.22 Superelevation Design

Figure 11.26 illustrates how the length of spiral (L_s) is used to change the pavement cross slope from normal crown to full superelevation. Figure 11.26(b) illustrates that the pavement can be revolved about the centerline, which is the usual case, or the pavement can be revolved about the inside or outside edges, a technique that is often encountered on divided four-lane highways where a narrow median restricts drainage profile manipulation.

Figures 11.28(b) and 11.29 clearly show the technique used to achieve pavement superelevation when revolving the pavement edges about the centerline (℄) profile. At points A and A', the pavement is at normal crown cross section—with both edges (inside and outside) of the pavement a set distance below the ℄ elevation. At points S.C. (D) and C.S. (D') the pavement is at full superelevation—with the outside edge a set distance above the ℄ elevation and the inside edge the same set distance below the ℄ elevation. The transition from normal crown cross section to full superelevation cross section proceeds as follows:

OUTSIDE EDGE

- From A to the T.S. (B), the outside edge rises at a ratio of 400:1—relative to the ℄ profile—and becomes equal in elevation to the ℄.
- From the T.S. (B), the outside edge rises (relative to the ℄ profile) at a uniform rate, from being equal to the ℄ elevation at the T.S. (B) until it is at full superelevation above the ℄ elevation at the S.C. (D).

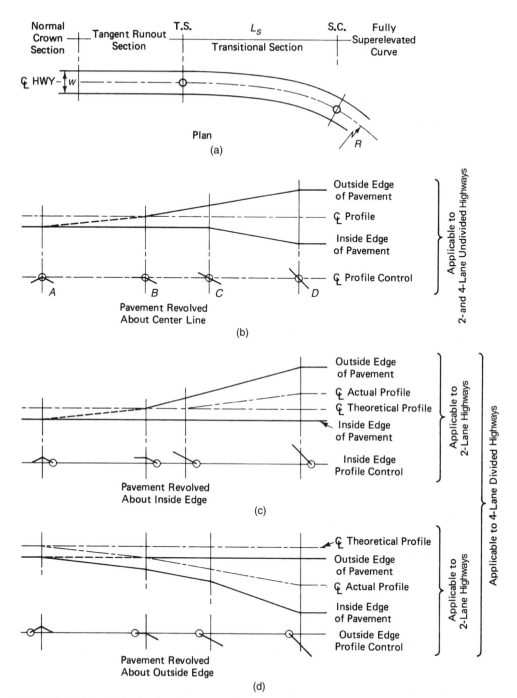

FIGURE 11.26 Methods of attaining superelevation for spiraled curves. (From *Geometric Design Standards for Ontario Highways*, Ministry of Transportation, Ontario)

- From A, through the T.S. (B), to point C, the inside edge remains below the ₵ profile at normal crown depth.
- From C to the S.C. (D), the inside edge drops at a uniform rate, from being at normal crown depth below the ₵ profile to being at full superelevation depth below the ₵ profile.

The transition from full superelevation at the C.S. (D') to normal cross section at A' proceeds in a manner reverse to that just described for the transition from A to S.C.

■ **EXAMPLE 11.9** *Superelevation Problem (Foot Units)*

See Figure 11.27 for vertical curve computations and Figures 11.28 and 11.29 for pavement superelevation computations. This example uses the horizontal curve data of Example 11.7.

You are given the following data:

$$V = 40 \text{ mph}$$
$$\Delta = 25°45' \text{ RT}$$
$$D = 9°$$
PI at 36 + 17.42
Two-lane highway, 24 ft wide, each lane 12 ft wide
PVI at 36 + 00
Elevation PVI = 450.00
$$g_1 = -1.5\%$$

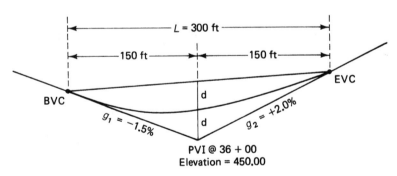

BVC = (36 + 00) − 150 = 34 + 50

EVC = (36 + 00) + 150 = 37 + 50

Elevation BVC = 450.00 + (150 × 0.015) = 452.25

Elevation EVC = 450.00 + (150 × 0.020) = 453.00

Midchord Elevation $= \dfrac{452.25 + 453.00}{2} = 452.63$

Tangent Offset: $d = \dfrac{452.63 - 450.00}{2} = 1.315$ ft

FIGURE 11.27 Vertical curve solution for the problem of Example 11.9.

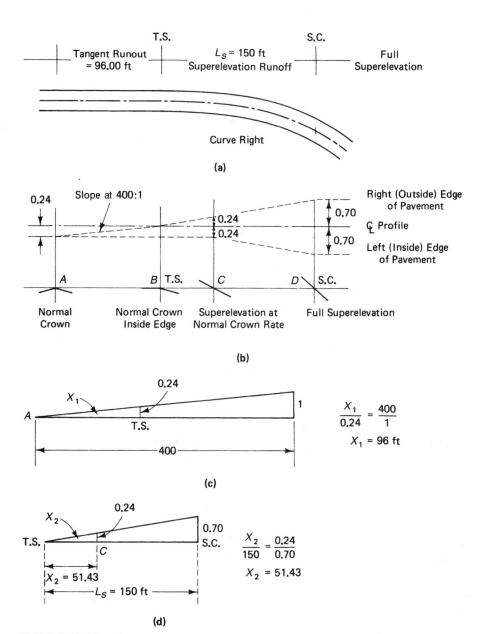

FIGURE 11.28 Sketches for the superelevation example of Section 11.22 and Example 11.6. (a) Plan. (b) Profile and cross sections. (c) Computation of tangent runout. (d) Location of point C.

$$g_2 = +2\%$$
$$L = 300 \text{ ft}$$
Tangent runout at 400:1
Normal crown at 2%
Pavement revolved about the ℄

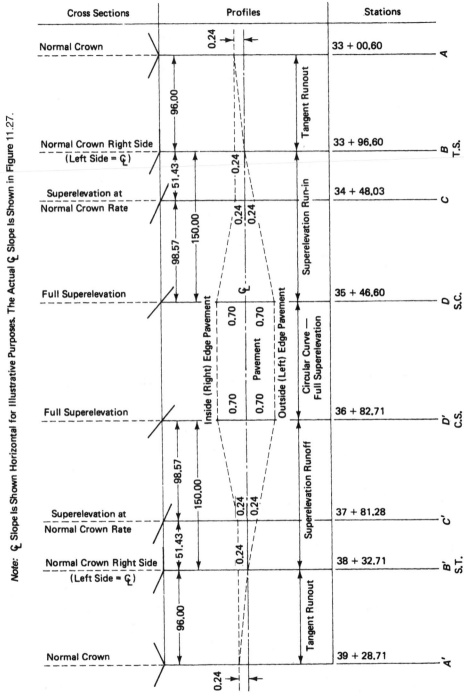

FIGURE 11.29 Superelevated pavement profiles and cross sections for the problem of Example 11.9.

From Table 11.4:

$$L_s = 150 \text{ ft}$$
$$e = 0.058$$

Chainage of low point:

$$X = \frac{-g_1 L}{A} = \frac{1.5 \times 300}{2 - (-1.5)} = 128.57 \text{ ft}$$

BVC	at	34	+	50
X	=	1		28.57
Low point	=	35	+	78.57

You must determine the ℄ and edge-of-pavement elevations for even 50-ft stations and all other key stations.

Solution
1. Compute the key horizontal alignment stations. (These stations have already been computed in Example 11.7).
2. Solve the vertical curve for ℄ elevations at 50-ft stations, the low point, *plus any horizontal alignment key stations that may fall between the BVC and EVC.*
3. Begin preparation of Table 11.11, showing all key stations for both horizontal and vertical alignment. List the ℄ grade elevation for each station.
4. Compute station *A*; see Figure 11.28(c):

$$\text{Cross fall at 2\% for 12 ft} = 0.02 \times 12 = 0.24 \text{ ft}$$
$$\text{Tangent runout} = \frac{400}{1} \times 0.24 = 96 \text{ ft}$$

T.S. =	33	+	96.60
X_1 (tangent runout) =			96.00
A =	33	+	00.60

Compute station *A'* (i.e., tangent runout at higher chainage spiral):

S.T. =	38	+	32.71
Tangent runout		+	96.00
A' =	39	+	28.71

5. Compute station *C*; see Figure 11.28(d):

$$\text{Cross fall at 5.8\% for 12-ft lane} = 0.058 \times 12 = 0.70 \text{ ft}$$
$$\text{Distance from T.S. to } C = 150 \times \frac{0.24}{0.70} = 51.43 \text{ ft}$$

T.S. =	33	+	96.60
X distance =			51.43
C =	34	+	48.03

Table 11.11 ℄ PAVEMENT ELEVATIONS

Station		Tangent Elevation	Tangent Offset $\left(\dfrac{X}{L/2}\right)^2 d$		℄ Elevation
A	33 + 00.60	454.49			454.49
	33 + 50	453.75			453.75
T.S.(B)	33 + 96.60	453.05			453.05
	34 + 00	453.00			453.00
C	34 + 48.03	452.28			452.28
BVC	34 + 50	452.25	$(0/150)^2 \times 1.32 =$	0	452.25
	35 + 00	451.50	$(50/150)^2 \times 1.32 =$	0.15	451.65
S.C.(D)	35 + 46.60	450.79	$(96.6/150)^2 \times 1.32 =$	0.55	451.35
	35 + 50	450.75	$(100/150)^2 \times 1.32 =$	0.58	451.33
Low point	35 + 78.57	450.32	$(128.6/150)^2 \times 1.32 =$	0.97	451.29
PVI	36 + 00	450.00	$(150/150)^2 \times 1.32 =$	1.32	451.32
	36 + 50	451.00	$(100/150)^2 \times 1.32 =$	0.58	451.58
C.S.(D')	36 + 82.71	451.65	$(67.3/150)^2 \times 1.32 =$	0.27	451.92
	37 + 00	452.00	$(50/150)^2 \times 1.32 =$	0.15	452.15
EVC	37 + 50	453.00	$(0/150)^2 \times 1.32 =$	0	453.00
C'	37 + 81.28	453.63			453.63
	38 + 00	454.00			454.00
S.T.(B')	38 + 32.71	454.65			454.65
	39 + 00	456.00			456.00
A'	39 + 28.71	456.57			456.57

Compute station C' (i.e., at higher chainage spiral):

$$\text{S.T.} = 38 + 32.71$$
$$X_2 \text{ distance} = 51.43$$
$$C' = 37 + 81.28$$

6. Figure 11.29 shows that right-side pavement elevations are 0.24 ft below ℄ elevation from A to C and from C' to A', and 0.70 ft below ℄ from S.C. to C.S. Right-side pavement elevations between C and S.C. and C.S. and C' must be interpolated. Figure 11.29 also shows that left-side pavement elevations must be interpolated between A and T.S., between T.S. and S.C., between C.S. and S.T., and between S.T. and A'. Between S.C. and C.S., the left-side pavement elevation is 0.70 higher than the corresponding ℄ elevations.

7. Fill in the left- and right-edge pavement elevations (Table 11.12), where the computation simply involves adding or subtracting normal crown (0.24) or full superelevation (0.70).

8. Perform the computations necessary to interpolate for the missing pavement-edge elevations in Table 11.12 (values are underlined). See Figures 11.30 and 11.31.

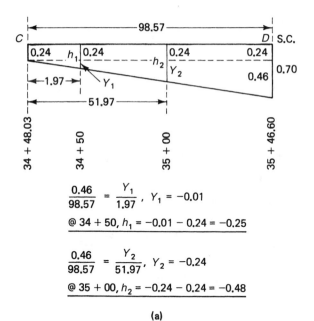

$$\frac{0.46}{98.57} = \frac{Y_1}{1.97}, \quad Y_1 = -0.01$$

@ 34 + 50, $h_1 = -0.01 - 0.24 = -0.25$

$$\frac{0.46}{98.57} = \frac{Y_2}{51.97}, \quad Y_2 = -0.24$$

@ 35 + 00, $h_2 = -0.24 - 0.24 = -0.48$

(a)

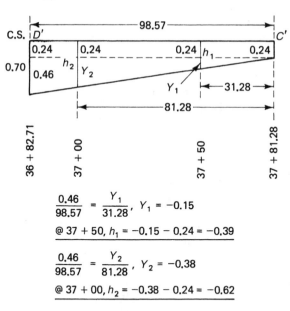

$$\frac{0.46}{98.57} = \frac{Y_1}{31.28}, \quad Y_1 = -0.15$$

@ 37 + 50, $h_1 = -0.15 - 0.24 = -0.39$

$$\frac{0.46}{98.57} = \frac{Y_2}{81.28}, \quad Y_2 = -0.38$$

@ 37 + 00, $h_2 = -0.38 - 0.24 = -0.62$

(b)

FIGURE 11.30 Right-edge pavement elevation interpolation for the problem of Example 11.9. (a) Tangent run-in. (b) Tangent runout.

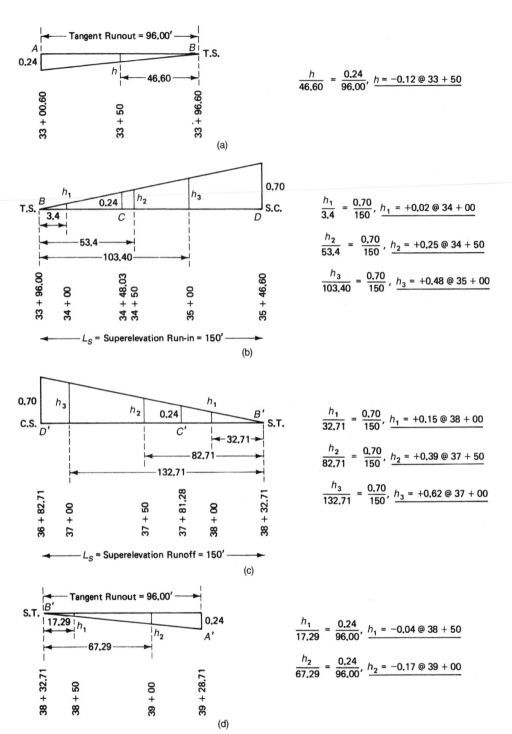

$$\frac{h}{46.60} = \frac{0.24}{96.00}, \quad \underline{h = -0.12 \ @ \ 33 + 50}$$

$$\frac{h_1}{3.4} = \frac{0.70}{150}, \quad \underline{h_1 = +0.02 \ @ \ 34 + 00}$$

$$\frac{h_2}{53.4} = \frac{0.70}{150}, \quad \underline{h_2 = +0.25 \ @ \ 34 + 50}$$

$$\frac{h_3}{103.40} = \frac{0.70}{150}, \quad \underline{h_3 = +0.48 \ @ \ 35 + 00}$$

$$\frac{h_1}{32.71} = \frac{0.70}{150}, \quad \underline{h_1 = +0.15 \ @ \ 38 + 00}$$

$$\frac{h_2}{82.71} = \frac{0.70}{150}, \quad \underline{h_2 = +0.39 \ @ \ 37 + 50}$$

$$\frac{h_3}{132.71} = \frac{0.70}{150}, \quad \underline{h_3 = +0.62 \ @ \ 37 + 00}$$

$$\frac{h_1}{17.29} = \frac{0.24}{96.00}, \quad \underline{h_1 = -0.04 \ @ \ 38 + 50}$$

$$\frac{h_2}{67.29} = \frac{0.24}{96.00}, \quad \underline{h_2 = -0.17 \ @ \ 39 + 00}$$

FIGURE 11.31 Left-edge pavement elevation interpolation for the problem of Example 11.9.

Table 11.12 PAVEMENT ELEVATIONS FOR EXAMPLE 11.9*

	Station	₵ Grade	Left-Edge Pavement† Above/Below ₵	Left-Edge Pavement† Elevation	Right-Edge Pavement† Below ₵	Right-Edge Pavement† Elevation
A	33 + 00.60	454.49	−0.24	454.25	−0.24	454.25
	33 + 50	453.75	−0.12	453.63	−0.24	453.51
T.S.(B)	33 + 96.60	453.05	0.00	453.05	−0.24	452.81
	34 + 00	453.00	+0.02	453.02	−0.24	452.76
C	34 + 48.03	452.28	+0.24	452.52	−0.24	452.04
BVC	34 + 50	452.25	+0.25	452.50	−0.25	452.00
	35 + 00	451.65	+0.48	452.13	−0.48	451.17
S.C.(D)	35 + 46.60	451.35	+0.70	452.05	−0.70	450.65
	35 + 50	451.33	+0.70	452.02	−0.70	450.62
Low pt.	35 + 78.57	451.29	+0.70	451.99	−0.70	450.59
PVI	36 + 00	451.32	+0.70	452.02	−0.70	450.62
	36 + 50	451.58	+0.70	452.28	−0.70	450.88
C.S.(D′)	36 + 82.71	451.92	+0.70	452.62	−0.70	451.22
	37 + 00	452.15	+0.62	452.77	−0.62	451.53
EVC	37 + 50	453.00	+0.39	453.39	−0.39	452.61
C′	37 + 81.28	453.63	+0.24	453.87	−0.24	453.39
	38 + 00	454.00	+0.15	454.15	−0.24	453.76
S.T.(B′)	38 + 32.71	454.65	0.00	454.65	−0.24	454.41
	38 + 50	455.00	−0.04	454.96	−0.24	454.76
	39 + 00	456.00	−0.17	455.83	−0.24	455.76
A′	39 + 28.71	456.57	−0.24	456.33	−0.24	456.33

*Interpolated values are shown underlined.
†Pavement revolved about the centerline (₵).

Review Questions

11.1. Why are curves used in roadway horizontal and vertical alignments?

11.2. Curves can be established by occupying ₵ or offset stations and then turning off appropriate deflection angles, or curves can be established by occupying a central control station and then turning off angles and measuring out distances—as determined through coordinates analyses. What are the advantages and disadvantages of each technique?

11.3. Why do chord and arc lengths for the same curve interval sometimes appear to be equal in value?

11.4. What characteristic do parabolic curves and spiral curves have in common?

11.5. Describe techniques that can be used to check the accuracy of the layout of a set of interchange curves using polar techniques (i.e., angle/distance layout from a central control station).

11.6. Why do construction surveyors usually provide offset stakes?

Problems

11.1. Given PI at $9 + 27.26$, $\Delta = 29°42'$, and $R = 700$ ft, compute the tangent (T) and the length of arc (L).

11.2. Given PI at $15 + 88.10$, $\Delta = 7°10'$, and $D = 8°$, compute the tangent (T) and the length of arc (L).

11.3. From the data in Problem 11.1, compute the stationing of the BC and EC.

11.4. From the data in Problem 11.2, compute the stationing of the BC and EC.

11.5. A straight-line route survey, which had PIs at $3 + 81.27$ ($\Delta = 12°30'$) and at $5 + 42.30$ ($\Delta = 10°56'$), later had 600-ft-radius circular curves inserted at each PI. Compute the BC and EC stationing (chainage) for each curve.

11.6. Given PI at $5 + 862.789$, $\Delta = 12°47'$, and $R = 300$ m, compute the deflections for even 20-m stations.

11.7. Given PI at $8 + 272.311$, $\Delta = 24°24'20''$, and $R = 500$ m, compute E (external), M (mid-ordinate), and the stations of the BC and EC.

11.8. Given PI at $10 + 71.78$, $\Delta = 36°10'30''$ RT, and $R = 1,150$ ft, compute the deflections for even 100-ft stations.

11.9. From the distances and deflections computed in Problem 11.6, compute the three key ℄ chord layout lengths, i.e., (1) the BC to the first 20-m station, (2) the chord distance for 20-m (arc) stations, and (3) from the last even 20-m station to the EC.

11.10. From the distances and deflections computed in Problem 11.8, compute the three key ℄ chord layout lengths, i.e., (1) the BC to the first 100-ft station, (2) the chord distance for 100-ft (arc) stations, and (3) from the last even 100-ft station to the EC.

11.11. From the distances and deflections computed in Problem 11.8, compute the chords (six) required for layout directly on offsets 50 ft right and 50 ft left of the ℄.

11.12. Two highway ℄ tangents must be joined with a circular curve of radius 1,000 ft (see Figure 11.32). The PI is inaccessible because its location falls in a river. Point A is established near the river on the back tangent, and point B is established near the river on the forward tangent. Distance AB is measured to be 615.27 ft. Angle $\alpha = 51°31'20''$, and angle $\beta = 32°02'45''$. Perform the calculations required to locate the BC and the EC in the field.

11.13. Two street curb lines intersect with $\Delta = 71°36'$ (see Figure 11.33). A curb radius must be selected so that an existing catch basin (CB) will abut the future curb. The curb side of the catch basin ℄ is located from point V: V to CB = 8.713 m and angle E-V-CB = $21°41'$. Compute the radius that will permit the curb to abut the existing catch basin.

11.14. Given the following compound curve data: $R_1 = 200$ ft, $R_2 = 300$ ft, $\Delta_1 = 44°26'$, and $\Delta_2 = 45°18'$, compute T_1 and T_2 (see Figure 11.15).

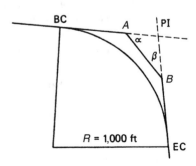

FIGURE 11.32 Sketch for Problem 11.12.

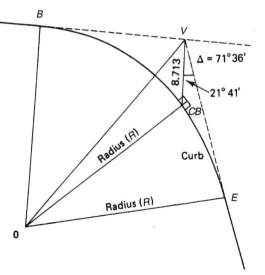

FIGURE 11.33 Sketch for Problem 11.13.

11.15. Given the following vertical curve data: PVI at $7 + 25.712$, $L = 100$ m, $g_1 = -3.2\%$, $g_2 = +1.8\%$, and elevation of PVI = 210.440, compute the elevations of the curve low point and even 20-m stations.

11.16. Given the following vertical curve data: PVI at $19 + 00$, $L = 500$ ft, $g_1 = +2.5\%$, $g_2 = -1\%$, and elevation at PVI = 723.86 ft, compute the elevations of the curve summit and even full stations (i.e., 100-ft even stations).

11.17. Given the following vertical curve data: $g_1 = +3\%$, $g_2 = -1\%$, design speed = 110 km/hr (from Table 11.3, $K = 90$ m), PVI at $0 + 360.100$, with an elevation of 156.663 m, compute the curve elevations at the high point (summit) and the even 50-m stations.

11.18. Given the following spiral curve data: $D = 8°$, $V = 40$ mph, $\Delta = 16°44'$, and PI at $11 + 66.18$, determine the value of each key spiral and circular curve component (L_s, R, P, q, LT, ST, X, Y, Δ_s, Δ_c, T_c, and L_c) and determine the stationing (chainage) of the T.S., S.C., C.S., and S.T.

11.19. Given the same data as in Problem 11.18, use the approximate equations (Equations 11.33 to 11.38) to compute X, Y, q, P, LT, and ST. Enter these values and the equivalent values determined in Problem 11.18 in a table to compare the results.

11.20. Use the data from Problem 11.18 to compute the deflections for the curve system (spirals and circular curves) at even 50-ft stations.

Chapter 12

Highway Construction Surveys

12.1 Preliminary (Preengineering) Surveys

Before the actual construction of a highway can begin, a great deal of investigative work has to be done. The final route chosen for the highway reflects costs due to topography (cuts and fills); relocating services; rail, highway, and water-crossing bridges; environmental impact; and a host of other considerations. Area information is first assembled from topographic maps, county maps, available aerial photography, and satellite imagery. The more likely routes may be flown and aerial photographs taken, with the resultant photogrammetric maps also being used to aid in the functional planning process. Figure 12.1 shows a simple stereo photo-analysis system that corrects distortions in the photos (rectifies) and permits the operator to digitize the horizontal and vertical (elevations) positions of all required points.

When the route has been selected, lower level aerial photography will probably form the basis for the preliminary or preengineering survey. Figure 12.2 shows part of a general plan prepared for a new highway location. The actual design drawings—which will show much more detail—are often drawn at 1 in.= 50 ft (1:500 metric). The proposed centerline is established in the field, with the stationing carried through from the initial point to the terminal point. Each time the tangent centerline changes direction, the PI is referenced, and the deflection angle is measured and recorded. Figure 12.3 shows a split-angle tie to a PI.

The highway designer can now insert horizontal curves at each PI, with the radius or degree of curve being controlled mostly by the design speed. At this point, the centerline stations are adjusted to account for the curve alignment—which is shorter, of course, than the tangent-only alignment. Horizontal control monuments, and both permanent and temporary benchmarks, are established along the route. Their placement intervals are usually less than 1,000 ft (300 m). Control for proposed aerial photography may require even smaller intervals. These control monuments are targeted so that their locations will show up on the aerial photographs.

In addition to referencing the PIs, centerline stations on tangent are referenced at regular intervals (500 to 1,000 ft) to aid the surveyor in reestablishing the centerline alignment

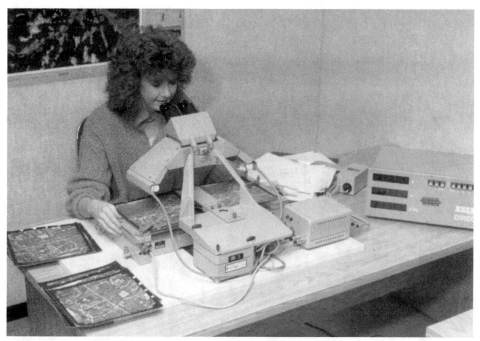

FIGURE 12.1 Zeiss G-2 Stereocord. Stereo-paired aerial photos are analyzed with horizontal and vertical data digitized for later plotting and analysis.

for additional preliminary surveys, numerous layout surveys, interim payment surveys, and finally the as-built survey. See Figure 12.4 for typical tie-in techniques. Swing ties are referenced to spikes driven into trees or to nails and washers driven into the roots of a tree. Strong swing ties intersect at angles near 120°. Three ties are used so that if one is lost, the point can still be reestablished from the remaining two. In areas of sparse tree cover, centerline ties are usually taken at right angles to the centerline station and can be cut crosses in rock outcrops. Iron bars (1″ reinforcing bars with brass caps) driven into the ground, aluminum monuments, and even concrete monuments can be used for this purpose.

The plan, usually prepared from aerial imaging, is upgraded in the field by the surveyor so that ambiguous or unknown features—that is, types of fences, dimensions of structures, types of manholes (storm or sanitary), road and drive surfaces, and the like—on the plan are properly identified. Soundings are taken at all water crossings, and soil tests are taken at bridge sites and areas suspected of instability. All drainage crossings, watercourses, ditches, and so on, are identified and tied in. Tree locations are fixed so that clearing and grubbing estimates can be prepared. Utility crossings (pipelines, conduits, and overhead cables) are located with respect to horizontal and vertical location. (Total station instruments are programmed to compute overhead clearances directly by remote object elevation—see Chapter 5.)

Railroad crossings require intersection angle and chainage, and track profiles a set distance right and left of the highway centerline. At-grade crossings require sight-line surveys for visibility patterns at set distances up the track ($^1/_4$ mile each way is not uncommon).

Sec. 12.1   Preliminary (Preengineering) Surveys   **445**

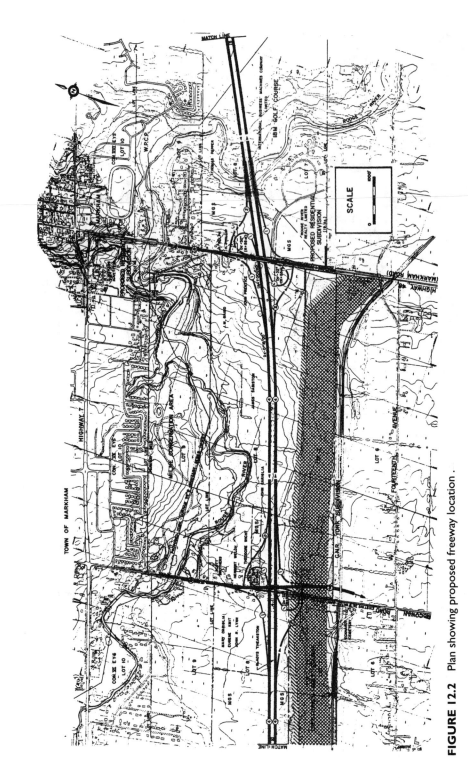

FIGURE 12.2 Plan showing proposed freeway location.

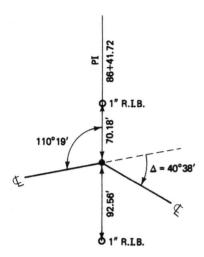

FIGURE 12.3 PI tie-in by split angle. Split angle in this example is $(180° + 40°38')/2 = 110°\ 19'$.

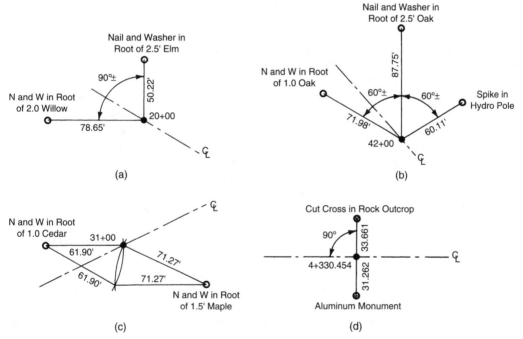

FIGURE 12.4 Centerline tie-ins. (a) Strong two-point swing tie. (b) Strong three-point swing tie. (c) Weak two-point swing tie, resulting in an ambiguous or poorly defined intersection. (d) 90° centerline tie-in (metric).

Road and other highway crossings require intersection angles and chainages, together with their centerline profiles left and right of the main centerline.

Cross sections are taken full width at regular stations, typically 25 ft (10 m) in rock, 50 ft (20 m) in cut, and 100 ft (30 m) in fill. In addition, extra cross sections are taken at any significant changes in slope that occur between regular stations. Rod readings are usually taken to the closest 0.1 ft (0.01 m), and rod location is tied in to the closest foot (0.1 m) for both chainage and offset distances. Stations are at 100-ft intervals (e.g., 0 + 00, 1 + 00) in foot units and are at 1,000-m intervals (e.g., 0 + 000, 1 + 000) in metric units.

12.2 Highway Design

The design of a highway depends on the type of service planned for the highway. Highways (and municipal streets) are classified as being locals, collectors, arterials, or freeways. The bulk of the highways, in mileage, are arterials that join towns and cities together in state or provincial networks. Design parameters, such as number of lanes, lane width, thickness and type of granular material, thickness and type of surface (asphalt or concrete), maximum climbing grades, minimum radius of curvature (related to design speed), and other items (including driver comfort and roadside aesthetics) vary from the highest consideration (for freeways) to the lowest consideration (for locals).

Local highways have lower design speeds, narrower lanes, and sharper curves to reflect the fact that the primary service function of locals is that of property access. Freeways, on the other hand, have mobility as their primary service function, and vehicle access is restricted to interchanges that are often many miles apart. Freeways have relatively high design speeds, which require flatter curves and wider (and perhaps more) traffic lanes. Flatter climbing grades (5 percent maximum is often used) enable trucks to maintain their operating speeds and thus help in keeping the overall operating speed for all traffic close to the speed for which the facility was designed.

The highway designer must keep these service-related parameters in mind as the highway's profiles, cross sections, and horizontal geometrics are incorporated. Figure 12.5 shows a typical design cross section for a two-lane arterial highway. Figure 12.6 is a plan and profile of an arterial two-lane highway, which shows the horizontal and vertical geometrics, toe of slope location, and cut-and-fill quantities. This plan and profile and the appropriate cross section (e.g., Figure 12.5) are all the surveyor normally needs to provide line and grade to the contractor. Figure 12.7 is a plan showing interchange ramps with full-width pavement elevations at 25-ft stations. This level of detail is commonly provided for all superelevated sections of pavement.

Complex interchanges require special attention from the surveyor. Figure 12.8 shows some of the curve data and chainage coordinates for an interchange. All key curve stations (PI, BC, EC, CPI, T.S., S.T., SPI, etc.) are established and referenced as shown in Figure 12.3 and Figure 12.4. The curves can be laid out by deflection angles (Chapter 11) or by polar layout (Section 5.7). If the interchange is to be laid out by using polar ties, then the surveyor must establish control monuments in protected areas, as discussed in Chapter 9.

Figure 12.8 also shows a creek diversion (Green Creek) necessitated by the interchange construction. Figure 12.8 gives the creek centerline alignment chainages and coordinates. Figure 12.9 shows some additional details for the creek diversion channel, including the centerline profile and cross sections, which can be used by the surveyor to provide line and grade for the channel.

12.3 Highway Construction Layout

When the decision is made to proceed with the highway construction, the surveyor goes back to the field to begin the stakeout. In some cases, months and even years may have

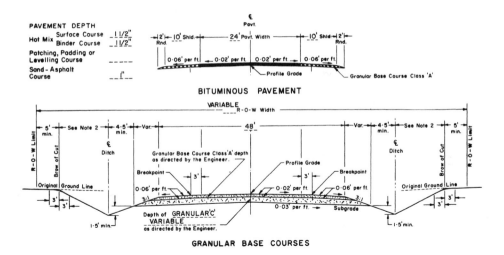

BITUMINOUS PAVEMENT

GRANULAR BASE COURSES

2 LANE HIGHWAYS
EARTH CUT SECTION

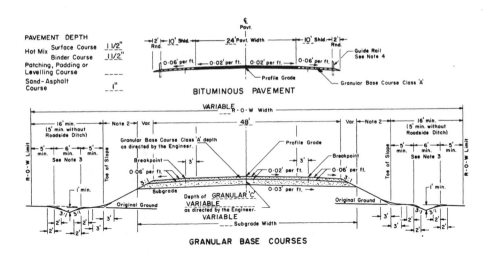

BITUMINOUS PAVEMENT

GRANULAR BASE COURSES

2 LANE HIGHWAYS
EARTH FILL SECTION

FIGURE 12.5 Typical cut-and-fill design cross sections for a two-lane arterial highway.

passed since the preliminary survey was completed, and some of the control monuments (as well as their reference points) may have been destroyed. However, if the preliminary survey has been referenced properly, sufficient horizontal and vertical control still exists for starting the stakeout. The centerline chainage is verified by measuring from referenced

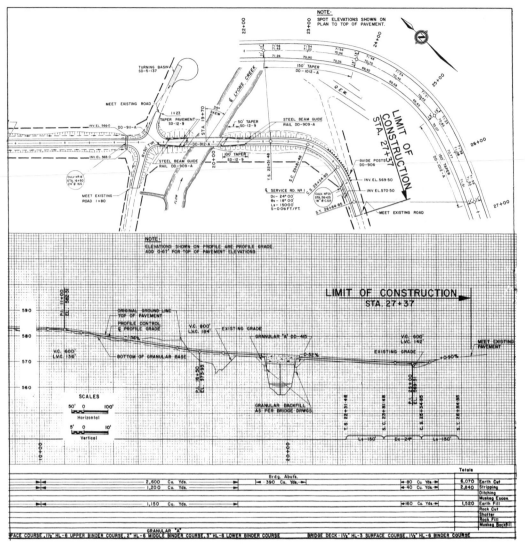

FIGURE 12.6 Plan and profile showing horizontal and vertical alignment, including pavement elevations through the superelevated section, cumulative quantities, and toe of slope locations.

stations, by checking cross-road intersections, by using GPS measurements, and by checking into any cross-road ties that were noted in the original survey.

Highways are laid out at 100-ft (30- or 40-m) stations, with additional stations being established at all changes in horizontal direction (e.g., BC, EC, T.S., S.T.) and all changes in vertical direction (e.g., BVC, ECV, low points, tangent runouts). The horizontal and vertical curve sections are often staked out at 50-ft (15- to 20-m) intervals to ensure that the finished product closely conforms to the design (25-ft intervals are often used on sharp-radius interchange ramps). When foot units are used, the full stations are at 100-ft intervals (e.g., 0 + 00, 1 + 00). Municipalities following the metric system use 100-m full-station

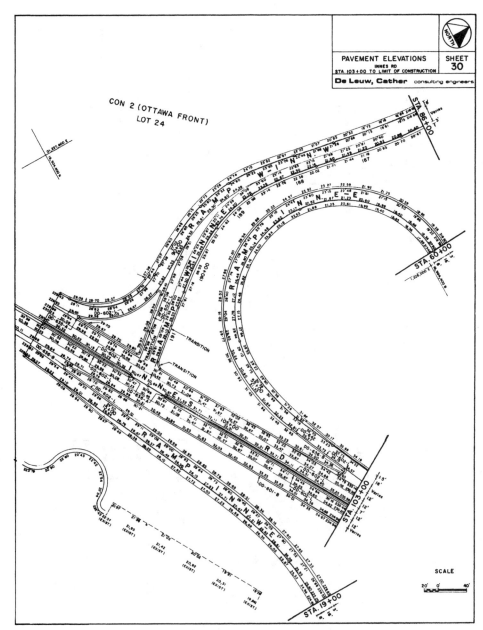

FIGURE 12.7 Interchange plan showing full-width pavement elevations at 25-ft station intervals.

intervals (0 + 00, 1 + 00), whereas most highway agencies use 1,000-m (km) intervals for full stations (e.g., 0 + 000, 0 + 100, . . . , 1 + 000).

The ℄ of construction is staked out by using a steel tape, GPS, or total station, with specifications ranging from 1/3,000 to 1/10,000 accuracy. The accuracy of ℄ layout can be

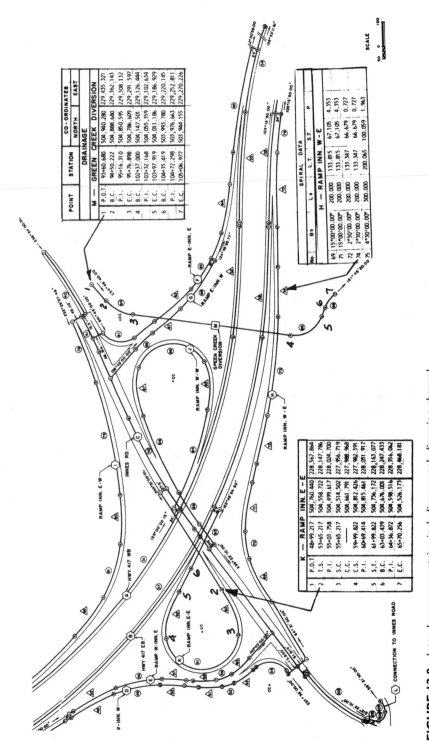

FIGURE 12.8 Interchange geometrics, including a creek diversion channel.

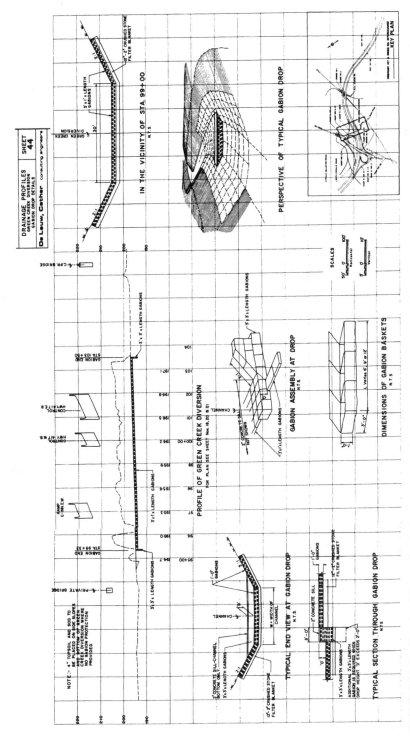

FIGURE 12.9 Details of Green Creek diversion channel, showing Gabion basket details, channel cross section, and profile. See Figure 12.8 for horizontal alignment.

verified at reference monuments, at road intersections, via GPS use, and the like. Highways can also be laid out from coordinated stations using theodolite/EDM instruments or total stations using polar layouts rather than rectangular layouts. Most interchanges are now laid out by polar methods, whereas most of the highways between interchanges are laid out with rectangular layout offsets. (The methods of polar layout were covered in Chapter 5.) GPS positioning techniques are now being used in many layout surveys, including highways.

The profile grade, which is shown on the contract drawing, can refer to the top of granular elevation, or it can refer to the top of asphalt (or concrete) elevation. The surveyor must ensure that he or she is using the proper reference before calculating subgrade elevations for the required cuts and fills.

12.4 Clearing, Grubbing, and Stripping Topsoil

Clearing and **grubbing** are the terms used to describe the cutting down of trees, and the removal of all stumps and litter, respectively. The full highway width is staked out, approximating the limits of cut and fill, so that the clearing and grubbing can be accomplished.

The first construction operation after clearing and grubbing is the stripping of topsoil. The topsoil is usually stockpiled for later use. In cut sections, the topsoil is stripped for full width, which extends to the points at which the far-side ditch slopes intersect the existing ground (EG)—also known as original ground (OG)—surface. See Figure 12.10. In fill sections, the topsoil is usually stripped for the width of the highway embankment (see Figure 12.11). Some highway agencies do not strip the topsoil where heights of fill exceed 4 ft (1.2 m) in the belief that this water-bearing material cannot damage the road base below that depth.

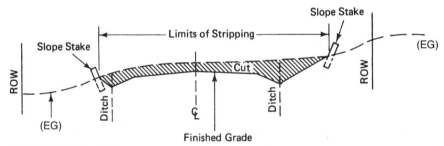

FIGURE 12.10 Highway cut section.

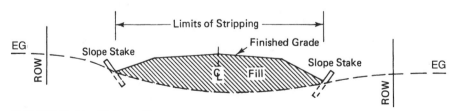

FIGURE 12.11 Highway fill section.

The bottom of fills (toe of slope) and the top of cuts (top of slope) are marked by slope stakes. These stakes, which are angled away from the ℄, not only delineate the limits of stripping but also indicate the limits for cut and fill. The latter operations take place immediately after the stripping operation. Lumber crayon (keel) or permanent markers are used on the stakes to show the station and the slope stake (s/s) identifications.

12.5 Placement of Slope Stakes

Figure 12.12 shows typical cut-and-fill sections in both foot and metric dimensions. The side slopes shown are 3:1, although most agencies use a steeper slope (2:1) for cuts and fills over 4 ft (1.2 m). To locate slope stakes, the difference in elevation between the profile grade at the ℄ and the invert of ditch (cut section) or the toe of embankment (fill section) must be determined first. In Figure 12.12(a), the difference in elevation consists of:

$$\text{Depth of granular} = 1.50 \text{ ft}$$

$$\text{Subgrade cross fall at 3\% over 24.5 ft} = 0.74 \text{ ft}$$

$$\text{Minimum depth of ditch} = \underline{1.50 \text{ ft}}$$

$$\text{Total difference in elevation} = 3.74 \text{ ft}$$

The ℄ of this minimum-depth ditch is 29.0 ft from the ℄ of construction. In cases when the ditch is deeper than minimum values, the additional difference in elevation and the additional distance from the ℄ of construction can be calculated easily from the same slope values.

In Figure 12.12(b), the difference in elevation consists of:

$$\text{Depth of granular} = 0.45 \text{ m}$$

$$\text{Fall at 3\% over 7.45 m} = 0.22 \text{ m}$$

$$\text{Minimum depth of ditch} = \underline{0.50 \text{ m}}$$

$$\text{Total difference in elevation} = 1.17 \text{ m}$$

The ℄ of this minimum-depth ditch is 8.95 m from the ℄ of construction. In these first two demonstrations, only the distance from the highway ℄ to the ditch has been determined. See Example 12.1 for additional practice.

In Figure 12.12(c), the difference in elevation consists of:

$$\text{Depth of granular} = 1.50 \text{ ft}$$

$$\text{Fall at 3\% over 24.5 ft} = \underline{0.74 \text{ ft}}$$

$$\text{Total difference in elevation} = 2.24 \text{ ft}$$

The difference from the ℄ of construction to the point where the subgrade intersects the side slope is 24.5 ft.

In Figure 12.12(d), the difference in elevation consists of:

$$\text{Depth of granular} = 0.45 \text{ m}$$

$$\text{Fall at 3\% over 7.45 m} = \underline{0.22 \text{ m}}$$

$$\text{Total difference in elevation} = 0.67 \text{ m}$$

The distance from the ℄ of construction to the point where the subgrade intersects the side slope is 7.45 m. In these last two demonstrations, the computed distance locates the slope stake.

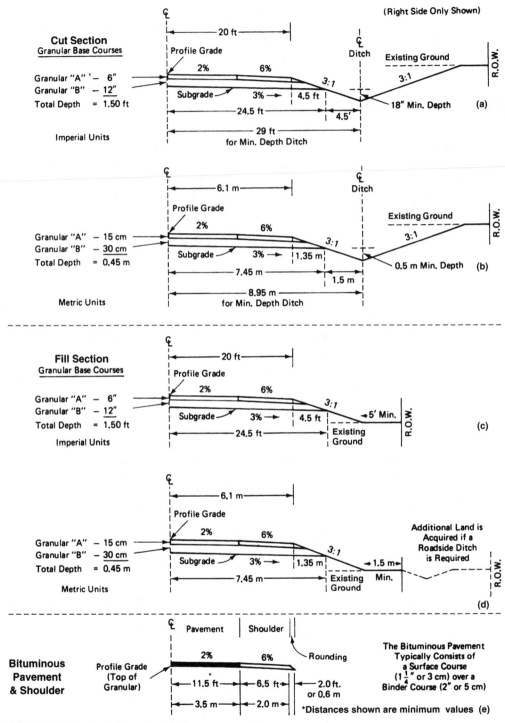

FIGURE 12.12 Typical two-lane highway cross section.

Figure 12.12(e) shows the pavement and shoulder cross section. The pavement and shoulder are built on top of the granular cross sections shown in Figure 12.12(a), (b), (c), and (d) in the final stages of construction.

■ **EXAMPLE 12.1** *Location of a Slope Stake in a Cut Section*

Refer to Figure 12.13. The profile grade (top of granular) is 480.00 and the HI is 486.28. Thus, we have the following information:

$$\text{Ditch invert} = 480.00 - 3.74 = 476.26$$
$$\text{Grade rod} = 486.28 - 476.26 = 10.02$$
$$\text{Depth of cut} = \text{grade rod} - \text{ground rod}$$

The following equation must be satisfied by trial-and-error ground rod readings:

$$X = (\text{depth of cut} \times 3) + 29.0$$

Solution

Using trial-and-error techniques, the rod holder moves from location to location away from the ₵ until the above equation is satisfied.

The rod holder, holding a fiberglass tape as well as the rod, estimates the desired location and gives a rod reading. For this example, assume that the rod reading is 6.0 ft at a distance of 35 ft from the ₵:

$$\text{Depth of cut} = 10.02 - 6.0 = 4.02$$
$$X = (4.02 \times 3) + 29.0 = 41.06 \text{ ft}$$

Because the rod holder is only 35 ft from the ₵, he or she must move farther out. At the next point, 43 ft from the ₵, a reading of 6.26 is obtained:

$$\text{Depth of cut} = 10.02 - 6.26 = 3.76$$
$$X = (3.76 \times 3) + 29.0 = 40.3$$

Because the rod holder is at 43 ft, he or she is too far out.

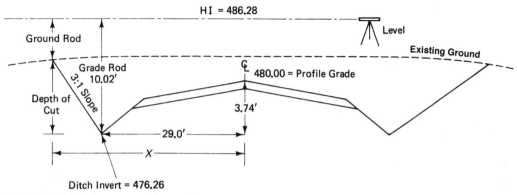

FIGURE 12.13 Location of slope stake in cut section (Example 12.1).

Sec. 12.5   Placement of Slope Stakes

457

The rod holder moves closer in and gives a rod reading of 6.10 at 41 ft from the
℄. Now we have:

$$\text{Depth of cut} = 10.02 - 6.10 = 3.92$$
$$X = (3.92 \times 3) + 29 = 40.8$$

This location is close enough for placing the slope stake. The error of 0.2 ft is not significant in this type of work. Usually two or three trials are required to locate the slope stake properly.

Figure 12.14 and 12.15 illustrate the techniques used when one is establishing slope stakes in fill sections. The slope stake distance from centerline can also be scaled from cross sections or topographic plans. In most cases, cross sections (see Chapter 17) are drawn at even stations (100 ft, or 30 to 40 m). The cross sections are necessary to calculate the volume estimates used in contract tendering. The location of the slope stakes can be scaled from this cross-section plot. In addition, highway contract plans are now usually developed photogrammetrically from aerial photos. These plans show contours that are precise enough for most slope-stake purposes. One can usually scale off the required distances from the ℄ by using either the cross-section plan or the contour plan to the closest 1.0 ft or 0.3 m.

The cost saving gained by determining this information in the office usually outweighs any resultant loss of accuracy. It is now possible to increase the precision through

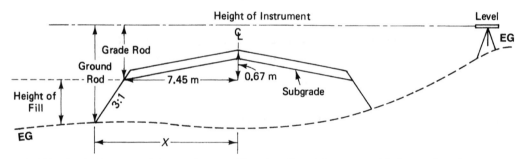

FIGURE 12.14 Location of slope stakes in a fill section. Case 1: instrument HI above subgrade.

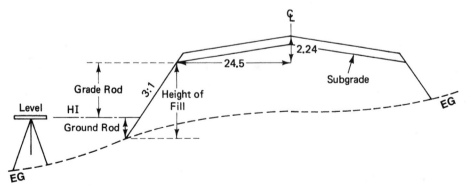

FIGURE 12.15 Location of a slope stake in a fill section. Case 2: instrument HI below subgrade.

advances in computers and photogrammetric equipment. Occasional field checks can be used to check these scale methods. If scale methods are employed, trigonometric leveling or total stations can be used to establish the horizontal distance from the ℄ to the slope stake. These methods are more accurate than using a fiberglass tape on deep cuts or high fills when breaking tape may be required several times.

12.6 Layout for Line and Grade

It is often possible in municipal work to put the grade stakes on offset, issue a grade sheet, and then continue with other work. The surveyor may be called back to replace the odd stake knocked over by construction equipment, but usually the layout is thought to be a one-time occurrence. In highway work, the surveyor must accept the fact that the grade stakes will be laid out several times. The chief difference between the two types of work is the large values for cut and fill. For the grade stakes to be in a safe location, they must be located beyond the slope stakes. This location is used for the initial layout, but as the work progresses, this distance back to the ℄ becomes too cumbersome to allow for accurate transfer of alignment and grade.

As a result, as the work progresses, the offset lines are moved ever closer to the ℄ of construction, until the final location for the offsets is 2 to 3 ft (1 m) from each edge of the proposed pavement. The number of times that the layout must be repeated is a direct function of the height of fill or the depth of cut involved. Machine guidance techniques (see Chapter 10) are proving to be much more economical when staking becomes repetitive.

In highway work, the centerline is laid out at the appropriate stations. The centerline points are then offset individually at convenient distances on both sides of the ℄. For the initial layout, the ℄ stakes, offset stakes, and slope stakes are all put in at the same time. The cuts and fills are written on the grade stakes and referenced either to the top of the stake or to a mark on the side of the stake that will give even foot (even decimeter) values. The cuts and fills are written on that side of the stake facing the ℄, whereas the stations are written on that side of the stake facing back to the 0 + 00 location, as previously noted.

The highway designer attempts to balance cuts and fills so that the overall costs are kept lower. Most highway projects employ scrapers (see Figure 12.16) first to cut the material that must be removed and then transport the material to the fill area, where it is discharged. The grade supervisor keeps checking the scraper cutting-and-filling operations by checking back to the grade stakes. (See Section 12.7 for grade-transfer techniques.) In rocky areas, blasting is required to loosen the rock fragments, which can then be loaded into haulers for disposal, either in fill areas or in shoulder or off-site areas. Figure 12.17 shows a typical dump hauler, larger versions of which can accommodate up to 140 yd^3 of material.

As the work progresses and the cuts and fills become more pronounced, care should be taken in breaking tape when laying out grade stakes so that the horizontal distance is maintained. The centerline stakes are offset by turning 90°, either with a right-angle prism or (more usually) by the swung-arm method. Cloth or fiberglass tapes are used to lay out the slope stakes and offset stakes. Once a ℄ station has been offset on one side, care is taken when offsetting to the other side to ensure that the two offsets and the ℄ stake are all in a straight line.

When the cut and/or fill operations have brought the work to the proposed subgrade (bottom of granular elevations), the subgrade must be verified by cross sections before the contractor is permitted to place the granular material. Usually a tolerance of 0.10 ft (30 mm) is allowed. Once the top of the granular profile has been reached, layout for pavement

FIGURE 12.16 Scraper (shown being assisted by a bulldozer) used for efficient transfer of cut material to fill areas. (Courtesy of Caterpillar, Inc.)

(sometimes a separate contract) can commence. Figure 12.18 shows a road grader shaping the crushed stone with the guidance of a rotating laser.

The final layout for pavement is usually on a very close offset (3 ft or 1 m). If the pavement is to be concrete, nails driven into the tops of the stakes provide more precise alignment.

When the highway construction has been completed, a final survey is performed. The final survey includes cross sections and locations that are used for final payments to the contractor and for a completion of the as-built drawings. The final cross sections are taken at the same stations used in the preliminary survey.

The description in this section refers to two-lane highways. The procedure for layout of a four-lane divided highway is very similar. The same control is used for both sections: grade stakes can be offset to the center of the median and used for both sections. When the lane separation becomes large and the vertical and horizontal alignment is different for each direction, the project can be approached as being two independent highways. The layout for elevated highways, often found in downtown urban areas, follows the procedures used for structures layout.

12.7 Grade Transfer

Grade stakes can be set so that the tops of the stakes are at grade. Stakes set to grade are colored red or blue on the top to differentiate them from all other stakes. This procedure is time consuming and often impractical except for final pavement layout. Generally, the larger the offset distance, the more difficult it is to drive the tops of the stakes to grade.

FIGURE 12.17 Euclid R35 hauler, capacity of 30.5 cubic yards. (Courtesy of Euclid-Hitachi Heavy Equipment, Ltd.)

As noted earlier, the cut and fill can refer to the top of the grade stake or to a mark on the side of the grade stake referring to an even number of feet (decimeters) of cut or fill. The mark on the side of the stake is located by sliding the rod up and down the side of the stake until a value is read on the rod that will give the cut or fill to an even foot (decimeter). This procedure of marking the side of the stake is best performed by two workers, one to hold the rod and the other to steady the bottom of the rod and then to make the mark on the stake. The cut or fill can be written on the stake or entered on a grade sheet, one copy of which is given to the contractor.

To transfer the grade (cut or fill) from the grade stake to the area of construction, a means of transferring the stake elevation horizontally is required. When the grade stake is close (within 6 ft or 2 m), the grade transfer can be accomplished using a carpenter's level set on a piece of sturdy lumber [see Figure 12.19(a)]. The recently developed torpedo (laser) level permits the horizontal reference (laser beam) to extend much beyond the actual location of the level itself (i.e., right across the grade). See Figure 12.19(b). When the grade stake is far from the area of construction, a string line level can be used to transfer the grade (cut or fill). See Figure 12.20. In this case, a fill of 1 ft 0 in. is marked on the grade stake (in addition to the offset distance). A guard stake has been placed adjacent to the grade stake, and the grade mark is transferred to the guard stake. A 1-ft distance is measured up the guard stake, and the grade elevation is marked. A string line is attached to the guard stake at the grade elevation mark. Then a line level is hung from the string (near the halfway mark), and the string is pulled taut to eliminate most of the sag (it is not possible to eliminate all the sag). The string line is adjusted up and down until the bubble in the line level is centered. With the bubble centered, the surveyor can see quickly at the ℄ how much more fill may be required to bring the highway, at that point, to grade. Figure 12.20 shows that

FIGURE 12.18 Rotating laser shown controlling fine grading operations by a road grader. (Courtesy of Leica Geosystems Inc.)

more fill is required to bring the total fill to the top of the subgrade elevation. The surveyor can convey this information to the grade inspector so that the fill can be increased properly. As the height of fill approaches the proper elevation (top of the subgrade), the grade checks become more frequent.

In the preceding example, the grade fill was 1 ft 0 in. Had the grade been cut 1 ft, the procedure with respect to the guard stake would have been the same: that is, measure up the guard stake 1 ft so that the mark now on the guard stake is 2 ft above the ℄ grade. The surveyor or inspector simply measures, at centerline, down 2 ft using a tape measure from the level string at the ℄. If the measurement down to the "present" height of fill exceeds 2 ft, it indicates that more fill is required; if the measurement down to the "present" height of fill is less than 2 ft, it indicates that too much fill has been placed and that an appropriate depth of fill must be removed.

Another method of grade transfer used when the offset is large and the cuts or fills are significant is the use of boning rods (batter boards). See Figure 12.21. In the preceding example, a fill grade is transferred from the grade stake to the guard stake. The fill grade is measured up the guard stake, and the grade elevation is marked. (Any even foot/decimeter cut or fill mark can be used, as long as the relationship to profile grade is clearly marked.) A crosspiece is nailed on the guard stake at the grade mark and parallel to the ℄. A similar

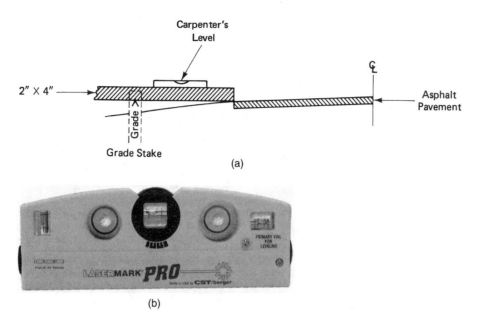

FIGURE 12.19 (a) Grade transfer using a conventional carpenter's level. (b) 80 laser torpedo level, featuring accuracy up to ¹/₄″ (0.029) at 100 ft (6 mm at 30 m); 3 precision glass vials—horizontal, vertical, and adjustable; and three AA cell batteries providing approximately forty hours of intermittent use. (Courtesy of CST/Berger, Illinois)

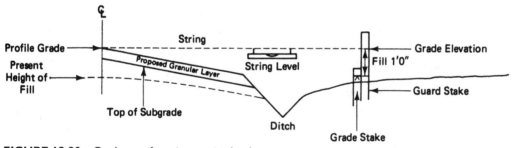

FIGURE 12.20 Grade transfer using a string level.

guard stake and crosspiece are established on the opposite side of the ℄. (In some cases, two crosspieces are used on each guard stake, the upper one indicating the ℄ profile grade and the lower one indicating the shoulder elevation.) The surveyor or grade inspector can then sight over the two crosspieces to establish a profile grade datum at that point. Another worker can move across the section with a rod, and the progress of the fill (cut) operation can be checked visually. Figure 12.22 shows a divided highway with grade stakes on either side of the ℄ swale, controlling grades on both sections of highway.

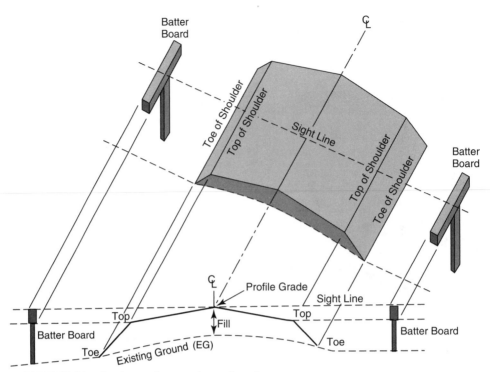

FIGURE 12.21 Grade transfer using batter boards.

12.8 Ditch Construction

The ditch profile often parallels the ℄ profile, especially in cut sections. When the ditch profile does parallel the ℄ profile, no additional grades are required to assist the contractor in construction. It is quite possible, however, to have the ℄ profile at one slope (even 0 percent) and the ditch profile at another slope (0.3 percent is often taken as a minimum slope to give adequate drainage). If the ditch grades are independent of the ℄ profile, the contractor must be given these cut or fill grades, either from the existing grade stakes or from grade stakes specifically referencing the ditch line. In the extreme case (e.g., a spiraled and superelevated highway going over the brow of the hill), the contractor may require five separate grades at one station (i.e., ℄, two edges of pavement, and two different ditch grades). It is even possible in this extreme case to have the two ditches flowing in opposite directions for a short distance.

Review Questions

12.1. What are the major differences between arterial highways and freeways?

12.2. What are two different techniques that can be used to tie in PI locations? Use sketches in your response.

12.3. How can slope-stake locations be determined prior to going out to the field?

12.4. What factors influence the choice of a route for a new highway?

FIGURE 12.22 A multilane divided highway, showing center drainage with grade stakes.

12.5. If it has been decided to locate features and elevations using aerial techniques, what field activities remain for the field surveyor?

12.6. Describe in detail how you would locate a railway crossing in the field.

12.7. What are the advantages of using machine control and guidance to construct highways? What are the disadvantages?

12.8. Describe three different ways of establishing a right angle to be used in a feature location.

Chapter 13

Municipal Street Construction Surveys

13.1 General Background

Preengineering surveys for a street or an area are requested when there is a strong likelihood that planned construction will be approved for the following year's budgeted works. The surveys manager must choose between aerial surveys and ground surveys for the preparation of the engineering drawings. On the basis of past experience, the surveys manager will know the cost per mile or kilometer for both ground and aerial surveys for various levels of urban density. More densely packed topographic and human-made detail can usually be picked up more efficiently using aerial methods. However, the use of total stations (described in Chapter 5) and satellite positioning techniques (described in Chapter 8) has made ground surveys competitive with aerial surveys for certain levels of detail density. A survey crew using a total station and two prism poles can capture as many as 1,000 points (X, Y, and Z coordinates) a day. In addition to this tremendous increase in data acquisition, we have the automatic transfer of data to the computer and the field note plot, which requires little additional work; thus, we have a highly efficient survey operation. See Tables 7.1 and 7.2 for typical scales for maps and plans used in municipal and highway projects.

As with highways, preliminary surveys for municipal streets include all topographic detail within the road allowance. Buried electric power lines, gas lines, telephone lines, and water services are usually staked out by the respective utility. Failure to request a stakeout can later (during construction) result in a broken utility and an expensive repair—not counting delays in the construction process.

Railroad crossings are profiled left and right of the street right-of-way (ROW), with additional sight-line data acquired as requested by the railroad or municipality. Water crossings and other road crossings are intersected, tied in, and profiled left and right of the street ROW with the same methods described for highways. Proposed connections to existing works (bridges, sewers, pipelines, streets, etc.) are surveyed carefully and precisely to ensure that the contractor makes an accurate connection.

In short, all topographic and works data that may have an effect on the cost estimates, the works design, or the construction of the proposed facility are tied in. Cross sections are taken full width at all regular stations and at any change in grade that may occur between them. Elevations are determined carefully for any connecting points (sewers, curbs, street centerlines, etc.). Reconstruction of existing roads may require drilling to determine the extent of any granular material that can be recycled. Road drilling is usually scheduled during the off-season.

Municipal benchmarks are utilized. Care must be taken always to check into adjacent or subsequent benchmarks to verify the starting elevation. As the preliminary survey proceeds, temporary benchmarks (TBMs) are established and carefully referenced to aid in subsequent preliminary and construction surveys.

13.2 Classification of Roads and Streets

The plan shown in Figure 13.1 depicts a typical municipal road pattern. The local roads shown have the primary purpose of providing access to individual residential lots.

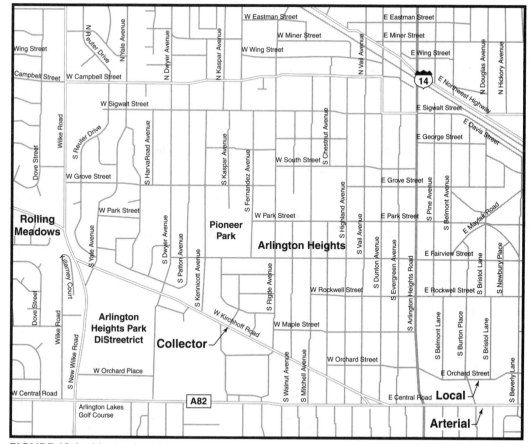

FIGURE 13.1 Municipal road pattern.

The collector roads (both major and minor) provide the dual service of lot access and traffic movement. The collector roads connect the local roads to arterial roads. The main purpose of the arterial roads is to provide a relatively high level of traffic movement.

Municipal works engineers base their road design on the level of service to be provided. The proposed cross sections and geometric alignments vary in complexity and cost for the fundamental local roads up to the more complex arterials. The highest level of service is given by the freeways, which provide high-velocity, high-volume routes with limited access (interchanges only) and ensure continuous traffic flow when design conditions prevail.

13.3 Road Allowances

The road allowance varies in width from 40 ft (12 m) for small locals to 120 ft (35 m) for major arterials. In parts of North America, including most of Canada, the local road allowances were originally 66 ft wide (one Gunter's chain). When widening was required due to increased traffic volumes, it was common to take 10-ft widenings on each side, initially resulting in an 86-ft road allowance for major collectors and minor arterials. Further widenings left major arterials at 100- and 120-ft widths.

13.4 Road Cross Sections

A full-service municipal road allowance usually has asphalt pavement, curbs, storm and sanitary sewers, water distribution pipes, hydrants, catch basins, and sidewalks. Additional utilities, such as natural gas pipelines, electrical supply cables, and cable television, are also often located on the road allowance. The essential differences between local cross sections and arterial cross sections are the widths of pavement and the quality and depths of pavement materials. The construction layout of sewers and pipelines is covered in Chapter 14. See Figure 13.2 for typical municipal road cross sections.

The cross fall (height of crown) used on the pavement varies from one municipality to another but is usually close to a 2 percent slope. The curb face is often 6 in. (150 mm) high except at driveways and crosswalks, where the height is restricted to about 2 in. (50 mm) for vehicle and pedestrian access. The slope on the boulevard from the curb to the street line usually rises at a 2 percent minimum slope, thus ensuring that roadway storm drainage does not run onto private property.

13.5 Plan and Profile

A typical plan and profile are shown in Figure 13.3. The **plan and profile,** which usually also show the cross-section details and construction notes, form the "blueprint" from which the construction is accomplished. The plan and profile, together with the contract specifications, spell out in detail precisely where and how the road (in this example) is to be built. The **plan** portion of the plan and profile gives the horizontal location of the facility, including curve radii, whereas the **profile** shows the key elevations and slopes along the road centerline, including vertical curve information. Both the plan and the profile relate all data to the project stationing established as horizontal control.

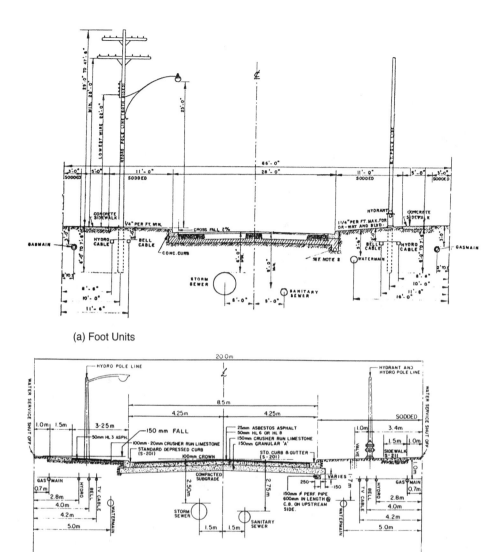

(a) Foot Units

(b) Metric Units

FIGURE 13.2 Typical cross section of a local residential road. Gas and water services shown under boule-vards are normally installed on only one side of the roadway; far-side house connections are located under the road and boulevard but above the sewer pipes. (a) Foot units. (b) Metric units.

13.6 Establishing Centerline (¢)

Let us use the case where a ditched residential road is to be upgraded to a paved and curbed road. The first job for the construction surveyor is to reestablish the centerline (¢) of the roadway. Usually this task entails finding several property markers delineating the street line ($). Fence and hedge lines can be used initially to guide the surveyor to the approximate locations of the property markers. When the surveyor finds one property marker, he

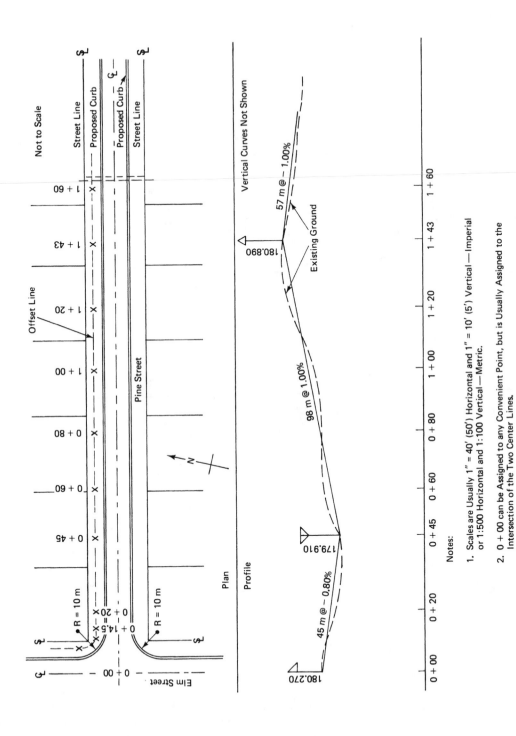

FIGURE 13.3 Plan and profile.

or she can measure out the frontage distances shown on the property plan (plat) to locate a sufficient number of additional markers. Usually the construction surveyor has the notes from the preliminary survey showing the location of property markers used in the original survey. If possible, the construction surveyor utilizes the same evidence used in the preliminary survey, taking the time, of course, to verify the resultant alignment. If the evidence used in the preliminary survey has been destroyed, as is often the case when a year or more has elapsed between the two surveys, the construction surveyor takes great care to ensure that the new results are not appreciably different from those of the original survey, unless, of course, a blunder occurred on the original survey.

The property markers can be square or round iron bars (including rebars) or round iron or aluminum pipes that are magnetically capped. The markers can vary from $1^1/_2$ to 4 ft in length. It is not unusual for the surveyor to have to use a shovel because the tops of the markers are often buried. The surveyor can use a magnetic or electronic metal detector to aid in locating buried markers. Sometimes even an exhaustive search of an area does not turn up a sufficient number of markers to establish the ₵. The surveyor must then extend the search to adjacent blocks or backyards to reestablish the missing markers. The surveyor can also approach the homeowners in the affected area and inquire about the existence of a mortgage survey plan (see Figure 13.4) for the specific property. Such a plan is required in most areas before a financial institution will provide mortgage financing. The mortgage survey plan shows dimensions from the building foundation to the street line and to both sidelines. Information thus gained can be used to narrow the search for a missing marker or can be used directly to establish points on the street line.

Once several points have been established on both sides of the roadway, the ₵ can be marked from each of these points by measuring (at right angles) half the width of the road allowance. The surveyor then sets up the theodolite on a ₵ mark near one of the project extremities and sights in on the ₵ marker nearest the other project extremity. He or she checks if all the markers line up in a straight line (assuming tangent alignment). See Figure 13.5. If all the markers do not line up, the surveyor checks the affected measurements. If discrepancies still occur (as is often the case), the surveyor makes the "best fit" of the available evidence. Depending on the length of roadway involved and the quantity of the ₵ markers established, the number of markers lining up perfectly will vary. Three well-spaced, perfectly aligned markers is the absolute minimum number required for the establishment of the ₵. The reason that all markers do not line up is that most lots are resurveyed over the years. Some lots may be resurveyed several times. Land surveyors' prime area of concern is that area of the plan immediately adjacent to their client's property, and they must ensure that the property stakeout is consistent for both evidence and plan intentions. Over several years, cumulative errors and mistakes can significantly affect the overall alignment of the ₵ markers.

If the ₵ is being marked on an existing road, as in this case, the surveyor uses nails with washers and red plastic flagging to establish the marks. The nails can be driven into gravel, asphalt, and (in some cases) concrete surfaces. The washers keep the nails from sinking below the road surface, and the red flagging helps in relocation.

If the project involves a new curbed road in a new subdivision, the establishment of the ₵ is much simpler. The recently set property markers would be intact, for the most part, and discrepancies between markers would be minimal (all markers would have been set in the same comprehensive survey operation). Wood stakes 2″ by 2″ or 2″ by 1″ and 18″ long would mark the ₵.

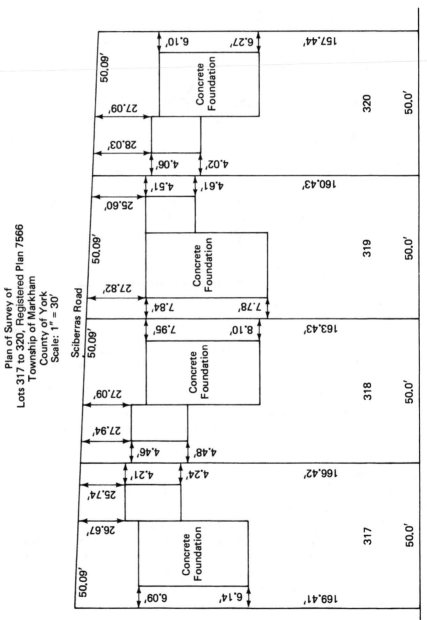

FIGURE 13.4 Mortgage survey plan (plat).

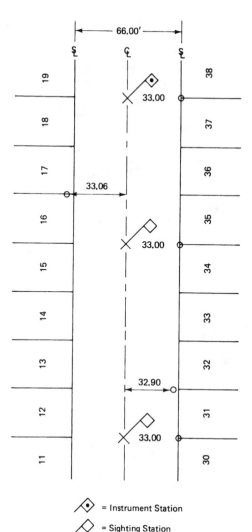

66.00'

19
18
17
33.06
16
15
14
13
32.90
12
11

38
37
36
35
34
33
32
31
30

33.00

33.00

33.00

◇ with dot = Instrument Station
◇ = Sighting Station

FIGURE 13.5 Property markers used to establish centerline.

13.7 Establishing Offset Lines and Construction Control

We continue here with the case where a ditched residential road is being upgraded to a paved and curbed road. The legal fabric of the road allowance, as given by the property markers, constitutes the horizontal control. Construction control consists of offset lines referenced to the proposed curbs with respect to line and grade. In the case of ditched roads and most highways, the offset lines are referenced to the proposed centerline with respect to line and grade. The offset lines are placed as close to the proposed location of the curbs as possible. It is essential that the offset stakes do not interfere with equipment and form work. It is also essential that the offset stakes be far enough removed so that they are not destroyed during cut or fill operations. Ideally, the offset stakes, once established, will

FIGURE 13.6 Slip-form concrete curber; line and grade are provided by the guide string (can also be guided by laser). Precision of $1/8$ in. or .01 ft. (Courtesy of Ausran)

remain in place for the duration of construction. This ideal can often be realized in municipal road construction, but it is seldom realized in highway construction because of the significant size of cuts and fills. If cuts and fills are not too large, offset lines for curbs can be 3 to 5 ft (1 to 2 m) from the proposed face of the curb. An offset line this close allows for very efficient transfer of line and grade (see Section 12.7). Figure 13.6 shows a concrete curb and gutter being installed by a slip-form concrete curber. The curber is kept on line and grade by keeping in contact (using a sonic detector) with a guide string or wire, which is established by measurements from the grade stakes.

In the case of a ditched gravel road being upgraded to a curbed paved road, the offset line must be placed far enough away on the boulevard to avoid the ditch-filling operation and any additional cut and fill that may be required. In the worst case, it may be necessary to place the offset line on the street line, an 18- to 25-ft (6- to 7-m) offset. Figure 13.7 shows a grade supervisor checking the status of excavation by sighting over cross rails referenced to the finished grade and set at right angles to the street. These sight rails (also called batter boards) are erected by measuring from the grade stakes placed by the surveyor.

When the street pavement is to be concrete, the pavement can be placed first, before the curbs are installed. Figure 13.8 shows a concrete-paving operation being controlled by a guide wire positioned precisely in line and grade by the surveyor working from previously placed grade stakes. After the pavement is in place, the curb (or curb and

FIGURE 13.7 Progress of street construction being monitored by the use of sight rails (batter boards).

gutter) can be placed adjacent to the new concrete pavement without the need of any further layouts.

When the street pavement is to be asphalt, the curbs (or curb and gutter) are always constructed first by measurements from the grade stakes. If the curb and gutter have been placed, the subsequent asphalt pavement is simply placed adjacent to the edge of the concrete gutter. If curb alone has been installed, the subsequent paving operation is guided by keel marks on the curb face a set distance below the top of the new curb (a 6-in. curb face is common). Figure 13.9 shows an asphalt-paving operation at an interchange. The asphalt is being placed adjacent to a barrier curb at the design profile marked on the curb face.

13.8 Construction Grades for a Curbed Street

The offset stakes (with nails or tacks for precise alignment) are usually placed at 50-ft (20-m) stations and at any critical alignment change points. The elevations of the tops of the stakes are determined by rod and level and are based on the vertical control established for the project. It is then necessary to determine the proposed elevation for the top of curb at each offset station. You can see in Figure 13.3 that elevations and slopes have been designed

FIGURE 13.8 Concrete paver controlled for line and grade by guide wire established from grade stakes. (Courtesy of Caterpillar Inc., Peoria, Illinois)

for a portion of the project. The plan and profile have been simplified for illustrative purposes, and offsets are shown for one curb line only.

Given the ₵ elevation data, the construction surveyor must calculate the proposed curb elevations. He or she can proceed by calculating the relevant elevations on ₵ and then adjusting for crown and curb height differential, or by applying the differential first and working out the curb elevations directly. To determine the difference in elevation between the ₵ and the top of curb, the surveyor must analyze the appropriate cross section. Figure 13.10 shows that the cross fall is $4.5 \times 0.02 = 0.090$ m (90 mm). The face on the curb is 150 mm; therefore, the top of curb is 60 mm above the ₵ elevation. A list of key stations (Table 13.1) is prepared, and the ₵ elevation at each station is calculated. The ₵ elevations are then adjusted to produce curb elevations. Because superelevation is seldom used in municipal design, it is safe to say that the curbs on both sides of the road are normally parallel in line and grade. A notable exception can occur when the intersections of collectors and arterials are widened to allow for turn lanes and both line and grade are affected.

The construction surveyor can then prepare a grade sheet (see Table 13.2), copies of which are given to the contractor and the project inspector. The tops of stake elevations, determined by level and rod, are assumed in this illustrative case. The grade sheet, signed by the construction surveyor, also includes the street name; date; limits of the contract; and, most important, the offset distance to the face of the curb.

FIGURE 13.9 Asphalt spreader laying asphalt to finished grade as marked on the curb face. (Courtesy of Caterpillar Inc., Peoria, Illinois)

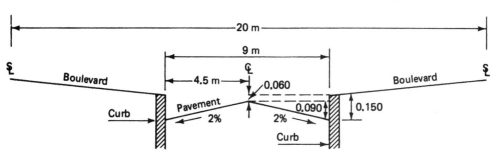

FIGURE 13.10 Cross section showing the relationship between the centerline and the top of the curb elevations.

Note that the construction grades (cut and fill) refer only to the vertical distance to be measured down or up from the top of the grade stake to locate the proposed elevation. Construction grades do not define with certainty whether the contractor is in a cut or fill situation at any given point. Refer to Figure 13.11. At 0 + 20, a construction grade of cut 0.145 is given, whereas the contractor is actually in a fill situation (i.e., the proposed top of curb is above the existing ground at that station). This lack of correlation between construction grades and the construction process can become more pronounced as the offset distance lengthens. For example, if the grade stake at station 1 + 40 had been located at the street line, the construction grade would have been cut, whereas the construction process is almost entirely in a fill operation.

Sec. 13.8   Construction Grades for a Curbed Street

Table 13.1 GRADE SHEET COMPUTATIONS

Station	℄ Elevation		Curb Elevation
0 + 00	180.270		
	−0.116		
BC 0 +14.5	180.154	+0.060	180.214
	−0.044		
0 + 20	180.110	+0.060	180.170
	−0.160		
0 + 40	179.950	+0.060	180.010
	−0.040		
0 + 45	179.910	+0.060	179.970
	+0.150		
0 + 60	180.060	+0.060	180.120
	+0.200		
0 + 80	180.260	+0.060	180.320
	+0.200		
1 + 00	180.460	+0.060	180.520
	+0.200		
1 + 20	180.660	+0.060	180.720
	+0.200		
1 + 40	180.860	+0.060	180.920
	+0.030		
1 + 43	180.890	+0.060	180.950
etc.			

Table 13.2 GRADE SHEET

Station	Curb Elevation	Stake Elevation	Cut	Fill
0 + 14.5	180.214	180.325	0.111	
0 + 20	180.170	180.315	0.145	
0 + 40	180.010	180.225	0.215	
0 + 45	179.970	180.110	0.140	
0 + 60	180.120	180.185	0.065	
0 + 80	180.320	180.320	On grade	
1 + 00	180.520	180.475		0.045
1 + 20	180.720	180.710		0.010
1 + 40	180.920	180.865		0.055
1 + 43	180.950	180.900		0.050
etc.				

When the layout is performed using foot units, the basic station interval is 50 ft. The dimensions are recorded and calculated to the closest one-hundredth (0.01) of a foot. All survey measurements are in feet and decimals of a foot, but for the contractor's purposes, the final cuts and fills are often expressed in feet and inches. The decimal-inch relationships are soon committed to memory by surveyors working in the construction field (see Table 13.3) Cuts and fills are usually expressed to the closest $1/_8$ in. for concrete, steel, and pipelines and to the closest $1/_4$ in. for highways, granular surfaces, and ditch lines. The grade sheet in

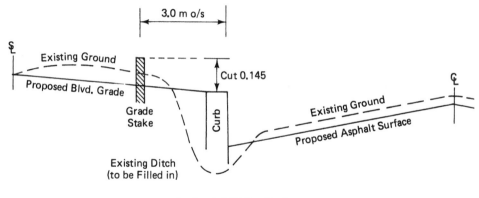

Station 0 + 20 Cut Grade

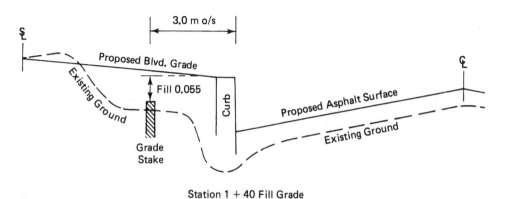

Station 1 + 40 Fill Grade

FIGURE 13.11 Cross sections showing cut-and-fill grades.

Table 13.3 DECIMAL FOOT-INCH CONVERSION

	$1'' = 1/12' = 0.083'$	
$1'' = 0.08(3)'$	$7'' = 0.58'$	$1/8'' = 0.01'$
$2'' = 0.17'$	$8'' = 0.67'$	$1/4'' = 0.02'$
$3'' = 0.25'$	$9'' = 0.75'$	$1/2'' = 0.04'$
$4'' = 0.33'$	$10'' = 0.83'$	$3/4'' = 0.06'$
$5'' = 0.42'$	$11'' = 0.92'$	
$6'' = 0.50'$	$12'' = 1.00'$	

Table 13.4 illustrates the foot-inch relationships. The first column and columns (6) and (7) are all that are required by the contractor, in addition to the offset distance, to construct the facility properly.

In some cases, the cuts and fills (grades) are written directly on the appropriate grade stakes. This information, written with lumber crayon (keel) or felt marker, is always written on the side of the stake facing the construction. The station is also written on each stake and is placed on that side of each stake facing the lower chainage (station).

Table 13.4 GRADE SHEET (FOOT UNITS)

(1) Station	(2) Curb Elevation	(3) Stake Elevation	(4) Cut	(5) Fill	(6) Cut	(7) Fill
0 + 30	470.20	471.30	1.10		$1'1\frac{1}{4}''$	
0 + 50	470.40	470.95	0.55		$0'6\frac{5}{8}''$	
1 + 00	470.90	470.90	On grade		On grade	
1 + 50	471.40	471.23		0.17		$0'2''$
2 + 00	471.90	471.46		0.44		$0'5\frac{1}{4}''$
2 + 50	472.40	472.06		0.34		$0'4\frac{1}{8}''$

13.9 Street Intersections

An intersection curb radius can range from 30 ft (10 m) for two local streets intersecting, to 60 ft (18 m) for two arterial streets intersecting. The angle of intersection is ideally 90° (for good sight lines); however, the range from 70° to 110° is often permitted for practical purposes. See Figure 13.12.

Street intersections require special attention from the surveyor for both line and grade. The curved curb lines (shown at the intersection of Pine Street and Elm Street in Figure 13.3) are often established in the field, not by the deflection angle technique of Chapter 11, but instead by first establishing the curve center and then establishing the arc by swinging off the curve radius. Also the profile of this curved section of proposed curb requires analysis because this section of curb does not fit the profile of Pine Street or of Elm Street. We can determine the curb elevation at the BC on Pine Street; it is 180.214 at 0 + 14.5 (Table 13.1).

At this point, the surveyor must obtain the plan and profile of Elm Street and determine the chainage at the EC of the curb coming from the BC at 0 + 14.5, on the north side of Pine Street (see Figure 13.3). Once this station has been determined, the proposed curb elevation at the EC on Elm Street can be computed. Let's assume this value to be 180.100 m. This procedure has to be repeated to determine the curb elevation on Elm Street at the EC of the curve coming from the BC on the south side of Pine Street (if that curb is also being constructed).

The surveyor, knowing the elevation of both the BC on Pine Street and the EC on Elm Street, can compute the length of arc between the BC and the EC and then compute the profile grade for that particular section of curved curb. The procedure is as follows. The curb elevation at the BC (180.214) is determined from the plan and profile of Pine Street (see Figure 13.3 and Table 13.1). The curb elevation at the EC is determined from the plan and profile of Elm Street (assume that the EC elevation = 180.100). The length of curb can be calculated using Equation 11.5:

$$L = \frac{\pi R \Delta}{180}$$
$$= 15.708 \text{ m}$$

The slope from BC to EC can be determined as follows:

$$180.214 - 180.100 = 0.114 \text{ m}$$

The fall is 0.114 m over an arc distance of 15.708 m, which is −0.73 percent.

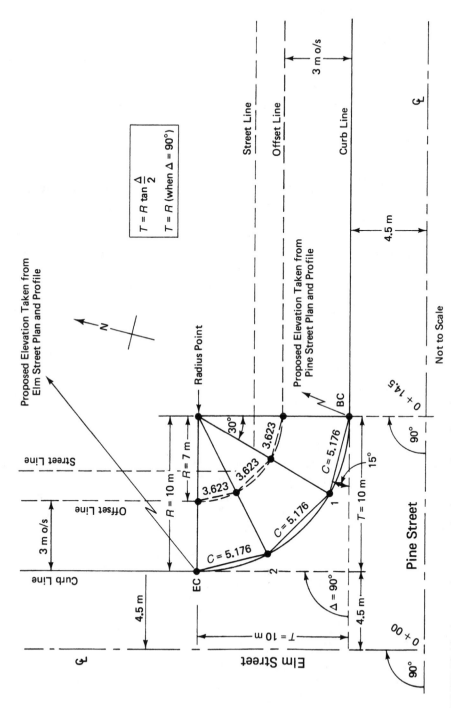

FIGURE 13.12 Intersection geometrics (one quadrant shown).

These calculations indicate that a satisfactory slope (0.5 percent is the usual minimum) joins the two points. The intersection curve is located by four offset stakes, BC, EC, and two intermediate points. In this case, 15.708/3 = 5.236 m, the distance measured from the BC to locate the first intermediate point, the distance measured from the first intermediate point to the second intermediate point, and the distance used as a check from the second intermediate point to the EC.

In actual practice, the chord distance is used rather than the arc distance. Because $\Delta/2 = 45°$ and we are using a factor of $^1/_3$, the corresponding deflection angle for one-third of the arc would be 15°, which we use in the following equation:

$$C = 2R \sin (\text{deflection angle})$$
$$= 2 \times 10 \times \sin 15° = 5.176 \text{ m}$$

These intermediate points can be deflected in from the BC or EC, or they can be located by the use of two tapes, with one surveyor at the radius point (holding 10 m, in this case) and another surveyor at the BC or intermediate point (holding 5.176), while the stake surveyor holds the zero point of both tapes. The latter technique is used most often on these small-radius problems. The only occasions when these curves are deflected in by theodolite occur when the radius point (curve center) is inaccessible (due to fuel pump islands, front porches, etc.).

The proposed curb elevations on the arc are computed as follows:

Station	Elevation
BC 0 +14.5	180.214
	−0.038
Stake #1	180.176
	−0.038
Stake #2	180.138
	−0.038
EC	180.100

The arc interval is 5.236 m. The difference in elevation = 5.236 × 0.0073 = 0.038. Grade information for the curve can be included on the grade sheet.

The offset curve can be established in the same manner after making allowances for the shortened radius (see Figure 13.12). For an offset (o/s) of 3 m, the radius becomes 7 m. The required chords can be calculated as follows:

$$C = 2R \sin (\text{deflection})$$
$$= 2 \times 7 \times \sin 15°$$
$$= 3.623 \text{ m}$$

13.10 Sidewalk Construction

The sidewalk is constructed adjacent to the curb or at some set distance from the street line. If the sidewalk is adjacent to the curb, no additional layout is required because the curb itself gives line and grade for construction. In some cases, the concrete for this curb and sidewalk is placed in one operation.

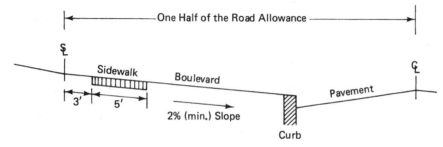

Note: The Sidewalk is Always Constructed so that it Slopes Toward the Road — Usually @ 1/4" per foot (2%).

FIGURE 13.13 Typical location of a sidewalk on the road allowance.

When the sidewalk is to be located at some set distance from the street line (⌀), a separate layout is required. Sidewalks located near the ⌀ give the advantages of increased pedestrian safety and boulevard space for the stockpiling of a winter's accumulation of plowed snow in northern regions. See Figure 13.13, which shows the typical location of a sidewalk on a road allowance.

Sidewalk construction usually takes place after the curbs have been built and the boulevard has been brought to sod grade. The offset distance for the grade stakes can be quite short (1 to 3 ft). If the sidewalk is located within 1 to 3 ft of the ⌀, the ⌀ is an ideal location for the offset line. In many cases, only the line is required for construction because the grade is already established by boulevard grading and the permanent elevations at ⌀ (existing elevations on private property are seldom adjusted in municipal work). The cross slope (toward the curb) is usually given as being $1/4$ in./ft (2 percent).

13.11 Site Grading

Every construction site (whether it be small, as for a residential house, or large, as for an airport) requires a site grading plan showing proposed (and existing) ground elevations. These ground elevations have been designed to (1) ensure proper drainage for storm-water runoff, (2) provide convenient pedestrian and vehicular access, and (3) balance cut and fill optimally. Typical municipal projects that require grading surveys include residential, commercial, or industrial developments; sanitary landfill sites; parks or greenbelts; boulevards; and the like.

Figure 13.14 shows part of a site grading plan for a residential development. The existing ground elevations are shown in brackets below each proposed elevation at all lot corners. With this plan, the surveyor can determine the runoff flow direction for each lot and set grade stakes at the lot corners to define finished ground grade. Figure 13.15 shows a backhoe/loader roughly shaping the ground behind a housing development, and Figure 13.16 shows another loader with a landscaping/pulverizing attachment bringing the ground to its final elevation just prior to sodding or seeding.

Grade control can be provided by rotating lasers or electronic levels (see Figure 12.18). Grade can also be given by grade stakes or grade stakes with batter boards. Large-scale

(83.45)—existing elevation

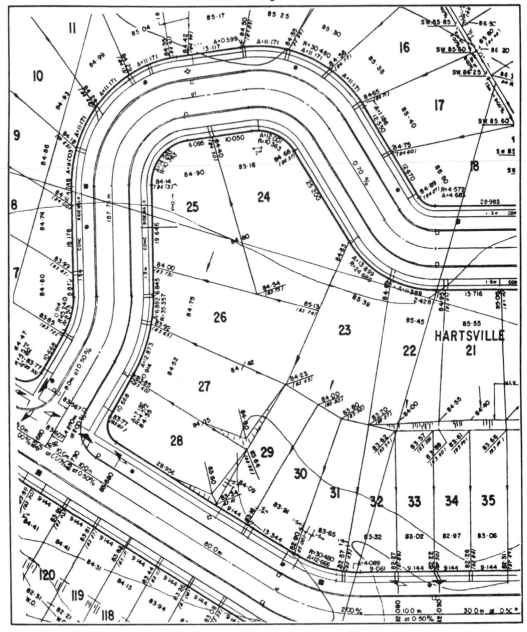

FIGURE 13.14 Site grading plan. Proposed elevation: 84.23. Existing elevation: (83.45).

FIGURE 13.15 Rough grading by loader/backhoe. (Courtesy of John Deere, Moline, Illinois)

projects, with deep cuts or fills, may require several layouts before finally being brought to grade, unless machine guidance and control techniques are being used (see Chapter 10).

Problems

13.1. A new road is to be constructed beginning at an existing road (℄ elevation = 472.70 ft) for a distance of 600 ft. The ℄ gradient is to rise at 1.18 percent. The elevations of the offset grade stakes are as follows: 0 + 00 = 472.60, 1 + 00 = 472.36, 2 + 00 = 473.92, 3 + 00 = 475.58, 4 + 00 = 478.33, 5 + 00 = 479.77, and 6 + 00 = 480.82. Prepare a grade sheet like the one in Table 13.4 showing the cuts and fills in feet and inches.

13.2. A new road is to be constructed to connect two existing roads. The ℄ intersection with the east road (0 + 00) is at an elevation of 210.666 m, and the ℄ intersection at the west road (1 + 32.562) is at an elevation of 209.446 m. The elevations of the offset grade stakes are as follows: 0 + 00 = 210.831, 0 + 20 = 210.600, 0 + 40 = 211.307, 0 + 60 = 210.114, 0 + 80 = 209.772, 1 + 00 = 209.621, 1 + 20 = 209.308, and 1 + 32.562 = 209.400. Prepare a grade sheet like the one in Table 13.2 showing cuts and fills in meters.

Refer to Figure 13.17 for Problems 13.3 to 13.8.

13.3. Figure 13.17 shows proposed curb locations, together with grade stakes offset from the north-side curb. The straight section of curb begins at 0 + 52 and ends at 8 + 38 (centerline stations). With the aid of a sketch, show how these two chainages were determined.

13.4. Compute the final centerline road elevations at the points of curve (PC)—that is, 0 + 52 and 8 + 38—and at all even 50-ft stations.

FIGURE 13.16 Fine grading by landscaping/pulverizing attachment on loader. (Courtesy of John Deere, Moline, Illinois)

13.5. Compute the top-of-curb elevations from 0 + 52 to 8 + 38. See Figure 13.17 (cross section) for the centerline/top-of-curb relationship.

13.6. Using the grade stake elevations shown in the chart below, prepare a grade sheet showing cuts and fills (ft and in.) for the proposed curb from 0 + 52 to 8 + 38.

		Grade Stake Elevations		
PC	0 + 52	504.71	4 + 50	508.46
	1 + 00	504.78	5 + 00	508.77
	1 + 50	504.93	5 + 55	510.61
	2 + 00	504.98	6 + 00	511.73
	2 + 50	505.82	6 + 50	512.00
	3 + 00	506.99	7 + 00	512.02
	3 + 50	507.61	7 + 50	512.11
	3 + 75	507.87	8 + 00	512.24
	4 + 00	508.26	PC 8 + 38	512.73

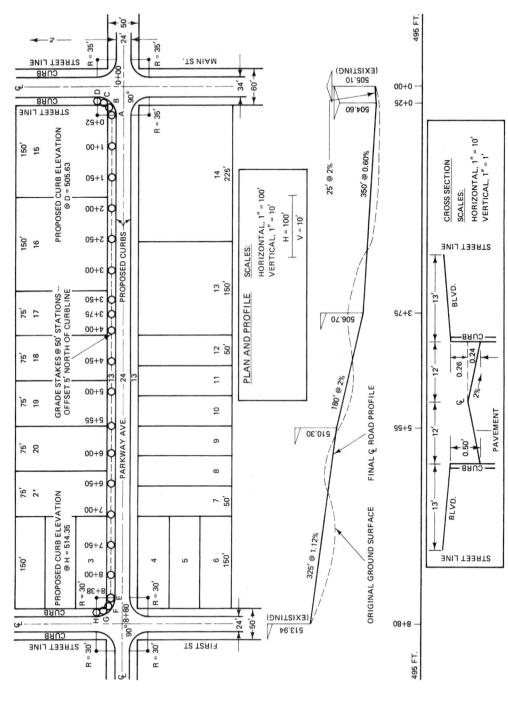

FIGURE 13.17 Plan, profile, and cross section for pavement and curbs, Parkway Avenue (foot units). See also Figure 14.24.

13.7. For the intersection curve (R = 35 ft) at Main Street, station A (0 + 52) to station D:
  (a) Compute the length of the arc.
  (b) Determine the curb-line gradient (percentage) from A to D.
  (c) Determine the proposed curb elevations at stations B and C. B and C divide the arc into three equal sections; that is, AB = BC = CD.
  (d) Using the grade stake elevations shown in the chart below, determine the cuts and fills (ft and in.) for stations B, C, and D.

Grade Stake Elevations			
A	504.71	C	506.37
B	506.22	D	506.71

13.8. For the intersection curve (R = 30 ft) at First Street, station E (8 + 38) to station H:
  (a) Compute the length of the arc.
  (b) Determine the curb-line gradient (percentage) from E to H.
  (c) Determine the proposed curb elevations at stations F and G. F and G divide the arc into three equal sections; that is, EF = FG = GH.
  (d) Using the grade stake elevations shown in the chart below, determine the cuts and fills (ft and in.) for stations F, G, and H.

Grade Stake Elevations			
E	512.73	G	512.88
F	512.62	H	513.27

Use Figure 13.18 for Problems 13.9 to 13.14.

13.9. Figure 13.18 shows proposed curb locations, together with offset stakes, for the north-side curb. The straight section of curb begins at station 0 + 15 and ends at station 2 + 10. With the aid of a sketch, show how these two chainages were determined.

13.10. Compute the final centerline road elevations at the points of curve (PC)—that is, 0 + 15 and 2 + 10—and at all even 20-m stations.

13.11. Compute the top-of-curb elevations from 0 + 15 to 2 + 10. See Figure 13.18 (cross section) for the centerline/top-of-curb relationship.

13.12. Using the grade stake elevations shown in the chart below, calculate the cuts and fills for the proposed curb from 0 + 15 to 2 + 10.

Grade Stake Elevations				
PC	0 + 15	186.720	1 + 20	188.025
	0 + 20	186.387	1 + 40	188.003
	0 + 40	185.923	1 + 60	187.627
	0 + 60	186.425	1 + 72	187.455
	0 + 72	186.707	1 + 80	187.907
	0 + 80	187.200	2 + 00	187.993
	1 + 00	187.527	PC 2 + 10	188.125

13.13. For the intersection curve (R = 10 m) at Elm Street, station A (0 + 15) to station D:
  (a) Compute the length of arc.

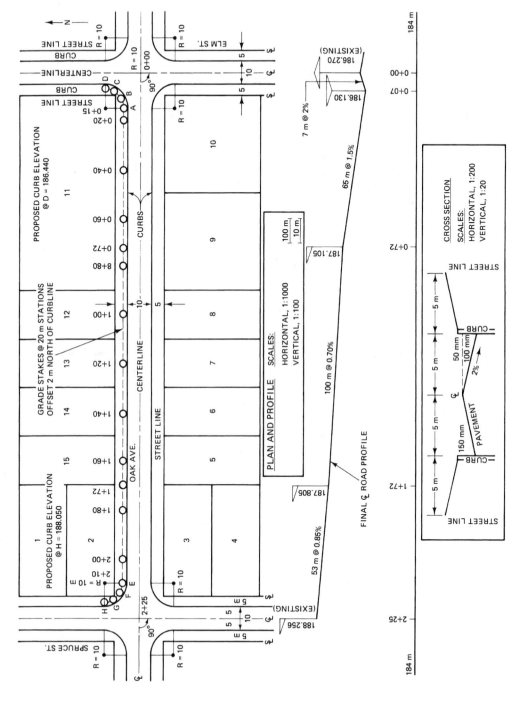

FIGURE 13.18 Plan, profile, and cross section for pavement and curbs, Oak Avenue (metric units). See also Figure 14.25.

(b) Determine the curb-line gradient (percentage) from A to D.

(c) Determine the proposed curb elevations at stations B, C, and D. B and C divide the arc into three equal sections; that is, AB = BC = CD.

(d) Using the grade stake elevations shown in the chart below, determine the cuts and fills for stations B, C, and D.

Grade Stake Elevations			
A	186.720	C	186.575
B	186.447	D	186.567

13.14. For the intersection curve (R = 10 m) at Spruce Street, station E (2 + 10) to station H:

(a) Determine the curb-line gradient (percentage) from E to H.

(b) Determine the proposed curb elevations at stations F, G, and H. F and G divide the arc into three equal sections; that is, EF = FG = GH.

(c) Using the grade stake elevations shown in the chart below, determine the cuts and fills for stations F, G, and H.

Grade Stake Elevations	
E	188.125
F	188.007
G	188.015
H	188.010

13.15. Refer to Figure 13.14 (site grading plan). A grade stake was set near the middle of Lot #26, and the stake-top elevation was determined to be 84.15 m. Compute the cut and fill at each of the four lot corners.

Chapter 14

Pipeline and Tunnel Construction Surveys

14.1 Pipeline Construction

Pressurized pipelines are designed to carry water, oil, natural gas, and sometimes sewage. The pipeline flow rates depend on the amount of pressure applied, the pipe size, and other factors. Because pressure systems do not require close attention to grade lines, the layout for pipelines can proceed at a much lower order of precision than is required for gravity pipes. Pipelines are usually designed so that the cover over the crown is adequate for the loading conditions expected and also adequate to prevent damage due to frost penetration, erosion, and the like.

The pipeline location can be determined from the contract drawings. The line-and-grade stakes are offset an optimal distance from the pipe ℄ and are placed at 50- to 100-ft (15- to 30-m) intervals. When existing ground elevations are not being altered, the standard cuts required can be simply measured down from the ground surface. (Required cuts in this case would equal the specified cover over the crown plus the pipe diameter plus the bedding, if applicable.) See Figure 14.1, Trench "A." In the case of proposed general cuts and fills, grades must be given to establish suitable crown elevation so that final cover is as specified. See Figure 14.1, Trench "B." Additional considerations and higher precisions are required at major crossings (e.g., rivers, highways, utilities).

Figure 14.2(a) shows a pipeline being installed a set distance below existing ground (as in Trench "A," Figure 14.1). Material cast from the trench is shown on the right side of the trench, and the installation equipment is shown on the left side of the trench. The surveyor, in consultation with the contractor, determines a location for the line-and-grade stakes that will not interfere with either the cast material or the installation equipment.

Figures 14.2(b) shows a pipe-jacking operation. The carrier pipe is installed at the same time that the tunnel is being excavated. A construction laser that has been set to the design slope and horizontal alignment provides line and grade. This type of operation is used where an open cut would be unacceptable—for example, across important roads or highways.

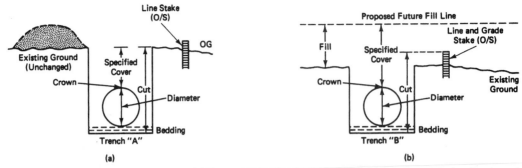

FIGURE 14.1 Pipeline construction. (a) Existing ground to be unchanged. (b) Existing ground to be altered.

(a)

FIGURE 14.2 (a) Pipeline installation. (Courtesy of Caterpillar, Inc.)

(continued)

Final surveys show the actual location of the pipe and appurtenances (valves and the like). As-built drawings, produced from final surveys, are especially important in urban areas, where it seems there is no end to underground construction.

(b)

FIGURE 14.2 *(continued)* (b) Pipe-jacking operation. (Courtesy of Astec Underground/American Augers)

14.2 Sewer Construction

Sewers are usually described as being in one of two categories: sanitary and storm. Sanitary sewers collect domestic and industrial liquid waste and convey these wastes (sewage) to a treatment plant. Storm sewers are designed to collect runoff from rainfall and to transport this water (sewage) to the nearest natural receiving body (e.g., creek, river, lake). The rainwater enters the storm-sewer system through ditch inlets or through catch basins located at the curb line on paved roads. The design and construction of sanitary and storm sewers are similar because the flow of sewage is usually governed by gravity. Because the sewer gradelines (flow lines) depend on gravity, it is essential that construction grades be precise.

Figure 14.3(a) shows a typical cross section for a municipal roadway. The two sewers are typically located 5 ft (1.5 m) on either side of the ℄. The sanitary sewer is usually deeper than the storm sewer because it must be deep enough to allow for all house

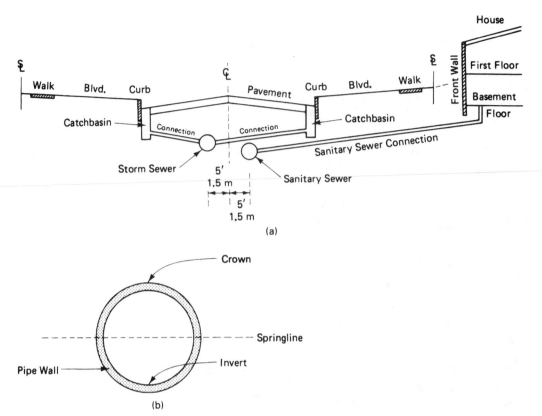

FIGURE 14.3 (a) Municipal road allowance showing typical service locations. (b) Sewer pipe section.

connections. The sanitary house connection is usually at a 2 percent (minimum) slope. If sanitary sewers are being added to an existing residential road, the preliminary survey must include the basement floor elevations. The floor elevations are determined by taking a rod reading on the window sill and then, after getting permission to enter the house, measuring from the window sill down to the basement floor. As a result of deep basements and long setbacks from the ₵, sanitary sewers often have to be at least 9 ft (2.75 m) below ₵ grade.

The minimum depth of storm sewers below ₵ grade depends, in the southern United States, on the traffic loading and, in the northern United States and most of Canada, on the depth of frost penetration. The minimum depth of storm sewers ranges from 3 ft (1 m) in some areas in the south to 8 ft (2.5 m) in the north. The design of the inlets and catch basins depends on the depth of the sewer and the quality of the effluent.

The minimum slope for storm sewers is usually 0.50 percent, whereas the minimum slope for sanitary sewers is often set at 0.67 percent. In either case, designers try to achieve self-cleaning velocity (i.e., a minimum of 2.5 to 3 ft/s or 0.8 to 0.9 m/s) to avoid excessive sewer maintenance costs. Manholes (MHs) are located at each change in direction, slope, or pipe size. In addition, manholes are located at 300- to 450-ft (100- to 140-m) maximum intervals.

Catch basins are usually located at 300-ft (100-m) maximum intervals. They are also located at the high side of intersections and at all low points. The 300-ft (100-m) maximum interval is reduced as the slope on the road increases.

For construction purposes, sewer layout is considered only from one manhole to the next. The stationing (0 + 00) commences at the first (existing) manhole (or outlet) and proceeds upstream only to the next manhole. If a second leg is also to be constructed, station 0 + 00 is assigned to the downstream manhole and proceeds upstream only to the next manhole. A unique manhole number (e.g., 1, 2, 3, . . . , 1A, 2A, 3A, . . .) describes each manhole. This type of numbering system avoids confusion with the stations for extensive sewer projects. Figure 14.3(b) shows a section of sewer pipe. The **invert** is the inside bottom of the pipe. The invert grade is the controlling grade for construction and design. The sewer pipes may consist of vitrified clay, steel, some of the newer plastics, or (as is usually the case) concrete.

The pipe wall thickness depends on the diameter of the pipe. For storm sewers, 12 in. (300 mm) is usually taken as the minimum diameter. The **springline** of the pipe is at the halfway mark, and connections are made above this reference line. The **crown** is the outside top of the pipe. Although the term *crown* is relatively unimportant (sewer cover is measured to the crown) for sewer construction, it is important for pipeline (pressurized pipes) construction because it gives the controlling grade for that type of construction.

14.3   Layout for Line and Grade

As in other construction work, offset stakes are used to provide line and grade for the construction of sewers. Grade can also be defined by the use of in-trench or surface lasers. Before deciding on the offset location, it is wise to discuss the matter with the contractor. The contractor will be excavating the trench and casting the material to one side or loading it into trucks for removal from the site. Also, the sewer pipe will be delivered to the site and positioned conveniently alongside its future location. The position of the offset stakes should not interfere with either of these operations.

The surveyor will position the offset line as close to the pipe centerline as possible, but seldom is it possible to locate the offset line closer than 15 ft (5 m) away. Station 0 + 00 is assigned to the downstream manhole or outlet, the chainage proceeding upstream to the next manhole. The centerline of construction is laid out, with stakes marking the location of the two terminal points of the sewer leg. The surveyor should use survey techniques giving 1:3,000 accuracy as a minimum for most sewer projects. Large-diameter (6-ft or 2-m) sewers require increased precision and accuracy.

The two terminal points on the ₵ are occupied by total station or theodolite, and right angles are turned to locate the terminal points precisely at the assigned offset distance. Station 0 + 00 on offset is occupied by a total station or theodolite, and a sight is taken on the other terminal offset point. Stakes are then located at 50-ft (20-m) intervals. Checking in at the terminal point verifies accuracy.

The tops of the offset stakes are surveyed and their elevations are determined, with the surveyor taking care to see that his or her leveling is accurate. The existing invert elevation of MH 3, shown on the contract plan and profile in Figure 14.4, is verified at the same time. The surveyor next calculates the sewer invert elevations for the 50-ft (20-m) stations. He or she then prepares a grade sheet showing the stations, stake elevations, invert grades, and cuts. The following cases illustrate the techniques used.

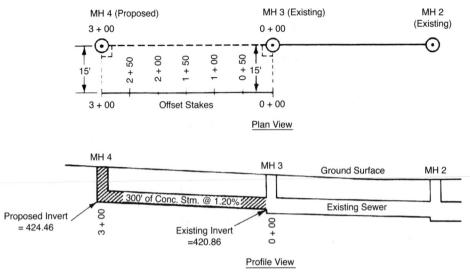

FIGURE 14.4 Plan and profile of a proposed sewer (foot units).

Table 14.1 SEWER GRADE SHEET: FOOT UNITS*

Station	Invert Elevation	Stake Elevation	Cut (ft)	Cut
MH 3 0 + 00	420.86	429.27	8.41	8′4⁷⁄₈″
0 + 50	421.46	429.90	8.44	8′5¹⁄₄″
1 + 00	422.06	430.41	8.35	8′4¹⁄₄″
1 + 50	422.66	430.98	8.32	8′3⁷⁄₈″
2 + 00	423.26	431.72	8.46	8′5¹⁄₂″
2 + 50	423.86	431.82	7.96	7′11¹⁄₂″
MH 4 3 + 00	424.46	432.56	8.10	8′1¹⁄₄″

*Refer to Table 13.3 for foot-inch conversion.

Assume that an existing sewer is to be extended from existing MH 3 to proposed MH 4 (see Figure 14.4). The horizontal alignment will be a straight-line production of the sewer leg from MH 2 to MH 3. The vertical alignment is taken from the contract plan and profile (Figure 14.4). The straight line is produced by setting up the total station or theodolite at MH 3, sighting MH 2, and double centering to the location of MH 4. The layout then proceeds as described previously. The stake elevations, as determined by differential leveling, are shown in Table 14.1.

At station 1 + 50, the cut is 8 ft 3⁷⁄₈ in. (see Figure 14.5). To set a cross-trench batter board at the next even foot, measure up 0 ft 8¹⁄₈ in. to the top of the batter board. The offset distance of 15 ft can be measured and marked at the top of the batter board over the pipe ℄ and a distance of 9 ft measured down to establish the invert elevation. This even foot measurement from the top of the batter board to the invert is known as the grade rod distance. A value can be picked for the grade rod so that it is the same value at each station. In this example, 9 ft

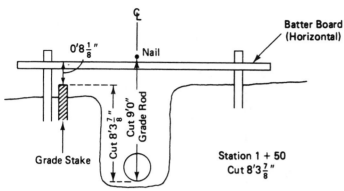

FIGURE 14.5 Use of cross-trench batter boards.

appears to be suitable for each station because it is larger than the largest cut. The arithmetic shown for station 1 + 50 is performed at each station so that the batter boards can be set at 50-ft intervals. The grade rod (a 2 in. by 2 in. length of lumber held by a worker in the trench) has a foot piece attached to the bottom at a right angle to the rod so that the foot piece can be inserted into the pipe. The grade rod foot piece allows measurement precisely from the invert.

This method of line-and-grade transfer has been shown first because of its simplicity; it has not been widely used in the field in recent years, however. With the introduction of larger and faster trenching equipment, which can dig deeper and wider trenches, this method would only slow the work down because it involves batter boards spanning the trench at 50-ft intervals. Many grade transfers are now accomplished by freestanding off-set batter boards or laser alignment devices. Using the data from the previous case, we can illustrate the technique of freestanding batter boards (see Figure 14.6).

The batter boards (3 ft or 1 m wide) are erected at each grade stake. As in the previous case, the batter boards are set at a height that will result in a grade rod that is an even number of feet (decimeters) long. With this technique, however, the grade rod distance will be longer because the top of the batter board should be at a comfortable eye height for the inspector. The works inspector usually checks the work while standing at the lower chainage stakes and sighting forward to the higher chainage stakes. The line of sight over the batter boards is a straight line parallel to the invert profile, and in Figure 14.6, the line of sight over the batter boards is rising at 1.20 percent.

As the inspector sights over the batter boards, he or she can include the top of the grade rod, which is being held on the most recently installed pipe length, in the field of view. The top of the grade rod has a horizontal board attached to it in a similar fashion to the batter boards. The inspector can determine visually whether the line over the batter boards and the line over the grade rod are in the same plane. The worker in the trench makes the necessary adjustment up or down, if one is required, and has the work rechecked. Grades can be checked to the closest $\frac{1}{4}$ in. (6 mm) in this manner. The preceding case is now worked out using a grade rod of 14 ft (see Table 14.2).

The grade rod of 14 ft requires an eye height of 5 ft $7\frac{1}{8}$ in. at 0 + 00. If this value is considered too high, a grade rod of 13 ft can be used, which results in an eye height of 4 ft $7\frac{1}{8}$ in. at the first batter board. The grade rod height is chosen to suit the needs of the inspector.

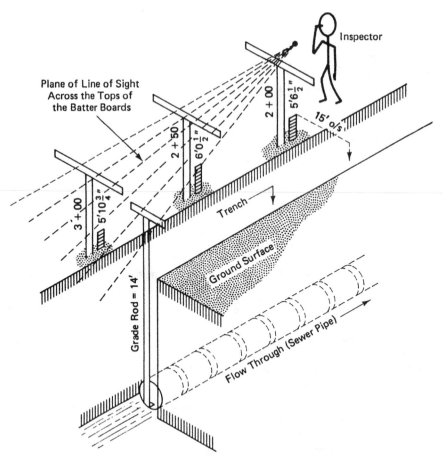

FIGURE 14.6 Freestanding batter boards.

Table 14.2 SEWER GRADE SHEET—WITH BATTER BOARDS (GRADE ROD = 14 FT)

Station	Invert Elevation	Stake Elevation	Cut	Stake to Batter Board	Stake to Batter Board
MH 3 0 + 00	420.86	429.27	8.41	5.59	5'7 1/8"
0 + 50	421.46	429.90	8.44	5.56	5'7 3/4"
1 + 00	422.06	430.41	8.35	5.65	5'7 3/4"
1 + 50	422.66	430.98	8.32	5.68	5'8 1/8"
2 + 00	423.26	431.72	8.46	5.54	5'6 1/2"
2 + 50	423.86	431.82	7.96	6.04	6'0 1/2"
MH 4 3 + 00	424.46	432.56	8.10	5.90	5'10 3/4"

In some cases, an additional station is put in before 0 + 00 (i.e., 0 − 50). The grade stake and batter board refer to the theoretical pipe ₵ and invert profile produced back through the first manhole. This batter board lines up, of course, with all the others and will be useful in checking the grade of the first few pipe lengths placed. Pipes are not installed unless a minimum of three batter boards (also known as sight rails) can be viewed simultaneously. Note that many agencies use 25-ft (10-m) stations, rather than the 50-ft (20-m) stations used in this case. The smaller intervals allow for much better grade control.

One distinct advantage to the use of batter boards in construction work is that an immediate check is available on all the survey work involved in the layout. The line of sight over the tops of the batter boards (which is actually a vertical offset line) must be a straight line. If, upon completion of the batter boards, all the boards do not line up precisely, it is obvious that a mistake has been made. The surveyor checks the work by first verifying all grade computations and then by releveling the tops of the grade stakes. Once the boards are in alignment, the surveyor can move on to other projects.

When the layout is performed in foot units, the basic station interval is 50 ft. The dimensions are recorded and calculated to the closest one-hundredth (0.01) of a foot. Although all survey measurements are in feet and decimals of a foot, for the contractor's purposes, the final cuts and fills are often expressed in feet and inches. The decimal-inch relationships are soon committed to memory by surveyors working in the construction field (see Table 13.3). Cuts and fills are usually expressed to the closest $1/8$ in. for concrete, steel, and pipelines and to the closest $1/4$ in. for highway granular surfaces and ditch lines.

Figure 14.7 includes a sketched plan and profile, design slope and inverts, and assumed elevations for the tops of the grade stakes. Also shown are the cuts and stake–to–batter board distances, based on a 5.0-m grade rod. All values are in metric (SI) units. Figures 14.8 and 14.9 show the installation of steel and concrete sewer pipes.

■ EXAMPLE 14.1 *Sewer Grade Sheet*

Figure 14.10 is a reproduction of a plan and profile sheet from a book of contract documents. The figure shows plan and profile data for a new storm sewer that terminates in a cross culvert at MH 44. The sewer runs up from the culvert at $418 + 50$ through MHs 42, 40, and 39. As the sewer proceeds downstream toward the culvert outlet, the pipes must be of increasingly larger diameter to accommodate the cumulative flow being collected by each additional leg of the sewer.

Figures 14.10 and 14.11 show that the sewer flowline drops as it passes through each manhole; that is, the west invert in each case is at a higher elevation than is the east invert. This flowline drop at each manhole provides additional head, which is utilized to overcome flow losses caused by turbulence in manholes.

When the sewer pipes increase in diameter, and as the sewer proceeds through a manhole, the required flowline drop is achieved by simply keeping the crowns of the incoming and outgoing pipes at the same elevation. That is, at MH 40, the incoming pipe's diameter is 27 in., whereas the outgoing pipe's diameter is 30 in. If the crowns are kept at the same elevation, the flowline inverts drop 3 in., or 0.25 ft.

Solution

To compute the sewer grade sheets, the surveyor can summarize pertinent plan and profile data from the contract document, as in Figure 14.10, onto a layout sketch, as shown in Figure 14.11. In this example, it is assumed that three legs of the sewer are being constructed: MH 44 (culvert) to MH 42, MH 42 to MH 40, and MH 40 to MH 39.

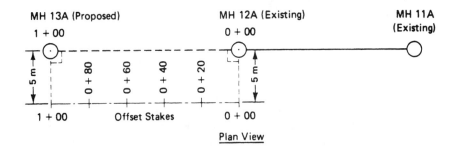

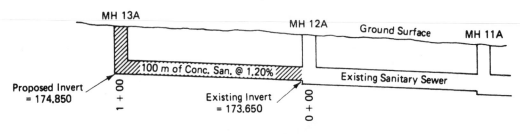

Station	Invert Elev.	Stake Elev. *	Cut	Stake to Batter Board, GR = 5.0 m
MH 12A 0 + 00	173.650	177.265	3.615	1.385
0 + 20	173.890	177.865	3.975	1.025
0 + 40	174.130	177.200	3.070	1.930
0 + 60	174.370	178.200	3.830	1.170
0 + 80	174.610	178.005	3.395	1.605
MH 13A 1 + 00	174.850	178.500	3.650	1.350

*Stake elevations and computations are normally carried out to the closest 5 mm.

FIGURE 14.7 Sewer construction example using metric units. An existing sanitary sewer is being extended from MH 12A to MH 13A. Five-meter offset (o/s) stakes were surveyed, with the resultant stake elevations shown in the Stake Elev. column.

The field surveyor first lays out the proposed sewer line manholes (42, 40, and 39). By setting up at each manhole stake, turning off 90°, and then measuring out the appropriate offset distance, the manholes are located on offset. The offset sewer line can then be staked out at 50-ft stations, from one manhole to the next. The elevations of the tops of the offset stakes are then surveyed with a rod and level. If possible, the surveyor checks into a second benchmark to ensure the accuracy of the work. In this example, the elevations of the tops of the stakes are assumed and are shown in Table 14.3.

The surveyor next calculates the pipe invert elevation at each station that has been staked out. Station 0 + 00 is taken to be the lower manhole station, with the chainage proceeding to the upper manhole of that leg. For the second leg, 0 + 00 is taken to be the station of the lower manhole, and once again the chainage proceeds upstream to the next manhole. This procedure is repeated for the length of the sewer.

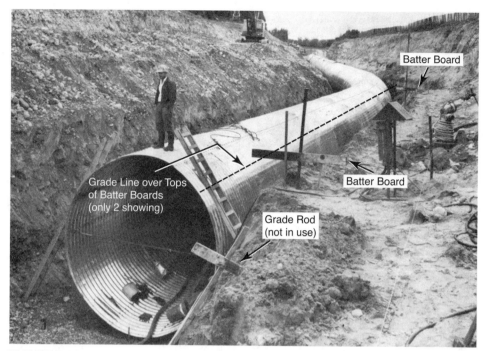

FIGURE 14.8 Corrugated steel sewer pipe, showing both tangent and curved sections. Only two batter boards are visible in this view, along with the grade rod leaning against the side of the trench. (Courtesy of Corrugated Steel Pipe Institute)

FIGURE 14.9 John Deere 792 Excavator shown installing concrete sewer pipe. (Courtesy of John Deere, Moline, Illinois)

Sewers are always constructed working upstream from the outlet (culvert, in this case) so that the trench can be drained during all phases of construction.

Table 14.3 shows the computed invert grades, the assumed stake elevations, and the resultant cut values. Using the technique of freestanding batter boards, as depicted

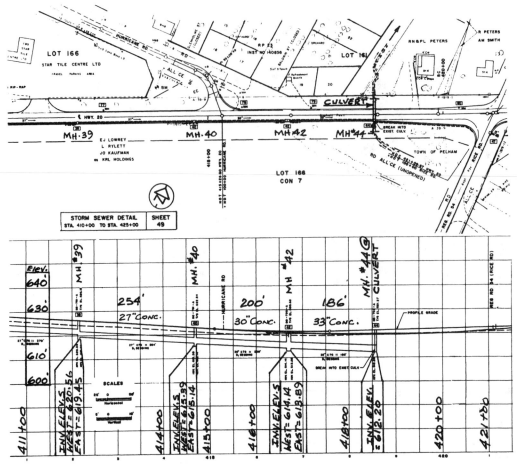

FIGURE 14.10 Plan and profile of Highway 20 (adapted).

in Figure 14.6, a grade rod is selected for each leg of the sewer that will give the works inspector a comfortable eye height—12 ft for the first two legs and 13 ft for the third leg. The stake–to–batter board dimension is computed for each station and then converted to feet and inches for use by the contractor.

14.3.1 Laser Alignment

Laser devices are widely used in most forms of construction work. Lasers are normally used in a fixed direction and slope mode or in a revolving horizontal pattern. One such device, shown in Figure 14.12, can be mounted in a sewer manhole, aligned for direction and slope, and used with a target for the laying of sewer pipe. The laser beam can be deflected by dust or high humidity, so care must be taken to overcome these factors. Blowers that remove the dusty or humid air can accompany lasers used in sewer trenches.

A rotating laser can also be used above grade with a signal-sensing target rod (the infrared signal sensor is not required when working with visible beam lasers). Working above

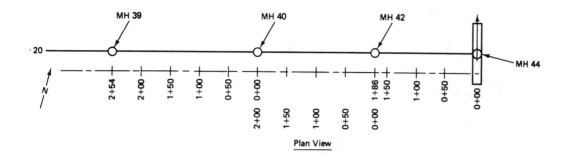

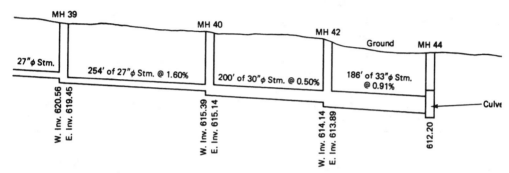

Plan View

FIGURE 14.11 Layout sketch for sewer construction. Adapted from plan and profile of Highway 20 (slope percentages computed from given inverts and distances).

Table 14.3 GRADE SHEETS FOR SEWER CONSTRUCTION (SEE FIGURES 14.10 AND 14.11)

Station	Pipe Elevation	Stake Elevation (Assumed)	Cut	Stake to Batter Board (ft)	Stake to Batter Board (ft and in.)	
MH 44 0 + 00	612.20	618.71	6.51	5.49	5'5 $\frac{7}{8}$''	
0 + 50	612.65	618.98	6.33	5.67	5' 8"	Grade
1 + 00	613.11	619.40	6.29	5.71	5'8 $\frac{1}{2}$"	rod
1 + 50	613.56	619.87	6.31	5.69	5'8 $\frac{1}{4}$"	= 12 ft
MH 42 1 + 86	613.89	620.33	6.44	5.55	5'6 $\frac{5}{8}$"	
MH 42 0 + 00	614.14	620.33	6.19	5.81	5'9 $\frac{3}{4}$"	
0 + 50	614.39	620.60	6.21	5.79	5'9 $\frac{1}{2}$"	Grade
1 + 00	614.64	620.91	6.27	5.73	5'8 $\frac{3}{4}$"	rod
1 + 50	614.89	621.63	6.74	5.26	5'3 $\frac{1}{8}$"	= 12 ft
MH 40 2 + 00	615.14	622.60	7.46	4.54	4'6 $\frac{1}{2}$"	
MH 40 0 + 00	615.39	622.60	7.21	5.79	5'9 $\frac{1}{2}$"	
0 + 50	616.19	623.81	7.62	5.38	5'4 $\frac{5}{8}$"	Grade
1 + 00	616.99	624.57	7.58	5.42	5'5"	rod
1 + 50	617.79	625.79	8.00	5.00	5'0"	= 13 ft
2 + 00	618.59	626.93	8.34	4.66	4'7 $\frac{7}{8}$"	
MH 39 2 + 54	619.45	627.55	8.10	4.90	4'10 $\frac{7}{8}$"	

FIGURE 14.12 (a) Small diameter pipe application. (b) Larger diameter pipe application. This pipeline laser features self-leveling and warnings when the instrument goes off-level or off-alignment; a gradient setting of +0.5% is shown in Figure 14.12(a). (Courtesy of Trimble)

ground not only eliminates the humidity factor but also allows for more accurate and quicker vertical alignment. These devices allow for setting slope within the limits of $-10°$ to $30°$. Some devices have automatic shutoff capabilities when the device is disturbed from its desired setting. Rotating lasers are used above ground as near as possible to the trench works. The slope of the sewer can be entered into the laser, and a rod reading can be taken at the existing manhole invert. That rod reading is then the controlling grade as the work proceeds upstream to the sewer leg terminal point (e.g., the next manhole).

14.4 Catch-Basin Construction Layout

Catch basins are constructed along with the storm sewer or at a later date, just prior to curb construction. The catch basin (CB) is usually located by two grade stakes—one on each side of the CB. The two stakes are on the curb line and are usually 5 ft (2 m) from the center of the catch basin. The cut or fill grade is referenced to the CB grate elevation at the curb face. The ₵ pavement elevation is calculated, and the crown height is subtracted from it to arrive at the top of the grate elevation (see Figures 14.13 and 14.14).

At low points, particularly at vertical curve low points, it is usual practice for the surveyor to lower the CB grate elevation arbitrarily to ensure that ponding does not occur on either side of the completed catch basin. We noted in Chapter 11 that the longitudinal slope at vertical curve low points is almost flat for a significant distance. The CB grate elevation can be arbitrarily lowered as much as 1 in. (25 mm) to ensure that the gutter drainage goes directly into the catch basin without ponding. The catch basin (which can be made of concrete poured in place but is more often prefabricated and delivered to the job site) is set below finished grade until the curbs are constructed. At the time of curb construction, the

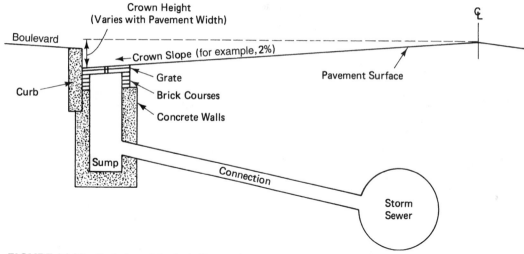

FIGURE 14.13 Typical catch basin (with sump).

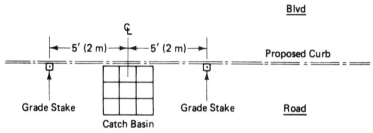

FIGURE 14.14 Catch basin layout, plan view.

finished grade for the grate is achieved by adding one or more courses of brick or concrete shim collars laid on top of the concrete walls.

14.5 Tunnel Construction

Tunnels are used in road, sewer, and pipeline construction when the cost of working at or near the ground surface becomes prohibitive. For example, sewers are tunneled when they must be at a depth that would make open cut too expensive (or not feasible operationally), or sewers may be tunneled to avoid disruption of services on the surface (as would occur if an open cut were put through a busy expressway). Roads and railroads are tunneled through large hills and mountains to maintain optimal grade lines. Control surveys for tunnel layouts are performed on the surface and join the terminal points of the tunnel. These control surveys use triangulation, precise traverse survey methods, or global positioning system (GPS) techniques and allow for the computation of coordinates for all key points (see Figure 14.15).

In the case of highway (railway) tunnels, the ℄ can be run directly into the tunnel and is usually located on the roof either at the ℄ or at a convenient offset (see Figure 14.16). If the tunnel is long, intermediate shafts can be sunk to provide access for materials, ventilation, and alignment verification. Conventional engineering theodolites are illustrated in

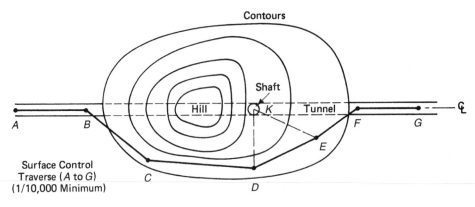

Contours

Shaft

Hill

Tunnel

A   B   F   G   ₵

Surface Control
Traverse (A to G)
(1/10,000 Minimum)   C   D   E

Direction and Length of B–F, D–K, and E–K
can be Computed from Traverse Data

(a)

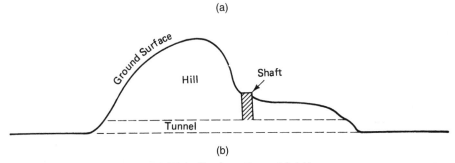

Ground Surface

Hill

Shaft

Tunnel

(b)

FIGURE 14.15   (a) Plan view and (b) profile view of tunnel location.

Figure 14.16; for cramped quarters, a suspension theodolite (Figure 14.17) can be used. Levels can also be run directly into the tunnel, and temporary benchmarks are established in the floor or roof of the tunnel. In the case of long tunnels, work can proceed from both ends and meet somewhere near the middle. Constant vigilance with respect to errors and mistakes is of prime importance.

In the case of a deep sewer tunnel, mining surveying techniques must be employed to establish line and grade (Figure 14.18). The surface ₵ projection AB is established carefully on beams overhanging the shaft opening. Plumb lines (piano wire) are hung down the shaft, and overaligning the total station or theodolite to the set points in the tunnel develops the tunnel ₵. A great deal of care is required in overaligning because this very short backsight will be produced over relatively long distances, thus magnifying any sighting errors.

The plumb lines usually employ heavy plumb bobs (capable of taking additional weights if required). Sometimes the plumb bobs are submerged in heavy oil to dampen the swing oscillations. If the plumb-line swing oscillations cannot be eliminated, the oscillations must be measured and then averaged. Some tunnels are pressurized to control groundwater seepage. The air locks associated with pressure systems cut down considerably on the clear dimensions in the shaft, making the plumbed-line transfer even more difficult.

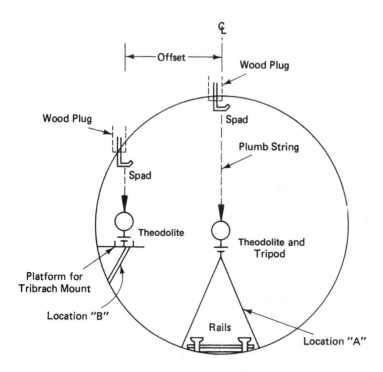

FIGURE 14.16 Establishing line in a tunnel.

Tunnel ₵ is Usually Located in the Roof (Location "A") and then can be
Offset (Location "B") to Provide Space for Excavation and Materials Movement.
Line and Grade can be Provided by a Single Laser Beam which has been
Oriented for Both Alignment and Slope.

Transferring the ₵ from the surface to underground locations by use of plumb lines is
an effective—although outdated—technique. Modern survey practice favors the use of precise optical plummets (see Figure 14.19) to accomplish the line transfer. These plummets are
designed for use in zenith or nadir directions, or in both zenith and nadir directions (as illustrated in Figure 14.19). The accuracy of this technique can be as high as 1 or 2 mm in 100 m.

Gyrotheodolites have also been used successfully for underground alignment control.
Several surveying equipment manufacturers produce gyro attachments for use with repeating
theodolites. Figure 14.20 shows a gyro attachment mounted on a 20-second theodolite. The
gyro attachment (also called a gyrocompass) consists of a wire-hung pendulum supporting a
high-speed, perfectly balanced gyromotor capable of attaining the required speed of 12,000 rpm
in 1 min. Basically the rotation of the earth affects the orientation of the spin axis of the
gyroscope so that the gyroscope spin axis orients itself toward the pole in an oscillating motion
that is observed and measured in a plane perpendicular to the pendulum. This north-seeking
oscillation, known as precession, is measured on the horizontal circle of the theodolite. Extreme left (west) and right (east) readings are averaged to arrive at the meridian direction.

The theodolite with a gyro attachment is set up and oriented approximately to north
through the use of a compass. The gyromotor is engaged until the proper angular velocity
has been reached (about 12,000 rpm for the instrument shown in Figure 14.20), and then

FIGURE 14.17 Breithaupt mining suspension theodolite. It can be used with a tripod or suspended from a steel punch in the ceilings of cramped galleries or drifts. (Courtesy of Keuffel & Esser Co.)

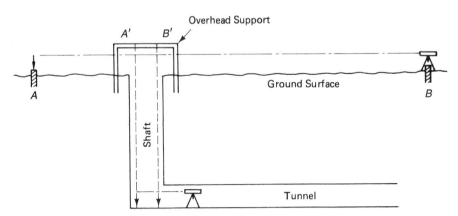

Tunnel ℄ AB is Carefully Marked on the Overhead Support at A′ and B′. These Two Marks are Set as Far Apart as Possible for Plumbing into the Shaft.

The theodolite in the Tunnel Over-aligns the Two Plumb Lines (Trial and Error Technique); ℄ is then Produced Forward using Double-Centering Techniques.

Precision can be Improved by Repeating this Process when the Tunnel Excavation has Progressed to the Point where a much Longer Backsight is Possible.

FIGURE 14.18 Transfer of surface alignment to the tunnel.

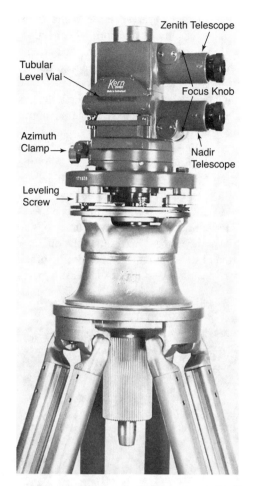

Zenith Telescope

Tubular
Level Vial

Kern

Focus Knob

Azimuth
Clamp

Nadir
Telescope

Leveling
Screw

FIGURE 14.19 Kern OL precise optical plummet. SE in 100 m for a single measurement (zenith or nadir) = ±1 mm (using a coincidence level). Used in high-rise construction, towers, shafts, and the like. (Courtesy of Leica Geosystems Inc.)

FIGURE 14.20 Gyro attachment, mounted on a 20-second theodolite. Shown with battery charger and control unit. (Courtesy of SOKKIA Corp.)

the gyroscope is released. The precession oscillations are observed through the gyro-attachment viewing eyepiece, and the theodolite is adjusted closer to the northerly direction, if necessary. When the theodolite is pointed to within a few minutes of north, the extreme precession positions (west and east) are noted in the viewing eyepiece and then recorded on the horizontal circle. As we noted earlier, the position of the meridian is the value of the averaged precession readings. This technique, which takes about a half hour to complete, is accurate to within 20″ of azimuth. These instruments can be used in most tunneling applications, where tolerances of 25 mm are common for both line and grade.

Lasers have been used for a wide variety of tunneling projects. Figure 14.21 shows a large-diameter boring machine that is kept aligned (both line and grade) by keeping the laser beam centered in the two targets mounted near the front and rear of the machine.

The techniques of prismless EDM (Sections 2.22) are used in the automated profile scanner shown in Figure 14.22. This system uses a Wild DIOR 3001 EDM that can measure distances from 0.3 m to 50 m (without using a reflecting prism) with an accuracy of 5 to 10 mm. An attached laser is used to mark the feature being measured so that the operator can verify the work. The profiler is set up at key tunnel stations so that a 360° profile of the tunnel can be measured and recorded on a memory card. The number of measurements

FIGURE 14.21 Laser-guided tunnel boring machine. (Courtesy of The Robbins Company, Solon, Ohio)

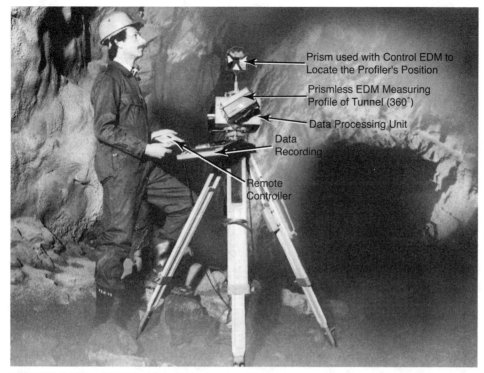

FIGURE 14.22 A.M.T. Profiler 2000 prismless EDM, used to measure tunnel profile at selected stations. (Courtesy of Amberg Measuring Technique)

taken as the profiler revolves through 360° can be preset in the profiler, or the remote controller can control it manually. The data on the memory card are then transferred to a microcomputer for processing. Figure 14.23 shows the station plot along with theoretical and excavated profiles. In addition to driving the digital plotter, the system software computes the area at each station and then computes the excavated volumes by averaging two adjacent areas and multiplying by the distance between them. (See Chapter 17 for these computational techniques.)

Problems

14.1. A storm sewer is to be constructed from existing MH 8 (invert elevation = 360.21) at + 1.32 percent for a distance of 240 ft to proposed MH 9. The elevations of the offset grade stakes are as follows: 0 + 00 = 368.75, 0 + 50 = 368.81, 1 + 00 = 369.00, 1 + 50 = 369.77, 2 + 00 = 370.22, and 2 + 40 = 371.91. Prepare a grade sheet like the one in Table 14.2 showing stake–to–batter board distances in ft and in. Use a 14-ft grade rod.

14.2. A sanitary sewer is to be constructed from existing MH 4 (invert elevation = 150.810) at +0.68 percent for a distance of 115 m to proposed MH 5. The elevations of the offset grade stakes are as follows: 0 + 00 = 152.933, 0 + 20 = 152.991, 0 + 40 = 153.626, 0 + 60 = 153.725, 0 + 80 = 153.888, 1 + 00 = 153.710, and 1 + 15 = 153.600. Prepare a grade sheet like the one in Figure 14.7 showing stake–to–batter board distances in meters. Use a 4-m grade rod.

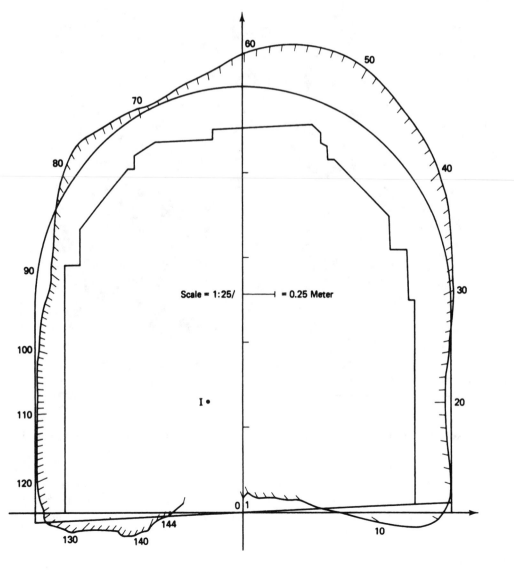

Scale = 1:25/ ⊢———⊣ = 0.25 Meter

RESULTS OF AREA COMPUTATION

Comp. Range	PT 1 − 144	
Range	Measure B1	Total B1+B2
Measured Area	24.17	24.59 m2
Theor. Area	22.18	22.71 m2
Overprofile	2.45	2.45 m2
Underprofile	0.46	0.57 m2

FIGURE 14.23 Computations and profile plot for profiler setup, where 144 prismless EDM readings were automatically taken and recorded. (Courtesy of Amberg Measuring Technique)

14.3. Refer to Figure 14.24 (plan and profile of Parkway Ave.). Compute the sewer invert elevations, at 50-ft stations, from MH 9 (0 + 05, ℄) to MH 1 (3 + 05, ℄). Here, ℄ refers to roadway centerline stationing (not the sewer).

14.4. Given the sewer grade stake elevations shown below, compute the cut distances at each 50-ft station for the section of sewer in Problem 14.3.

MH 9	0 + 00	503.37	1 + 50	504.09
	0 + 50	503.32	2 + 00	504.10
	1 + 00	503.61	2 + 50	504.77
			MH 1 3 + 00	504.83

14.5. Select a realistic grade rod, and prepare a grade sheet showing the stake–to–batter board dimensions both in feet and in feet and inches for the section of sewer in Problems 14.3 and 14.4.

14.6. Refer to Figure 14.24 (plan and profile of Parkway Ave.). For the sections of sewer from MH 1 (3 + 05, ℄) to MH 2 (5 + 05, ℄) and from MH 2 (5 + 05, ℄) to MH 3 (6 + 55, ℄), determine the following:

(a) The invert elevations of 50-ft stations.

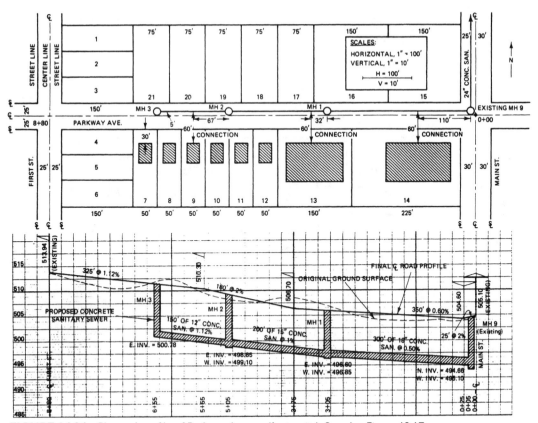

FIGURE 14.24 Plan and profile of Parkway Avenue (foot units). See also Figure 13.17.

(b) The cut distances from the top of the grade stake to invert at each 50-ft station.

(c) After selecting suitable grade rods, determine the stake–to–batter board distances at each stake.

Use the following grade stake elevations:

MH 1 0 + 00	504.83		MH 2 0 + 00	507.26
0 + 50	505.21		0 + 50	507.43
1 + 00	505.30		1 + 00	507.70
1 + 50	506.17		MH 3 1 + 50	507.75
MH 2 2 + 00	507.26			

14.7. Refer to Figure 14.25 (plan and profile of Oak Ave.). Compute the sewer invert elevations at 20-m stations from MH 13 (0 + 05, ₵) to MH 1 (1 + 05, ₵) and from MH 1 (1 + 05, ₵) to MH 2 (1 + 85, ₵).

14.8. Given the grade stake elevations shown below, compute the cut distances from stake to invert at each stake for both sections of sewer from Problem 14.7.

MH 13 0 + 00	186.713		MH 1 0 + 00	187.255
0 + 20	186.720		0 + 20	187.310
0 + 40	186.833		0 + 40	187.333

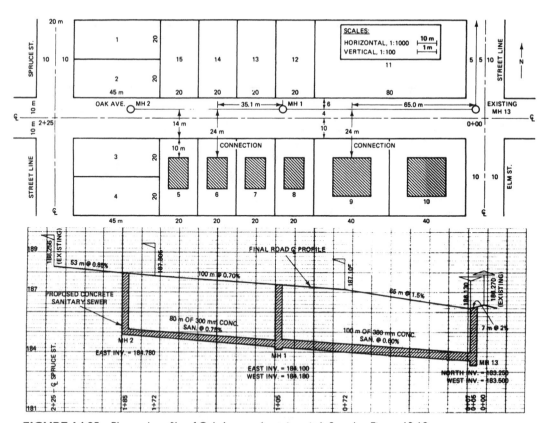

FIGURE 14.25 Plan and profile of Oak Avenue (metric units). See also Figure 13.18.

0 + 60	186.877		0 + 60	187.340
0 + 80	186.890	MH 2	0 + 80	187.625
MH 1 1 + 00	187.255			

14.9. Using the data in Problem 14.8, select suitable grade rods for both sewer legs, and compute the stake–to–batter board distance at each stake.

14.10. Refer to Section 14.2 and Figure 14.3 and 14.24. Determine the minimum invert elevations for the sanitary-sewer building connections at the front-wall building lines for Lots 9, 13, and 14. Use a minimum slope of 2 percent for sewer connection pipes.

14.11. Refer to Section 14.2 and Figures 14.3 and 14.25. Determine the minimum invert elevations for the sanitary-sewer building connections at the front-wall building lines for Lots 6 and 9. Use a minimum slope of 2 percent for sewer connection pipes.

14.12. Figure 14.26 shows typical excavation equipment that can be used in pipeline and other construction projects. Write a report describing the types of construction projects in which each of the equipment can be utilized effectively. Describe why some excavation equipment is well suited for some specific roles and not suitable at all for others. Data can be obtained from the library, trade and professional journals, equipment dealers or manufacturers, and construction companies.

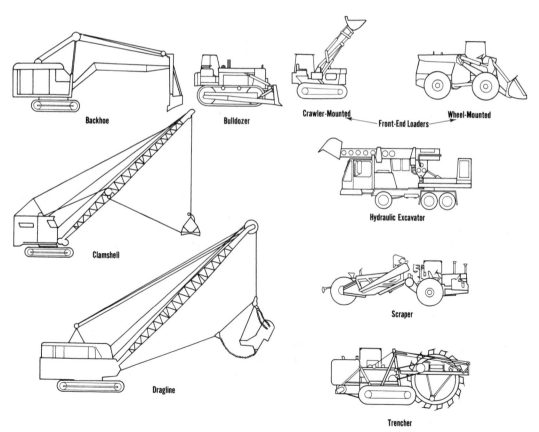

FIGURE 14.26 Typical excavation equipment. (Courtesy of American Concrete Pipe Association)

Chapter 15

Culvert and Bridge Construction Surveys

15.1 Culvert Construction

The plan location and invert grade of culverts are shown on the construction plan and profile. The intersection of the culvert ℄ and the highway ℄ is shown on the plan and is identified by its highway stationing (chainage). In addition, when the proposed culvert is not perpendicular to the highway ℄, the skew number or skew angle is shown (see Figure 15.1). The construction plan shows the culvert location ℄ chainage, skew number, and length of the culvert. The construction profile shows the inverts for each end of the culvert. One grade stake is placed on the ℄ of the culvert, offset a safe distance from each end (see Figure 15.2). The grade stake references the culvert ℄ and gives the cut or fill to the top of the footing for open footing culverts, to the top of the slab for concrete box culverts, or to the invert of the pipe for pipe culverts. If the culvert is long, intermediate grade stakes may be required. The stakes may be offset 6 ft (2 m) or longer distances if site conditions warrant. When concrete culverts are laid out, it is customary to place two offset line stakes to define the end of the culvert, in addition to placing the grade stakes at either end of the culvert. These end lines are normally parallel to the ℄ of construction or perpendicular to the culvert ℄. See Figure 15.3 for types of culverts.

15.2 Culvert Reconstruction

Intensive urban development creates an increase in impervious surfaces—for example, roads, walks, drives, parking lots, and roofs. Prior to development, rainfall had the opportunity to seep into the ground and eventually the water table, until the ground became saturated. After saturation, the rainfall ran off to the nearest watercourse, stream, river, or lake. The increase in impervious surfaces associated with urban development can result in a significant increase in surface runoff, which causes flooding where the culverts, and sometimes bridges, are now no longer capable of handling the increased flow. Some

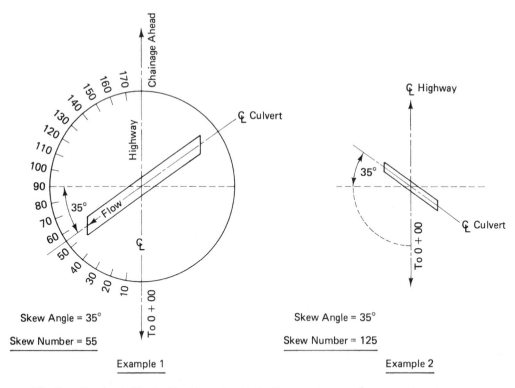

The Skew Number is Obtained by Measuring Clockwise to the Nearest 5°, the Angle Between the Back Tangent Ç of the Highway and the Ç of the Culvert.

FIGURE 15.1 Culvert skew numbers, showing the relationship between the skew angle and the skew number.

municipalities demand that developers provide detention and storage facilities (e.g., ponds and oversized pipes) to keep increases in site runoff to a minimum.

Figure 15.4 shows a concrete box culvert being added to an existing box culvert, resulting in what is called a twin-cell culvert. The proposed centerline grade for the new culvert will be the same as for the existing culvert. The outside edge of the concrete slab is laid out on close offset (o/s), with the construction grade information (cuts to floor slab elevation) referenced also from the o/s layout stakes. In addition, one edge (inside, in this case) of the wing wall footing is laid out on close o/s, with the alignment and grade information referenced to the same o/s stakes. Wing walls are used to retain earth embankments adjacent to the ends of the culvert.

Figure 15.5 shows a situation where road improvements require the replacement of a cross culvert. The new culvert is skewed at #60 to fit the natural stream orientation better, it is longer (140 ft) to accommodate the new road width, and it is larger to provide increased capacity for present and future developments. Figure 15.5 shows the plan and cross section of a detour that will permit the culvert to be constructed without closing the highway. (The detour centerline curve data were described in detail in Chapter 11).

The suggested staging for the construction is shown in eighteen steps. Essentially the traffic is kept to the east side of the highway while the west half of the old culvert is removed and the west half of the new culvert is constructed. Once the concrete in the west

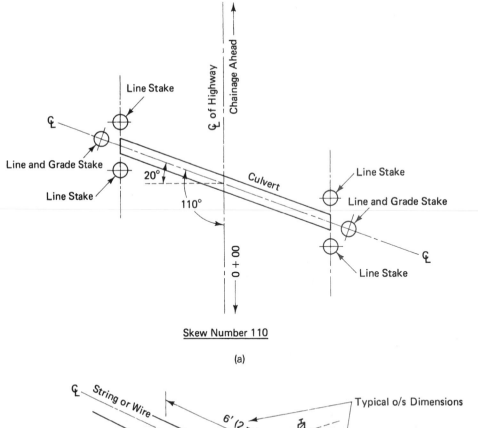

Line Stake

℄ of Highway

Chainage Ahead

Line Stake

Line and Grade Stake

20°

Line Stake

110°

Line and Grade Stake

℄

Line Stake

Culvert

℄

0 + 00

Line Stake

Skew Number 110

(a)

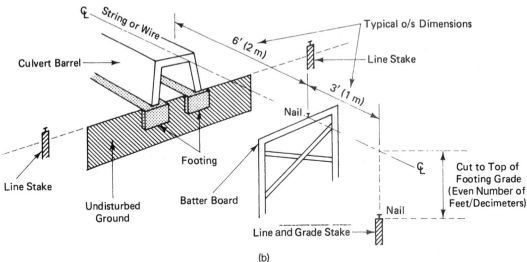

℄ String or Wire

Typical o/s Dimensions

Culvert Barrel

6' (2 m)

Line Stake

3' (1 m)

Nail

Line Stake

Footing

℄

Cut to Top of
Footing Grade
(Even Number of
Feet/Decimeters)

Batter Board

Undisturbed
Ground

Nail

Line and Grade Stake

(b)

FIGURE 15.2 Line and grade for culvert construction. (a) Plan view. (b) Perspective view. (Courtesy of the Ministry of Transportation, Ontario)

half of the new culvert has gained sufficient strength (usually in about thirty days), the detour can be constructed as shown in Figure 15.5. With the traffic now diverted to the detour, the east half of the old culvert is removed, and the east half of the new culvert is constructed. The improved road cross section can now be built over the completed culvert, and the detour and other temporary features can be removed.

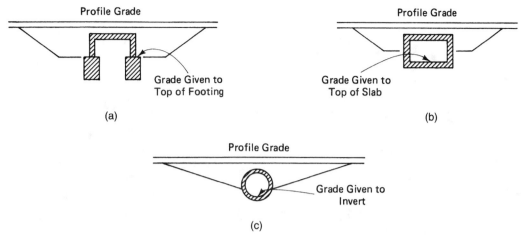

FIGURE 15.3 Types of culverts. (a) Open footing culvert. (b) Concrete box culvert. (c) Circular, arch, etc., culvert.

15.3 Bridge Construction: General Background

Accuracy requirements for structure construction are generally of the highest order for survey layouts. Included in this topic are bridges, elevated expressways, and so on. Accuracy requirements for structure layouts range from 1/3,000 for residential housing, to 1/5,000 for bridges and 1/10,000 for long-span bridges. The accuracy requirements depend on the complexity of the construction, the type of construction materials specified, and the ultimate design use of the facility. The accuracy required for any individual project can be specified in the contract documents, or it can be left to the common sense and experience of the surveyor.

Preliminary surveys for bridges include bore holes drilled for foundation investigation. Bridge designers indicate the location of a series of bore holes on a highway design plan. The surveyor locates the bore holes in the field by measuring centerline chainages, offsets, and ground elevations. As the bridge design progresses, the surveyor may have to go back to the site several times to establish horizontal and vertical control for additional bore holes, which may be required for final footing design. Figure 15.6 shows the plan and profile location for a series of bore holes at abutment, pier, and intermediate locations. In addition, the coordinates of each bore hole location are shown, permitting the surveyor to establish the field points by polar ties from coordinated monuments. The surveyor may also establish the points by the more traditional centerline chainage and offset measurements, as previously noted, or by GPS techniques. The establishment of permanent, well-referenced construction control (as outlined in Chapters 8 and 9) ensures that all aspects of the project—preliminary tie-ins and cross sections, bore holes, staged construction layouts, and final measurements—are all referenced to the same control net.

15.4 Contract Drawings

Contract drawings for bridge construction typically include general arrangement, foundation layout, abutment and pier details, beam details, deck details, and reinforcing steel

Existing Culvert

Telephone Cables
Temporarily Supported

Forms and Reinforcing Steel
for Outside Culvert Wall

Concrete Slab

Survey Layout Line
Outside of
Culvert Wall

Survey Layout Line
Inside Edge of
Retaining Wall Footing

FIGURE 15.4 Concrete culvert construction addition, showing floor slab, wing wall footing, and culvert walls—with reinforcing steel and forms.

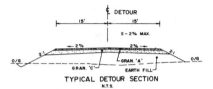

DETOUR & STAGING
FOR CULVERT @ STA. 62+26

CULVERT STATION 62+26

SUGGESTED CONSTRUCTION STAGING

1.   Construct road widening on east side of existing highway.
2.   Switch traffic to east half of highway.
3.   Place sheet piling roadway protection Stage I.
4.   Remove west end of existing concrete culvert.
5.   Place twin 30" C.S.P. culverts.
6.   Construct Stage I temporary ditch relocation and divert ditch.
7.   Construct west part of new concrete culvert.
8.   Place sheet piling road protection Stage II.
9.   Construct detour.
10. Switch traffic to detour.
11. Remove sheet piling roadway protection Stage I.
12. Construct Stage II temporary ditch relocation and divert ditch.
13. Remove remaining part of existing concrete culvert.
14. Construct east part of new concrete culvert.
15. Construct ditch relocation and divert ditch through new concrete culvert.
16. Construct east side highway embankment.
17. Shift traffic to east side of highway.
18. Remove detour, sheet piling roadway protection Stage II and C.S.P. culverts.

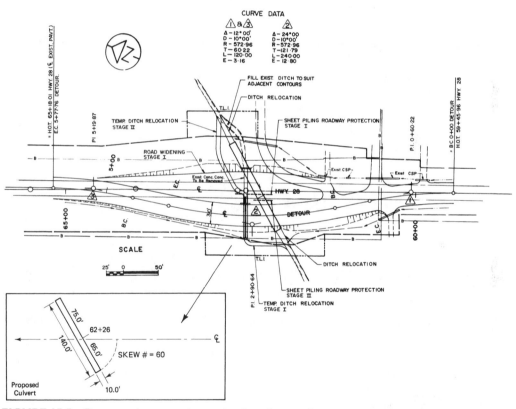

FIGURE 15.5 Detour and construction staging for culvert replacement and realignment.

Sec. 15.4   Contract Drawings

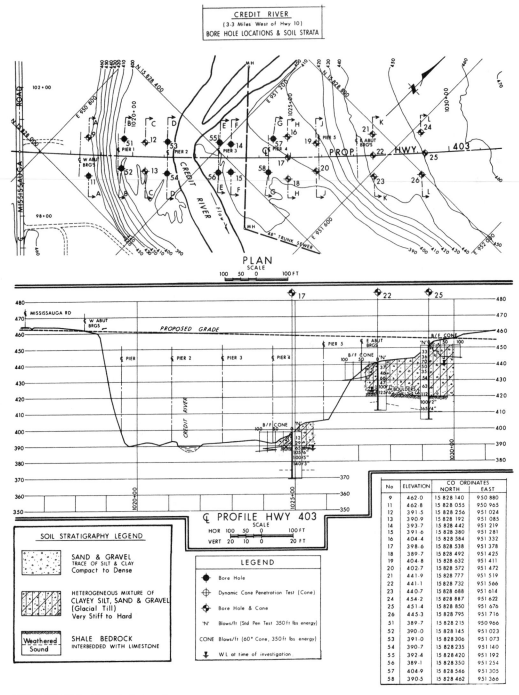

FIGURE 15.6 Bore hole locations.

schedules. These contract drawings can also include details on railings, wing walls, and the like. Whereas the contractor must utilize all drawings in the construction process, the construction surveyor is concerned only with those drawings that facilitate the construction layout. The entire layout can usually be accomplished by using only the general arrangement drawing (Figures 15.7 and 15.8) and the foundation layout drawing Figure 15.9. These plans are analyzed to determine which of the many dimensions shown are to be utilized for construction layout.

The key dimensions are placed on foundation layout sketches (Figures 15.10 and 15.11), which will be the basis for the layout. The sketches show the location of the piers and the abutment bearing $\mathbb{C}$, the location of each footing, and (in this case) the skew angle. Although most bridges have symmetrical foundation dimensions, the bridge in this case is asymmetrical because of the spiraled pavement requirements. This lack of symmetry does not pose any problems for the actual layout, but it does require additional work in establishing dimension check lines in the field.

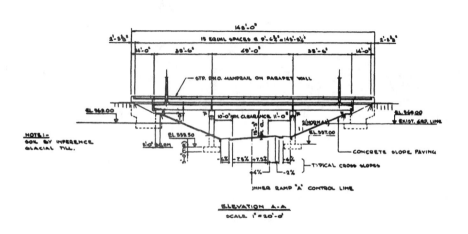

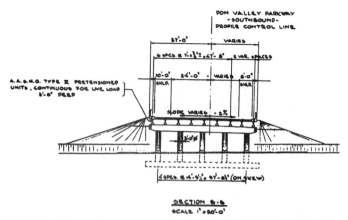

FIGURE 15.7 General arrangement plan for Bridge No. 5, Don Valley Parkway, and Highway 401 East off ramp.

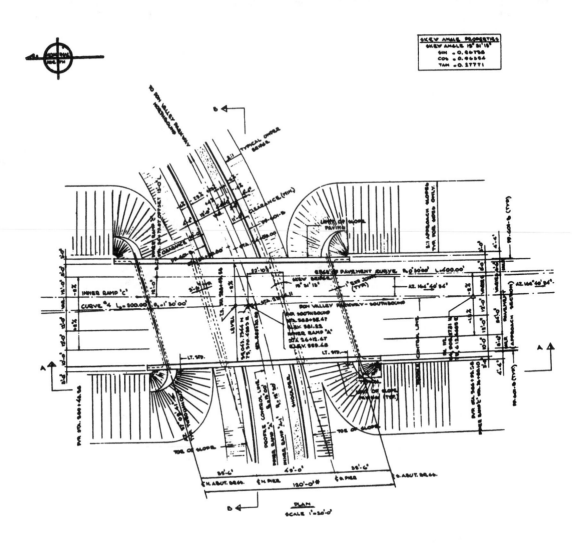

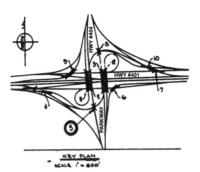

FIGURE 15.8 Key plan and general layout for Bridge No. 5, Don Valley Parkway, and Highway 401 East off ramp. *See same dimension, Figure 15.9.

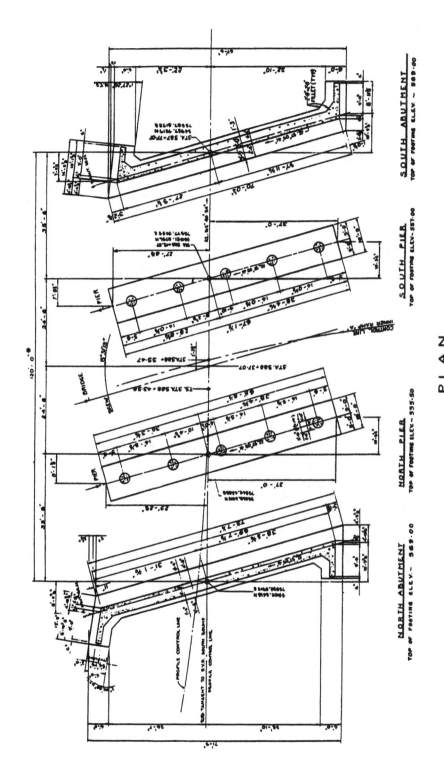

FIGURE 15.9 Foundation layout plan for bridge No. 5, Don Valley Parkway. *See same dimension, Figure 15.8.

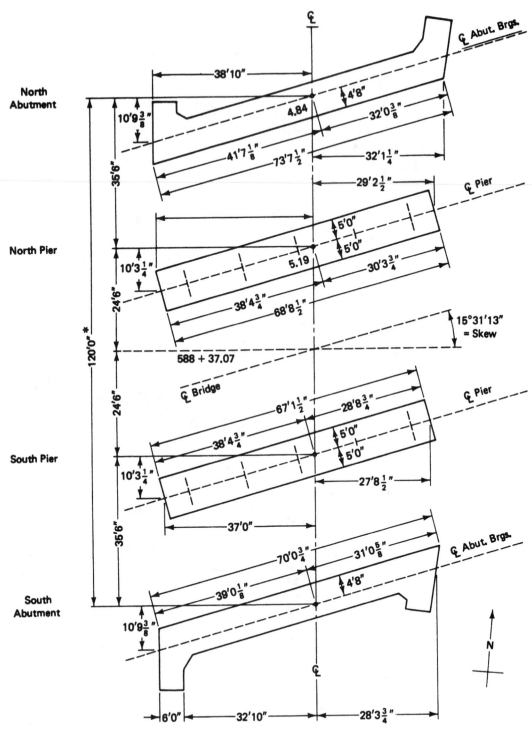

FIGURE 15.10 Foundation layout sketch. *See same dimension on Figures 15.8 and 15.9.

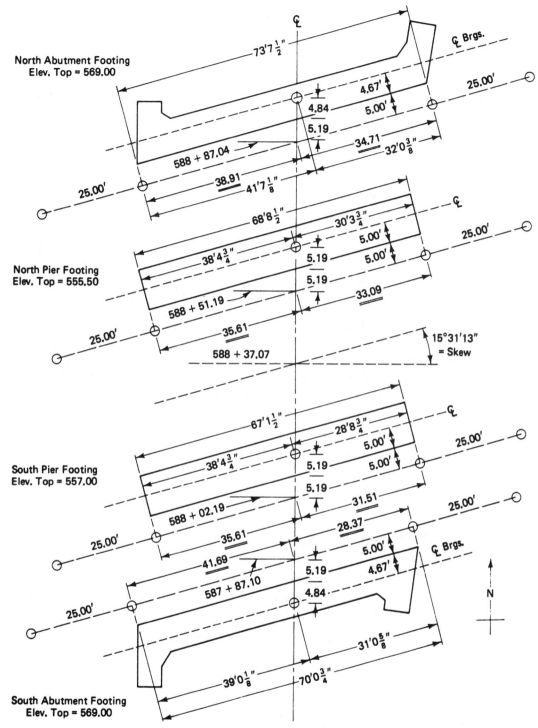

FIGURE 15.11 Foundation layout sketch and offsets sketch. Circles shown are ℄ and o/s layout bars.

15.5 Layout Computations

The following information shows the computations used in establishing the key bridge component stations. See Figures 15.9 and 15.10.

Chainage at ₵ of bridge	588 + 37.07	Chainage at ₵ of bridge	588 + 37.07
	+ 24.50		− 24.50
₵ north pier	588 + 61.57	₵ south pier	588 + 12.57
	+ 35.50		− 35.50
₵ north abut. brg.	588 + 97.07	₵ south abut. brg.	587 + 77.07

$$(588 + 97.07) - (587 + 77.07) = 120.00 \text{ ft} \quad \text{Check}$$

For this case, assume that the offset lines are 5 ft south from the south footing faces of the north abutment and both piers, and 5 ft north from the north face of the south abutment. The offset stakes (1-in. reinforcing steel, 2 to 4 ft long) are placed opposite the ends of the footings and an additional 25 ft away on each offset line. The stakes can be placed by using right-angle ties from ₵ or by locating the offset line parallel to the footings. For the latter option, it is necessary to compute the ₵ stations for each offset line. You can see in Figure 15.10 that the distance from the abutment ₵ of the bearings to the face of the footing is 4 ft 8 in. (4.67 ft). Along the N–S bridge ₵, this dimension becomes 4.84 ft (4.67 × sec 15°31′13″). Similarly the dimension from the pier ₵ to the face of the pier footing is 5.00 ft; along the N–S bridge ₵, this dimension becomes 5.19 ft (5.00 × sec 15°31′13″). Along the N–S bridge ₵, the 5-ft offset distance also becomes 5.19 ft. Accordingly, the stations for the offset lines at the N–S bridge ₵ are as follows (refer to Figure 15.11):

$$\text{North abutment offset} = 588 + 97.07 - 10.03 = 588 + 87.04$$
$$\text{North pier offset} = 588 + 61.57 - 10.38 = 588 + 51.19$$
$$\text{South pier offset} = 588 + 12.57 - 10.38 = 588 + 02.19$$
$$\text{South abutment offset} = 587 + 77.07 + 10.03 = 587 + 87.10$$

15.6 Offset Distance Computations

The stations at which the offset lines intersect the bridge ₵ are each occupied with a transit (theodolite). The skew angle (15°31′13″) is turned and doubled (minimum), and the appropriate offset distances are measured on each side of the bridge ₵. For the north abutment offset line, the offset distance left is:

$$41.598 \, (41'7^{1}/_{8}'') - (10.03 \times \sin 15°31'13'') = 38.91'$$

The offset distance right is 32.03 + 2.68 = 34.71 ft. See Figures 15.10 and 15.11. The following computation is a check of the answer:

$$38.91 + 34.71 = 73.62' = 73'7^{1}/_{2}'' \quad \text{Check}$$

For the north pier offset line, the offset distance left is:

$$38.39 \, (38'4^{3}/_{4}'') - (10.38 \sin 15°31'13'') = 35.61'$$

The offset distance right is 30.31 + 2.78 = 33.09 ft (see Figures 15.10 and 15.11):

$$35.61 + 33.09 = 68.70 = 68'8^1/_2'' \quad \text{Check}$$

For the south pier offset line, the offset distance left is:

$$38.39 - (10.38 \times \sin 15°31'13'') = 35.61'$$

The offset distance right is 28.73 ($28'8^3/_4$) + 2.78 = 31.51 ft (see Figures 15.10 and 15.11):

$$35.61 + 31.51 = 67.12' = 67'1^1/_2'' \quad \text{Check}$$

For the south abutment offset line, the offset distance left is:

$$39.01 (39'0^1/_8'') + (10.03 \times \sin 15°31'13'') = 41.69'$$

The offset distance right is 31.05 − 2.68 = 28.37 ft (see Figures 15.10 and 15.11):

$$41.69 + 28.37 = 70.06' = 70'0^3/_4'' \quad \text{Check}$$

All these distances are shown double underlined in Figure 15.11. Once these offsets and the stakes placed 25 ft farther on each offset line have been accurately located, the next step is to verify the offsets by some independent means.

As an alternative to the direct layout techniques described here, key layout points may be coordinated on the computer, with all coordinates then uploaded into total stations and/or GPS receiver computers. The layout can then proceed using the polar layout and positioning techniques described in Chapters 5 and 8.

15.7 Dimension Verification

For this type of analysis, the technique described in Section 6.10 (omitted measurements) is especially useful. Bearings are assumed that will provide the simplest solution. These values are shown in Figure 15.12. The same techniques can be used to calculate any other series of diagonals if sight lines are not available for the diagonals shown. In addition to diagonal check measurements, or even in place of them, check measurements can consist of right-angle ties from the ℄, as shown in Figure 15.10. Additional calculations are required for the chainage and distance to the outside stakes. Tables 15.1 and 15.2 show these calculations.

As in all construction layout work, great care must be exercised when the grade and alignment stakes are placed. The grade and alignment stakes (iron bars) should be driven flush with the surface (or deeper) to prevent deflection if heavy equipment were to cross over them. Placing substantial guard stakes adjacent to them protects the grade and alignment stakes. The surveyor must bear in mind that, for many reasons, it is more difficult to place a marker in a specific location than it is to locate a set marker by field measurements. For example, it is difficult to drive a 3- or 4-ft steel bar into the ground while keeping it in its proper location with respect to line and distance. As the top of the bar nears the ground surface, it often becomes evident that the top is either off line or off distance. The solution to this problem is either to remove the bar (which is difficult to do) and replace it or to wedge it into its proper location by driving a piece of rock into the ground adjacent to the bar. The latter solution often proves unsuccessful, however, because the deflected bar can, in time, push the

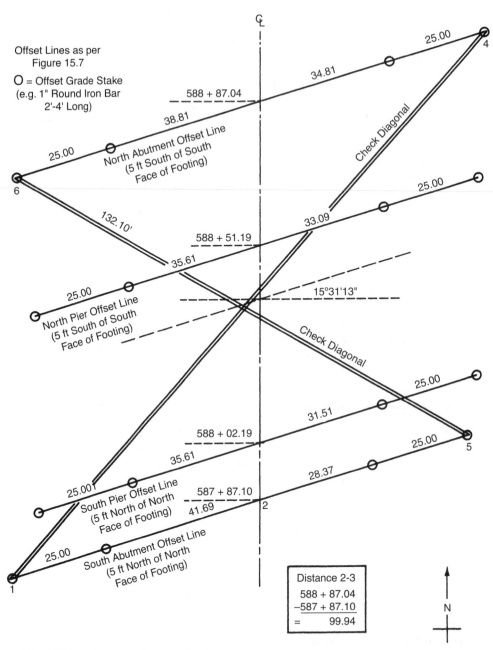

FIGURE 15.12 Layout verification sketch.

rock aside and return to near its incorrect position. These problems can be avoided by taking great care during the original placement; continuous line and distance checks are advised.

In this bridge construction case, an offset line was established 5 ft from the face of the footings. Foundation excavations are usually excavated precisely ("neat"), so the value

Table 15.1 COMPUTATION FOR CHECK DIAGONAL 1–4

Line	Bearing	Distance	Latitude	Departure
1-2	N 74° 28′47″ E (90° skew angle chosen for convenience)	66.69	+17.84	+64.26
2-3	Due north	99.94	+99.94	+0.0
3-4	N 74°28′47″ E	59.81	+16.00	+57.63
1-4			+133.78	+121.89

Distance 1–4 $= \sqrt{133.78^2 + 121.89^2} = 180.98$ ft

Allowable error $= \pm 0.03$ ft, for an accuracy ratio of $\dfrac{1}{7,000}$

Table 15.2 COMPUTATION FOR CHECK DIAGONAL 5–6

Line	Bearing	Distance	Latitude	Departure
5-2	S 74°28′47″ W	53.37	−14.28	−51.42
2-3	Due north	99.94	+99.94	−0.0
3-6	S 74°28′47″ W	63.81	−17.07	−61.48
5-6			+68.59	−112.90

Distance 5–6 $= \sqrt{68.59^2 + 112.90^2} = 132.10$ ft

Allowable error $= \pm 0.02$ ft, for an accuracy ratio of $\dfrac{1}{7,000}$

of 5 ft is quite realistic; however, the surveyor, in consultation with the contractor, must choose an offset that is optimal. As in all construction work, the shorter the offset distance, the easier it is to transfer line and grade accurately to the structure.

15.8 Vertical Control

The 5-ft offset line in this case can be used not only to place the footings but also to place the abutment walls and pier columns as the work proceeds above ground. The form work can be checked easily because the offset lines will be clear of all construction activity. The proposed top-of-footing elevations are shown on the general arrangements plan (Figure 15.7) and in Figure 15.11. The elevations of the grade stakes opposite each end of the footings are determined by differential leveling, and the resultant cuts are given to the contractor in the form of either a grade sheet or batter boards.

A word of caution: A benchmark is often located quite close to the work, and probably one instrument setup is sufficient to take the backsight, grade stake sights, and the foresight back to the benchmark. However, two potential problems should be considered. One problem is that, after the grade stake sights have been determined, the foresight back to the benchmark is taken. The surveyor, knowing the value of the backsight taken previously, may be influenced in reading the foresight because he or she expects to get the same value. In this case, the surveyor may exercise less care in reading the foresight, using it only as a quick check reading. Blunders have been known to occur in this situation. The second potential problem involves the use of automatic (self-leveling) levels. Sooner or later, all automatic levels become inoperative due to failure of the compensating device. This device is used to keep the line of sight horizontal and often relies on wires or bearings to achieve its purpose. If a wire breaks, the compensating device will produce a line of sight that is seriously in error. If the surveyor is unaware that the level is not functioning properly, the leveling work will be unacceptable, in spite of the fact that the final foresight agrees with the backsight.

The only safe way to deal with these potential problems is always to use two benchmarks (BMs) when setting out grades. Start at one BM and check into the second BM. Even if additional instrument setups are required, the extra work is a small price to pay for this accuracy check. Each structure location should have three benchmarks established prior to construction so that, if one is destroyed, a minimum of two will remain for the completion of the project.

The walls and piers are constructed above ground, so the surveyor can check the forms (poured concrete bridges) for plumb by occupying one of the 25-ft offset stakes (Figures 15.11 and 16.8), sighting the other 25-ft offset stake, and then raising the telescope to sight-check measurements to the form work with the vertical cross hair. The 5-ft offset line should be far enough from the finished wall to allow for the line of sight to be clear of the concrete forms and other falsework supports. (See the concrete forms and other falsework in Figures 15.4 and 16.8.) Realizing that steep vertical sightings can accentuate some instrument errors, most surveyors will not take check sightings higher than 45° without double centering.

For high bridges, instrument stations are moved farther away (farther than 25 ft, in this case) so that check sights with the theodolite are kept to relatively small vertical angles. Additional offset stations are established, clear of the work, so that sights can also be taken (in this case) to the 90° adjacent walls. Figure 15.13 shows typical locations used for vertical control stations for the south pier of the bridge example.

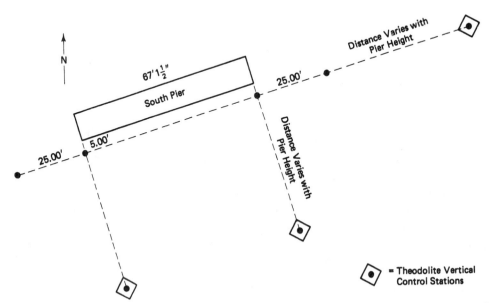

FIGURE 15.13 Theodolite stations that can be used for plumb checks as the construction rises above the ground. See also Figure 16.8.

15.9 Cross Sections for Footing Excavations

Original cross sections must be taken prior to construction. As the excavation work proceeds, cross sections are taken. The structural excavations (because of their higher cost) are kept separate from other cut-and-fill operations for the bridge site. When all work is completed, final cross sections are used for payments and final records. The structural excavation quantities are determined by taking preliminary and final cross sections on each footing. The offset line for each footing is used as a baseline for individual footing cross sections.

Review Questions

15.1. Explain the difference between skew angle and skew number.

15.2. Why would it be inappropriate to use wood stakes as alignment and grade markers in the layout of a high-level bridge?

15.3. When laying out structures, why is it important to measure check diagonals?

15.4. Why is it recommended that two or more benchmarks be located on a construction site?

15.5. A new bridge may require twenty to thirty drawings in the contract drawings package. Which of those drawings does the construction surveyor typically consult for the data needed to provide the survey layout?

Chapter 16

Building Construction Surveys

16.1 Building Construction: General Background

All buildings must be located with reference to the property limits. Accordingly, the initial stage of the building construction survey involves the careful retracing and verification of the property lines. Once the property lines are established, the building is located according to plan, with all corners marked in the field. Large-scale building projects have horizontal control points established. These points are based on a state plane grid or a transverse Mercator grid. These horizontal control monuments are tied into the project property lines as well as the state or provincial grid (see Section 9.2).

Temporary benchmarks are surveyed onto all major sites from the closest benchmark, and then the work is verified by closing the survey into another independent benchmark. The surveyor establishes a minimum of three temporary benchmarks at each site to ensure that, if one is destroyed, at least two will be available for all layout work (see Section 15.8).

16.2 Single-Story Construction

Construction surveys for single-story buildings may entail survey layouts only for the building footings and main floor elevation. The contractor can tie in the rest of the building components to the footings and the first floor elevation for both line and grade. Of course, site grading surveys are also required in most circumstances.

Figure 16.1(a) shows a block plan, which illustrates the general area of the construction site. Figure 16.1(b) shows a site plan, which gives existing and proposed elevations, key building dimensions (16 m × 20 m), setbacks from front and side lines (5 m and 7.5 m, respectively), parking areas, walks, and so on. Figure 16.1(c) shows a ground floor plan, which gives the basic dimensions, as well as first-floor elevations and column locations. Figure 16.1(d) shows an elevation and section; both show footing elevations, first-floor elevations, and roof elevations.

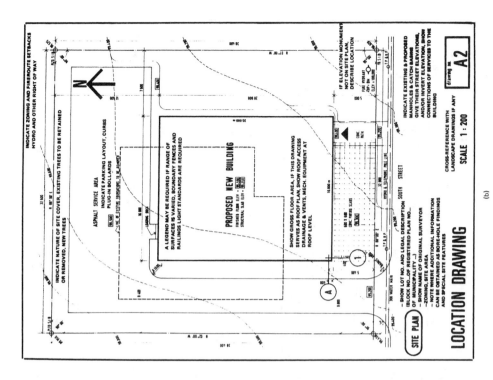

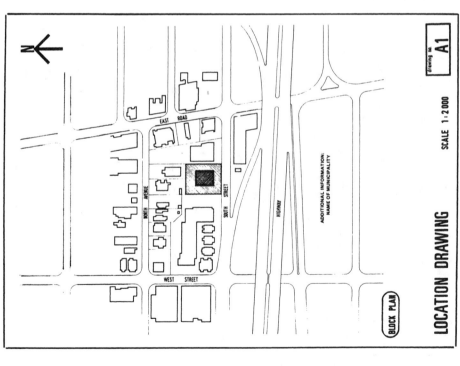

FIGURE 16.1 Location drawing for a single-story building. (a) Block plan. (b) Site plan.

(continued)

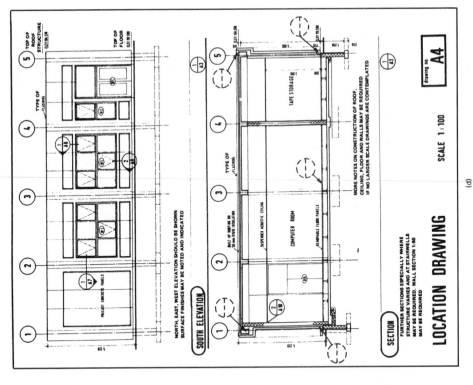

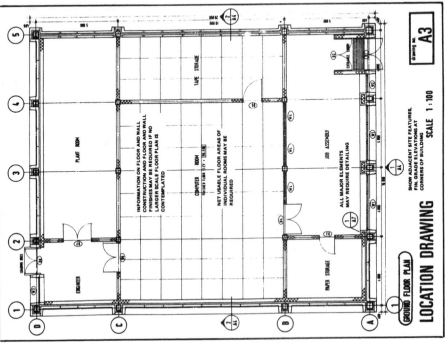

FIGURE 16.1 (*continued*) (c) Ground floor plan. (d) Elevation and section.

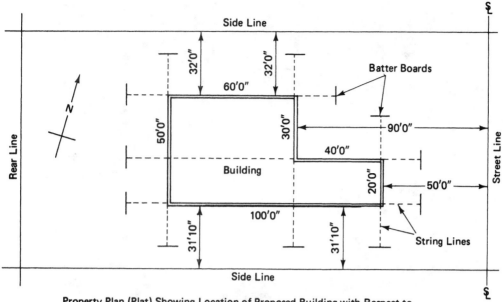

Property Plan (Plat) Showing Location of Proposed Building with Respect to the Property Lines.

This Plan also Shows the Location of the Batter Boards and the String Lines for Each Building Wall Footing.

FIGURE 16.2 Building layout.

The surveyor sets up the property lines as shown on the site plan (a licensed surveyor will be required at this stage to establish the legal lines). The building corners are then set out by measuring the setbacks from the property lines. After the corners have been staked, diagonal distances (on line or offsets) can be measured to verify the layout— that is, for the building shown (16.000 m × 20.000 m), the diagonals should be $\sqrt{16.000^2 + 20.000^2} = 25.612$ m. At an accuracy requirement of 1:5,000, the diagonal tolerance would be +0.005 m (i.e., 0.005/25.612 ≈ 1:5,000). Once the corners have been laid out and verified, the offsets and batter boards can be laid out. Figure 16.2 shows the location of batter boards and string lines used for control of an L-shaped building. The batter boards and string lines are usually set at the first-floor elevation.

Figure 16.3 shows a combination transit and level that can be found on many small construction sites. Although this transit-level is not as precise as many of the instruments previously introduced in this text, its precision is well suited for small construction sites, where the instrument sights are relatively short—typically less than 150 ft or 50 m.

Figure 16.4 shows a backhoe beginning to excavate a basement and footings for a house; one set of corner batter boards can be seen in the photograph. Figure 16.5 shows a small bulldozer shaping the site to conform to the site grading plan. This aspect of the building survey usually takes place only after the services have been trenched in from the street and the house has been erected. Figure 16.6 shows the steel work for a single-story commercial building. The columns shown have been bolted to the footing anchor bolts previously laid out by the surveyor. Figure 16.7 shows two different types of connections: Figure 16.7(a)

FIGURE 16.3 David White LT8-300 level-transit. (Courtesy of CST/Berger-David White)

shows the bolt connection, and Figure 16.7(b) shows the pocket connection. In both cases, a template can be utilized to align the bolt patterns with the column centerlines properly. The tolerance is usually set at $\frac{1}{8}$ inch (3 or 4 mm) for each set of column bolts, with the overall tolerance for the length of the building set at $\frac{1}{4}$ inch (6 mm). The columns can be plumbed with the aid of a theodolite, although for single-story construction, the columns can be plumbed satisfactorily with a spirit level (carpenter's level).

FIGURE 16.4 Excavation for house basement and footings. (Courtesy of John Deere, Moline, Illinois)

FIGURE 16.5 Site grading with a bulldozer. (Courtesy of John Deere, Moline, Illinois)

Once steel is in place, the surveyor can mark the columns a set distance above the floor grade. In North America, the offset marks are set at 4 or 5 ft above the floor slab. Masonry or concrete walls can be checked for plumb by setting the theodolite on an offset line, sighting a target set on the same offset line, and then checking all parts of the wall or concrete forms with the aid of precut offset boards, either held against the wall or nailed directly to the form. See Figure 16.8.

FIGURE 16.6 Single-story building columns and column footings.

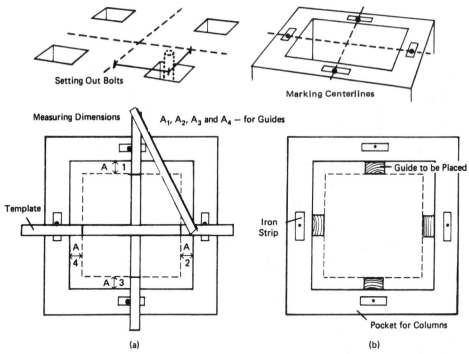

FIGURE 16.7 Connection of columns to footings. (a) Bolt connection. (b) Pocket connection. (Courtesy of National Swedish Institute for Building Research, FIG Report #69)

16.2.1   Construction Lasers and Electronic Levels

The word **laser** is an acronym for **l**ight **a**mplification by **s**timulated **e**mission of **r**adiation. Construction lasers, which have been used for the past generation, are made of helium-neon gas and are considered safe when in normal use. Lasers were introduced in Chapter 14 for control of line and grade in sewers [Figures 14.2(b) and 14.10] and for control of line and grade for a boring machine in tunnels (Figure 14.21). In both examples, a laser beam, fixed in slope and horizontal alignment, controlled the construction work. Figure 12.18 showed a rotating laser guiding grading operations. In addition to being used to guide and control machines as described in Section 12.3, visible beam and infrared lasers can be used with a leveling rod to duplicate the conventional leveling process. As noted in Section 12.3, green-beam visible lasers are now used outdoors as well as indoors.

In building construction, a rotating instrument defining a horizontal or vertical plane is used to control the construction. Figure 16.9(a) shows a surveyor marking the steel columns at the 4-ft mark with the aid of an electronic level; the rotating beam in this instrument is an infrared beam emitted by a laser diode. Figures 16.10 and 16.11 show rotating lasers controlling horizontal and vertical building components, respectively. A detection device, such as that shown in Figure 16.9(b), is used to capture the rotating beam. A detection device is required for the electronic level because the infrared beam is invisible, and such a device may also be required for the visible-beam rotating laser when used outdoors because it is not easily visible to the eye when it is used in bright sunlight.

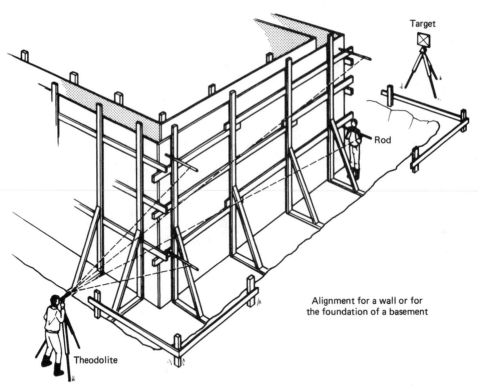

FIGURE 16.8 Vertical alignment of walls. (Courtesy of National Swedish Institute for Building Research, FIG Report #69)

The visible-beam laser instrument can be used in darkened areas indoors without a beam detector because the now-visible beam itself can be used as a reference line. For example, when false ceilings are installed, the column-mounted, rotating laser beam makes a laser mark on the ceiling hanger as it strikes its surface. The ceiling hanger is then marked or bent at the laser mark, ensuring a quick and horizontal installation for the ceiling channels.

Many rotating lasers and electronic levels can be battery-driven for a 10-hr operation and require overnight recharging. These instruments are also self-leveling when first set up to within 4° of level. These automatic instruments stop rotating if knocked off level. The rotating speed can be varied from stop to 420 rpm. The instrument itself (Figure 16.9) can be set up on both threaded and bolt-type tripods and also on any flat surface. The accuracy of the instrument shown in Figure 16.9 is $^1/_8$ inch (0.01 ft) at 100 ft and $^3/_{16}$ inch at 200 ft, with decreasing accuracy up to a limit of about 600 ft. The adjustment of the instrument can be checked by comparing results with a good-quality automatic level. If the electronic level is off by more than $^1/_8$ inch in 100 ft, it is returned to the dealer for recalibration.

When used for setting 4-ft marks, 1-m marks, or other floor reference marks, this technique is quite cost effective. If a level and rod are used for this purpose, it will take two surveyors (one for the instrument and one for the rod) to do the job. When the electronic or

FIGURE 16.9 Electronic level. (a) Marking vertical control (4-foot marks on a steel column). (Courtesy of CST/Berger-David White)

(continued)

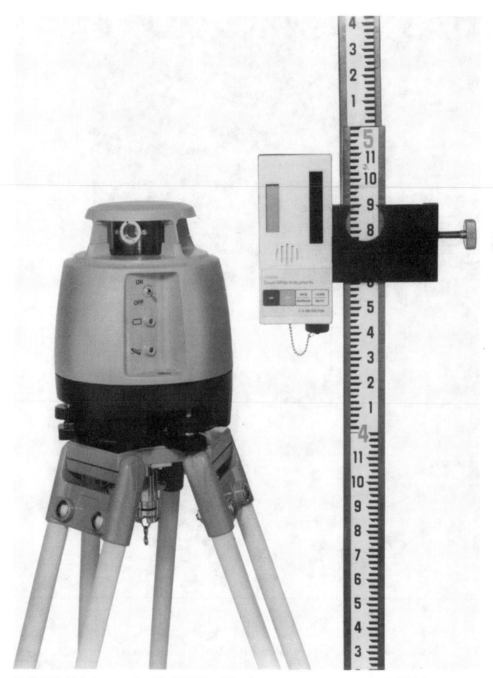

FIGURE 16.9 (*continued*) (b) AEL300 self-leveling, automatic, rotating level (300-ft radius range), with leveling rod and laser detector (C6). (Courtesy of CST/Berger-David White)

FIGURE 16.10 Spectra precision rotating laser level and an electronic receiver attached to a grade rod, shown being used to set concrete forms to the correct elevation. (Courtesy of Trimble)

laser level is used, the survey crew is reduced to just one surveyor, which eliminates communications and the chance for communication errors. According to some reports, it has resulted in a 50-percent increase in work accomplished, along with a 50-percent decrease in the work force. Figure 16.12 illustrates the ease with which one worker can check concrete footings prior to the concrete-block wall construction.

16.3 Multistory Construction

Multistory construction demands a high level of precision from the surveyor. As the building rises many stories, cumulative errors can cause serious delays and expense. Multistory

FIGURE 16.11 Rotating laser positioned to check plumb orientation of construction wall. (Courtesy of Leica Geosystems Inc.)

columns are laid out by intersecting precisely established column lines, and the distances between columns are then checked in all directions. Templates are used to position anchor bolts or pockets, as shown in Figure 16.7.

A theodolite sighted on premarked column centers or on the column edge checks for plumb, or verticality. Column edges are sighted from stations set on offsets a distance of one-half the column width. Shim plates (up to $^1/_8$ in.) are permitted in some specifications when one is plumbing columns mounted on anchor bolts. See Figures 16.7 and 16.13. Wedges are used to help plumb columns in pockets. Connecting beams later hold the columns in their plumbed positions.

16.3.1 Vertical Control

Elevations are usually set two floors at a time. If the stairs are in place, the elevations can be carried up by differential leveling, as shown in Figure 16.14. Many building surveyors, however, prefer to transfer elevations upward by taking simultaneous readings from a properly tensioned, fully graduated steel tape, which can be hung down from the upper floor. The lower-floor surveyor, set up with a known height of instrument (HI), takes a reading on the suspended steel tape. At the same time, the second surveyor, who has a level set up a floor or two higher, also takes a reading on the tape. (Radios can be used for synchronization.)

FIGURE 16.12 Autolaser 300, shown checking the footing elevations prior to construction of the concrete-block walls. (Courtesy of CST/Berger-David White)

The difference in readings is added to the lower HI to give the upper HI; temporary benchmarks are then set on the upper floor.

16.3.2 Horizontal Positioning

Horizontal control can be extended upward by downward plumbing through small (6 in.-square) throughfloor ports onto the lower-floor control stations, or onto marks offset from control stations (see Figure 16.15). The plumbing operation is often accomplished by using optical plummets, as shown in Figure 14.19. Upward plumbing is dangerous because material may be accidentally discharged through the floor port onto the surveyor below. In addition to using optical plummets, the surveyor can use heavy plumb bobs, with swing oscillations dampened in oil or water, as described in Section 14.5.

New layouts are constantly checked back to previously set layout points for both line and grade. All angles are doubled (face right and face left), and all distance tolerances are in the 5-mm (0.02-ft) range. As noted above, upper-floor control is checked back to control

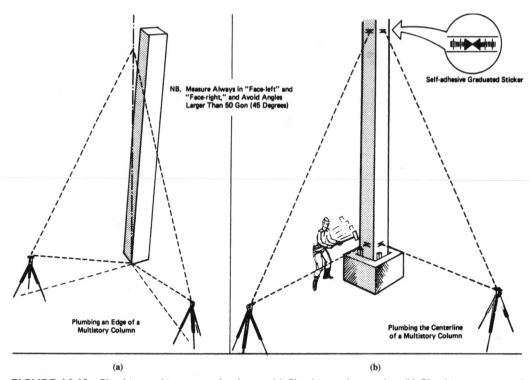

FIGURE 16.13 Plumbing multistory steel columns. (a) Plumbing column edge. (b) Plumbing pretargeted column centerlines. (Courtesy of the National Swedish Institute for Building Research, FIG Report #69)

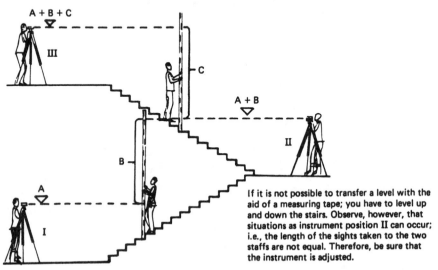

FIGURE 16.14 Transferring elevations in a multistory building. (Courtesy of the National Swedish Institute for Building Research, FIG Report #69)

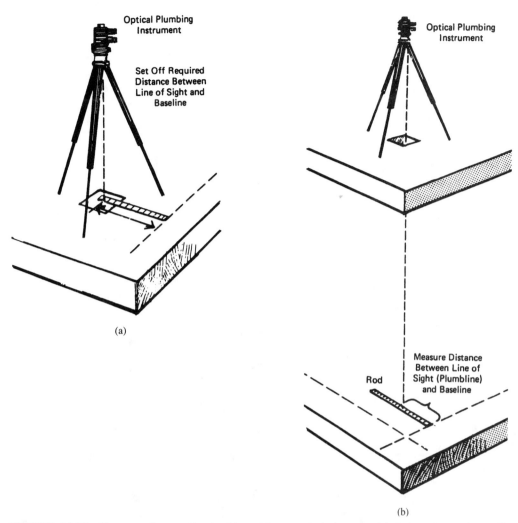

FIGURE 16.15 Downward eccentric plumbing with an optical plummet. (a) Orientation to lower floor. (b) Transfer lower floor position to instrument station floor. See Figure 14.19 for optical plummet. (Courtesy of the National Swedish Institute for Building Research, FIG Report #69)

previously set on lower floors. When it becomes necessary to transfer control to higher floors from the ground, for steel structures, the survey work must be done early in the morning or on cloudy days so that the sun cannot heat and thus deform various parts of the structure, which would cause an erroneous layout.

Figure 16.16 illustrates the case where a theodolite is set up by interlining (wiggling in) between targets (see Section 4.11) set on adjacent structures. These permanent targets are originally set from ground control before the project starts. An alternative to this technique is to set up on a control station located on a high building, sight a permanent target on another building, and then transfer this line down onto each new floor of the building as it is constructed.

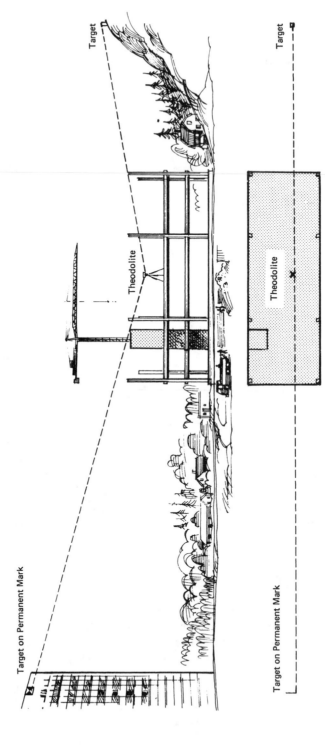

FIGURE 16.16 Establishing control line by bucking-in or wiggling in. See also Figure 4.21. (Courtesy of the National Swedish Institute for Building Research, FIG Report #69)

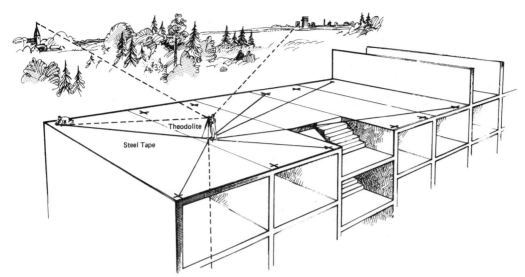

FIGURE 16.17 Optical plumbing combined with reference azimuth points for polar layouts. See also Figure 14.19. (Courtesy of the National Swedish Institute for Building Research, FIG Report #69)

16.3.3   Horizontal Alignment—Reference Azimuth Points

Figure 16.17 shows a survey station located on an upper floor by downward optical plumbing to a previously set control mark. Alignment is provided by sighting reference azimuth points (RAPs)—points of known position (see Section 9.8). Knowing the coordinates of the instrument station and the coordinates of at least two RAPs, the surveyor can quickly determine the azimuth and distance to any of the coordinated building layout points (a third RAP is sighted to provide an accuracy check). If a programmed total station is being used, the coordinates of the layout points, the RAPs, and the instrument station can be loaded into the data collector, with the required layout distances and angles (polar ties) calculated automatically.

16.3.4   Horizontal Alignment—Free Stations

Free stationing, which can be used for most layout surveys (not just building surveys), has several advantages: (1) the theodolite (or total station) can be set up to avoid obstacles, (2) setup centering errors are eliminated, (3) fewer control stations are required, (4) the setup station can be located close to the current layout work, and (5) the layout work is greatly accelerated if total stations are employed. Figure 16.18 illustrates resection techniques being used to provide the coordinates of the theodolite station. In this case, the theodolite or total station is set up at any convenient location (called a **free station**). Angles are then taken to a minimum of three coordinated control stations; a fourth control station can be taken to provide a second computation of the instrument station as an accuracy check. If both angles and distances are to be taken to the control stations, only two, or preferably three, stations are required. By using trigonometric relationships (identities) to solve the resulting triangles, the coordinates of the instrument station are computed. Most total stations are now programmed to solve resection problems. Once the coordinates of the free station

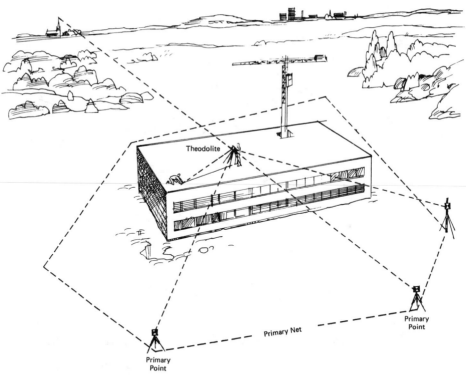

FIGURE 16.18 Location and alignment of theodolite by resection (free stationing). (Courtesy of the National Swedish Institute for Building Research, FIG Report #69)

are solved and the total station is backsighted to a known reference point, the total station programs can then compute the polar ties to any layout point.

Review Questions

16.1. What are the steps required to provide the horizontal layout for a building? List the steps in point form.

16.2. What are the steps required to provide a vertical location layout for (a) a one-story building or (b) a multistory building? List the steps in point form.

16.3. What are the advantages of using a rotating laser rather than a rod and level on a single-story building site?

16.4. What is the advantage of using a visible-beam laser instrument rather than an infrared-beam laser instrument (electronic level) on a construction site?

Chapter 17

Quantity and Final Surveys

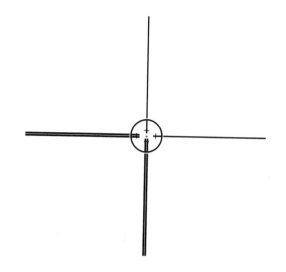

17.1 Construction Quantity Measurements: General Background

After the line and grade for a project have been established, the construction surveyor's next concern is to supply the project supervisor with survey measurements—and the resultant quantities—that reflect the progress achieved by the contractor. Progress payments, which are based on quantities supplied by the surveyor and the construction inspector, are processed either at the end of a regular time period (e.g., monthly) or at the completion of previously agreed-upon project stages. Usually, the contractor also employs a surveyor to provide similar data; this practice ensures that questionable quantities are quickly discovered and remeasured.

Some construction projects are bid on a lump-sum basis, where the contractor's one-price bid covers all the work required to complete a project. The demolition of a structure is a good example of a situation where a lump-sum bid is appropriate. Here, the owner (e.g., private company, municipality, state, or province) simply wants the structure removed; the demolition technique is of little importance to the owner (blasting would be an exception). The owner simply wants the job done as cheaply and expeditiously as possible.

Most construction projects, however, are bid on a unit-price basis, where all the facets of the job are defined in detail. The owner (e.g., municipality, highway department, developer, railroad) specifies the line and grade (and cross section) of the facility. Also specified are the types of construction materials to be used, the qualities of the finished product (e.g., compressive strength of concrete), the compaction level of earth and granular fills, and the appearance of the finished product.

The owner lists all categories (units) of materials and operations and prepares an estimate for the total quantities for each item—that is, total cut and fill, total length of fence,

Table 17.1 TYPICAL CONSTRUCTION CATEGORIES WITH TENDERING UNITS

Lineal (ft, m)	Area (ft², m², acres, hectares)	Volume (ft³, m³)	Weight (tons, tonnes)*
Curb	Clearing	Concrete-in structures	Granular material
Curb and gutter	Grubbing	Cuts	Crushed stone
Sewer pipe	Sodding	Fills	Steel
Pipeline	Seeding	Borrow material	Asphalt
Pipe jacking	Mulching	Water	
Height of manhole	Road surfaces	Dust control chemicals	
Depth of pile		Various excavation operations	
Guide rail		Riprap	
Noise barrier		Gabions	
		Blasting (rock)	

*1 ton = 2,000 lbs = 0.907 metric tons (tonnes).

total volume of concrete placed, total area of sod or seeding, and the like. See Table 17.1 for other typical construction categories, with tendering units included. The contractor bids a price against each unit item; all items are then added to compute the total unit bid. Finally, these individual total bids are summed to produce the grand total for the contract bid. The contract is often awarded to the qualified contractor who submits the lowest bid. The contractor is paid for the work as it is completed (progress payments) and when all the work is completed (final payment). Usually, the owner holds a certain percentage of the final payment until after the guarantee period (two years is typical) has elapsed.

Table 17.2 shows typical measuring precision, in both feet and meters, for layout and quantity surveys. Surveyors, working for both the owner and the contractor, record the progress payment measurements in their field notes. These notes must be accurate, complete, and unambiguous. Figures 17.1, 17.2, 17.3, and 17.4 illustrate typical survey notes for quantity surveys.

Lineal units, as shown in Table 17.1, are added to obtain final quantities. Simply adding the scale tickets for the various items totals weight units. Both the area and the volume units require additional work from the surveyor, however, before the unit totals can be determined. The following sections illustrate the more commonly used techniques for both area and volume computations.

17.2 Area Computations

Areas enclosed by closed traverses can be computed using the coordinate method (Section 6.12). Figure 17.5 illustrates two additional area computation techniques.

17.2.1 Trapezoidal Technique

The area in Figure 17.5 was measured using a fiberglass tape for the offset distances. A common interval of 15 ft was chosen to delineate the riverbank suitably. Had the riverbank been more uniform, a larger interval could have been used; had it been even more irregular, a smaller interval would have been appropriate. The trapezoidal technique assumes that the

Table 17.2 TYPICAL MEASUREMENT PRECISION FOR VARIOUS CONSTRUCTION QUANTITIES

Activity	Foot	Metric
Cross Sections		
Backsight and foresight readings to be taken to the nearest:	0.01 ft	1 mm
Maximum sight distance with level	300 ft	90 m
Maximum allowable error between adjacent benchmarks	0.08 ft	20 mm
Intermediate rod readings to be taken to the nearest:	0.10 ft	10 mm
Intervals:		
Earth cut	100 ft	25 m
Rock cut	50 ft	10 m
Rock cut with overburden	50 ft	10 m
Muskeg (bog) excavation	100 ft	25 m
Fills with stripping, subexcavation, or ditching	100 ft	25 m
Transition from cut to fill	100 ft	25 m
Fills	100 ft	25 m
Earth or rock fills	100 ft	25 m
Borrow pits	50 ft	25 m
Maximum transverse interval for cross-section elevations:		
Earth	100 ft	25 m
Rock	50 ft	10 m
Borrow	50 ft	25 m
Offset distances to be measured to the nearest:	1.0 ft	0.1 m
Grade and Superelevation Calculations		
Calculate grade percentage:	To two decimal places	To three decimal places
Calculate grade elevation to the nearest:	0.01 ft	1 mm
Calculate full superelevation cross fall to the nearest:	0.0001 ft	0.0001 m
Calculate the rate of rise or fall to the nearest:	0.00001 ft/ft	0.0001 m/m
Chainage of T.S., S.C., C.S., S.T. to be recorded to the nearest:	0.01 ft	1 mm
Layout/intervals (with the exception of plus sections, layout is normally at the same interval as the cross sections/grade calculations):		
Rock:	50 ft	10 m
Earth:	100 ft	25 m
Maximum interval for setting structure footing grades	25 ft	10 m
Structure grades to be set to an accuracy of:	0.01 ft	1 mm
Adjustment to slope stake distances to allow for grubbing losses	1 ft	300 mm
Set grades for earth grading to the nearest:	0.10 ft	10 mm
Set grades for granular to the nearest:	0.01 ft	5 mm
Layout stake offset for curb and gutter (ideal)	6 ft	2 m
Layout stake interval for curb and gutter	50/25 ft	20/10 m
Set curb and gutter grades to the nearest:	0.01 ft	1 mm
Maximum staking interval for layout of a radius (intersections)	10 ft	3 m
Layout stake offset for concrete pavement (ideal)	6 ft	2 m
Set grades for concrete pavement to an accuracy of:	0.01 ft	1 mm
Calculate individual cleaning and grubbing areas to the nearest:	ft^2	0.10 m^2
Convert the total clearing area to the nearest:	0.01 acre	0.01 ha
Calculate earth, rock, or borrow end areas to the nearest:	ft^2	0.10 m^2
Convert the total volume per cut to pit to the nearest:	yd^3	m^3
The quantity of blast cover, paid by box measurement, will not exceed:	1500 yd^3	1000 m^3
Truck box volumes shall be calculated to the nearest:	0.10 yd^3	0.10 m^3
The item total for "water" shall be taken to the nearest:	1,000 gal	m^3
Water tank volumes shall be calculated to the nearest:	gal	0.01 m^3
Item totals for calcium chloride shall be taken to the nearest:	0.01 ton	0.01 ton
Culvert and structures excavation end areas shall be calculated to the nearest:	ft^2	0.10 m^2
Volumes per culvert and structure site shall be calculated to the nearest:	yd^3	m^3

(continued)

Table 17.2 (*continued*)

Activity	Foot	Metric
Linear dimensions for the calculation of culvert and structure concrete volumes shall be to the closest:	0.01 ft	1 mm
Volume calculations for concrete culverts and structures shall be to the nearest:	0.50 ft^3	0.01 m^3
Concrete culvert and structure items totals shall be taken to the nearest:	0.10 yd^3	0.10 m^3
Pipe culvert placement shall be measured to the nearest:	ft	100 mm
Pipe removal shall be measured to the nearest:	ft	100 mm
Sewer trench measurements shall be taken at maximum intervals of:	100 ft	30 m
The height of manholes and catch basins shall be measured to:	0.10 ft	10 mm
Concrete in manholes and catch basins shall be calculated to the nearest:	0.01 yd^3	0.01 m^3
Excavation and backfill for sewer systems shall be calculated to the nearest:		
(a) End area	ft^2	0.10 m^2
(b) Item total	yd^3	m^3
Sewer pipe and subdrains shall be measured to the nearest:	ft	100 mm
Concrete pavement, sidewalk area (placing and removal) shall be calculated to the nearest:	ft^2	0.10 m^2
The item total for concrete pavements, sidewalk area (placing and removal) shall be rounded to the nearest:	yd^2	m^2
Curb and gutter, fence, guide rail (placing and removal) shall be measured to the nearest:	ft	100 mm
And the item totaled to the nearest:	ft	1.0 m
Seeding and mulching, and sodding shall be calculated to the nearest:	ft^2	0.10 m^2
And the item totaled to the nearest:	yd^2	1.0 m^2
Topsoil stockpile and areas shall be calculated to the nearest:	ft^2	0.10 m^2
And the stockpile volume calculated to the nearest:	yd^3	1.0 m^3
Riprap		
Depth to be measured to the nearest:	0.10 ft	10 mm
Length and width measured to the nearest:	ft	100 mm
Volume calculated to the nearest:	ft^3	0.10 m^3
Item total to the nearest:	yd^3	1.0 m^3
Reinforcing steel shall be totaled to the nearest:	0.01 ton	0.01 ton
The length of each pile driven shall be measured to the nearest:	in.	10 mm
The pile driving item total shall be taken to the nearest:	ft	0.10 m
The area of restored roadway surface shall be calculated to the nearest:	ft^2	0.10 m^2
And the item totaled to the nearest:	yd^2	1.0 m^2
Width measurements shall be taken to the nearest pavement width plus:	1 ft	300 mm
And length to the nearest:	1 ft	300 mm
Field measurements of length and width for clearing and grubbing shall be taken to the nearest:	ft.	100 mm
Measurements to establish truck box volume shall be taken to the nearest:	0.10 ft	10 mm
Measurements to establish water truck tank volume shall be taken to the nearest:	0.10 ft	10 mm
Measurements to establish boulder volume shall be taken to the nearest:	0.10 ft	10 mm
Field measurements for concrete pavement and sidewalk shall be taken to the nearest:	0.10 ft	10 mm
Field measurements for seeding, mulching, and sodding shall be taken to the nearest:	0.1 ft	100 mm
Field measurements for asphalt sidewalk and hot mix miscellaneous shall be taken to the nearest:	Foot length 0.10 ft width	10-mm length 10 mm width

Source: Adapted from *Construction Manual*, Ministry of Transportation, Ontario.

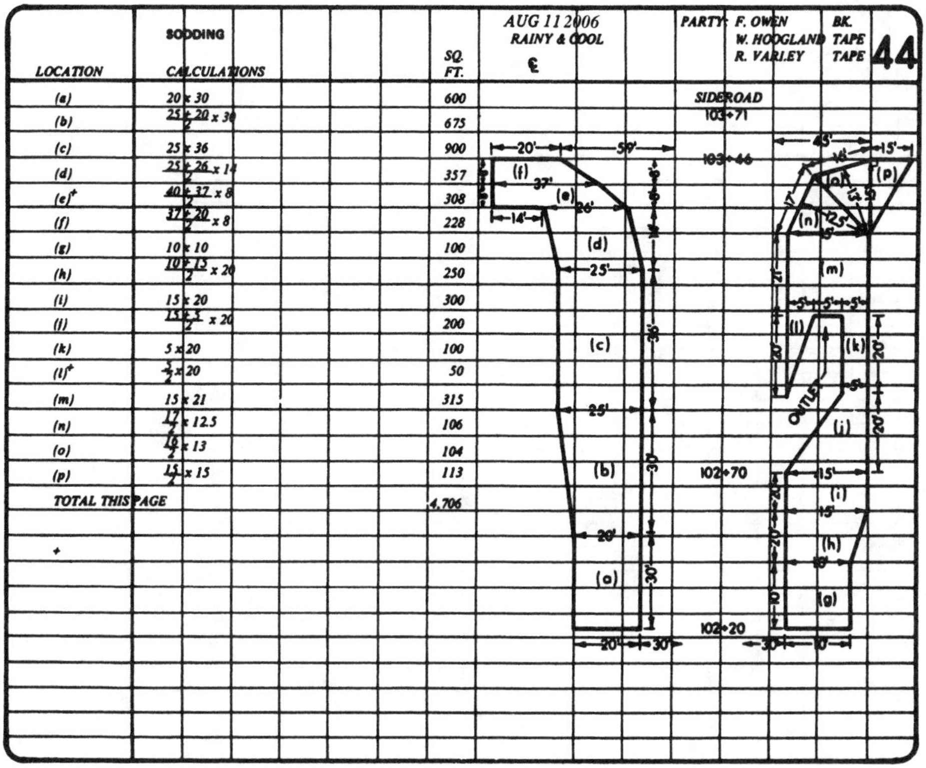

LOCATION	SODDING CALCULATIONS	SQ. FT.
(a)	20 x 30	600
(b)	(25 + 20)/2 x 30	675
(c)	25 x 36	900
(d)	(25 + 26)/2 x 14	357
(e)*	(40 + 37)/2 x 8	308
(f)	(37 + 20)/2 x 8	228
(g)	10 x 10	100
(h)	(10 + 15)/2 x 20	250
(i)	15 x 20	300
(j)	(15 + 5)/2 x 20	200
(k)	5 x 20	100
(l)*	5/2 x 20	50
(m)	15 x 21	315
(n)	17/2 x 12.5	106
(o)	16/2 x 13	104
(p)	15/2 x 15	113
TOTAL THIS PAGE		4,706

AUG 11 2006
RAINY & COOL
PARTY: F. OWEN
W. HOOGLAND
R. VARLEY
BK.
TAPE
TAPE
44
SIDEROAD
103+71
103+46
102+70
102+20
OUTLET

FIGURE 17.1 Example of the method for recording sodding payment measurements. (From *Construction Manual*, Ministry of Transportation, Ontario)

lines joining the ends of each offset line are straight lines (the smaller the common interval, the more valid this assumption).

The end sections can be treated as triangles:

$$A = \frac{(8.1 \times 26.1)}{2} = 106 \text{ ft}^2$$

and

$$A = \frac{(11.1 \times 20.0)}{2} = \frac{111 \text{ ft}^2}{217 \text{ ft}^2} = \text{subtotal}$$

The remaining areas can be treated as trapezoids. The trapezoidal rule is stated as follows:

$$\text{Area} = X\left[\frac{(h_1 + h_n)}{2} + h_2 + \cdots + h_{n-1}\right] \tag{17.1}$$

where X = common interval between the offset lines
h = offset measurement
n = number of offset measurements

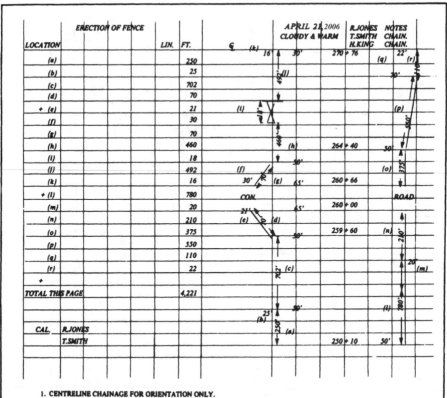

FIGURE 17.2 Field notes for fencing payment measurements. (From *Construction Manual*, Ministry of Transportation, Ontario)

From Figure 17.5:

$$A = 15\left(\frac{26.1 + 20.0}{2} + 35.2 + 34.8 + 41.8 + 45.1 + 40.5 + 30.3 + 25.0\right)$$

$$= 4{,}136 \text{ ft}^2$$

Total area $= 4{,}136 + 217 = 4{,}353 \text{ ft}^2$

17.2.2 Simpson's One-Third Rule

This technique gives more precise results than the trapezoidal technique and is used where one boundary is irregular, like that shown in Figure 17.5. The rule assumes that an odd number of offsets are involved and that the lines joining the ends of three successive offset lines are parabolic in configuration. Simpson's one-third rule is stated mathematically as follows:

$$A = \frac{1}{3} \times \text{interval} \times (h_1 + h_n + 2\Sigma h \text{ odd} + 4\Sigma h \text{ even}) \tag{17.2}$$

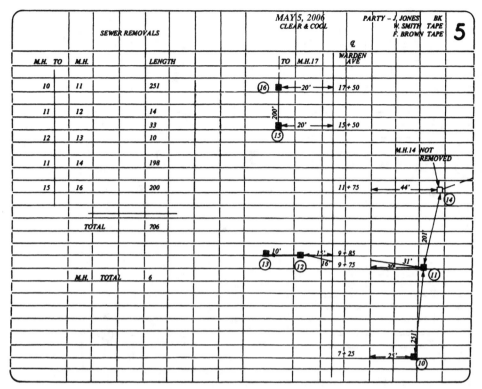

FIGURE 17.3 Example of field-book entries regarding removal of sewer pipe, etc. (From *Construction Manual*, Ministry of Transportation, Ontario)

We can also say that the area is equal to one-third of the common interval times the sum of the first and last offsets ($h_1 + h_n$) plus twice the sum of the other odd-numbered offsets (Σh odd) plus four times the sum of the even-numbered offsets (Σh even). We can use data from Figure 17.5 to calculate Equation 17.2:

$$A = \frac{15}{3} [26.1 + 20.0 + 2(34.8 + 45.1 + 30.3)$$
$$+ 4(35.2 + 41.8 + 40.5 + 25.0)]$$
$$= 4{,}183 \text{ ft}^2$$
$$\text{Total area} = 4{,}183 + 217 \text{ (from preceding example)}$$
$$= 4{,}400 \text{ ft}^2$$

If a problem is encountered with an even number of offsets, the area between the odd numbers of offsets is determined by Simpson's one-third rule, with the remaining area being determined by using the trapezoidal technique. The discrepancy between the trapezoidal technique and Simpson's one-third rule is 47 ft^2 in a total of 4,400 ft^2 (about 1 percent in this example).

PILE NUMBER	LENGTH	CUT OFF	IN PLACE	SPLICES	REMARKS	DATE DRIVEN	DRIVING TUBE PILES WEST ABUTMENT BRIDGE NO. 2	5
1	20'	5'	15'			03/14/06	CONTRACTOR USING DELMAG D-12	
2	20'	5'-6"	14'-6"					
3	20'	4'	16'			"		
4	20'	3'	17'			"		
5	20'	5'	15'			"		
6	20'	5'-6"	14'-6"			"		
7	20'	4'	16'			"		
8	20'	2'	18'			03/18/06		
9	20'	1'	19'			"		
10	20'+20'/40'	15'	25'	1		"		
11	20'+20'/40'	12'	28'	1		"		
12	20'+20'/40'	10'	30'	1		"		
13	20'+20'/40'	15'	25'	1		"		
14	20'+20'/40'	16'	24'	1		"		
15	20'+20'/40'	16'	24'	1		"		
16	20'	–	20'		DRIVEN TO GRADE	03/22/06		
17	20'	–	20'		"	"		
18	20'	4'	16'			"		
19	20'	3'	17'			"		
20	20'	2'	18'			"		
	520'	128'	392'	6 TOTAL				

CONTINUED ON PAGE 6

WHEN SHEET PILES ARE DRIVEN – SHOW NUMBER AND TYPE ALONG EACH SIDE – NUMBERED CONTINUOUSLY AROUND THE PERIMETER.

FIGURE 17.4 Example of field notes for pile driving. (From *Construction Manual,* Ministry of Transportation, Ontario)

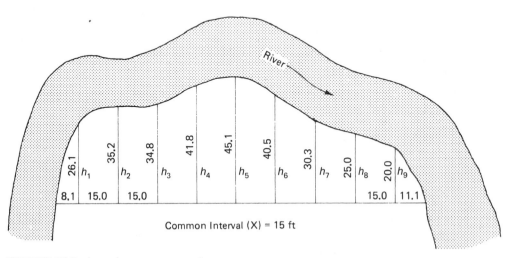

FIGURE 17.5 Irregular area computation.

560   Chap. 17   Quantity and Final Surveys

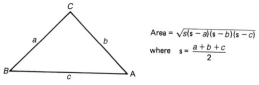

$$\text{Area} = \sqrt{s(s-a)(s-b)(s-c)}$$

$$\text{where} \quad s = \frac{a+b+c}{2}$$

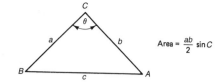

$$\text{Area} = \frac{ab}{2}\sin C$$

FIGURE 17.6 Areas by trigonometric formulas.

17.2.3 Areas by Trigonometric and Geometric Formulas

Some construction quantities can be separated into geometric figures, which can then be analyzed by trigonometric and geometric formulas. These formulas are given in Figures 17.6 and 17.7 and are further illustrated by the worked-out examples in Figure 17.8.

17.3 Area by Graphical Analysis

We have seen that areas can be determined very precisely using coordinates (Chapter 6), and less precisely by using the somewhat approximate methods illustrated by the trapezoidal rule and Simpson's one-third rule. Areas can also be determined by analyzing plotted data on plans and maps. For example, if a transparent sheet is marked off in grid squares to some known scale, an area outlined on a map can be determined by placing the grid paper over (sometimes under) the map and counting the number of squares and partial squares within the boundary limits shown on the map. The smaller the squares, the more precise the result.

Another method of graphic analysis involves the use of a planimeter (Figures 17.9 and 17.10). A planimeter consists of a graduated measuring drum attached to an adjustable or fixed tracing arm, which itself is attached to a pole arm. One end of the pole arm is anchored to the working surface by a needle. The graduated measuring drum gives partial revolution readings, while a disk keeps count of the number of full revolutions as the area outlined is traced.

Areas are determined by placing the pole-arm needle in a convenient location, setting the measuring drum and revolution counter to zero (some planimeters require recording an initial reading), and then tracing (using the tracing pin) the outline of the area being measured. As the tracing proceeds, the drum (which is also in contact with the working surface) revolves, measuring a value that is proportional to the area being measured. Some planimeters measure directly in square inches, while others can be set to map scales. When in doubt, or as a check on planimeter operation, the surveyor can measure out a scaled figure [e.g., 4-in. (100-mm) square] and then measure the area (16 in.^{2}) with a planimeter so that the planimeter area can be compared with the actual area laid off by scale. If the planimeter gives a result in square inches—say, 51.2 in.^{2}—and the map is at a scale of 1 in. = 100 ft, the actual ground area portrayed by 51.2 in.^{2} would be $51.2 \times 100^2 = 512{,}000$ ft^2 = 11.8 acres.

The planimeter is normally used with the pole-arm anchor point outside the area being traced. If it is necessary to locate the pole-arm anchor point inside the area being measured, as in the case of a relatively large area, the area of the zero circle of the planimeter must be added to the planimeter readings. This constant is supplied by the manufacturer or can be deduced by simply measuring a large area twice, once with the anchor point outside the area and once with the anchor point inside the area.

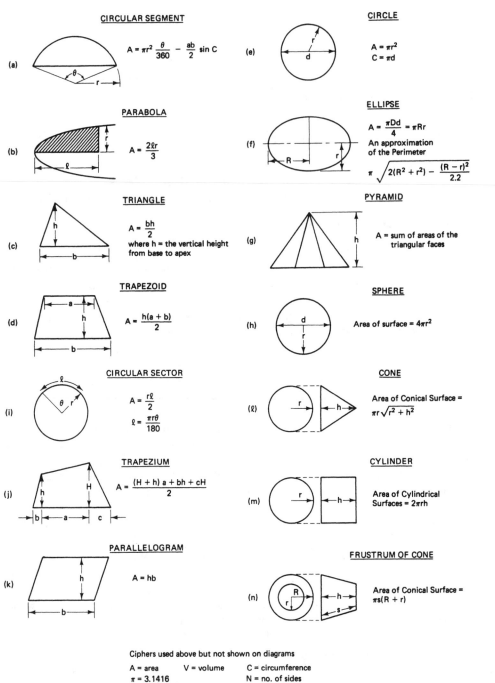

FIGURE 17.7 Areas by geometric formulas. (a) Circular segment. (b) Parabola. (c) Triangle. (d) Trapezoid. (e) Circle. (f) Ellipse. (g) Pyramid. (h) Sphere. (i) Circular sector. (j) Trapezium. (k) Parallelogram. (l) Cone. (m) Cylinder. (n) Frustum of cone. (From *Construction Manual*, Ministry of Transportation, Ontario)

(a) Area of triangle:
given 3 sides

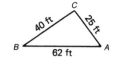

Area $= \sqrt{s(s - a)(s - b)(s - c)}$

$s = \dfrac{40 + 25 + 62}{2} = 63.5$ ft

$A = \sqrt{63.5(63.5 - 40)(63.5 - 25)(63.5 - 62)}$

$A = \sqrt{63.5 \times 23.5 \times 38.5 \times 1.5}$

$A = \sqrt{86177} = \underline{293.6}$ ft^2

(b) Area of triangle:
given 2 sides and
the contained
angle

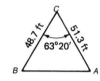

Area $= \dfrac{ab}{2} \sin C$

$A = \dfrac{48.7 \times 51.3}{2} \sin 63°20'$

$A = \underline{1116.3}$ ft^2

(c) Area of triangle:
given the base and
vertical height

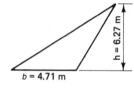

Area $= \dfrac{bh}{2}$

Area $= 4.1 \times 6.27$

Area $= \underline{29.53}$ m

(d) Area of circle:
given radius (r)

Area $= \pi r^2$

$= \pi \times 18^2$

$= 3.1416 \times 324$

$= \underline{1017.9}$ ft^2

(e) Area of circular sector:
given radius and central
angle.

$\theta = 100°$

$\ell = ?$

$r = 38.2$ m

Area $= \dfrac{r\ell}{2}$

where $\ell = 2\pi r \times \dfrac{\theta}{360}$

$\ell = 3.1416 \times 38.2 \times \dfrac{100}{180} = 66.67$ m

$A = \dfrac{38.2 \times 66.67}{2} = \underline{1273.4}$ m^2

or

$A = \pi r^2 \times \dfrac{\theta}{360}$

$= 3.1416 \times 38.2 \times \dfrac{100}{360}$

$A = \underline{1273.4}$ m^2

FIGURE 17.8 Area examples.

(continued)

Sec. 17.3   Area by Graphical Analysis

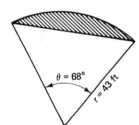

(f) Area of circular segment: given θ and r

$$\text{Area} = \left(\pi r^2 \frac{\theta}{360}\right) - \left(\frac{ab}{2}\sin C\right)$$

$$A = \left(3.1416 \times 43^2 \times \frac{68}{360}\right) - \left(\frac{43 \times 43}{2}\sin 68°\right)$$

$$A = 1097.2 - 857.2$$

$$A = \underline{\underline{240.0\ \text{ft}^2}}$$

(g) Area of sphere: given r, the radius

$$\text{Area} = 4\pi r^2$$

$$A = 4 \times 3.1416 \times 18^2$$

$$A = \underline{\underline{4071.5\ \text{ft}^2}}$$

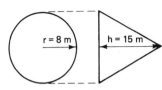

(h) Area of the surface of a cone: given radius (r), and height (h)

$$\text{Area} = \pi r\sqrt{r^2 + h^2}$$

$$A = 3.1416 \times 8\sqrt{8^2 + 15^2}$$

$$A = \underline{\underline{427.3\ \text{m}^2}}$$

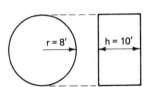

(i) Area of cylinder: given radius (r) and height (h)

$$\text{Area} = 2\pi rh$$

$$A = 2 \times 3.1416 \times 8 \times h$$

$$A = \underline{\underline{502.7\ \text{ft}^2}}$$

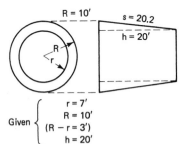

(j) Area of the conical surface of the frustum of a cone: given radius of both top (r) and bottom (R), and length of side (s) or height (h)

$$\text{Area} = \pi s(R + r)$$

$$s = \sqrt{(R - r)^2 + h^2}$$

$$s = \sqrt{3^2 + 20^2}$$

$$s = 20.2'$$

$$A = 3.1416 \times 20.2 \times 17$$

$$A = \underline{\underline{1078.8\ \text{ft}^2}}$$

$$\text{Given}\begin{cases} r = 7' \\ R = 10' \\ (R - r = 3') \\ h = 20' \end{cases}$$

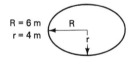

(k) Area of an ellipse: given the two diameters or radii

$$\text{Area} = \pi Rr$$

$$A = 3.1416 \times 6 \times 4$$

$$A = \underline{\underline{75.40\ \text{m}}}$$

FIGURE 17.8 (continued)

(ℓ) Area of a parabolic
   segment: given
   radius (r) and
   length (ℓ)

ℓ = 8.27 m
r = 3.5 m

Area = $\frac{2}{3}$ ℓr

A = $\frac{2}{3}$ × 8.27 × 3.5

A = <u>19.30</u> m²

(m) Area of a parallelogram
    or rectangle: given
    base (b) and height (h)

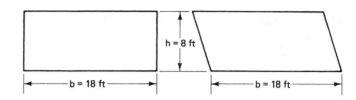

h = 8 ft

b = 18 ft   b = 18 ft

Area = bh = 18 × 8 = <u>144</u> ft²

(n) Area of a trapezoid:
    given the lengths of
    the two parallel sides
    (a and b), and the
    distance (h) between
    them

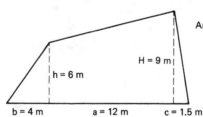

a = 20'
h = 10'
b = 28'

Area = $\frac{a + b}{2}$ × h

A = $\frac{20 + 28}{2}$ × 10

A = <u>240</u> ft²

(o) Area of a trapesium:
    given the base and
    the heights (H and h)
    to the ends of the
    opposite side

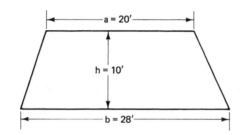

H = 9 m
h = 6 m
b = 4 m   a = 12 m   c = 1.5 m

Area = $\frac{bh + (h + H)a + cH}{2}$

A = $\frac{4 × 6 + (6 + 9)12 + 1.5 × 9}{2}$

A = <u>108.75</u> m

(p) Area of the surface
    of a pyramid: given
    the base (b) and the
    height (h)

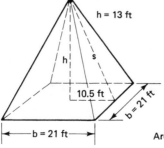

h = 13 ft
h   s
10.5 ft
b = 21 ft
b = 21 ft

Area = sum of triangular faces

Area of triangle = $\frac{b × s}{2}$

s = $\sqrt{10.5^2 + 13^2}$

s = 16.7 ft

Area of triangle = $\frac{21 × 16.7}{2}$

A = 350.7 ft²

Area of 4-sided pyramid = 4 × 350.7 = <u>1402.8</u> ft²

FIGURE 17.8 (continued)

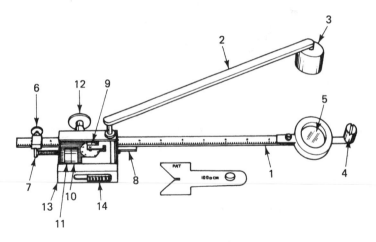

Part 1 Tracer Arm
Part 2 Pole Arm
Part 3 Pole Weight
Part 4 Hand Grip
Part 5 Tracing Magnifier
Part 6 Clamp Screw
Part 7 Fine Movement Screw

Part 8 Tracer Arm Vernier
Part 9 Revolution Recording Dial
Part 10 Measuring Wheel
Part 11 Measuring Wheel Vernier
Part 12 Idler Wheel
Part 13 Carriage
Part 14 Zero Setting Slide Bar

FIGURE 17.9 Polar planimeter.

FIGURE 17.10 Area takeoff by polar planimeter.

FIGURE 17.11 Area takeoff by an electronic planimeter.

Planimeters are particularly useful in measuring end areas (Section 17.4) used in volume computations. For the measuring of watershed areas, planimeters are also used effectively as a check on various construction quantities (e.g., areas of sod, asphalt), and as a check on areas determined by coordinates. Electronic planimeters (Figure 17.11) measure larger areas in less time than traditional polar planimeters. Computer software is available for highways and other earthworks applications (e.g., cross sections) and for drainage basin areas. The planimeter shown in Figure 17.11 has a 36″ by 30″ working-area capability, with a measuring resolution of 0.01 in.^{3} (0.02-in.^{2} accuracy).

17.4 Construction Volumes

In highway construction, designers try to balance cut-and-fill volumes optimally, for economic reasons. Cut and fill cannot be balanced precisely because of geometric and aesthetic design considerations and because of the unpredictable effects of shrinkage and swell on excavated material. Shrinkage occurs when a cubic yard (meter) is excavated and then placed while being compacted. The same material formerly occupying 1 yd^3 (m^3) volume now occupies a smaller volume. **Shrinkage** reflects an increase in density of the excavated material when it is placed and compacted and is obviously greater for silts, clays, and loams than it is for granular materials, such as sand and gravel. **Swell** occurs when the shattered (blasted) rock is placed in the roadbed. Obviously 1 yd^3 (m^3) of solid rock will expand significantly when shattered. Swell is usually in the 15 to 20 percent range, whereas shrinkage is in the 10 to 15 percent range—although values as high as 40 percent are possible with organic material in wet areas.

To keep track of cumulative cuts and fills as the profile design proceeds, the cumulative cuts (plus) and fills (minus) are shown graphically in a mass diagram. The mass diagram is an excellent method of determining waste or borrow volumes and can be adapted to show haul (transportation) considerations (see Figure 17.12). The total cut-minus-fill is plotted at each station directly below the profile plot.

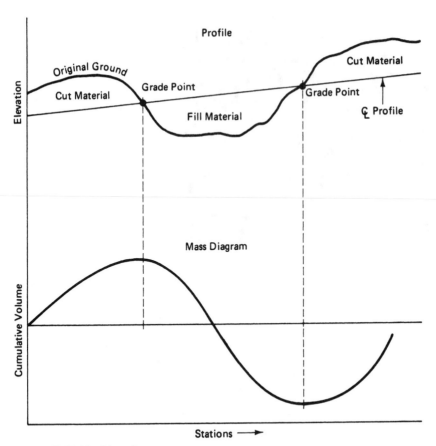

FIGURE 17.12 Mass diagram.

Large fills require borrow material, usually taken from a nearby borrow pit. Borrow-pit leveling procedures are described in Section 3.9 and Figure 3.23. The borrow pit in Figure 3.23 was laid out on a 50-ft grid. The volume of a grid square is the average height $[(a + b + c + d)/4]$ times the area of the base (50^2). The partial grid volumes (along the perimeter of the borrow pit) can be computed by forcing the perimeter volumes into regular geometric shapes (wedge shapes or quarter-cones).

When high precision is less important, volumes can be determined by analysis of contour plans; the smaller the contour interval, the more precise the result. The areas enclosed by a contour line can be taken off by planimeter. Electronic planimeters are very useful for this purpose. The computation is shown in the following equation:

$$V = \frac{I[C_1 + C_2]}{2} \tag{17.3}$$

V is the volume (ft^3 or m^3) of earth or water, C_1 and C_2 are areas of adjacent contours, and I is the contour interval (ft or m). The prismoidal formula (see Section 17.6) can be used if m is an intervening contour (C_2) between C_1 and C_3. This method is also well suited for many water-storage volume computations.

Perhaps the most popular present-day volume computation technique involves the use of computers and any one of a large number of available software programs. The computer programmer uses techniques similar to those described here, but the surveyor's duties may end with proper data entry into the computer (see Chapter 5).

17.5 Cross Sections, End Areas, and Volumes

Cross sections establish ground elevations at right angles to a proposed route. Cross sections can be developed from a contour plan (like profiles were in the previous section), although it is common to have cross sections taken by field surveys. Chapter 5 introduced a computerized system that permits generation of profiles and cross sections from a general database.

Cross sections are useful in determining quantities of cut and fill in construction design. If the original ground cross section is plotted and then the as-constructed cross section is also plotted, the end area at that particular station can be computed. In Figure 17.13(a), the proposed road at station $7 + 00$ is at an elevation below existing ground (EG), which indicates a cut situation (i.e., the contractor will cut out that amount of soil shown between the proposed section and the original section). In Figure 17.13(b), the proposed road elevation at station $3 + 00$ is above the existing ground, which indicates a fill situation (i.e., the contractor will bring in or fill in that amount of soil shown). Figure 17.13(c) shows a transition section between cut-and-fill sections.

When the end areas of cut or fill have been computed for adjacent stations, the volume of cut or fill between those stations can be computed by simply averaging the end areas and multiplying the average end area by the distance between the end-area stations. Figure 17.14 illustrates this concept, and Equation 17.4 gives the general case for volume computation:

$$V = \frac{[A_1 + A_2]}{2} \times L \tag{17.4}$$

A_1 and A_2 are the end areas of two adjacent stations, and L is the distance (feet or meters) between the stations. To give the answer in cubic yards, divide the answer, in cubic feet, by 27; when metric units are used, the answer is left in cubic meters.

The average end-area method of computing volumes is entirely valid only when the area of the midsection is, in fact, the average of the two end areas. This situation is seldom the case in actual earthwork computations; however, the error in volume resulting from this assumption is insignificant for the usual earthwork quantities of cut and fill. For special earthwork quantities (e.g., expensive structures excavation) or for higher-priced materials (e.g., concrete in place), a more precise method of volume computation, the prismoidal formula (discussed in Section 17.6), must be used.

■ **EXAMPLE 17.1** *Volume Computations by End Areas*

Figure 17.15(a) shows a pavement cross section for a proposed four-lane curbed road. The total pavement depth is 605 mm, the total width is 16.30 m, the subgrade is sloping to the side at 2 percent, and the top of the curb is 20 mm below the elevation of ₵. This proposed cross section is shown in Figure 17.15(b), along with the existing ground cross section at station $0 + 340$. You can see that all subgrade elevations were derived from the proposed cross section, together with the ₵ design elevation of

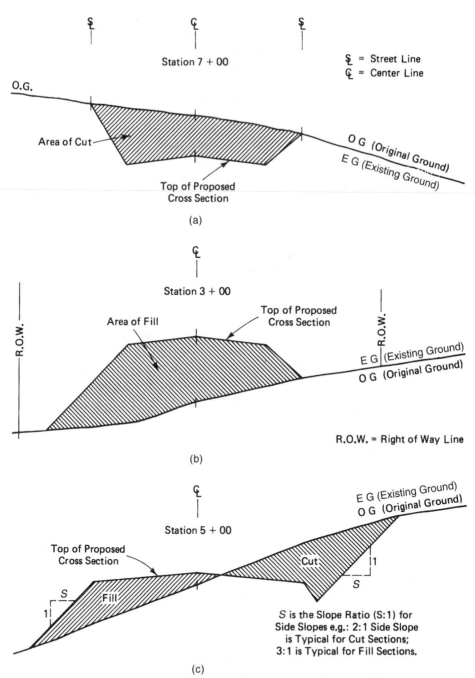

FIGURE 17.13 End areas. (a) Cut section. (b) Fill section. (c) Transition section (that is, both cut and fill).

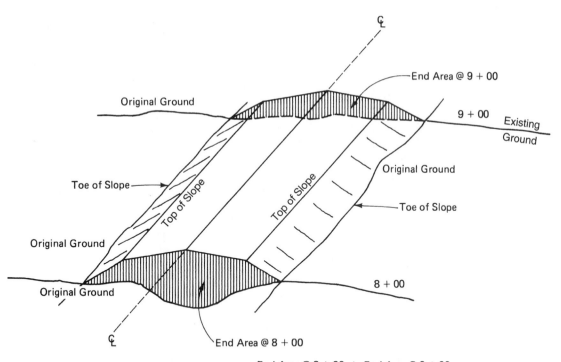

Volume of Fill Between 8 + 00 and 9 + 00 = $\dfrac{\text{End Area @ 8 + 00} + \text{End Area @ 9 + 00}}{2}$ × 100

FIGURE 17.14 Fill volume computations using end areas.

221.43. The desired end area is the area shown below the original ground plot and above the subgrade plot.

Solution

At this point, an elevation datum line is arbitrarily chosen (220.00). The datum line chosen can be any elevation value rounded to the nearest foot, meter, or 5-ft value that is lower than the lowest elevation in the plotted cross section. Figure 17.16 illustrates that end-area computations involve the computation of two areas:

1. Area between the ground cross section and the datum line.
2. Area between the subgrade cross section and the datum line.

The desired end area (cut) is area 1 minus area 2. For fill situations, the desired end area is area 2 minus area 1. The end-area computation can be determined as shown in the following chart:

Station	Plus	Subarea	Minus	Subarea
0 + 340	$\dfrac{(1.55 + 1.50)}{2} \times 4.5 =$	6.86	$\dfrac{(1.55 + 1.41)}{2} \times 2.35 =$	3.48
	$\dfrac{(1.50 + 2.00)}{2} \times 6.0 =$	10.50	$\dfrac{(0.66 + 0.82)}{2} \times 8.15 =$	6.03

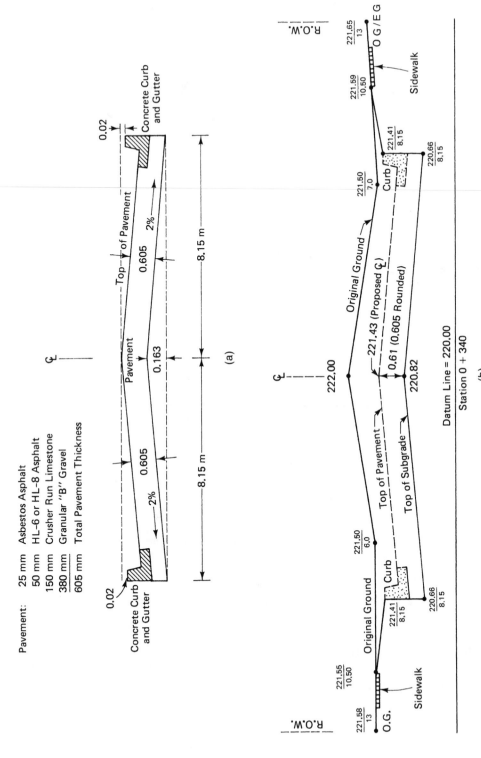

FIGURE 17.15 End-area computation, general. (a) Typical arterial street cross section. (b) Survey and design data plotted to show both the original ground and the design cross sections.

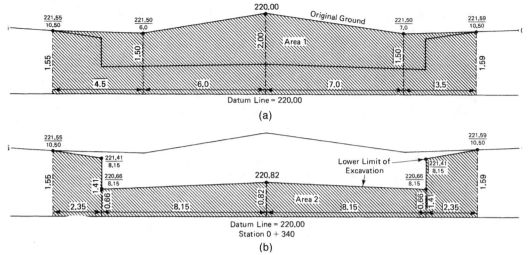

FIGURE 17.16 End-area computations for cut areas. (a) Area between ground cross section and datum line. (b) Area between subgrade cross section and datum line.

$$\frac{(2.00 + 1.50)}{2} \times 7.0 = 12.25 \qquad \frac{(0.82 + 0.66)}{2} \times 8.15 = 6.03$$

$$\frac{(1.50 + 1.59)}{2} \times \underline{3.5} = \underline{5.41} \qquad \frac{(1.41 + 1.59)}{2} \times \underline{2.35} = \underline{3.53}$$

Check: 21 m   **35.02 m^2**   Check: 21 m   **19.07 m^2**

End area = 35.02 − 19.07 = 15.95 m^2

Assuming that the end area at $0 + 300$ has been computed to be 18.05 m^2, we can now compute the volume of cut between $0 + 300$ and $0 + 340$ using Equation 17.4:

$$V = \frac{(18.05 + 15.95)}{2} \times 40 = 680 \text{ m}^3$$

17.6 Prismoidal Formula

If values more precise than end-area volumes are required, the prismoidal formula can be used. A prismoid is a solid with parallel ends joined by planes or continuously warped surfaces. The prismoidal formula is shown in Equation 17.5:

$$V = L\frac{(A_1 + 4A_m + A_2)}{6} \text{ ft}^3 \text{ or m}^3 \qquad (17.5)$$

where A_1 and A_2 are the two end areas, A_m is the area of a section midway between A_1 and A_2, and L is the distance from A_1 to A_2. A_m is not the average of A_1 and A_2 but is derived from distances that are the average of corresponding distances required for A_1 and A_2 computations.

The formula shown in Equation 17.5 is also used for other geometric solids (e.g., truncated prisms, cylinders, and cones). To justify its use, the surveyor must refine the field measurements to reflect the increase in precision being sought. A typical application of the prismoidal formula is the computation of in-place volumes of concrete. The difference in cost between a cubic yard or meter of concrete and a cubic yard or meter of earth cut or fill is sufficient reason for the increased precision.

■ **EXAMPLE 17.2** *Volume Computations by Prismoidal Formula*

See Figure 17.17 for the problem parameters. Compute the volume using the prismoidal formula in Equation 17.5.

Solution

$$\text{Volume}(V) = \frac{L}{6}(A_1 + 4A_m + A_2) \qquad A_1 = \frac{(0.75 + 4.75)}{2} \times 7 = 19.25 \text{ ft}^2$$

$$= \frac{17}{6}(19.25 + 4[11.25] + 5.25) \qquad A_m = \frac{(0.75 + 3.75)}{2} \times 5 = 11.25 \text{ ft}^2$$

$$= \frac{17}{6}(19.25 + 45.00 + 5.25) \qquad A_2 = \frac{(0.75 + 2.75)}{2} \times 3 = 5.25 \text{ ft}^2$$

$$= \frac{17}{6} \times 69.50$$

$$= \frac{1,181.50}{6}$$

$$= 196.92 \text{ ft}^3$$

$$= 7.29 \text{ yd}^3$$

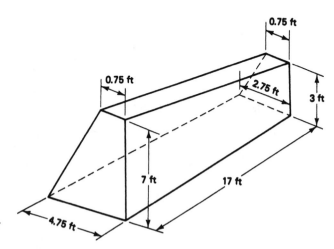

FIGURE 17.17 Sketch of prism for Example 17.2.

17.7 Volume Computations by Geometric Formulas

Some construction items can be broken into solid geometric figures for analysis and quantity computations. Structural concrete quantities, material stockpiles, and construction liquids (e.g., dust control) are some quantities that lend themselves to this sort of analysis. Figures 17.18 and 17.19 illustrate some of these formulas and their use.

17.8 Final (As-Built) Surveys

Similar to preliminary surveys, final surveys tie in *features that have just been built*. These measurements provide a record of construction and a check that the construction has proceeded according to the design plan.

The final plan, drawn from the as-built survey, is quite similar to the design plan except for the revisions made to reflect the changes invariably required during the construction process. Design changes occur when problems are encountered. Such problems are usually apparent only after the construction has commenced—for example, unexpected underground pipes, conduits, or structures that interfere with the designed facility. It is difficult, especially in complex projects, to plan for every eventuality. If the preliminary surveyor, the construction surveyor, and the designer have all done their jobs well, however, the design plan and the as-built drawing will be quite similar with respect to the horizontal

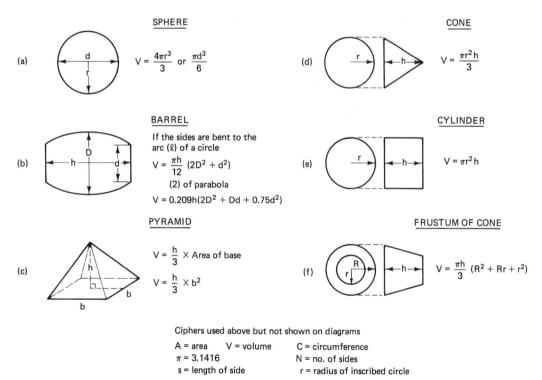

FIGURE 17.18 Volumes by geometric formulas. (a) Sphere. (b) Barrel. (c) Pyramid. (d) Cone. (e) Cylinder. (f) Frustum of cone. (From *Construction Manual*, Ministry of Transportation, Ontario)

(a) Volume of a sphere:
   given the radius (r)

$$\text{Volume} = \frac{4\pi r^3}{3}$$

$$V = \frac{4 \times 3.1416 \times 10^3}{3}$$

$$V = \underline{4189} \text{ ft}^3$$

(b) Volume of a cone:
   given the radius (r)
   and the height (h)

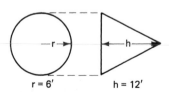

r = 6′   h = 12′

$$\text{Volume} = \frac{\pi r^2 h}{3}$$

$$V = \frac{3.1416 \times 6^2 \times 12}{3}$$

$$V = \underline{452.4} \text{ ft}^3$$

(c) Volume of a cylinder:
   given radius of base (r),
   and height or length (h)
   of the cylinder

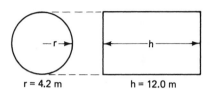

r = 4.2 m   h = 12.0 m

$$\text{Volume} = \pi r^2 h$$

$$V = 3.1416 \times 4.2^2 \times 12$$

$$V = \underline{665.0} \text{ m}^3$$

(d) Volume of frustum of
   cone: given the radius
   of both top (r) and
   bottom (R), and the
   height (h)

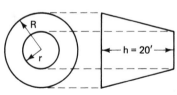

Given: r = 7′, R = 10′, h = 20′

$$\text{Volume} = \frac{\pi h}{3}(R^2 + Rr + r^2)$$

$$V = \frac{3.1416 \times 20}{3}(10^2 + 70 + 7^2)$$

$$V = \underline{4587} \text{ ft}^3$$

(e) Volume of a pyramid:
   given base (b) and
   height (h)

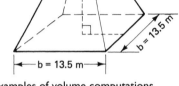

h = 27.4 m

b = 13.5 m   b = 13.5 m

$$\text{Volume} = \frac{h}{3} \times \text{area of base}$$

$$V = \frac{27.4}{3} \times 13.5^2$$

$$V = \underline{1664.6} \text{ m}^3$$

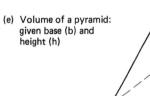

FIGURE 17.19 Examples of volume computations.

and vertical position of all detail. When machine control or guidance has been utilized in the construction process (see Chapter 10), the continual updating of the project's digital data file, in the in-cab computer, results in the final survey being already in the computerized data file at the conclusion of the construction process. When more traditional surveying methods are used, however, the as-built field survey is usually begun after all the construction work has been completed.

The horizontal and vertical control points used in the preliminary and construction surveys are reestablished and referenced again, if necessary. The cross sections taken for the final payment survey are incorporated into the as-built survey. All detail is tied in, as in a preliminary survey. Pipelines and sewers are surveyed before the trenches are backfilled. It is particularly important at this stage to tie in (horizontally and vertically) any unexpected pipes, conduits, or structures encountered in the trench. As-built drawings, produced from final (as-built) surveys, are especially valuable in urban areas, where it seems there is no end to underground construction and reconstruction. After the completion of the contracted work and the expiration of the guarantee period, the as-built drawing is often transferred to CD/DVD or other secure computerized storage.

Problems

17.1. The field measurements (in feet) for an irregularly shaped sodded area are shown in Figure 17.20. Calculate the area of sod to the closest square yard, using (a) the trapezoidal technique and (b) Simpson's one-third rule.

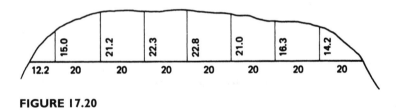

FIGURE 17.20

17.2. The field measurements (in feet) for an irregularly shaped paved area are shown in Figure 17.21. Calculate the area of pavement to the closest square yard.

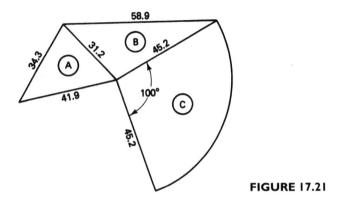

FIGURE 17.21

17.3. Refer to Figure 17.22. Calculate the end areas at stations 3 + 60 and 3 + 80, and then determine the volume (m³) of cut required in that section of road. For the end area at 3 + 60, use 180 m as the datum elevation; for the end area at 3 + 80, use 181 m as the datum elevation.

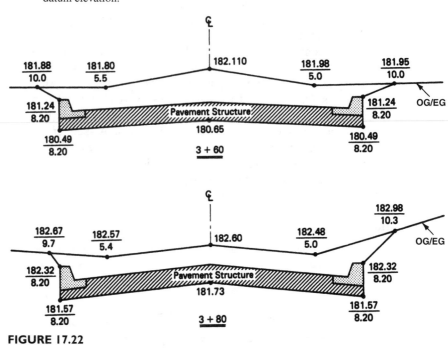

FIGURE 17.22

17.4. Use the prismoidal formula to compute the volume (yd³) of concrete contained in Figure 17.23.

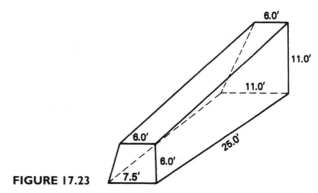

FIGURE 17.23

17.5. A stockpile of crushed stone is shaped into a conical figure. The base has a diameter of 72 ft, and the height of the stockpile is 52.5 ft. Compute the volume (yd³) of stone in the stockpile.

17.6. Compute the volume (yd³) contained in the footing and abutment shown in Figure 17.24.

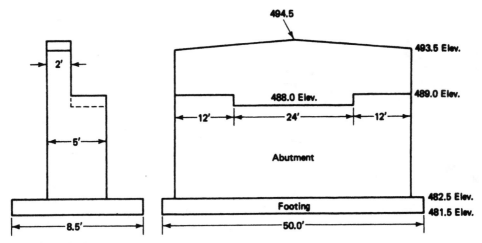

FIGURE 17.24

Appendix A

Trigonometry and Coordinate Geometry Review

A.1 Trigonometric Definitions and Identities

A.1.1 Right Triangles

Basic Function Definitions (See Figure A.1)

$$\sin A = \frac{a}{c} = \cos B \tag{A.1}$$

$$\cos A = \frac{b}{c} = \sin B \tag{A.2}$$

$$\tan A = \frac{a}{b} = \cot B \tag{A.3}$$

$$\sec A = \frac{c}{b} = \operatorname{cosec} B \tag{A.4}$$

$$\operatorname{cosec} A = \frac{c}{a} = \sec B \tag{A.5}$$

$$\cot A = \frac{b}{a} = \tan B \tag{A.6}$$

Derived Relationships

$$a = c \sin A = c \cos B = b \tan A = b \cot B = \sqrt{c^2 - b^2}$$
$$b = c \cos A = c \sin B = a \cot A = a \tan B = \sqrt{c^2 - b^2}$$
$$c = \frac{a}{\sin A} = \frac{a}{\cos B} = \frac{b}{\sin B} = \sqrt{a^2 + b^2}$$

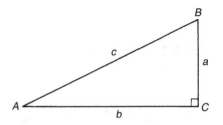

FIGURE A.1 Right triangle.

Note: The quadrant numbers in Figure A.2 reflect the traditional geometry approach (counterclockwise) to quadrant analysis. In surveying, the quadrants are numbered 1 (N.E.), 2 (S.E.), 3 (S.W.), and 4 (N.W.). The analysis of algebraic signs for the trigonometric functions (as shown) remains valid. Handheld calculators automatically provide the correct algebraic sign if the angle direction is entered in the calculator in its azimuth form.

A.1.2 Oblique Triangles

Refer to Figure A.3 for the following relationships.
Sine Law

$$\frac{a}{\sin A} = \frac{b}{\sin B} = \frac{c}{\sin C} \tag{A.7}$$

Cosine Law

$$a^2 = b^2 + c^2 - 2bc \cos A \tag{A.8}$$

$$b^2 = a^2 + c^2 - 2ac \cos B \tag{A.9}$$

$$c^2 = a^2 + b^2 - 2ab \cos C \tag{A.10}$$

Given	Required	Formulas
A, B, a	C, b, c	$c = 180 - (A + B); b = \dfrac{a}{\sin A} \sin B; c = \dfrac{a}{\sin A} \sin C$
A, b, c	a	$a^2 = -b^2 + c^2 - 2bc \cos A$
a, b, c	A	$\cos A = \dfrac{b^2 + c^2 - a^2}{2\,bc}$
a, b, c	Area	Area $= \sqrt{s(s-a)(s-b)(s-c)}$ where $s = \frac{1}{2}(a + b + c)$
C, a, b	Area	Area $= \frac{1}{2}ab \sin C$

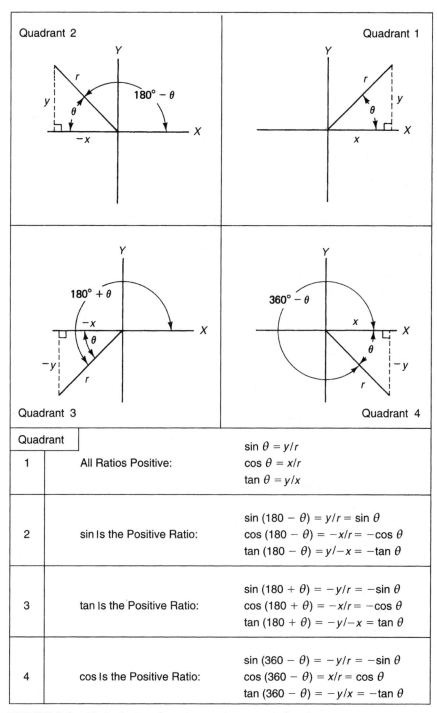

Quadrant		
1	All Ratios Positive:	$\sin \theta = y/r$ $\cos \theta = x/r$ $\tan \theta = y/x$
2	sin Is the Positive Ratio:	$\sin (180 - \theta) = y/r = \sin \theta$ $\cos (180 - \theta) = -x/r = -\cos \theta$ $\tan (180 - \theta) = y/-x = -\tan \theta$
3	tan Is the Positive Ratio:	$\sin (180 + \theta) = -y/r = -\sin \theta$ $\cos (180 + \theta) = -x/r = -\cos \theta$ $\tan (180 + \theta) = -y/-x = \tan \theta$
4	cos Is the Positive Ratio:	$\sin (360 - \theta) = -y/r = -\sin \theta$ $\cos (360 - \theta) = x/r = \cos \theta$ $\tan (360 - \theta) = -y/x = -\tan \theta$

FIGURE A.2 Algebraic signs for primary trigonometric functions.

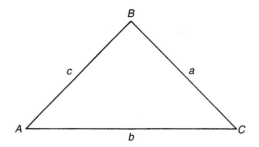

FIGURE A.3 Oblique triangle.

A.1.3 General Trigonometric Formulas

$$\sin A = 2 \sin \tfrac{1}{2} A \cos \tfrac{1}{2} A = \sqrt{1 - \cos^2 A} = \tan A \cos A \tag{A.11}$$

$$\cos A = 2\cos^2 \tfrac{1}{2} A - 1 = 1 - 2\sin^2 \tfrac{1}{2} A$$

$$= \cos^2 \tfrac{1}{2} A - \sin^2 \tfrac{1}{2} A = \sqrt{1 - \sin^2 A} \tag{A.12}$$

$$\tan A = \frac{\sin A}{\cos A} = \frac{\sin 2A}{1 + \cos 2A} = \sec^2 A - 1 \tag{A.13}$$

ADDITION AND SUBTRACTION IDENTITIES

$$\sin (A \pm B) = \sin A \cos B \pm \sin B \cos A \tag{A.14}$$

$$\cos (A \pm B) = \cos A \cos B \pm \sin A \sin B \tag{A.15}$$

$$\tan (A \pm B) = \frac{\tan A \mp \tan B}{1 \mp \tan A \tan B} \tag{A.16}$$

$$\sin A + \sin B = 2 \sin \tfrac{1}{2}(A + B) \sin \tfrac{1}{2}(A - B) \tag{A.17}$$

$$\sin A - \sin B = 2 \cos \tfrac{1}{2}(A + B) \sin \tfrac{1}{2}(A - B) \tag{A.18}$$

$$\cos A + \cos B = 2 \cos \tfrac{1}{2}(A + B) \cos \tfrac{1}{2}(A - B) \tag{A.19}$$

$$\cos A - \cos B = 2 \sin \tfrac{1}{2}(A + B) \sin \tfrac{1}{2}(A - B) \tag{A.20}$$

DOUBLE-ANGLE IDENTITIES

$$\sin 2A = 2 \sin A \cos A \tag{A.21}$$

$$\cos 2A = \cos^2 A - \sin^2 A = 1 - 2 \sin^2 A = 2 \cos^2 A - 1 \tag{A.22}$$

$$\tan 2A = \frac{2 \tan A}{1 - \tan^2 A} \tag{A.23}$$

$$\sin \frac{A}{2} = \sqrt{\frac{1 - \cos A}{2}} \tag{A.24}$$

$$\cos \frac{A}{2} = \sqrt{\frac{1 + \cos A}{2}} \tag{A.25}$$

$$\tan \frac{A}{2} = \frac{\sin A}{1 + \cos A} \tag{A.26}$$

A.2 Coordinate Geometry

Coordinate geometry was introduced in Section 6.11, where traverse station coordinates were computed along with the area enclosed by the traverse. This section (Section A.2) describes the basics of coordinate geometry along with some applications. An understanding of these concepts will help you understand the fundamentals underlying coordinate geometry software programs that are now used to process most electronic surveying field data. See Chapter 5.

A.2.1 Geometry of Rectangular Coordinates

Figure A.4 shows two points $P_1(X_1, Y_1)$ and $P_2(X_2, Y_2)$ and their rectangular relationships to the X and Y axes.

$$\text{Length } P_1 P_2 = \sqrt{(X_2 - X_1)^2 + (Y_2 - Y_1)^2} \tag{A.27}$$

$$\tan \alpha = \frac{(X_2 - X_1)}{(Y_2 - Y_1)} \tag{A.28}$$

where α is the bearing or azimuth of $P_1 P_2$. Also:

$$\text{Length } P_1 P_2 = \frac{(X_2 - X_1)}{\sin \alpha} \tag{A.29}$$

$$\text{Length } P_1 P_2 = \frac{(Y_2 - Y_1)}{\cos \alpha} \tag{A.30}$$

Use the equation having the larger numerical value of $(X_2 - X_1)$ or $(Y_2 - Y_1)$.

It is clear from Figure A.4 that $(X_2 - X_1)$ is the departure of $P_1 P_2$ and that $(Y_2 - Y_1)$ is the latitude of $P_1 P_2$. In survey work, the Y value (latitude) is known as the northing, and the X value (departure) is known as the easting.

From analytic geometry, the slope of a straight line is $m = \tan (90 - \alpha)$, where $(90 - \alpha)$ is the angle of the straight line with the X axis (Figure A.4); that is:

$$m = \cot \alpha$$

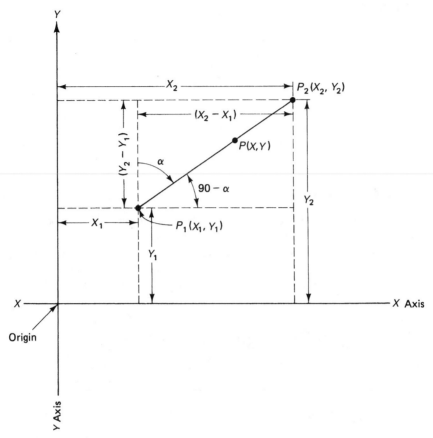

FIGURE A.4 Geometry of rectangular coordinates.

From coordinate geometry, the equation of straight line $P_1 P_2$, where the coordinates of P_1 and P_2 are known, is:

$$\frac{y - y_1}{y_2 - y_1} = \frac{x - x_1}{x_2 - x_1} \tag{A.31}$$

Equation A.31 can be written as:

$$y - y_1 = \frac{y_2 - y_1}{x_2 - x_1}(x_2 - x_1)$$

where $(y_2 - y_1)/(x_2 - x_1) = \cot \alpha = m$ (from analytic geometry), the slope. Analysis of Figure A.4 also shows that slope = rise/run = $(y_2 - y_1)/(x_2 - x_1) = \tan(90 - \alpha)$.

When the coordinates of one point (P_1) and the bearing or azimuth of a line are known, the equation becomes:

$$y - y_1 = \cot \alpha \, (x - x_1) \tag{A.32}$$

where α is the azimuth or bearing of the line through P_1 (x_1, y_1). Also from analytical geometry:

$$y - y_1 = \frac{-1}{\cot \alpha}(x - x_1) \tag{A.33}$$

which represents a line *perpendicular* to the line represented by Equation A.32; that is, the slopes of perpendicular lines are negative reciprocals.

Equations for circular curves are quadratics in the following form:

$$(x - H)^2 + (y - K)^2 = r^2 \tag{A.34}$$

where r is the curve radius, (H, K) are the coordinates of the center, and (x, y) are the coordinates of point P, which locates the circle (see Figure A.5). When the circle center is at the origin, the equation becomes:

$$x^2 + y^2 = r^2 \tag{A.35}$$

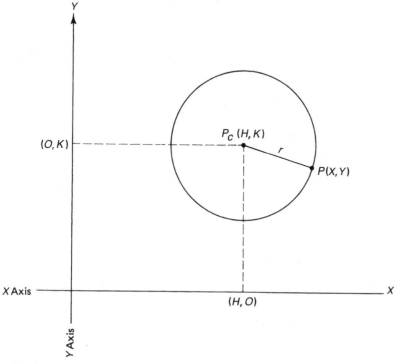

FIGURE A.5 Circular curve coordinates.

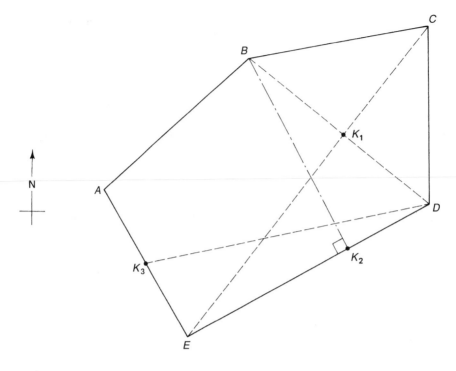

Station	Coordinates	
	North	East
A	1,000.00	1,000.00
B	1,250.73	1,313.61
C	1,302.96	1,692.14
D	934.77	1,684.54
E	688.69	1,160.27

FIGURE A.6 Coordinates for traverse problem.

A.2.2 Illustrative Problems in Rectangular Coordinates

■ EXAMPLE A.1

From the information shown in Figure A.6, calculate the coordinates of the point of intersection (K_1) of lines EC and DB. From Equation A.31, the equation of EC is:

$$\frac{y - 688.89}{1302.96 - 688.69} = \frac{x - 1160.27}{1692.14 - 1160.27} \tag{1}$$

$$y - 688.69 = \frac{1302.96 - 688.69}{1692.14 - 1160.27}(x - 1160.27)$$

$$y - 688.69 = 1.15492508x - 1340.025$$

$$y = 1.15492508x - 651.335 \qquad (1A)$$

The equation of *DB* is:

$$\frac{y - 934.77}{1250.73 - 934.77} = \frac{x - 1684.54}{1313.61 - 1684.54} \qquad (2)$$

$$y - 934.77 = \frac{1250.73 - 934.77}{1313.61 - 1684.54}(x - 1684.54)$$
$$y - 934.77 = -0.8518049x + 1434.899$$
$$y = -0.8518049x + 2369.669 \qquad (2A)$$

Substitute for *y*:

$$1.15492508x - 651.335 = -0.08518049x + 2369.669$$

$$x = 1505.436$$

Substitute the value of *x* in either (1A) or (2A):

$$y = 1087.33$$

Therefore, the coordinates of point of intersection K_1 are (1087.33 N, 1505.44 E).

Note: Coordinates are shown as (N, E) or (E, N), depending on local practice.

■ EXAMPLE A.2

From the information shown in Figure A.6, calculate (a) the coordinates of K_2, the point of intersection of line *ED*, and a line *perpendicular to ED* running through station *B*; and (b) distances K_2D and K_2E.

(a) From Equation A.31, the equation of *ED* is:

$$\frac{y - 688.69}{934.77 - 688.69} = \frac{x - 1160.27}{1684.54 - 1160.27} \qquad (1)$$
$$y - 688.69 = \frac{246.08}{524.27}(x - 1160.27) \qquad (1A)$$

From Equation A.32, the equation of BK_2 is:

$$y - 1250.73 - \frac{524.27}{246.08}(x - 1313.61) \qquad (2)$$

Simplifying, we find that these equations become (use five decimals to avoid rounding errors):

$$ED: 0.46938x - y = -144.09 \qquad (1B)$$

$$BK_2: 2.13049\,x + y = 4049.36 \qquad\qquad (2B)$$
$$2.59987x = 3905.27 \qquad\qquad (1B + 2B)$$
$$x = 1502.102$$

Substitute the value of x in Equation 1A, and check the results in Equation 2:

$$y = 849.15$$

Therefore, the coordinates of K_2 are (849.15 N, 1502.10 E).

(b) Figure A.7 shows the coordinates for stations E and D and intermediate point K_2 from Equation A.27:

$$\text{Length } K_2D = \sqrt{85.62^2 + 182.44^2} = 201.53$$

and:

$$\text{Length } K_2E = \sqrt{160.46^2 + 341.83^2} = 377.62$$
$$K_2D + K_2E = ED = 579.15$$

Check:

$$\text{Length } ED = \sqrt{246.08^2 + 524.27^2} = 579.15$$

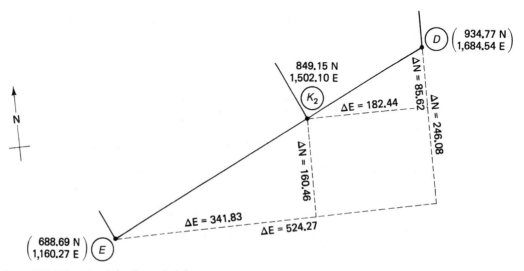

FIGURE A.7   Sketch for Example A.2.

■ EXAMPLE A.3

From the information shown in Figure A.6, calculate the coordinates of the point of intersection (K_3) on the line EA of a *line parallel to CB* running from station D to line EA. From Equation A.31:

$$CB: \frac{y - 1302.96}{1250.73 - 1302.96} = \frac{x - 1692.14}{1313.16 - 1692.14}$$

$$y - 1302.96 = \frac{-52.23}{-378.53}(x - 1692.14)$$

$$\text{Slope (cot } \alpha) \text{ of } CB = \frac{-52.23}{-378.53}$$

Because DK_3 is parallel to BC:

$$\text{Slope (cot } \alpha) \text{ of } DK_3 = \frac{-52.23}{-378.53}$$

$$DK_3: y - 934.77 = \frac{52.23}{378.53}(x - 1684.54) \tag{1}$$

$$EA: \frac{y - 688.69}{1000.00 - 688.69} = \frac{x - 1160.27}{1000.00 - 1160.27} \tag{2}$$

$$DK_3: 0.13798x - y = -702.34 \tag{1A}$$

$$EA: 1.94241x + y = +2942.41 \tag{2A}$$

$$2.08039x = +2240.07 \tag{1A + 2A}$$

$$x = 1076.7547$$

Substitute the value of x in Equation 1A, and check the results in Equation 2A:

$$y = 850.91$$

Therefore, the coordinates of K_3 are (850.91 N, 1076.75 E).

■ EXAMPLE A.4

From the information shown in Figure A.8, calculate the coordinates of the point of intersection (L) of the ₵ of Fisher Road with the ₵ of Elm Parkway.

The coordinates of station M on Fisher Road are (4,850,277.101 N, 316,909.433 E), and the bearing of the Fisher Road ₵ (ML) is S 75°10′30″ E. The coordinates of the center of the 350 M radius highway curve are (4,850,317.313 N, 317,112.656 E). The coordinates here are referred to a coordinate grid system having 0.000 m north at the equator and 304,800.000 m east at longitude 79°30′W.

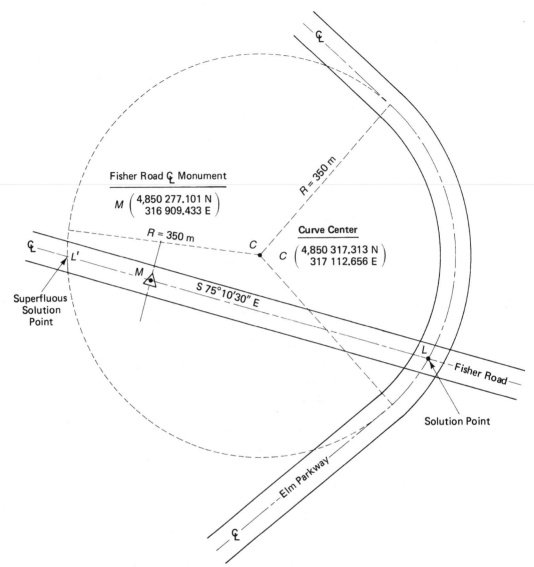

Fisher Road ℄ Monument

$M \left(\begin{array}{c} 4{,}850\ 277.101\ N \\ 316\ 909.433\ E \end{array} \right)$

$R = 350\ m$

$R = 350\ m$

Curve Center

$C \left(\begin{array}{c} 4{,}850\ 317.313\ N \\ 317\ 112.656\ E \end{array} \right)$

S 75°10'30" E

Superfluous
Solution
Point

Solution Point

Fisher Road

Elm Parkway

FIGURE A.8 Intersection of a straight line with a circular curve. (See Example A.4.)

The coordinate values are, of necessity, very large and would cause significant rounding error if they were used in calculator computations. Accordingly, an auxiliary set of coordinate axes will be used, allowing the values of the given coordinates to be greatly reduced for the computations; the amount reduced will later be added to give the final coordinates. The summary of coordinates is shown next:

	Grid Coordinates		Reduced Coordinates	
Station	y	x	$y'(y-4{,}850{,}000)$	$x'(x-316{,}500)$
M	4,850,277.101	316,909.433	277.101	409.433
C	4,850,317.313	317,112.656	317.313	612.656

From Equation A.32, the equation of the Fisher Road ℄ (*ML*) is:

$$y' - 277.101 = \cot 75°10'30'' \, (x' - 409.433) \tag{1}$$

From Equation A.34, the equation of the Elm Parkway ℄ is:

$$(x' - 612.656)^2 + (y' - 317.313)^2 = 350.000^2 \tag{2}$$

Simplify Equation 1 to:

$$y' - 277.101 = -0.2646782x' + 108.368$$

$$y' = -0.2646782x' + 385.469 \tag{1A}$$

Substitute the value of y' into Equation 2:

$$(x' - 612.656)^2 + (-0.2646782x' + 385.469 - 317.313)^2 - 350.000^2 = 0$$
$$(x' - 612.656)^2 + (-0.2646782x' + 68.156)^2 - 350.000^2 = 0$$
$$1.0700545x'^2 - 1261.391x' + 257492.61 = 0$$

This quadratic of the form $ax^2 + bx + c = 0$ has the following roots:

$$x = \frac{-b \pm \sqrt{b^2 - 4ac}}{2a}$$

$$x' = \frac{1261.3908 \pm \sqrt{1{,}591{,}107.30 - 1{,}102{,}124.50}}{2.140109}$$

$$= \frac{1261.3908 \pm 699.27305}{2.140109}$$

$$= 916.1514 \text{ or } x' = 262.658$$

Solve for y' by substituting in Equation 1A:

$$y' = 142.984 \text{ or } y' = 315.949$$

When these coordinates are now enlarged by the amount of the original reduction, the following values are obtained:

Station	Reduced Coordinates		Grid Coordinates	
	y'	x'	$y(y' + 4{,}850{,}000)$	$x(x' + 316{,}500)$
L	142.984	916.151	4,850,142.984	317,416.151
L'	315.949	262.658	4,850,315.949	316,762.658

Analysis of Figure A.8 is required in order to determine which of the two solutions is the correct one. The sketch shows that the desired intersection point L is south and east of station M; that is, L (4,850,142.984 N, 317 416.151 E) is the set of coordinates for the intersection of the ℄ of Elm Parkway and of Fisher Road. The other intersection point (L') solution is superfluous. See Figure A.8.

Appendix B

Surveying and Mapping Web Sites

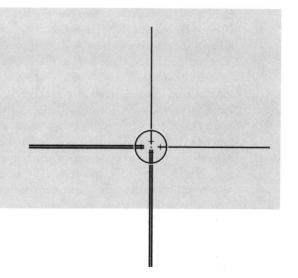

The web sites listed here cover surveying, global positioning systems (GPSs), photogrammetry, geographic information systems (GISs), and mapping. Some sites include web links to various related sites. Although the web sites were verified at the time of publication, some changes are inevitable. Revised web addresses may be accessed by searching the links shown at other sites listed here or by conducting an on-line search via a search engine.

ACAD Tutorial   www.cadtutor.net/acad/index.html
American Association of State Highway and Transportation Officials (AASHTO)
    http://www.aashto.com
American Congress on Surveying and Mapping (ACSM)   http://www.acsm.net
American Society for Photogrammetry and Remote Sensing   http://www.asprs.org
ARCINFO tutorial home page   http://boris.qub.ac.uk/shane/arc/ARChome.html
Australian Surveying and Land Information Group
    http://www.ga.gov.au/acres/reference/about_facts.jsp
Beadle's (John) Introduction to GPS Applications   http://ares.redsword.com/gps/apps
Bennett (Peter) NMEA-0183 and GPS information   http://vancouver-webpages.com/
    pub/peter/
Berntsen International, Inc. (surveying markers)   http://www.berntsen.com
Canada Centre for Remote Sensing   http://www.ccrs.nrcan.gc.ca/
Canada Centre for Remote Sensing (tutorial)
    www.ccrs.nrcan.gc.ca/ccrs/resource/index_e.php#tutor
Canadian Geodetic Survey   http://www.geod.nrcan.gc.ca
Centre for Topographic Information (Canada)   www.cits.nrcan.gc.ca
CORS information   http://www.ngs.noaa.gov/CORS/cors-data.html
Dana (Peter H.) GPS Overview
    www.colorado.edu/geography/gcraft/notes/gps/gps_ftoc.html
Flatirons Surveying Site (general surveying information)   http://www.flatsurv.com

Galileo   http://europa.eu.int/comm/dgs/energy_transport/galileo/index_en.htm

Glonass home page (Russian Federation)   http://www.glonass-center.ru/nagu.txt

GPS World magazine   http://gpsworld.com

Homeland Security, United States Coast Guard (USCG) Navigation Center (use search box for information on DGPS)   http://www.uscg.mil/uscg.shtm

International Earth Rotation Service (IERS)   http://www.iers.org

Land Surveying and Geomatics   http://surveying.mentabolism.org/

Leica Geosystems Inc.   http://www.leica-geosystems.com/

Leick GPS GLONASS GEODESY home page   http://www.spatial.maine.edu/~leick/

MAPINFO (mapping)   http://www.mapinfo.com

MicroSurvey (surveying and design software)   http://www.microsurvey.com/

NASA, EROS Data Center   http://edcwww.cr.usgs.gov

National Map   http://nationalmap.usgs.gov

Natural Resources Canada   www.nrcan.gc.ca

NGS home page   www.ngs.noaa.gov/

Nikon (surveying instruments)   http://www.nikon-trimble.com

Optech (Lidar—airborne laser terrain mapper)   www.optech.ca

POB *Point of Beginning* magazine   http://www.pobonline.com/

Professional Surveyor magazine   http://www.profsurv.com

Sokkia   http://www.sokkia.com/

Spectra Precision (Trimble)   http://www.contractorstools.com/spectra.html

Thales Navigation   products.thalesnavigation.com/en

Topcon GPS (including tutorial)   http://www.topcongps.com

Topcon Instrument Corporation   http://www.topcon.com/home.html

Trimble   http://www.trimble.com

Trimble (GIS)   www.trimble.com/mgis.shtml

Trimble GPS Tutorial   www.trimble.com/gps/index.htm

Tripod Data Systems (Trimble)   http://www.tdsway.com

United States Bureau of Land Management (BLM) geographic coordinate database www.blm.gov/gcdb

United States Bureau of Land Management (BLM) land survey information www.lsi.blm.gov

United States Bureau of Land Management (BLM) National Integrated Lands System (NILS)   www.blm.gov/nils

United States Coast Guard (USCG) Navigation Center   http://www.navcen.uscg.gov

United States Geological Survey (USGS)   http://www.usgs.gov/

United States maps and data   www.geodata.gov/

Appendix C

Glossary

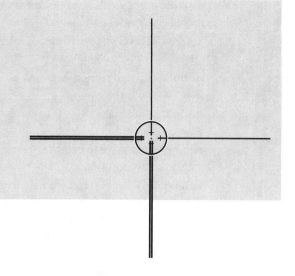

absolute positioning   The direct determination of a station's coordinates by taking GPS observations on a minimum of four GPS satellites. Also known as *point positioning*.

absorption   The process by which radiant energy is retained by a substance. The absorbing medium itself may emit energy, but only after an energy conversion has taken place.

abutment   The part of a bridge substructure that supports the end of the superstructure and retains the approach fill.

accuracy   The conformity of a measurement to the "true" value.

accuracy ratio   The error in a measurement divided by the overall value of the measurement; that is, an error of 1 ft in 3,000 ft would result in an accuracy ratio of 1/3,000. Also known as *error ratio*.

aerial survey   Preliminary and final surveys using traditional aerial photography and aerial imaging. Aerial imagery includes the use of digital cameras, multispectral scanners, lidar, and radar.

alignment   The location of the centerline of a survey or a facility.

ambiguity   The integer number of carrier cycles between the GPS receiver and a satellite.

antenna reference height (ARH)   The vertical height of a GPS receiver antenna above a control station.

arithmetic check   A check on the reductions of differential leveling involving the sums of the backsights and the foresights.

arterial road   A (highway) road mainly designed for traffic mobility—with some property access consideration.

as-built survey A postconstruction survey that confirms design execution and records in-progress revisions. Also known as *final survey*.

automatic level A surveyors' level that has the line of sight automatically maintained in the horizontal plane once the instrument is roughly leveled.

azimuth The direction of a line given by an angle, measured clockwise from a north (usually) meridian.

backfill Material used to fill an excavation.

backsight (BS) A sight taken with a level to a point of known elevation, thus permitting the surveyor to compute the elevation of the HI. In theodolite work, the backsight is a sighting taken to a point of known position to establish a reference direction.

baseline A line of reference for survey work; often the centerline, the street line, or the centerline of construction is used, although any line could be arbitrarily selected or established.

batter board A horizontal crosspiece on a grade stake or grade rod that refers to the proposed elevations.

bearing The direction of a line given by the acute angle from a meridian and accompanied by a cardinal compass direction.

bearing plate A plate that is secured to the abutment seat or pier top on which rest the beams, girders, and the like.

benchmark (BM) A fixed solid reference point with a precisely determined published elevation.

board measure A standard unit of timber measure: 1 board ft equals 1 ft square by 1 in. thick.

borrow pit A source of granular fill material that is located off the right of way.

break line A line that joins points defining significant changes in ground surface slope, such as toe and top of slope, top and bottom of ditches and swales, creek centerlines, and the like.

bucking in A trial-and-error technique of establishing a theodolite on a line between two points that themselves are not intervisible. Also known as *wiggling in* and *interlining*.

catch basin A structure designed to collect surface water and transfer it to a storm sewer.

central meridian A reference meridian in the center of the zone covered by the plane coordinate grid—at every 6° of longitude in the UTM grid.

chainage *See* station.

circular curve A curve with a constant radius.

clearing The cutting and removal of trees from a construction site.

COGO Coordinate geometry. Software programs that facilitate coordinate geometry computations used in surveying and civil engineering.

collector road A (highway) road designed to provide property access with some traffic mobility; it connects local roads to arterials.

compass rule Used in traverse balancing, an adjustment that distributes the errors in latitude and departure for each traverse course in the same proportion as the course distance is to the traverse perimeter.

compound curve Two or more circular arcs turning in the same direction and having common tangent points and different radii.

construction survey A survey to provide line and grade to a contractor for the construction of a facility.

continuously operating reference station (CORS) A GPS control station that continuously receives satellite signals and compares the time-stamped updated position coordinates to the known station coordinates. This differential positioning data can be used in postprocessing solutions or in real-time positioning by transmitting the differential data to single-receiver surveyors and navigators to permit the higher-precision differential positioning normally found only in two-receiver (multireceiver) surveys.

contour A line on a map joining points of similar elevation.

control survey A survey used to establish reference points and lines for preliminary and construction surveys.

coordinate geometry computer programs *See* COGO.

coordinates A set of numbers (X, Y) defining the two-dimensional position of a point given by the distances measured north and east of an origin reference point having coordinates of (0, 0).

CORS *See* continuously operating reference station.

cross section A profile of the ground, or the like, that is taken at right angles to a reference line.

crown The uppermost point on a road, pipe, or cross section. The rate of cross fall on pavement.

culvert A structure designed to provide an opening under a road, or the like, usually for the transportation of storm water.

cut In construction, the excavation of material. The measurement down from a grade mark.

cutoff angle *See* mask angle.

cycle slip A temporary loss of lock on satellite carrier signals, causing a miscount in carrier cycles. Lock must be reestablished to continue GPS positioning solutions.

data collector An electronic field book designed to accept field data—both measured and descriptive.

datum An assumed or a fixed reference plane.

deck The floor of a bridge.

deflection angle The angle between the prolongation of the back survey line measured right (R) or left (L) to the forward survey line.

departure (dep) The departure distance is the change in easterly displacement of a line. It is computed by multiplying the distance times the sine of its azimuth or bearing.

differential leveling   A technique for determining the differences in elevation between points using a surveyor's level.

differential positioning   Obtaining measurements at a known base station to correct simultaneous measurements made at rover receiving stations.

double centering   A technique of turning angles or producing straight lines involving a minimum of two sightings with a theodolite—once with the telescope direct and once with the telescope inverted.

drainage   The collection and transportation of ground and storm water.

DTM Digital terrain model.   A three-dimensional depiction of a ground surface—usually produced by software programs [sometimes referred to as a digital elevation model (DEM) with breaklines].

DXF Drawing exchange format.   An industry standard format that permits graphical data to be transferred among various data collector, CAD, GIS, and soft-copy photogrammetry applications programs.

EDM   Electronic distance measurement.

EFB   Electronic field book. *See* data collector.

elevation   The distance above, or below, a given datum. Also known as *orthometric height*.

elevation factor   The factor used to convert ground distances to sea-level distances.

engineering surveys   Preliminary and layout surveys used for engineering design and construction.

EOS   Earth observing system. NASA's study of the earth, scheduled to cover the period 2000 to 2015, in which a series of small to intermediate earth observation satellites will be launched to measure global changes. The first satellite (experimental) in the series (TERRA), was launched in 1999.

epoch   An observational event in time that forms part of a series of GPS observations.

error   The difference between a measured, or observed, value and the "true" value.

ETI$^+$ Enhanced thematic mapper. An eight-band, multispectral scanning radiometer, on-board Landsat 7, that is capable of providing relatively high resolution (15-m) imaging information about the earth's surface.

existing ground (EG)   The position of the ground surface just prior to construction. *See also* original ground.

external distance   The distance from the midcurve to the PI in a circular curve.

Father Point   A general adjustment of Canadian-U.S.-Mexican leveling observations resulted in the creation of the North American Vertical Datum of 1988 (NAVD88) in 1991. The adjustment held fixed the height of the primary tidal benchmark located at Father Point, Rimouski, Quebec, on the south shore of the St. Lawrence River.

fiducial marks   Reference marks on the edges of aerial photos, used to locate the principal point on the photo.

fill   Material used to raise the construction level. The measurement up from a grade mark.

final survey   *See* as-built survey.

footing   The part of the structure that is placed in, or on, the ground upon which the main structure rests.

forced centering   The interchanging of theodolites, prisms, and targets into tribrachs that have been left in position over the station.

foresight (FS)   In leveling, a sight taken to a BM or TP to obtain a check on a leveling operation or to establish a transfer elevation.

foundation   The portion of the structure that rests on the footing.

four-foot mark   A reference mark, used in building construction that is 4 ft above the finished first-floor elevation.

free station   A conveniently located instrument station used for construction layout, the position of which is determined after occupation, through resection techniques.

freeway   A highway designed for traffic mobility, with access restricted to interchanges with arterials and other freeways.

gabion   A wire basket filled with fragmented rocks or concrete, often used in erosion control.

general dilution of precision (GDOP)   A value that indicates the relative uncertainty in position, using GPS observations, caused by errors in time (GPS receivers) and satellite vector measurements. A minimum of four widely spaced satellites at high elevations usually produce good results (i.e., lower GDOP values).

geocoding   The linking of entity and attribute data to a specific geographic location.

geodetic datum   In North America, a precisely established and maintained series of benchmarks referenced to mean sea level (MSL) tied to the vertical control station Father Point, which is located in the St. Lawrence River Valley.

geodetic height (h)   The distance from the ellipsoid surface to the ground surface.

geodetic survey   A survey that reflects the curved (ellipsoidal) shape of the earth.

geographic information system (GIS)   A spatially and relationally referenced database.

geographic meridian   A line on the surface of the earth joining the poles; that is, a line of longitude.

geoid   A surface that is approximately represented by mean sea level (MSL) and is, in fact, the equipotential surface of the earth's gravity field. This surface is everywhere normal to the direction of gravity.

geoid height   *See* geoid undulation.

geoid separation   *See* geoid undulation.

geoid undulation (N)   The difference in elevation between the geoid surface and the ellipsoid surface. N is negative if the geoid surface is below the ellipsoid surface (also known as the *geoid height* and *geoid separation*).

geomatics　A term used to describe the science and technology dealing with earth measurement data, including collection, sorting, management, planning and design, storage, and presentation. It has applications in all disciplines and professions that use earth-related spatial data, for example, planning, geography, geodesy, infrastructure engineering, agriculture, natural resources, environment, land division and registration, project engineering, and mapping.

geostationary orbit　A satellite orbit such that it appears the satellite is stationary over a specific location on earth. A formation of geostationary satellites presently provides communications services worldwide. Also known as *geosynchronous orbit*.

GIS　*See* geographic information system.

global positioning system　*See* GPS.

gon　A unit of angular measure in which 1 revolution = 400 gon and 100.000 gon = a right angle. Also known as *grad*.

GPS　Global positioning system. A ground positioning (Y, X, and Z) technique based on the reception and analysis of NAVSTAR satellite signals.

grade sheet　A construction report giving line and grade, that is, offsets and cuts/fills at each station.

grade stake　A wood stake with a cut/fill mark referring to that portion of a proposed facility adjacent to the stake.

grade transfer　A technique of transferring cut/fill measurements to the facility. Typical techniques include carpenter's level, stringline level, laser, and batter boards.

gradient　The slope of a gradeline.

grid distance　A distance on a coordinate grid.

grid factor　A factor used to convert ground distances to grid distances.

grid meridian　A meridian parallel to a central meridian on a coordinate grid.

ground distance　A distance as measured on the ground surface.

grubbing　The removal of stumps, roots, and debris from a construction site.

Gunter's chain　Early (1800s) steel measuring device consisting of 100 links, 66 ft long.

haul　The distance that 1 cubic yard (meter) of cut material is transported to a fill location in highway construction.

head wall　A vertical wall at the end of a culvert that is used to keep fill material from falling into the creek or watercourse.

hectare　10,000 square meters.

height of instrument (HI)　In leveling, the height of the line of sight of the level above a datum.

height of instrument (hi)　In total station and theodolite work, the height of the instrument's optical axis above the instrument station.

horizontal line　A straight line perpendicular to a vertical line.

image rectification   The extraction of planimetric and elevation data from stereo-paired aerial photographs for the preparation of topographic maps.

interlining   A trial-and-error technique of establishing the theodolite or total station on a line between two points that are not intervisible. Also known as *bucking in* and *wiggling in*.

intermediate sight (IS)   A sight taken by a level or theodolite to determine a feature elevation and/or location.

invert   The inside bottom of a pipe or culvert.

ionosphere   That section of the earth's atmosphere that is about 50 km to 1,000 km above the earth's surface.

ionospheric refraction   In GPS, the impedance in the velocity of signals as they pass through the ionosphere.

laser   An acronym for *l*ight *a*mplification by *s*timulated *e*mission of *r*adiation. Construction lasers employ either visible light or infrared beams to provide construction control for both line and grade.

laser alignment   The horizontal and/or vertical alignment given by a fixed or rotating laser.

latitude (lat)   When used in reference to a traverse course, the latitude distance is the change in northerly displacement (ΔN) of a line. It is computed by multiplying the distance by the cosine of its azimuth or bearing. In geographic terms, the latitude is an angular distance measured northerly or southerly, at the earth's center, from the equator.

layout survey   A construction survey.

level line   A line in a level surface.

lidar   Light detection and ranging. This laser technique, used in airborne and satellite imagery, utilizes laser pulses that are reflected from ground features to obtain topographic and DTM mapping detail.

line   A GIS term describing the joining of two points and having one dimension.

line and grade   The designed horizontal and vertical position of a facility.

linear error of closure   The line of traverse misclosure representing the resultant of the measuring errors.

local road   A (highway) road designed for property access, connected to arterials by collectors.

longitude   An angular distance measured, at earth's center, east or west, in the plane of the equator, from the reference meridian through Greenwich, England. Lines of longitude are shown on globes as meridians.

magnetic meridian   A line parallel to the direction taken by a freely moving magnetized needle, as in a compass.

manhole (MH)   A structure that provides access to underground services.

mask angle   The vertical angle below which satellite signals are not recorded or processed. A value of 10° or 15° is often used. Also known as the *cutoff angle*.

mass diagram   A graphic representation of cumulative highway cuts and fills.

mean sea level (MSL)   A reference datum for leveling.

meridian   A north-south reference line.

midordinate distance   The distance from the midchord to the midcurve in a circular curve.

mistake   A poor result due to carelessness or a misunderstanding.

monument   A permanent reference point for horizontal and vertical positioning.

MSL   *See* mean sea level.

multispectral scanner   Scanning device used for satellite and airborne imagery to record reflected and emitted energy in two or more bands of the electromagnetic spectrum.

nadir   A vertical angle measured from the nadir direction (straight down) upward to a point.

NAVSTAR   A set of about thirty orbiting satellites used in navigation and positioning.

normal tension   The amount of tension required in taping to offset the effects of sag.

original ground (OG)   The position of the ground surface just prior to construction. *See also* existing ground.

orthometric height (H)   The distance (measured along a plumb line) from the geoid surface to the ground surface. Also known as *elevation*.

orthophoto maps   *See* planimetric maps.

page check   An arithmetic check of leveling notes.

parabolic curve   A curve used in vertical alignment to join two adjacent gradelines.

parallax   An error in sighting that occurs when the objective and/or the cross hairs are improperly focused.

photogrammetry   The science of taking accurate measurements from aerial photographs.

pier   A vertical column supporting beams, girders, and the like, which help to span the distance between abutments.

plan   Bird's-eye view of a route location, the same as if you are in an aircraft looking straight down. Gives the horizontal location of a facility, including curve radii.

plan and profile   Form the "blueprint" from which the construction is accomplished; usually also show the cross-section details and construction notes.

plane survey   A survey that ignores the curvature of the earth.

planimeter   A mechanical or electronic device used to measure area by tracing the outline of the area on a map or plan.

planimetric maps   Scaled maps showing the scaled horizontal locations of both natural and cultural feautures. Also known as *orthophoto maps*.

point   A GIS term describing a single spatial entity and having zero dimension.

polar coordinates The coordinates that locate a feature by distance and angle (r, θ).

polygon A GIS term for a closed chain of points representing an area.

precision The degree of refinement with which a measurement is made.

preengineering survey A preliminary survey that forms the basis for engineering design.

preliminary survey *See* preengineering survey.

profile A series of elevations taken along a baseline at some specified repetitive station interval. A side view or elevation of a route in which the longitudinal surfaces are highlighted.

property survey A survey to retrace or establish property lines or to establish the location of buildings with respect to property limits.

pseudorange The uncorrected distance from a GPS satellite to a ground receiver, determined by comparing the code transmitted from the satellite to the receiver's on-board replica code. To convert to the actual range (distance from the satellite to the receiver), corrections must be made for clock errors as well as natural and instrumental errors.

random errors Errors associated with the skill and vigilance of the surveyor.

real-time positioning A survey technique that requires a base station to measure the satellites' signals, process the baseline corrections, and then broadcast the corrections (differences) to any number of roving receivers that are simultaneously tracking the same satellites. Also known as real-time kinematic (RTK) positioning.

rectangular coordinates Two distances, 90° opposed, that locate a feature with respect to a baseline.

relative positioning The determination of position through the combined computations of two or more receivers simultaneously tracking the same satellites, thus resulting in the determination of the baseline vector (X, Y, Z) joining the two receivers.

remote object elevation The determination of the height of an object by a total station sighting and utilizing on-board applications software.

remote sensing Geodata collection and interpretive analysis for both airbone and satellite imagery.

resection The solution of the coordinate determination of an occupied station by the angle sighting of three or more coordinated reference stations (two or more stations if both angles and distances are measured).

retaining wall A wall built to hold back the embankment adjacent to a structure.

right of way (ROW) The legal property limits of a utility or access route.

route surveys Preliminary, control, and construction surveys that cover a long, but narrow area—as in highway and railroad construction.

ROW *See* right of way.

sag The error caused when a measuring tape is supported only at the ends, with insufficient tension.

scale factor The factor used to convert sea-level distances to plane grid distances.

sea-level correction factor The factor used to convert ground distances to sea-level-equivalent distances.

shaft An opening of uniform cross section joining a tunnel to the surface. It is used for access and ventilation.

shrinkage The decrease in volume that occurs when excavated material is placed on site under compaction.

skew number A clockwise angle (closest 5°) turned from the back tangent to the centerline of a culvert or bridge.

slope stake A stake placed to identify the top or bottom of a slope.

soft-copy (digital) stereoplotting The analysis (largely automated) of digitized images through the use of sophisticated algorithms.

span The unsupported length of a structure.

spiral curve A transition curve of constantly changing radius placed between a high-speed tangent and a central curve. It permits a gradual speed adjustment.

springline The horizontal bisector of a sewer pipe, above which connections may be made.

station A point on a baseline that is a specified distance from the point of commencement. The point of commencement is identified as $0 + 00$; 100 ft, 100 m, or 1,000 m (in some highway applications) are known as full stations; $1 + 45.20$ ($0 + 45.20$ in some highways applications) identifies a point 145.20 ft (m) distant from the point of commencement.

stripping The removal of topsoil from a construction site.

superelevation The banking of a curved section of road to help overcome the effects of centrifugal force.

survey drafting A term that covers a wide range of scale graphics, including both manual and digital plotting.

surveying The art and science of taking field measurements on or near the surface of the earth.

swell The increase in volume when blasted rock is placed on site.

systematic errors Errors whose magnitude and algebraic sign can be determined.

tangent A straight line, often referred to with respect to a curve.

temporary benchmark (TBM) A semipermanent reference point of known elevation.

three-wire leveling A more precise technique of differential leveling in which rod readings are taken at the stadia hairs in addition to the main cross hair in a telescope.

toe of slope The bottom of a slope.

topographic map The maps showing spot elevations and contours.

total station   An electronic theodolite combined with an EDM instrument, a central processor, and an electronic data collector. It is used to read and record angles and distances and to perform various surveying computations.

traverse   A continuous series of measured lines (angles and distances are measured).

triangulation   A control survey technique involving (1) a precisely measured baseline as a starting side for a series of triangles or chain of triangles; (2) the measurement of each angle in the triangle using a precise theodolite, which permits the computation of the lengths of each side; and (3) a check on the work made possible by precisely measuring a side of a subsequent triangle (the spacing of check lines depends on the desired accuracy level).

trilateration   The control surveying solution technique of measuring only the sides in a triangle.

turning point (TP)   A solid point used in leveling where an elevation is established temporarily so that the level may be relocated.

universal transverse Mercator system (UTM)   A worldwide grid system based on sixty zones, each 6° of longitude wide. The grid covers from 80° north latitude to 80° south latitude. Each central meridian, at the middle of the zone, has an x (easterly) value of 500,000.000 m, and the equator has a y (northerly) value of 0.000 m. The scale factor at the central meridian is 0.9996.

vertical angle   An angle in the vertical plane measured up ($+$) or down ($-$) from horizontal.

vertical curve   A parabolic curve joining two gradelines.

vertical line   A line from the surface of the earth to the earth's center. Also known as *plumb line* or *line of gravity*.

waving the rod   The slight waving of the leveling rod to and from the instrument, which permits the surveyor to take a more precise (lowest) rod reading.

wing wall   An abutment extension designed to retain the approach fill.

zenith   A point on a celestial sphere (i.e., a sphere of infinitely large radius, with its center at the center of the earth) that is directly above the observer.

zenith angle   A vertical angle measured downward from the zenith (upward plumbline) direction. This is the vertical angle measured by most modern theodolites and total stations.

Appendix D

Typical Field
Projects

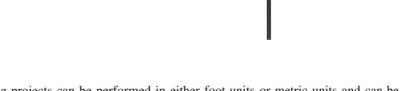

The following projects can be performed in either foot units or metric units and can be adjusted in scope to fit the available time.

D.1 Field Notes

Survey field notes can be entered into a bound field book or on loose-leaf field notepaper. If a bound field book is used, be sure to leave room for a title, index, and diary at the front of the book, if they are required by your instructor. Instructions and sample layout for completing a field book are listed below:

1. Write your name in ink on the outside cover.
2. Page numbers (for example, 1 to 72) are to be on the right side only. See Figure D.1 for examples of pages 1–7.
3. All entries should be in pencil, 2H or 3H.
4. All entries should be printed (uppercase, if permitted).
5. Calculations should be checked and initialed.
6. Sketches should be used to clarify field notes. Orient the sketch so that the included north arrow points toward the top of the page.
7. All field notes should be brought up to date each day.
8. Show the word *copy* at the top of all copied pages.
9. Field notes should be entered directly in the field book, not on scraps of paper to be copied later.
10. Mistakes in entered data are to be lined out, not erased.
11. Spelling mistakes, calculation mistakes, and the like, should be erased and reentered correctly.

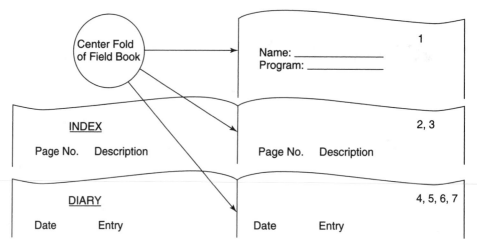

FIGURE D.I Typical field book layout.

12. Lettering is to be read from the bottom of the page or the right side.
13. The first page of each project should show the date, temperature, project title, crew duties, and so on.
14. The diary should show absentees, weather, description of work, and so on.

D.2 Project 1: Building Measurements

Description Measure the selected walls of an indicated campus building with a nylon-clad steel or fiberglass tape. Record the measurements on a sketch in the field book, as directed by the instructor (see Figure D.2 for sample field notes).

Equipment Nylon-clad steel tape or fiberglass tape (100 ft or 30 m).

Purpose To introduce you to the fundamentals of note keeping and survey measurement.

Procedure Use the measuring techniques described in class prior to going out.

- One crew member should be appointed to take notes for this first project. At the completion of the project (same day), the other crew members should copy the notes (ignoring erroneous data) into their field books. Include diary and index data. (Crew members should take equal turns acting as note keeper over the length of the program).
- Using a straightedge, draw a large sketch of the selected building walls on the right-hand (grid) side of your field book. Show the walls as they would appear in a plan view; for example, ignore overhangs, or show them as dashed lines. Keep the tape taut to remove sag, and try to keep the tape horizontal. If the building wall is longer than one tape length, make an intermediate mark (do not deface the building), and proceed from that point.

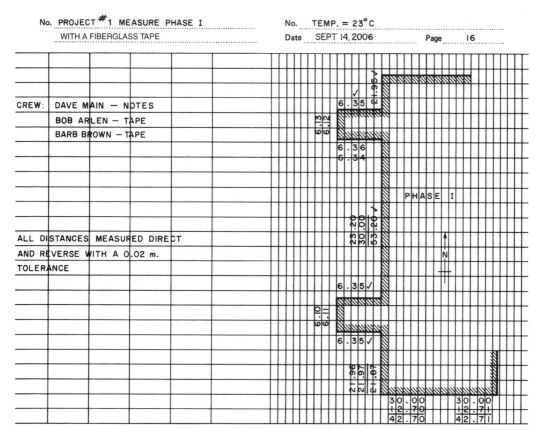

CREW: DAVE MAIN — NOTES
        BOB ARLEN — TAPE
        BARB BROWN — TAPE

PHASE I

ALL DISTANCES MEASURED DIRECT
AND REVERSE WITH A 0.02 m.
TOLERANCE

N

6.35 ✓

6.35 ✓

30.00   30.00
12.70   12.71
42.70   42.71

FIGURE D.2 Sample taping field notes for Project 1 (building dimensions).

• After completing all the measurements in one direction, start from the terminal point and remeasure all the walls. If the second measurement agrees with the first measurement, put a check mark beside the entered data. If the second measurement agrees acceptably (e.g., within ±0.10 ft or ±0.02 m), enter that measurement directly above or below the first entered measurement. If the second measurement disagrees with the first, (e.g., by more than 0.10 ft or 0.02 m—or other value given by your instructor), enter that value on the sketch and measure that dimension a third time. Discard the erroneous measurement by drawing a line (using a straightedge) through the erroneous value.

Discussion   If the class results are summarized on the chalkboard, it will be clear that all crews did not obtain the same results for specified building wall lengths. There will be much more agreement among crews on the lengths of the shorter building walls than on the lengths of the longer building walls (particularly the walls that were longer than one tape length). Discuss and enumerate the types of mistakes and errors that could account for measurement discrepancies among survey crews working on this project.

D.3 Project 2: Experiment to Determine Normal Tension

Description   Determine the tension required to eliminate errors due to tension and sag for a 100-ft (or 30-m) steel tape supported only at the ends. This tension is called normal tension.

Equipment   Steel tape (100.00 ft. or 30.000 m), two plumb bobs, and a graduated tension handle.

Purpose   To introduce you to measurement techniques requiring the use of a steel tape and plumb bob, and to demonstrate the "feel" of the proper tension required when using a tape that is supported only at the ends (the usual case).

Procedure

- With a 100-ft or 30-m tape fully supported on the ground and under a tension of 10 lbs or 50 N (standard tension) as determined by use of a supplied tension handle, measure from the initial mark and place a second mark at exactly 100.00 ft (or 30.000 m).
- Check this measurement by repeating the procedure (while switching personnel) and correcting if necessary. (If this initial measurement is not performed correctly, much time will be wasted.)
- Raise the tape off the ground and keep it parallel to the ground to eliminate slope errors.
- Using plumb bobs and the tension handle, determine how many pounds or Newtons of tension (see Table F.1) are required to force the steel tape to agree with the previously measured distance of 100.00 ft or (30.000 m).
- Repeat the process while switching crew personnel. (Acceptable tolerance is ±2 lbs.)
- Record the normal tension results (at least two) in the field book, as described in the classroom.
- Include the standard conditions for the use of steel tapes in your field notes (see Table D.1). For this project, you can assume that the temperature is standard, 68°F or 20°C.

Discussion   If the class results are summarized (e.g., on the chalkboard), it will be clear that not all survey crews obtained the same average value for normal tension. Discuss the reasons for the tension measurement discrepancies and agree on a "working" value for normal tension for subsequent class projects.

Table D.1   STANDARD CONDITIONS FOR THE USE OF STEEL TAPES

English System	or	Metric System
Temperature = 68°F		Temperature = 20°C
Tension = 10 lbs		Tension = 50 N (11.2 lbs)
Tape is fully supported.		Tape is fully supported.

D.4 Project 3: Field Traverse Measurements with a Steel Tape*

Description Measure the sides of a five-sided closed field traverse using techniques designed to permit a precision closure ratio of 1:5,000 (see Table 2.1). The traverse angles can be obtained from Project 5. See Figure D.3 for sample field notes.

Equipment Steel tape, two plumb bobs, hand level, range pole or plumb bob string target, and chaining pins or other devices to mark position on the ground.

Purpose To enable you to develop some experience in measuring with a steel tape and plumb bobs (or electronically) in the first stage of a traverse closure exercise (see also Project 5).

Procedure

- Each course of the traverse is measured twice—forward (direct) and then immediately back (reverse)—with the two measurements agreeing to within 0.03 ft or 0.008 m. If the two measurements do not agree, repeat them until they do and before the next course of the traverse is measured.
- When measuring on a slope, the high-end surveyor normally holds the tape directly on the mark, and the low-end surveyor uses a plumb bob to keep the tape horizontal.

No. PROJECT #3

TRAVERSE DISTANCES

No.

Date MARCH 29, 2006   Page   10

COURSE	DIRECT	REVERSE	MEAN	MEAN (C_T)
111–112	164.96	164.94	164.95	164.97
112–113	88.41	88.43	88.42	88.43
113–114	121.69	121.69	121.69	121.70
114–115	115.80	115.78	115.79	115.80
115–111	68.36	68.34	68.35	68.36

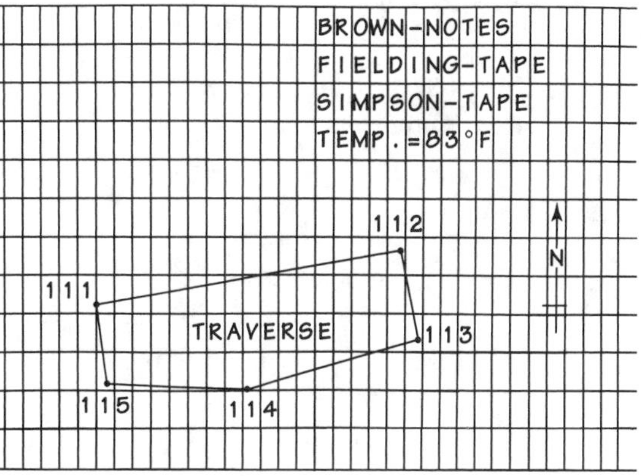

BROWN–NOTES

FIELDING–TAPE

SIMPSON–TAPE

TEMP. = 83°F

TRAVERSE

FIGURE D.3 Sample field notes for Project 3 (traverse distances).

*Projects 3 and 5 can be combined, using EDM-equipped theodolites, or total stations, and reflecting prisms. The traverse courses can be measured using an EDM instrument, or total station, and a prism (pole-mounted or tribrach-mounted). Each station can be occupied with a theodolite-equipped EDM instrument, or total station, and each pair of traverse courses can be measured at each setup. Traverse computations can use the mean distances thus determined and the mean angles obtained from each setup. Reference: Chapters 5 and 6.

- The low-end surveyor uses the hand level to keep the tape approximately horizontal by sighting the high-end surveyor and noting how much lower she or he is in comparison. The plumb bob can then be set to that height differential. Use chaining pins or other markers to mark intermediate measuring points temporarily on the ground. Use scratch marks or concrete nails on paved surfaces.
- If a range pole is first set behind the far station, the rear surveyor can keep the tape aligned properly by sighting at the range pole and directing the forward surveyor on line.
- Book the results as shown in Figure D.3. Then repeat the process until all five sides have been measured and booked. When booked erroneous measurements are to be discarded, strike them out using a straightedge—don't erase.
- If the temperature is something other than standard, correct the mean distance for temperature. Use C_T; that is:

$$C_T = .00000645(T - 68)L_\text{ft} \quad \text{or} \quad C_T = .0000116\,(T - 20)L_\text{m}$$

Reference   Chapter 2.

D.5   Project 4: Differential Leveling

Description   Use the techniques of differential leveling to determine the elevations of temporary benchmark (TBM) 33 and of the intermediate stations (if any) identified by the instructor. See Figure D.4 for sample field notes.

Equipment   Survey level, rod, and rod level (if available).

Purpose   To give you experience in the use of levels and rods and in properly recording all measurements in the field book.

Procedure

- Start at the closest municipal or college benchmark (BM) (description given by the instructor), and take a backsight (BS) reading to establish a height of instrument (HI).
- Insert the description of the BM (and all subsequent TPs), in detail, under Description in the field notes.
- Establish a turning point (TP 1) generally in the direction of the defined terminal point (TBM 33) by taking a foresight (FS) on TP 1.
- When you have calculated the elevation of TP 1, move the level to a convenient location and set it up again. Take a BS reading on TP 1, and calculate the new HI.
- The rod readings taken on any required intermediate points (on the way to or from the terminal point) are booked in the IS (intermediate sight) column—unless some of those intermediate points are also being used as turning points (TPs) (e.g., see TP 4 in Figure D.4).

STA	BS	HI	IS	FS	ELEV	DESCRIPTION
						BROWN-INST.
						SMITH-ROD
						TEMP.=65°F
BM 21	0.54	182.31			181.77	BM. BRONZE PLATE ON E. WALL OF S.E.
						STAIRWELL OF PHASE 1 BLDG., ABOUT
						1 m ABOVE THE GROUND.
TP 1	0.95	175.04		8.22	174.09	N. LUG ON TOP FLANGE OF HYD.@
						E/SIDE OF BUS SHELTER
TP 2	0.80	168.76		7.08	167.96	SPIKE IN S/SIDE OF HP @ 237 FINCH
						AVE.
TP 3	0.55	160.20		9.11	159.65	SPIKE IN S/SIDE OF HP @ 245 FINCH
						AVE.
111			4.22			TRAVERSE STATION I.B.
112			4.71			TRAVERSE STATION I.B.
113			2.03			TRAVERSE STATION I.B.
114			1.22			TRAVERSE STATION I.B.
TP 4	3.77	163.45		0.52	159.68	TOP OF I.B. @ STA. 115
TBM 33				1.18	162.27	BRASS CAP ON CONC. MON.-CONTROL
						STATION 1102
TBM 33	1.23	163.50			162.27	
TP 4	2.71	162.39		3.82	159.68	
TP 3	8.88	168.53		2.74	159.65	
TP 5	11.86	177.38		3.01	165.52	TOP OF N.E. CORNER OF CONCRETE
						STEP @ 233 FINCH AVE.
TP 1	10.61	184.72		3.27	174.11	
BM 21				2.94	181.78	(e=+0.01)
ΣBS = 41.90			ΣFS = 41.89			
ΣBS,	41.90	− ΣFS,	41.89 =	0.01		181.77 +.01 = 181.78, CHECK

FIGURE D.4 Sample field notes for Project 4 (differential leveling).

- If you can't see the terminal point (TBM 33) from the new instrument location, establish additional turning points (TP 2, TP 3, etc.) by repeating the steps above until you can take a reading on the terminal point.
- After you have taken a reading (FS) on the terminal point (TBM 33) and calculated its elevation, move the level slightly, and set it up again. Now take a BS on the terminal point (TBM 33), and prepare to close the level loop back to the starting benchmark.
- Repeat the leveling procedure until you have surveyed back to the original BM. If you use the original TPs on the way back, book them by their original numbers (you do not have to describe them again). If you use new TPs on the way back, describe each TP in detail under Description and assign each one a new number.

- If the final elevation of the starting BM differs by more than 0.04 ft (or 0.013 m) from the starting elevation (after the calculations have been checked for mistakes by performing an arithmetic check—also known as a page check), repeat the project. Your instructor may give you a different closure allowance, depending on the distance leveled and/or the type of terrain surveyed.

Notes

- Keep BS and FS distances from the instrument roughly equal.
- Wave the rod (or use a rod level) to ensure a vertical reading (see Figure 3.13).
- Eliminate parallax for each reading.
- Use only solid (steel, concrete, or wood) and well-defined features for TPs. If you cannot describe a TP precisely, do not use it!
- Perform an arithmetic check on the notes before assessing closure accuracy.

Reference   Chapter 3.

D.6   Project 5: Traverse Angle Measurements and Closure Computations

Description   Measure the angles of a five-sided field traverse using techniques consistent with the desired precision ratio of 1/5,000 (see also Project 3).

Equipment   Theodolite or total station and a target device (range pole or plumb bob string target, or prism).

Purpose   To introduce you to the techniques of setting a theodolite or total station over a closed traverse point, turning and doubling interior angles, and checking your work by calculating the geometric angular closure. Then compute (using latitudes and departures) a traverse closure to determine the precision ratio of your field work.

Procedure   If you are using a total station with a traverse closure program, use that program to check your calculator computations. For traverse computation purposes, assume a direction for one of the traverse courses, or use one supplied by your instructor. See Figure D.5 for sample field notes.

- Using the same traverse stations that you used for Project 3, measure each of the five angles (direct and double).
- Read all angles from left to right. Begin the first (of two) angles at 0°00'00" (direct), and begin the second angle (double) with the value of the first angle reading.
- Transit the telescope between the direct and double readings.

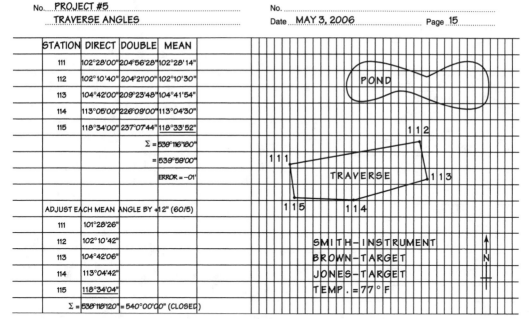

STATION	DIRECT	DOUBLE	MEAN
111	102°28'00"	204°56'28"	102°28'14"
112	102°10'40"	204°21'00"	102°10'30"
113	104°42'00"	209°23'48"	104°41'54"
114	113°05'00"	226°09'00"	113°04'30"
115	118°34'00"	237°07'44"	118°33'52"
		Σ = 539°116'180"	
		= 539°59'00"	
		ERROR = −01'	
ADJUST EACH MEAN ANGLE BY +12" (60/5)			
111	101°28'26"		
112	102°10'42"		
113	104°42'06"		
114	113°04'42"		
115	118°34'04"		
Σ = 539°118'120" = 540°00'00" (CLOSED)			

FIGURE D.5 Sample field notes for Project 5 (traverse angles).

- Divide the double angle by 2 to obtain the mean angle. If the mean angle differs by more than 30″ (or another value given by your instructor) from the direct angle, repeat the steps above.
- When all the mean angles have been booked, add the angles to determine the geometric angular closure.
- If the geometric closure exceeds 01′ (30″ $\sqrt{N}$), find the error.
- Combine the results from Projects 3 and 5 to determine the precision closure of the field traverse (1/5,000 or better is acceptable). Use an assumed direction for one of the sides (see above).

D.7   Project 6: Topographic Survey

Topographic field surveys can be accomplished in several ways. For example, you can use:

- Cross sections and tie-ins with a manual plot of the tie-ins, cross sections, and contours.
- Theodolite/EDM with a manual plot of the tie-ins and contours.
- Total station with a computer-generated plot on a digital plotter.

Purpose   Each type of topographic survey shown in this section is designed to give you experience in collecting field data (location details and elevations) using different specified surveying equipment and surveying procedures. The objective for each approach to topographic surveying is the same—that is, the production of a scaled map or plan showing all

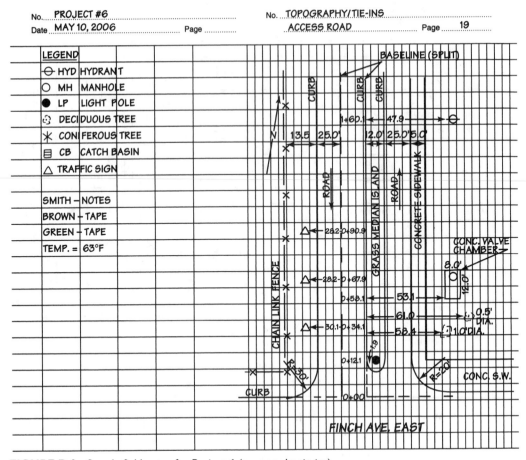

FIGURE D.6   Sample field notes for Project 6 (topography tie-ins).

relevant details and height information (contours and spot elevations) of the area surveyed. Time and schedule constraints normally limit most programs to include just one or two of these approaches.

D.7.1   Cross Sections and Tie-Ins Topographic Survey

Description   Using the techniques of right-angle tie-ins and cross sections (both referenced to a baseline), locate the positions and elevations of selected features on the designated area of the campus. See Figures D.6, D.7, and D.8).

Equipment   Nylon-clad steel tape or fiberglass tape and two plumb bobs. A right-angle prism is optional.

No. PROJECT #6 | No. TOPOGRAPHY/CROSS SECTIONS
Date MAY 17, 2006 Page | FINCH AVE. E. TO 3+00N. Page 23

STA.	B.S.	H.I.	I.S.	F.S.	ELEV.	DESCRIPTION
BM#3	8.21	318.34			310.13	S.W. CORNER OF CONC. VALVE CHAMBER
						@ 0+53.1
0+00			3.34		315.00	℄, ON ASPH.
			0.03		318.31	75.0' LT, ON ASPH.
			7.35		310.99	50.0' RT, ON ASPH.
0+50			6.95		311.39	℄, ON ASPH.
			7.00		311.34	25' LT, BOT. CURB
			0.3		318.0	38.5' LT, @ FENCE, ON GRASS
			7.5		310.8	6' RT, ℄ OF ISLAND, ON GRASS
			8.32		310.02	12' RT, BOT. CURB
			8.41		309.93	37.0' RT, BOT. CURB
			8.91		309.43	37.0' RT, TOP OF CONC. WALK
			9.01		309.33	42.0' RT, TOP OF CONC. WALK
			9.3		309.0	46.9' RT, TOP OF HILL, ON GRASS
			11.7		306.6	56.6' RT, @ BUILDING WALL, ON GRASS

FIGURE D.7 Sample field notes for Project 6 (topography cross sections).

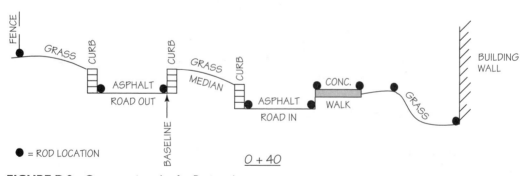

FIGURE D.8 Cross-section plot for Project 6.

Procedure

- Establish your own baseline using wood stakes or pavement nails, or use a curb line, as shown in Figure D.6, as the survey baseline. Use keel or other nonpermanent markers to mark the baseline.
- The station 0 + 00 is the point of intersection of your baseline with some other line or point, as defined by your instructor.
- Measure the baseline stations (e.g., 50 ft or 20 m) precisely with a nylon-clad steel tape. Mark them clearly on the ground with a keel marker, nails, or wood stakes.

- Determine the baseline stations of all features left and right of the baseline by estimating 90° (swung-arm technique) or by using a right-angle prism.
- When you have booked the baseline stations of all the features in a 50/100-ft or 20/30-m interval (the steel tape can be left lying on the ground on the baseline), determine and book the offset (o/s) distances left and right of the baseline to each feature. Tie in all detail to the closest 0.10 ft or 0.03 m. A fiberglass tape can be used for these measurements.
- Do not begin the measurements until the sketches have been made for the survey area.
- Elevations are determined using a level and a rod. The level is set up in a convenient location where a benchmark (BM) and a considerable number of intermediate sights (ISs) can be seen. See Figure D.7.
- Hold the rod on the baseline at each station and at all points where the ground slope changes (e.g., the top and bottom of the curb, edge of the walk, top of the slope, bottom of the slope, limit of the survey). See the typical cross section shown in Figure D.8.
- When all the data (that can be seen) have been taken at a station, the rod holder then moves to the next (50-ft or 20-m) station and repeats the process.
- When the rod can no longer be seen, the instrument operator calls for the establishment of a turning point (TP). After taking a foresight to the new TP, the instrument is moved closer to the next stage of work and a backsight is then taken to the new TP before continuing with the cross sections. In addition to cross sections at the even station intervals (e.g., 50 ft, 20 m), you must take full or partial sections between the even stations if the lay of the land changes significantly.

D.7.2   Theodolite/EDM Topographic Survey

Description   Using electronic distance measurement (EDM) instruments and optical or electronic theodolites, locate the positions and elevations of all topographic detail and a sufficient number of additional elevations to draw a representative contour drawing of the selected areas. See the sample field notes in Figure D.9.

Equipment   Theodolite, EDM, and one or more pole-mounted reflecting prisms.

Procedure

- Set the theodolite at a control station (northing, easting, and elevation known), and backsight on another known control station.
- Set an appropriate reference angle (or azimuth) on the horizontal circle (e.g., 0°00′00″ or some assigned azimuth).
- Set the height of the reflecting prisms (hr) on the pole equal to the height of the optical center of the theodolite/EDM (hi). If the EDM is not coaxial with the theodolite, set the height of the target (target/prism assembly) on the pole equal to the optical center of the instrument (see the left-hand illustration in Figure 2.28). Take all vertical angles to the prism target, or to the center of the prism if the EDM is coaxial with the theodolite.
- Sketch the area to be surveyed.

- Begin taking readings on the appropriate points. Enter the data in the field notes (see Figure D.9) and enter the shot number in the appropriate spot on the accompanying field-note sketch. Keep shot numbers sequential (begin perhaps with 1,000). Work is expedited if two prisms are employed. While one prism-holder walks to the next shot location, the instrument operator can take a reading on the other prism-holder.

- When all field shots (horizontal and vertical angles and horizontal distances) have been taken, sight the reference backsight control station again to verify the angle setting. Also verify that the height of the prism is unchanged.

- Reduce the field notes to determine station elevations and course distances, if required.

- Plot the topographic features and elevations at 1″ = 40′, or 1:500 metric (or at other scales as given by your instructor).

- Draw contours over the surveyed areas. See Chapter 7.

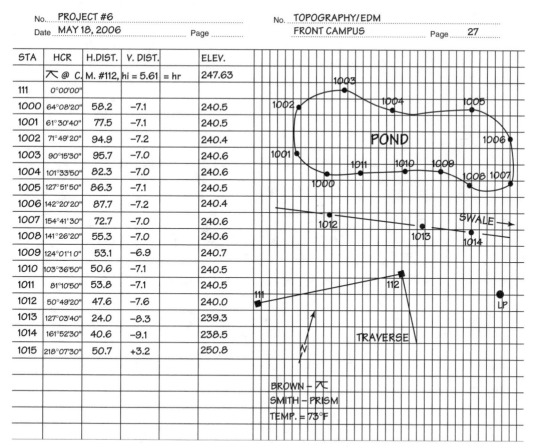

FIGURE D.9 Sample field notes for Project 6 (topography by theodolite/EDM).

D.7.3 Total Station Topographic Survey

Description Using a total station and one or more pole-mounted reflecting prisms, tie in all topographic features and any additional ground shots (including break lines) that are required to define the terrain accurately. See Section 7.4 and Figure D.10).

Equipment Total station and one (or more) pole-mounted reflecting prisms.

Procedure

- Set the total station over a known control point (northing, easting, and elevation known).[*] Turn on the instrument, and index the circles, if necessary, by transiting the telescope and revolving the instrument 360°. (Some newer total stations do not require this initializing operation.)

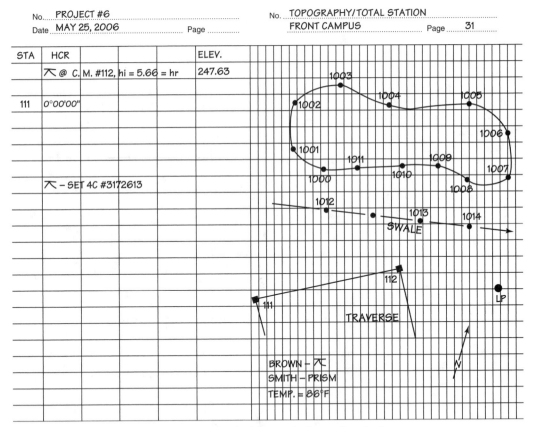

FIGURE D.10 Sample field notes for Project 6 (topography by total station).

---

[*]The total station can also be set in any convenient location with its position determined using the on-board resection program and after sighting the required number of visible control stations.

- Set the program menu to the type of survey (topography) being performed and to the required instrument settings. Select the type of field data to be stored (e.g., N, E, and Z; or E, N, and Z; etc.). Set the temperature and pressure settings if required.
- Check the configuration settings, for example, tilt correction, coordinate format, zenith vertical angle, angle resolution (e.g., 5″), $c + r$ correction (e.g., no.), units (ft/m, degree, mm Hg), and auto power off (say, 20 minutes).
- Identify the instrument station from the menu. Insert the date, station number coordinates, elevation, and hi. It may be possible to upload all control station data prior to going out to the field. In that case, scan through the data and select the appropriate instrument station and backsight station(s). Enter the height of instrument (hi), and store or record all the data.
- Backsight to one or more known control point(s) (point number, north and east coordinates, and elevation known). Set the horizontal circle to 0°00′00″ or to some assigned reference azimuth for the backsight reference direction. Store or record the data.
- Set the initial topography point number in the instrument (e.g., 1,000), and set for automatic point number incrementation. Adjust the height of the reflecting prism (hr) to equal the instrument hi.
- Begin taking intermediate sights. Provide an attribute code (consistent with the software code library) for each reading. Table D.2 shows a typical code library. Some software programs enable attribute codes to provide automatically for feature stringing (e.g., curb1, edge of water1, fence3), whereas other software programs require the surveyor to prefix the code with a character (e.g., Z) that turns on the stringing command. (See Section 5.5 and Figures 5.17 and 5.18). Most total stations have an automatic mode for topographic surveys, where one push of a button measures and stores all the point data as well as the code and attribute data. The code and attribute data of the previous point are usually presented to the surveyor as a default setting. If the code and attribute data are the same for a series of readings, the surveyor can press Enter and not enter all the identical data repeatedly.
- Put all or some selected point numbers on the field sketch. This information will be of assistance later in the editing process if mistakes have occurred in the numbering or coding.
- When all required points have been surveyed, check back into the control station originally backsighted to ensure that the instrument orientation is still valid.
- Transfer the field data into a properly labeled computer file.
- After opening the data-processing program, import the field data file and begin the editing process and the graphics generation process. (The graphics generation process is automatic for many programs.)
- Create the TIN and contours.
- Either finish the drawing with the working program or create a DXF file for transfer to a CAD program and then finish the drawing.
- Prepare a plot file and then plot the data (to a scale assigned by your instructor) on the lab digital plotter.

Table D.2 TYPICAL CODE LIBRARY

Control		Utilities	
TCM	Temporary control monument	HP	Hydro pole
CM	Concrete monument	LP	Lamp pole
SIB	Standard iron bar	BP	Telephone pole
IB	Iron bar	GS	Gas valve
RIB	Round iron bar	WV	Water valve
NL	Nail	CABLE	Cable
STA	Station		
TBM	Temporary benchmark		

Municipal		Topographic	
℄	Centerline	GND	Ground
RD	Road	TB	Top of bank
EA	Edge of asphalt	BB	Bottom of bank
BC	Beginning of curve	DIT	Ditch
EC	End of curve	FL	Fence line
PC	Point on curve	POST	Post
CURB	Curb	GATE	Gate
CB	Catch basin	BUSH	Bush
DCB	Double catch basin	HEDGE	Hedge
MH	Manhole	BLD	Building
STM	Storm sewer manhole	RWALL	Retaining wall
SAN	Sanitary sewer manhole	POND	Pond
INV	Invert	STEP	Steps
SW	Sidewalk	CTREE	Coniferous tree
HYD	Hydrant	DTREE	Deciduous tree
RR	Railroad		

D.8   Project 7: Building Layout

Description   Lay out the corners of a building and reference the corners with batter boards. See Figure D.11.

Equipment   Theodolite, steel tape, plumb bobs, wood stakes, light lumber for batter boards, C clamps, keel or felt pen, level, and rod. A theodolite/EDM (or total station) and prism can be used in place of the theodolite and steel tape.

Purpose   To give you experience in laying out the corners of a building according to dimensions taken from a building site plan, in constructing batter boards, and in referencing both the line and grade of the building walls and floor.

Procedure

- After the front and side lot lines have been defined by your instructor and after the building dimensions have been given, set stakes X and Y on the front lot line, as shown in Figure D.11.

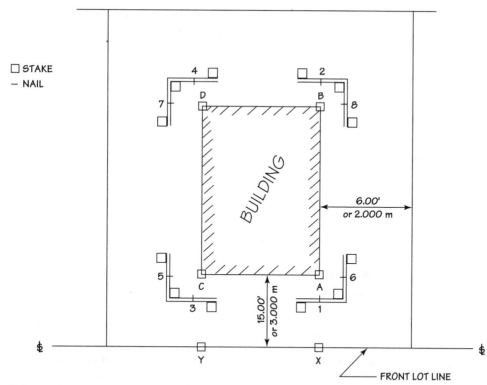

FIGURE D.11 Sample field notes for Project 7 (building layout).

- Set up the theodolite at X, sight on Y, turn 90° (double), and place stakes at A and B.
- Set up the theodolite at Y, sight on X, turn 90° (double), and place stakes at C and D.
- Measure the building diagonals to check the accuracy of the layout. Adjust and re-measure if necessary.
- After the building corners have been set, offset the batter boards a safe distance (e.g., 6 ft or 2 m), and set the batter boards at the first-floor elevation, as given by your instructor.
- After the batter boards have been set, place line nails on the top of the batter boards as follows:

  (a) Set up on A, sight B, place nail 2, transit the telescope, and set nail 1.

  (b) From setup on A, sight C, place nail 5, transit the telescope, and place nail 6.

  (c) Set up on D, and place nails 3, 4, 7, and 8 in a similar fashion.

D.9 Project 8: Horizontal Curve

Description Given the centerline alignment of two intersecting tangents (including a station reference stake), calculate and lay out a horizontal curve.

Equipment   Theodolite, steel tape, plumb bobs, wood stakes, and range pole or string target. A theodolite/EDM (or total station) and prism can be used in place of the theodolite and steel tape.

Purpose   To give you experience in laying out a circular curve at specified station intervals after first calculating all the necessary layout measurements from the given radius and the measured location of the PI and the Δ field angle.

Procedure

- Intersect the two tangents to create the PI.
- Measure the station of the PI.
- Measure (and double) the tangent deflection angle (Δ).
- After receiving the radius value from your instructor, compute T and L, and then compute the station of the BC and EC. (See Section 11.3.)
- Compute the deflections for even stations—at 50-ft or 20-m intervals. (See Section 11.4.) Compute the equivalent chords. (See Section 11.5.)
- Set the BC and EC by measuring out T from the PI along each tangent.
- From the BC, sight the PI, and turn the curve deflection angle ($\Delta/2$) to check the location of the EC. If the line of sight does not fall on the EC, check the calculations and measurements for the BC and EC locations, and make any necessary adjustments.
- Using the calculated deflection angles and appropriate chord lengths, stake out the curve.
- Measure from the last even station stake to the EC to verify the accuracy of the layout.
- Walk the curve, looking for any anomalies (e.g., two stations staked at the same deflection angle). The symmetry of the curve should be such that even minor mistakes will be obvious in a visual check.

Reference   Chapter 11.

D.10   Project 9: Pipeline Layout

Description   Establish an offset line and construct batter boards for line-and-grade control of a proposed storm sewer from MH 1 to MH 2. Stakes marking those points will be given for each crew in the field.

Equipment   Theodolite, steel tape, wood stakes, and light lumber and C clamps for the batter boards. A theodolite/EDM (or total station) and prism can be used in place of the theodolite and steel tape.

Purpose   To give you experience in laying out offset line-and-grade stakes for a proposed pipeline. You will learn how to compute a grade sheet and construct batter boards and how to check the accuracy of your work by sighting across the constructed batter boards.

Procedure

- Set up at the MH 1 stake, and sight the MH 2 stake.
- Turn off 90°, measure the offset distance (e.g., 10 ft or 3 m), and establish MH 1 on offset.
- Set up at the MH 2 stake, sight the MH 1 stake, and establish MH 2 on offset. Refer to Figure 14.7 for guidance.
- Give the MH 1 offset stake a station of 0 + 00. Measure out to establish grade stakes at the even stations (50 ft or 20 m). Finally, check the distance from the last even station to the MH 2 stake to check that the overall distance is accurate.
- Using the closest benchmark (BM), determine the elevations of the tops of the offset grade stakes. Close back to the benchmark within the tolerance given by your instructor.
- Assume that the invert of MH 1 is 7.97 ft or 2.430 m below the top of the MH 1 grade stake (or other assumed value given by your instructor).
- Compute the invert elevations at each even station. Then complete a grade sheet similar to those shown in Table 14.2 and Figure 14.7. Select a convenient height for the grade rod.
- Using the stake–to–batter board distances in the grade sheet, use the supplied light lumber and C clamps to construct batter boards similar to those shown in Figure 14.6. Use a small carpenter's level to keep the crosspieces horizontal. Check to see that all crosspieces line up in one visual line—a perfect visual check (all batter boards line up behind one another) is a check on all the measurements and calculations.

Reference   Chapter 14.

Appendix E

Answers to Selected Problems

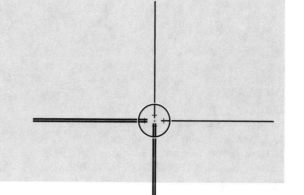

CHAPTER 2

2.3(a). 1,578.72 ft; 481.19 m **2.3(c).** 806.52 ft;
245.83 m **2.7.** 22.735 m **2.11.** H = 3,959.78 ft,
elevation = 476.47 ft **2.13.** Elevation at
L = 415.81 ft; elevation at M = 154.13 ft;
elevation at N = 232.55 ft **2.14.** H = 386.504 m,
elevation B = 139.077 m

CHAPTER 3

3.1(a). 0.010 ft m **3.1(b).** 0.19 ft **3.1(f).** 1.688 m
3.2(a)ii. 1.86 ft **3.2(b)iii.** 1.040 m **3.3.** 26.40 mi
3.7. Error = 0.005. Allowable for second order—
U.S. = 0.005; allowable second order—Canada =
0.007 Qualifies for second order in both countries.
3.14(a). Elevation = 269.76 ft **(b).** Station at 112 +
09.58 **3.16(b).** BM K110 at 165.957 (adjusted)

CHAPTER 6

6.1. 116°38′ **6.3(a).** S 8°23′ W
6.3(e). S 81°50 W **6.4(a).** 352°09′
6.4(e). 322°42′ **6.12(d).** E = 0.79 ft,
precision = 1/3,200. **6.15.** A = 7.66 acres
6.19. A = 5.27 hectares (ha)
6.20. AD = 1,851.44 ft **6.21.** CD = 852.597 m;
bearing DE = S 74°30′23″ W **6.25.** A = 2,430 m²

CHAPTER 7

7.9(a). H = SR.f = 20,000 × .153 = 3,060 m
Altitude = 3060 + 180 = 3,240 m
7.9(b). SR = 20,000 × 12 = 240,000

H = SR.f = 240,000 × 6.022/12 = 120,440 ft
Altitude = 120,400 + 520 = 120,920 ft
7.10(a). Photo scale = 23.07 × 1:50,000/4.75 =
1:10,295 **7.10(b).** Photo scale = 6.20 ×
1:100,000/1.85 = 1:29,839

CHAPTER 9

9.1. Grid bearing AB = S 79°29′00.4″W,
AB = 290.722 m **9.2.** B = 186°21′09.1″
9.4. Geodetic bearing of AB = S 74°35′18.2″W
9.5. B = 186°21′01.3″

CHAPTER 11

11.2. T = 44.85 ft, L = 89.58 ft
11.4. BC at 15 + 43.25 ft, EC at 16 + 32.83 ft
11.6. BC

BC	5 + 829.183	0°00′00″
	5 + 840	1°01′59″
	5 + 860	2°56′35″
	5 + 880	4°51′11″
EC	5 + 896.116	6°23′31″

11.12. A to BC = 565.07 ft, B to EC = 408.89 ft
11.13. R = 29.958 m **11.14.** T_1 = 228.74 ft,
T_2 = 270.02 ft **11.16.** Summit at 20 + 07.14,
elev. = 722.06 ft

11.17.			
	BVC	0 + 180	151.263
		0 + 200	151.841
		0 + 250	153.091
		0 + 300	154.063
		0 + 350	154.757
	PVI	0 + 360	154.863

	0 + 400	155.134
High Point	0 + 450	155.313
	0 + 500	155.174
EVC	0 + 540	154.863

11.18. T.S. = 9 + 85.68, S.C. = 11 + 35.68, C.S. = 11 + 94.85, S.T. = 13 + 44.85

CHAPTER 13

13.1. At 4 + 00, cut 0'10 7/8" **13.2.** At 0 + 60, "on grade" **13.4.** At 4 + 50, ₵ elev. = 508.20 ft
13.5. At 4 + 50, curb elev. = 508.46 ft
13.6. At 4 + 50, "on grade" **13.7(a).** L = 54.98 ft
13.7(b). Slope at 1.11% **13.7(c).** and **13.7(d).**

Station	Curb Elev.	Stake Elev.	Cut	Fill
A	505.02	504.71		0'3³/₄"
B	505.22	506.22	1'0"	
C	505.44	506.37	0'11¹/₈"	
D	505.63	506.71	1'1"	

CHAPTER 14

14.1. At 1 + 50, cut 7.58 ft; for grade rod of 14 ft, stake to batter board distance of 6'50"
14.2. At 0 + 80, cut 2.534 m; for a grade rod of 4 m, stake to batter board distance of 1.466 m
14.3. At 1 + 00, 495.60 **14.4.** At 1 + 00, cut 8.01 ft
14.5. GR = 14 ft; at 1 + 00, 5'11⁷/₈"
14.7. First leg, 0 + 40 = 183.740 m
14.8. First leg, 0 + 40, cut 3.093 m
14.9. First leg, GR = 5 m; at 0 + 40, cut 1.907
14.10. At lot 13, sewer invert opposite 13 = 496.85
+ (32 × 0.01) = 497.17
springline elevation = 497.17 + (1.25/2) = 497.80
60 ft of sewer connection pipe at 2% = + 1.20
minimum invert elevation of connection = 499.00

CHAPTER 17

17.1. Trapezoidal technique, 289 yd²; Simpson's one-third rule, 290 yd² **17.2.** A = 334 yd²
17.3. V = 401 m³
17.4. V = 60.7 yd³ **17.5.** V = 2639 yd³
17.6. V = 88.6 yd³

APPENDIX F

F.1. $C_s = -w^2L^3/24p^2 = -0.32^2 \times 48.888^3/24 \times 100^2$
$= -.050$ m
Corrected distance = 48.888 − .050 = 48.838 m

F.3. $C_s = -w^2L^3/[24 \times 24^2] = -0.018^2 \times 100^3/$
$[24 \times 24^2] = -0.023 \times 4 = -0.094$ ft
$C_s = -0.018^2 \times 71.16^3/[24 \times 24^2] = 0.008$
$\Sigma C_s = -(0.008 + 0.094) = -0.010$ ft
Corrected distance = 471.16 − 0.10 = 471.06 ft

APPENDIX G
G.1.

Theodolite Station

Station	Rod Inter-val	Vertical Angle	Hori-zontal Distance	Eleva-tion Differ-ence	Eleva-tion
					371.21
1	3.48	+0°58'	347.9	+5.9	371.1
2	0.38	−3°38'	37.8	−2.4	368.8

G.2.

Theodolite Station

Station	Rod Inter-val	Vertical Angle	Hori-zontal Distance	Eleva-tion Differ-ence	Eleva-tion
	hi 5 1.83				207.96
1	0.041	+ 2°19'	4.09	+0.17	208.13
2	0.072	+ 1°57'	7.19	V = +0.24	208.60
	on 1.43				
3	0.555	0°00'	55.5	V = 0	207.08
	on 2.71				

G.3(a).
V = 100 × 1.31 × cos 3°12' sin 3°12' = 7.30 ft
Elevation K + hi (5.36) + V (7.30) − RR (4.27) = elevation L (293.44)
Elevation K = 285.05 ft
G.3(b). H = 100 × 1.31 × cos²3°12' = 130.59 ft
G.5.

Station	Rod Inter-val	Hori-zontal Angle	Ver-tical Angle	Hori-zontal Dis-tance	Eleva-tion Differ-ence	Eleva-tion
	at K, hi = 1.82					167.78
L		0°00'				
0+00 ₵	0.899	34°15'	−19°08'	80.24	−27.84	139.94
S ditch	0.851	33°31'	−21°58'	73.19	−29.52	138.26
N ditch	0.950	37°08'	−20°42'	83.13	−31.41	136.37

399.92

Appendix F

Steel Tape Corrections

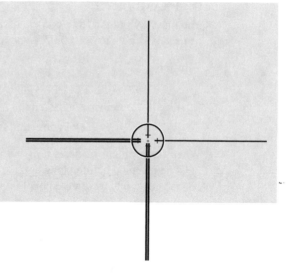

F.1 Erroneous Tape-Length Corrections

For all but precise work, new tapes supplied by the manufacturer are considered to be correct under standard conditions. As a result of extensive use, however, tapes become kinked, stretched, and repaired—perhaps imprecisely. When the tape length becomes something other than that specified, the tape must be corrected, or the measurements taken with the erroneous tape must be corrected.

■ EXAMPLE F.1

A measurement is recorded as 171.278 m with a 30-m tape that is found to be only 29.996 m under standard conditions. What is the corrected measurement?

Solution

$$\text{Correction per tape length} = -0.004$$

$$\text{Number of tape lengths} = \frac{171.278}{30}$$

$$\text{Total correction} = -0.004 \times \frac{171.278}{30}$$

$$\text{Total correction} = -0.023 \text{ m}$$

$$\text{Corrected distance} = 171.278 - 0.023$$

$$= 171.255 \text{ m}$$

$$\text{Corrected distance} = \frac{29.996}{30} \times 171.278 = 171.255$$

■ EXAMPLE F.2

You must lay out the front corners of a building, a distance of 210.08 ft. The tape to be used is known to be 100.02 ft under standard conditions.

Solution

$$\text{Correction per tape length} = +0.02\,\text{ft}$$
$$\text{Number of tape lengths} = 2.1008$$
$$\text{Total correction} = 0.02 \times 2.1008 = +0.04\,\text{ft}$$

When the problem involves a *layout distance,* the algebraic sign of the correction must be reversed before being applied to the layout measurement. We must find the distance that, when corrected by $+0.04$, will give 210.08 ft:

$$\text{Layout distance} = 210.08 - 0.04 = 210.04\,\text{ft}$$

You will discover that four variations of this problem are possible: correcting a measured distance using (1) a long tape or (2) a short tape and precorrecting a layout distance using (3) a long tape or (4) a short tape. To minimize confusion about the sign of the correction, consider the problem with the distance reduced to that of only one tape length (100 ft or 30 m).

In Example F.1, a recorded distance of 171.278 m was measured with a tape only 29.996 m long. The total correction was found to be 0.023 m. If doubt exists about the sign of 0.023, ask yourself what the procedure would be for correcting only one tape length. In this example, after one tape length has been measured, it is recorded that 30 m (the nominal length of the tape) has been measured. If the tape is only 29.996 m long, then the field book entry of 30 m must be corrected by -0.004; if the correction for one tape length is minus, then the corrections for the total distance must also be minus.

The magnitude of a tape error can be determined precisely by having the tape compared with a tape that has been certified by the National Bureau of Standards in the United States or the National Research Council in Canada. In field practice, if a tape is found to be incorrect, it is usually either repaired precisely or discarded.

F.2   Temperature Corrections

In Section 2.7, we noted that the standard temperature for steel tapes is 68°F, or 20°C. Most measurements taken with a steel tape occur at some temperature other than standard (68°F or 20°C). When the temperature is warmer or cooler than standard, the steel tape expands or contracts and thus introduces an error into the measurement. The coefficient of thermal expansion and contraction of steel is 0.00000645 per unit length per degree Fahrenheit (°F), or 0.0000116 per unit length per degree Celsius (°C).

Foot Unit Temperature Corrections

$$C_t = 0.00000645\,(T - 68)L \tag{F.1}$$

where C_t = the correction, in feet, due to temperature
T = the temperature (°F) of the tape during measurement
L = the distance measured in feet

Metric Unit Temperature Corrections

$$C_t = 0.0000116\,(T - 20)\,L \qquad \text{(F.2)}$$

where C_t = the correction, in meters, due to temperature

T = the temperature (°C) of the tape during measurement

L = the distance measured in meters

■ EXAMPLE F.3

A distance is recorded as being 471.37 ft at a temperature of 38°F. What is the corrected distance?

Solution

$$C_t = 0.00000645\,(38 - 68)\,471.37$$
$$= -0.09 \text{ ft}$$

The corrected distance is $471.37 - 0.09 = 471.28$ ft.

■ EXAMPLE F.4

You must lay out two points in the field that will be exactly 100.000 m apart. Field conditions indicate that the temperature of the tape will be 27°C. What distance will be laid out?

Solution

$$C_t = 0.0000116\,(27 - 20)\,100.000$$
$$= +0.008 \text{ m}$$

Because this is a layout (precorrection) problem, the correction sign must be reversed (i.e., we are looking for the distance that, when corrected by +0.008 m, will give us 100.000 m):

$$\text{Layout distance} = 100.000 - 0.008 = 99.992 \text{ m}$$

Accuracy demands for most surveys do not require high precision in determining the temperature of the tape. It is usually sufficient to estimate the air temperature. However, for more precise work (say, 1:10,000 and higher), care is required in determining the actual temperature of the tape, which can be significantly different from the temperature of the air.

F.2.1 Invar Steel Tapes

High-precision surveys may require the use of steel tapes with a low coefficient of thermal expansion. The invar tape is a nickel–steel alloy with a coefficient of thermal expansion of 0.0000002 to 0.00000055 per degree Fahrenheit. In the past, invar tapes were used to measure baselines for triangulation control surveys. Currently, EDM instruments measure baselines more efficiently. Invar tapes can still be used to advantage, however, in situations where high precision is required over short distances.

F.3 Tension and Sag Corrections

If a steel tape is fully supported and a tension other than standard (10 lb, foot system; 50 new-tons, metric system) is applied, a tension (pull) error exists. The tension correction formula is:

$$C_P = \frac{(P - P_s)L}{AE} \tag{F.3}$$

If a tape has been standardized while fully supported and is being used without full support, an error called sag will occur (see Figure F.1.) The force of gravity pulls the center of the unsupported section downward in the shape of a caternary, thus creating an error B′B. The sag correction formula is:

$$C_S = \frac{-w^2 L^3}{24 P^2} = \frac{-W^2 L}{24 P^2} \tag{F.4}$$

where $W^2 = w^2 L^2$

W = the weight of the tape between supports
w = the weight of the tape per unit length
L = length of the tape between supports

Table F.1 further defines the terms in these two formulas. You can see in the table that 1 newton is the force required to accelerate a mass of 1 kg by 1 m/s^2.

$$\text{Force} = \text{mass} \times \text{acceleration}$$

$$\text{Weight} = \text{mass} \times \text{acceleration due to gravity } (g)$$

$$g = 32.2 \text{ ft/s}^2 = 9.807 \text{ m/s}^2$$

FIGURE F.I Example of steel tape error called sag.

Table F.1 CORRECTION FORMULA TERMS DEFINED (FOOT, METRIC, AND SI UNITS)

Unit	Description	Foot	Old Metric	Metric (SI)
C_P	Correction due to tension per tape length	ft	m	m
C_s	Correction due to sag per tape length	ft	m	m
L	Length of tape under consideration	ft	m	m
P_s	Standard tension	lb (force)	kg (force)	N (newtons)
	Typical standard tension	10 lb	4.5–5 kg	50 N
P	Applied tension	lb	kg	N
A	Cross-sectional area	in.2	cm^2	m^2
E	Average modulus of elasticity of steel tapes	29×10^6 lb/in.^{2}	21×10^5 kg/cm^2	20×10^{10} N/m^2
E	Average modulus of elasticity in invar tapes	21×10^6 lb/in.^{2}	14.8×10^5 kg/cm^2	14.5×10^{10} N/m^{2}m
w	Weight of tape per unit length	lb/ft	kg/m	N/m
W	Weight of tape	lb	kg	N

App. F   Steel Tape Corrections

In SI units, a mass of 1 kg has a weight of $1 \times 9.807 = 9.807$ N (newtons); that is:

$$1 \text{kg (force)} = 9.807 \text{N}$$

Some spring balances are graduated in kilograms, and tape manufacturers give standard tension in newtons, so surveyors must be prepared to work in both old metric and SI units. Examples F.5 through F.7 give you practice in correcting for tension. Examples F.8 and F.9 give you practice in correcting for sag.

■ **EXAMPLE F.5** *Tension Correction*

A 100-ft steel tape with a standard tension of a 10-lb force is used with a 20-lb force pull. If the cross-sectional area of the tape is 0.003 in.^{2}, what is the tension error for each tape length used?

Solution

$$C_P = \frac{(20 - 10)\ 100}{(29{,}000{,}000 \times 0.003)} = +0.011 \text{ft}$$

To illustrate the use of this correction, consider that a field distance of 421.22 ft was recorded under the conditions described above. The total correction is $4.2122 \times 0.011 = +0.05$ ft. The corrected distance is 421.27 ft.

■ **EXAMPLE F.6** *Tension Correction*

A 30-m steel tape, with a standard tension of 50 N, is used with a 100-N force. If the cross-sectional area of the tape is 0.02 cm^2, what is the tension error per tape length?

Solution

$$C_P = \frac{(100 - 50)\ 30}{(0.02 \times 21 \times 10^5 \times 9.087)} = +0.0036 \text{ m}$$

To illustrate the use of this correction, consider that a field distance of 182.716 m was recorded under the conditions described above. The total correction is $182.716/30 \times 0.0036 = +0.022$ m. The corrected distance is 182.738 m.

The cross-sectional area of a tape can be calculated from micrometer readings, taken from the manufacturer's specifications, or determined from the following expression:

Tape length $\times$ tape area $\times$ specific weight of tape steel $=$ weight of tape

or

$$\text{Tape area} = \frac{\text{weight of tape}}{\text{length} \times \text{specific weight}}$$

■ **EXAMPLE F.7** *Tension Correction*

A tape weighs 1.95 lb. The overall length of the 100-ft steel tape (end to end) is 102 ft. The specific weight of steel is 490 lb/ft^2. What is the cross-sectional area of the tape?

Solution

$$\frac{102 \text{ ft} \times 12 \text{ in.} \times \text{area (in.}^2)}{1{,}728 \text{ in}^3} \times 490 \text{ lb/ft}^3 = 1.95 \text{ lb}$$

$$\text{Area} = \frac{1.95 \times 1728}{102 \times 12 \times 490} = 0.0056 \text{ in.}^2$$

Tension errors are usually quite small and so are relevant only for very precise surveys. Even for precise surveys, it is seldom necessary to calculate tension corrections because the availability of a tension spring balance allows the surveyor to apply standard tension and thus eliminates the necessity of calculating a correction.

■ **EXAMPLE F.8** *Sag Correction*

A 100-ft steel tape weighs 1.6 lb and is supported only at the ends with a force of 10 lb. What is the sag correction?

Solution

$$C_S = \frac{-1.6^2 \times 100}{24 \times 10^2} = -0.11 \text{ ft}$$

If the force is increased to 20 lb, the sag correction is reduced to:

$$C_S = \frac{-1.6^2 \times 100}{24 \times 20^2} = -0.03 \text{ ft}$$

■ **EXAMPLE F.9**

Calculate the length between two supports if the recorded length is 50.000 m, the mass of the tape is 1.63 kg, and the applied tension is 100 N.

Solution

$$C_S = \frac{-(1.63 \times 9.807)^2 \times 50.000}{24 \times 100^2}$$

$$= -0.053 \text{ m}$$

Therefore, the length between the supports is $50.000 - 0.053 = 49.947$ m.

Problems

F.1. A 100- ft steel tape, known to be only 99.98 ft long under standard conditions, was used to record a measurement of 398.36 ft. What is the distance corrected for the erroneous tape length?

F.2. A 30-m steel tape, known to be 30.004 m under standard conditions, was used to record a measurement of 271.118 m. What is the distance corrected for the erroneous tape length?

F.3. A rectangular commercial building must be laid out 200.00 wide and 300.00 ft long. If the steel tape used is 100.02 ft long under standard conditions, what distance would be laid out?

F.4. A survey distance of 338.12 ft was recorded when the field temperature was 96°F. What is the distance, corrected for temperature?

F.5. Station 3 + 54.67 must be marked in the field. If the steel tape to be used is only 99.98 under standard conditions, and if the temperature will be 87°F at the time of the measurement, how far from the existing station mark at 0 + 79.23 will the surveyor have to measure to locate the new station?

F.6. The point of intersection of the centerline of Elm Road with the centerline of First Street was originally recorded (using a 30-m steel tape) as being at 6 + 71.225. The temperature is −6°C, and the tape used for measuring is 29.995 under standard conditions. How far from existing station mark 5 + 00 on First Street would a surveyor have to measure along the centerline to reestablish the intersection point under these conditions?

F.7. A 50-m tape is used to measure the distance between two points. The average weight of the tape per meter is 0.320 N. If the measured distance is 48.888 m with the tape supported only at the ends and with a tension of 100 N, find the corrected distance.

F.8. A 30-m tape has a mass of 544 g and is supported only at the ends with a force of 80 N. What is the sag correction?

F.9. A 100-ft steel tape weighing 1.8 lb and supported only at the ends with a tension of 24 lb is used to measure a distance of 471.16 ft. What is the distance corrected for sag?

F.10. A distance of 72.55 ft is recorded; a steel tape supported only at the ends with a tension of 15 lb and weighing 0.016 lb per ft is used. Find the distance corrected for sag.

Appendix G

Early Surveying

G.1 Evolution of Surveying

Surveying is a profession with a very long history. Since the beginning of property ownership, boundary markers have been required to differentiate one property from another. Historical records dating to almost 3000 B.C. show evidence of surveyors' work in China, India, Babylon, and Egypt. The Egyptian surveyor, called the *harpedonapata* (rope-stretcher), was in constant demand because the Nile River flooded more or less continuously, destroying boundary markers in those fertile farming lands. These surveyors used ropes with knots tied at set graduations to measure distances.

Ropes were also used to lay out right angles. The early surveyors discovered that the 3:4:5 ratio provided right-angle triangles. To lay out XZ at 90 degrees to XY, a twelve-unit rope would have knots tied at unit positions 3 and 7, as shown in Figure G.1. One surveyor held the three-unit knot at X; the second surveyor held the seven-unit knot at Y; and the third surveyor, holding both loose ends of the rope, stretched the rope tightly, resulting in the location of point Z. These early surveyors knew that multiples of 3:4:5 (e.g., 30:40:50) would produce more accurate positioning.

Another ancient surveying instrument consisted of three pieces of wood in the form of an isosceles triangle with the base extended in both directions (see Figure G.2). A plumb bob suspended from the apex of the frame would line up with a notch cut in the midpoint of the base only when the base was level. These levels came in various sizes, depending on the work being done.

It is presumed that the great pyramids were laid out with knotted ropes, levels as described here, and various forms of water-trough levels for the foundation layout. These Egyptian surveying techniques were empirical solutions that were field proven. It remained for the Greeks to provide the mathematical reasoning and proofs to explain why the field techniques worked.

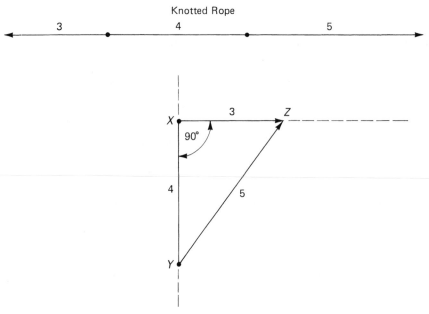

FIGURE G.1   Rope knotted at 3:4:5 ratio—used to place point Z at 90 degrees to point X from line XY.

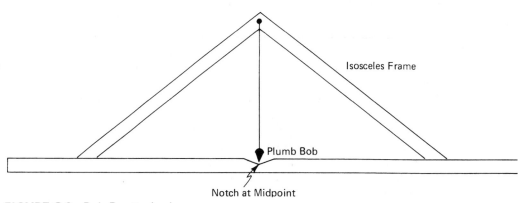

FIGURE G.2   Early Egyptian level.

Pythagoras is one of many famous Greek mathematicians. He and his school developed theories regarding geometry and numbers about 550 B.C. They were also among the first to deduce that the earth was spherical by noting the shape of the earth's shadow cast on the moon. The term *geometry* is Greek for "earth measurement," clearly showing the relationship between mathematics and surveying. In fact, the history of surveying is closely related to the history of mathematics and astronomy.

By 250 B.C., Archimedes recorded in a book known as the *Sand Reckoner* that the circumference of the earth was 30 myriads of stadia (i.e., 300,000 stadia). He had received some support for this computation from his friend Eratosthenes, who was a mathematician

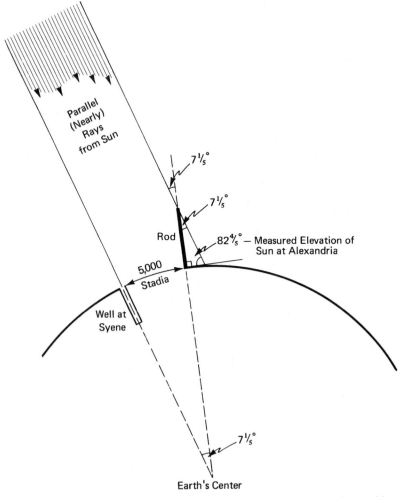

FIGURE G.3 Illustration of Eratosthenes' technique for computing the earth's circumference.

and a librarian at the famous library of Alexandria in Egypt. According to some reports, Eratosthenes knew that a town called Syene was 5,000 stadia south of Alexandria. He also knew that at summer solstice (around June 21) the sun was directly over Syene at noon because there were no shadows. He observed this fact by noting that the sun's reflection was exactly centered in the well water.

Eratosthenes assumed that at the summer solstice, the sun, the towns of Syene and Alexandria, and the center of the earth all lay in the same plane (see Figure G.3). At noon on the day of the summer solstice, the elevation of the sun was measured at Alexandria as being $82\frac{4}{5}°$. The angle from the top of the rod to the sun was then calculated as being $7\frac{1}{5}°$. Because the sun is such a long distance from the earth, it can be assumed that the sun's rays are parallel as they reach the earth. With that assumption, it can be deduced that the angle from the top of the rod to the sun is the same as the angle at the earth's center: $7^1/_5°$. Because $7^1/_5°$ is one-fiftieth of 360°, it follows that the circular arc subtending $7^1/_5°$ (the distance from Syene to Alexandria) is one-fiftieth of the circumference of the earth. The circumference of the earth is thus determined to be 250,000 stadia. If the stadia being used were $^1/_{10}$ of a mile (different values for the stadium existed, but one value was roughly $^1/_{10}$ of our mile), then it is possible that Eratosthenes had calculated the earth's circumference to be 25,000 miles. Using the Clarke ellipsoid, with a mean radius of 3,960 miles, the circumference of the earth would actually be $C = 2 \times 3.1416 \times 3,960 = 24,881$ miles.

After the Greeks, the Romans made good use of practical surveying techniques for many centuries to construct roadways, aqueducts, and military camps. Some Roman roads and aqueducts exist to this day. For leveling, the Romans used a *chorobate*, a 20-foot (approximate) wooden structure with plumbed end braces and a 5-foot (approximate) groove for a water trough (see Figure G.4). Linear measurements were often made with wooden poles 10 to 17 ft long. With the fall of the Roman Empire, surveying and most other intellectual endeavors became lost arts in the Western world.

Renewed interest in intellectual pursuits may have been fostered by the explorers' need for navigational skills. The lodestone, a naturally magnetized rock (magnetite), was used to locate magnetic north. The compass would later be used for navigation on both land and water. In the mid-1500s, the surveyors' chain was first used in the Netherlands. An Englishman, Thomas Digges, first used the term *theodolite* to describe an instrument that had a circle graduated in 360° and used to measure angles. By 1590, the plane table (a combined positioning and plotting device) was created by Jean Praetorius; it wasn't a great deal different from the plane tables used in the early 1900s. The telescope, which was invented in

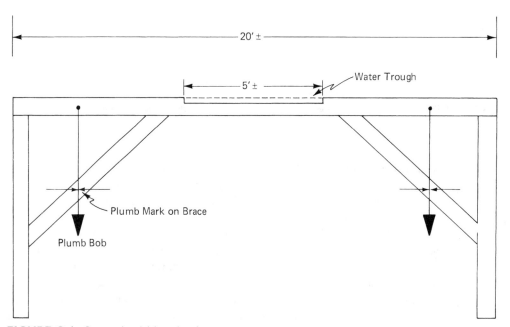

FIGURE G.4 Roman level (chorobate).

1609 by Galileo (among others), could be attached to a quadrant (angle-measuring device), thus permitting the technique of triangulation (a simpler method of determining long distances). See Section 9.1. Jean Picard (1620–1682) was apparently the first to use a spider-web cross hair in a telescope. He also used vernier scales to improve the precision of angular measurement. James Watt, the inventor of the steam engine, is also credited with being the first to install stadia hairs in the survey telescope.

The first dumpy levels were created in the first half of the 1700s by combining a telescope with a bubble level (see Figures G.5 and G.6). The repeating style of theodolite (see Sections 4.8 and G.3) was seen in Europe in the mid-1800s, but soon lost favor because scale imperfections caused large cumulative errors. Direction theodolites (see Section 4.10) were favored because high accuracy could be achieved by reading angles at different positions on the scales, thus eliminating the effect of scale imperfections. Refinements to theodolites continued over the years with better optics, micrometers, coincidence reading, lighter-weight materials, and so on. Heinrich Wild is credited with many significant improvements in the early 1900s that greatly affected the design of most European survey instruments produced by the Wild, Kern, and Zeiss companies.

Meanwhile, in the United States, William J. Young of Philadelphia is credited with being one of the first to create the transit (transiting theodolite) in 1831 (see Figure G.8). The transit differed from the early theodolites because the telescope was shortened so that it could be revolved (transited) on its axis. This simple but brilliant adaptation permitted the surveyor to produce straight lines accurately by simply sighting the backsight and transiting the telescope forward (see Section 4.15). When this technique was done twice—once with the telescope normal and once with the telescope inverted—most of the potential errors (e.g., scale graduation imperfections, cross-hair misalignment, standards misalignment) in the instrument could be removed by averaging.

Also, when using a repeating instrument, angles can be quickly and accurately accumulated (see Section 4.6.1). The transit proved to be superior for North American surveying needs. The emphasis in European surveying was on precise control surveys, but the emphasis in North America was on enormous projects in railroad and canal construction and vast projects in public land surveys, which were occasioned by the influx of large numbers of immigrants, the resulting increase in population, and westward expansion. The American repeating transit was fast, practical, and accurate, and thus was a significant factor in the development of the North American continent. European and Japanese optical theodolites, which replaced the traditional American vernier transit beginning in the 1950s, have now themselves been replaced by electronic theodolites and total stations.

Electronic distance measurement (EDM) was first introduced by Geodimeter, Inc., in the 1950s. This technique replaced triangulation for control survey distance measurements and the steel tape for all but short distances in boundary and engineering surveys. GPS surveys are now used for most control survey point positioning.

Aerial surveys became very popular after World War II. This technique is a very efficient method of performing large-scale topographic surveys and accounts for the majority of such surveys, although total station surveys are now competitive at lower levels of detail density.

In this text, we refer to airborne and satellite positioning in Chapters 7, 10, 12 and 13. Airborne and satellite photography, lidar imaging, and spectral scanning can be processed to produce plans of very large geographic areas at a much reduced cost compared with field surveying techniques. The images captured on aerial surveys can be converted to scaled plans through the techniques of photogrammetry and digital image processing. Satellite imagery has been used for many years to assess crop inventories, flood damage, migration patterns, and other large-scale geographic projects. The resolution of satellite images is rapidly improving, and remotely sensed data will be of use in many more applications, including civil engineering projects. It has been predicted that satellite imagery will soon be routinely available at a resolution of 0.3 m or better (see Chapter 7).

In the late 1980s, the total station instrument was thought to be the ultimate in surveying instrumentation. It provides electronic data collection of angles, distances, and descriptive data, with transfer to the computer and plans drawn by digital plotter, all on the same day. Now, many new applications (including construction stakeout and machine guidance) are being found for GPS techniques (see Chapters 8 and 10). In 2005, equipment manufacturers began to marketing total stations with integrated GPS receivers and with all ground and satellite data processed by the same integrated controller.

Our world is changing rapidly and technological innovations are leading the way. Traditional surveying practices have given way to electronic (digital) practices. The following sections describe equipment and techniques that have become almost obsolete during the current technological revolution. They are included to provide interested students some perspective on our rapidly changing field of study.

G.2   Dumpy Level

The dumpy level (see Figure G.5) was at one time used extensively on all leveling surveys. Although this simple instrument has been replaced mostly by more sophisticated instruments, it is shown here in some detail to aid in the introduction of the topic. For purposes of description, the level can be analyzed with respect to its three major components: telescope, level tube, and leveling head.

The telescope assembly is illustrated in Figure G.6. These schematics also describe the telescopes used in theodolites and transits. Rays of light pass through the objective (1) and form an inverted image in the focal plane (4). The image thus formed is magnified by the eyepiece lenses (3) so that the image can be seen clearly. The eyepiece lenses also focus

FIGURE G.5   Dumpy level. (Courtesy of Keuffel & Esser Co.)

the cross hairs, which are located in the telescope in the principal focus plane. The focusing lens (negative lens; 2) can be adjusted so that images at varying distances can be brought into focus in the plane of the reticle (4). In most telescopes designed for use in North America, additional lenses are included in the eyepiece assembly so that the inverted image

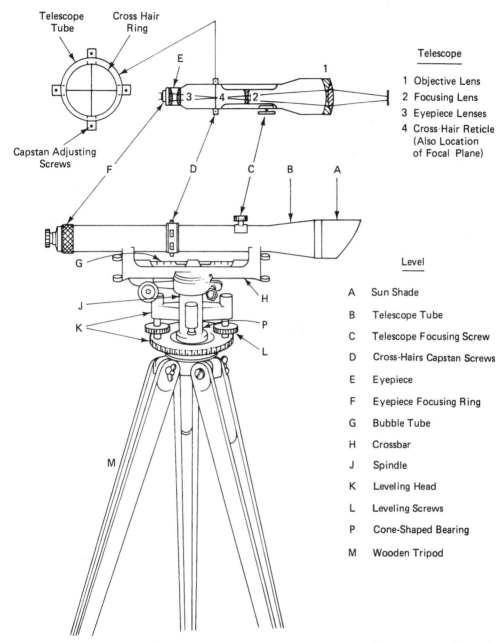

Telescope

1 Objective Lens
2 Focusing Lens
3 Eyepiece Lenses
4 Cross-Hair Reticle
  (Also Location
  of Focal Plane)

Level

A   Sun Shade
B   Telescope Tube
C   Telescope Focusing Screw
D   Cross-Hairs Capstan Screws
E   Eyepiece
F   Eyepiece Focusing Ring
G   Bubble Tube
H   Crossbar
J   Spindle
K   Leveling Head
L   Leveling Screws
P   Cone-Shaped Bearing
M   Wooden Tripod

FIGURE G.6   Dumpy level. (Adapted from *Construction Manual*, Ministry of Transportation, Ontario)

can be viewed as an upright image. The minimum focusing distance for the object ranges from 1 to 2 m, depending on the instrument.

The line of collimation (line of sight) joins the center of the objective lens to the intersection of the cross hairs. The optical axis is the line taken through the center of the objective lens and perpendicular to the vertical lens axis. The focusing lens (negative lens) is moved by focusing screw C (Figure G.6) and has the same optical axis as the objective lens.

The cross hairs (Figure G.6, 4) can be thin wires attached to a cross-hair ring, or (as is more usually the case) cross hairs are lines etched on a circular glass plate enclosed by a cross-hair ring. The cross-hair ring, which has a slightly smaller diameter than does the telescope tube, is held in place by four adjustable capstan screws. The cross-hair ring (and the cross hairs) can be adjusted left and right or up and down simply by loosening and then tightening the two appropriate opposing capstan screws.

In the case of the dumpy level, four leveling foot screws set the telescope level. The four foot screws surround the center bearing of the instrument (Figure G.6) and are used to tilt the telescope level, using the center bearing as a pivot.

Figure G.7 illustrates how the telescope is positioned during the leveling process. The telescope is first positioned directly over two opposite foot screws. The two screws are kept only snugly tightened (overtightening makes rotation difficult and could damage the threads). The screws are rotated in opposite directions until the bubble is centered in the level tube. Loose foot screws are an indication that the rotations have not proceeded uniformly; at worst, the foot screw pad can rise above the plate, making the telescope wobble.

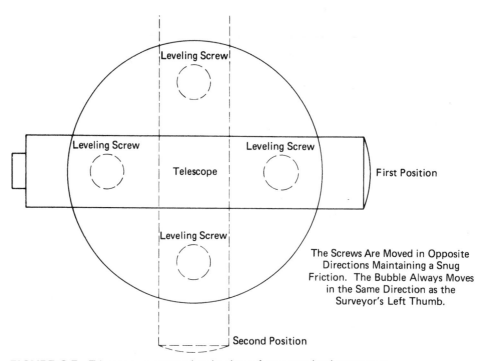

FIGURE G.7 Telescope positions when leveling a four-screw level instrument.

The solution for this condition is to turn the loose screw until it again contacts the base plate and provides a snug friction when turned in opposition to its opposite screw.

The telescope is first aligned over two opposing screws, and the screws are turned in opposite directions until the bubble is centered in the level tube. The telescope is then turned 90° to the second position, which is over the other pair of opposite foot screws, and the leveling procedure is repeated. This process is repeated until the bubble remains centered in those two positions. When the bubble remains centered in these two positions, the telescope is then turned 180° to check the adjustment of the level tube.

If the bubble does not remain centered when the telescope is turned 180°, it indicates that the level tube is out of adjustment. The instrument can still be used, however, by simply noting the number of divisions that the bubble is off center and by moving the bubble half the number of those divisions. For example, if upon turning the leveled telescope 180°, you note that the bubble is four divisions off center, the instrument can be leveled by moving the bubble to a position of two divisions off center. The bubble will remain two divisions off center, no matter which direction the telescope is pointed. It should be emphasized that the instrument is, in fact, level if the bubble remains in the same position when the telescope is revolved, regardless of whether or not that position is in the center of the level vial. See Section 3.11 for adjustments used to correct this condition.

G.3 The Engineer's Vernier Transit

G.3.1 General Background

Prior to the mid-1950s, and before the development and/or widespread use of electronic and optical theodolites, most engineering surveys for topography and layout were accomplished using the engineer's transit (see Figure G.8). This instrument has open circles for horizontal and vertical angles; angles are read with the aid of vernier scales. This four-screw instrument is positioned over the survey point by using a slip-knotted plumb bob string, which was attached to the chain hook hanging down from the instrument. Figure G.9 shows the three main assemblies of the transit. The upper assembly, called the *alidade*, includes the standards, telescope, vertical circle and vernier, two opposite verniers for reading the horizontal circle, plate bubbles, compass, and upper tangent (slow-motion) screw.

The spindle of the alidade fits down into the hollow spindle of the circle assembly. The circle assembly includes the horizontal circle that is covered by the alidade plate, except at the vernier windows, the upper clamp screw, and the hollow spindle. The hollow spindle of the circle assembly fits down into the leveling head. The leveling head includes the four leveling screws, the half-ball joint (about which opposing screws are manipulated to level the instrument), a threaded collar that permits attachment to a tripod, the lower clamp and slow-motion screw, and a chain with an attached hook for attaching the plumb bob.

The upper clamp tightens the alidade to the circle, whereas the lower clamp tightens the circle to the leveling head. These two independent motions permit angles to be accumulated on the circle for repeated measurements. Transits that have these two independent motions are called *repeating* instruments. Instruments with only one motion (upper) are called *direction* instruments. The circle on older direction instruments cannot be previously zeroed, so angles are usually determined by subtracting the initial setting from the final value. It is not possible to accumulate or repeat angles with a direction theodolite.

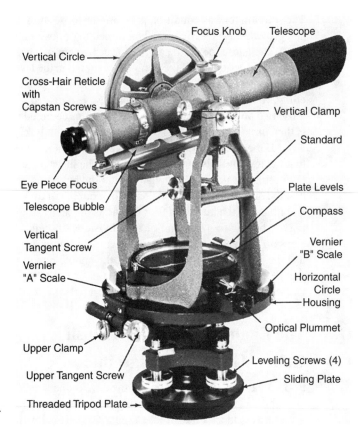

FIGURE G.8 Engineer's transit.
(Courtesy of Keuffel & Esser Co.)

G.3.2 Circles and Verniers

The horizontal circle is usually graduated into degrees and half-degrees or 30 minutes (see Figure G.10), although it is not uncommon to find the horizontal circle graduated into degrees and one-third degrees (20 minutes). To determine the angle value more precisely than the least count of the circle (i.e., 30 or 20 minutes), vernier scales are employed.

Figure G.11 shows a double vernier scale alongside a portion of a transit circle. The left vernier scale is used for clockwise circle readings (angles turned to the right), and the right vernier scale is used for counterclockwise circle readings (angles turned to the left). To avoid confusion about which vernier (left or right) scale should be used, recall that the vernier to be used is the one whose graduations are increasing in the same direction as are the circle graduations. The vernier scale is constructed so that thirty vernier divisions cover the same length of arc as do twenty-nine divisions (half-degrees) on the circle. The width of one vernier division is $(29/30) \times 30' = 29'$ on the circle. Therefore, the space difference between one division on the circle and one division on the vernier represents $01'$.

In Figure G.11, the first division on the vernier (left or right of the index mark) fails to line up exactly with the first division on the circle (left or right) by $01'$. The second division on the vernier fails to line up with the corresponding circle division by $02'$, and so on.

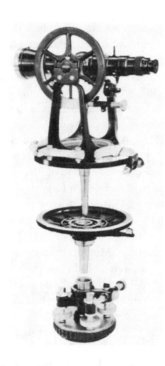

THE **ALIDADE ASSEMBLY**, WHICH
INCLUDES

    TELESCOPE
    VERTICAL CIRCLE
    STANDARDS
    VERNIERS
    VERNIER COVER
    PLATE LEVELS
    INNER CENTER
    UPPER TANGENT

THE **CIRCLE ASSEMBLY**, WHICH
INCLUDES

    HORIZONTAL LIMB
    OUTER CENTER
    UPPER CLAMP

THE **LEVELING HEAD ASSEMBLY**,
WHICH INCLUDES

    LEVELING HEAD
    LEVELING SCREWS
    SHIFTING PLATE
    FRICTION PLATE
    HALF BALL
    TRIPOD PLATE
    LOWER CLAMP
    LOWER TANGENT

FIGURE G.9 Three major assemblies of the transit. (Courtesy of SOKKIA Corp.)

THE INNER CENTER OF THE ALIDADE ASSEMBLY FITS INTO THE OUTER CENTER OF THE CIRCLE ASSEMBLY AND CAN BE ROTATED IN THE OUTER CENTER. THE OUTER CENTER FITS INTO THE LEVELING HEAD AND CAN BE ROTATED IN THE LEVELING HEAD.

FIGURE G.10 Part of a transit circle showing a least count of 30 minutes. The circle is graduated in both clockwise and counterclockwise directions, permitting the reading of angles turned to both the left and the right.

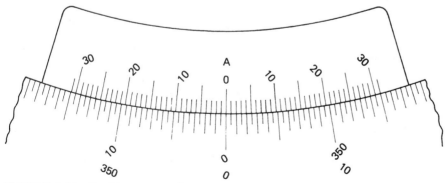

FIGURE G.11 Double vernier scale set to zero on the horizontal circle.

If the vernier were moved so that its first division lined up exactly with the first circle division (30′ mark), the reading would be 01′. If the vernier were moved again the same distance of arc (1′), the second vernier mark would now line up with the appropriate circle division line, indicating a vernier reading of 02′. Generally, the vernier is read by finding which vernier division line coincides exactly with any circle line, and then by adding the value of that vernier line to the value of the angle obtained from reading the circle to the closest 30′ (which is the case in this example).

In Figure G.12(a), the circle is divided into degrees and half-degrees (30′). Before even looking at the vernier, we know that its range will be 30′ (left or right) to cover the least count of the circle. Inspection of the vernier shows that thirty marks cover the range of 30′, indicating that the value of each mark is 01′. (Had each of the minute marks been further subdivided into two or three intervals, the angle could then have been read to the closest 30″ or 20″.)

If we consider the clockwise circle readings (field angle turned left to right), we see that the zero mark is between 184° and 184°30′; the circle reading is therefore 184°. Now to find the value to the closest minute, we use the left-side vernier. Moving from the zero mark, we look for the vernier line that lines up exactly with a circle line. In this case, the 08′ mark lines up, which is confirmed by noting that both the 07′ and 09′ marks do not line up with their corresponding circle mark, both by the same amount. The angle for this illustration is 184° + 08′ = 184°08′.

If we consider the counterclockwise circle reading in Figure G.12(a), we see that the zero mark is between 175°30′ and 176°; the circle reading is therefore 175°30′. To that value, we add the right-side vernier reading of 22′ to give an angle of 175°52′. As a check, the sum of the clockwise and counterclockwise readings should be 360°00′.

All transits are equipped with two double verniers (A and B) located 180° apart. Although increased precision can theoretically be obtained by reading both verniers for each angle, usually only one vernier is employed. Furthermore, to avoid costly mistakes, most surveying agencies favor use of the A vernier at all times.

As noted earlier, the double vernier permits angles to be turned to the right (left vernier) or to the left (right vernier). By convention, however, field angles are normally turned only to the right. Exceptions occur when deflection angles are employed, as in route surveys, or when construction layouts necessitate angles to the left, as in some curve deflections. There are a few more specialized cases (e.g., star observations) when it is advantageous to turn angles to the left, but as stated earlier, the bulk of surveying experience favors angles turned to the right. This type of consistency provides the routine required to foster a climate in which fewer mistakes occur, and in which mistakes that do occur can be recognized readily and eliminated.

The graduations of the circles and verniers as illustrated were in wide use in the survey field. However, there are several variations to both circle graduations and vernier graduations. Typically, the circle is graduated to the closest 30′ (as illustrated), 20′, or 10′ (rarely). The vernier has a range in minutes covering the smallest division on the circle (30′, 20′, or 10′), and it can be further graduated to half-minute (30″) or one-third minute (20″) divisions. A few minutes spent observing the circle and vernier graduations of an unfamiliar transit will easily disclose the proper technique required for reading [see also Figure G.12(b) and (c)].

The use of a magnifying glass (5×) is recommended for reading the scales, particularly for the 30′ and 20′ verniers. Optical theodolites replaced vernier transits in the 1970s and 1980s; optical theodolites were replaced in the 1990s by electronic transits/theodolites and total stations.

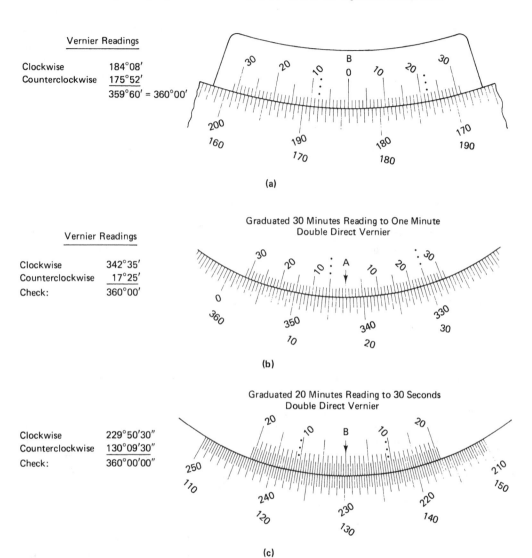

Vernier Readings

Clockwise	184°08′
Counterclockwise	175°52′
	359°60′ = 360°00′

(a)

Graduated 30 Minutes Reading to One Minute
Double Direct Vernier

Vernier Readings

Clockwise	342°35′
Counterclockwise	17°25′
Check:	360°00′

(b)

Graduated 20 Minutes Reading to 30 Seconds
Double Direct Vernier

Clockwise	229°50′30″
Counterclockwise	130°09′30″
Check:	360°00′00″

(c)

FIGURE G.12 Sample vernier readings. Triple dots identify aligned vernier graduations. *Note:* Appropriate vernier scale graduation numerals are angled in the same direction as the referenced circle graduation numerals.

G.3.3 Telescope

The telescope (see Figure G.8 in the transit is somewhat shorter that than in a level and has a reduced magnifying power (26× versus the 30× often used in the level). The telescope axis is supported by the standards, which are of sufficient height to permit the telescope to be revolved (*transited*) 360° about the axis. A level vial tube is attached to the telescope so that it can be used as a level. (Also see Figure 4.15 and Figure 4.16.)

The telescope level has a sensitivity of 30″ to 40″ per 2-mm graduation, compared to a level sensitivity of about 20″ for a dumpy level. When the telescope is positioned so that the level tube is under the telescope, it is said to be in the *direct* (normal) position; when the level tube is on top of the telescope, the telescope is said to be in a *reversed* (inverted) position. The eyepiece focus ring is always located at the eyepiece end of the telescope, whereas the object focus knob can be located on the telescope barrel just ahead of the eyepiece focus, midway along the telescope, or on the horizontal telescope axis at the standard.

G.3.4 Leveling Head

The leveling head supports the instrument; proper manipulation of the leveling screws allows the horizontal circle and telescope axis to be placed in a horizontal plane, which forces the alidade and circle assembly spindles to be placed in a vertical direction. When the leveling screws are loosened, the pressure on the tripod plate is removed, permitting the instrument to be shifted laterally a short distance ($^3/_8$ in.). This shifting capability permits the surveyor to position the transit center precisely over the desired point by noting the location of the hanging plumb bob.

G.3.5 Transit Adjustments

See also the theodolite adjustments in Section 4.13 of this text.

G.3.5.1 Standards Adjustment
The horizontal axis should be perpendicular to the vertical axis. The standards are checked for proper adjustment by first setting up the theodolite and then sighting a high (at least 30° altitude) point (see Figure 4.17). After clamping the instrument in that position, the telescope is depressed and point *B* is marked on the ground. The telescope is then transited (plunged), a lower clamp is loosened, and the transit is turned and once again set precisely on point *A*. The telescope is again depressed, and if the standards are properly adjusted, the vertical cross hair will fall on point *B*; if the standards are not properly adjusted, a new point *C* is established. The discrepancy between *B* and *C* is double the error resulting from the standards maladjustment. Point *D*, which is now established midway between *B* and *C*, will be in the same vertical plane as point *A*. The error is removed by sighting point *D* and then elevating the telescope to *A′*, adjacent to *A*. The adjustable end of the horizontal axis is then raised or lowered until the line of sight falls on point *A*. When the adjustment is complete, care is taken in retightening the upper friction screws so that the telescope revolves with proper tension.

G.3.5.2 Telescope Bubble
If the transit is to be used for leveling work, the axis of the telescope bubble and the axis of the telescope must be parallel. To check this relationship, the bubble is centered with the telescope clamped, and the peg test is performed (see Section 3.11). When the proper rod reading has been determined at the final setup, the horizontal cross hair is set on that rod reading by moving the telescope with the vertical tangent (slow-motion) screw. The telescope bubble is then centered by means of the capstan screws located at one (or both) end(s) of the bubble tube.

G.3.5.3 Vertical Circle Vernier

When the transit has been carefully leveled (via the plate bubbles), and the telescope bubble has been centered, the vertical circle should read zero. If a slight error (index error) exists, the screws holding the vernier are loosened, the vernier is tapped into its proper position, and then the screws are retightened so that the vernier is once again just touching, without binding, the vertical circle.

G.3.6 Plate Levels

Transits come equipped with two plate levels set at 90° to each other. Plate levels have a sensitivity range of 60″ to 80″ per 2-mm division on the level tube, depending on the overall precision requirements of the instrument.

G.3.7 Transit Setup

The transit is removed from its case, held by the standards or leveling base (never by the telescope), and placed on a tripod by screwing the transit snugly to the threaded tripod top. When carrying the transit indoors or near obstructions (e.g., tree branches), cradle it under the arm, with the instrument forward, where it can be seen. Otherwise, the transit and tripod can be carried on the shoulder. Total stations (see Chapter 5) should always be removed from the tripod and carried by the handle or in the instrument case.

The instrument is placed roughly over the desired point and the tripod legs are adjusted so that (1) the instrument is at a convenient height and (2) the tripod plate is nearly level. Usually, two legs are placed on the ground, and the instrument is roughly leveled by manipulation of the third leg. If the instrument is to be set up on a hill, the instrument operator faces uphill and places two of the legs on the lower position; the third leg is placed in the upper position and then manipulated to level the instrument roughly. The wing nuts on the tripod legs are tightened, and a plumb bob is attached to the plumb bob chain, which hangs down from the leveling head. The plumb bob is attached by means of a slip knot, which allows placement of the plumb bob point immediately over the mark. If it appears that the instrument placement is reasonably close to its final position, the tripod legs are pushed into the ground without jarring the instrument.

If necessary, the length of plumb bob string is adjusted as the setting-up procedure advances. If the instrument is not centered after pushing the tripod legs, one leg is either pushed in farther or pulled out and repositioned until the plumb bob is very nearly over the point or until it becomes obvious that manipulation of another leg would be more productive. When the plumb bob is within $1/4$ in. of the desired location, the instrument can then be leveled.

Two adjacent leveling screws are now loosened so that pressure is removed from the tripod plate and the transit can be shifted laterally until it is precisely over the point. If the same two adjacent leveling screws are retightened, the instrument will return to its level (or nearly so) position. Final adjustments to the leveling screws at this stage will not be large enough to displace the plumb bob from its position directly over the desired point.

The leveling procedure for a transit is a faster operation than that for a dumpy level. The transit has two plate levels, which means that the transit can be leveled in two directions, 90° opposed, without rotating the instrument. When both bubbles have been carefully centered, the instrument should be turned through 180° to check the adjustment of the plate bubbles. If one (or both) bubble(s) do not center after turning 180°, the discrepancy is noted and the bubble is brought to half the discrepancy by means of the leveling screws. If this procedure has been done correctly, the bubbles will remain in the same position while the instrument is revolved, indicating that the instrument is level.

G.3.8   Measuring Angles by Repetition (Vernier Transit)

Now that the instrument is over the point and is level, the following procedure is used to turn and double an angle. Turning the angle at least twice permits the elimination of mistakes and increases precision because of the elimination of most of the instrument errors. It is recommended that only the A vernier scale be used.

1. **Set the scales to zero.** Loosen both the upper and lower motion clamps. While holding the alidade stationary, revolve the circle by pushing on the circle underside with the fingertips. When the zero on the scale is close to the vernier zero, tighten (snug) the upper clamp. With the aid of a magnifying glass, turn the upper tangent (slow-motion) screw until the zeros are set precisely. It is good practice to make the last turn of the tangent screw against the tangent screw spring so that spring tension is assured.

2. **Sight the initial point.** See Figure 4.8. Assume that the instrument is at station A, and that the angle from B to E is to be measured. With the upper clamp tightened and the lower clamp loose, turn and point at station B, and then tighten the lower clamp. Check the eyepiece focus and object focus to eliminate parallax (see Section 3.7). If a range pole or even a pencil is giving the sight, always sight as close to the ground level as possible to eliminate plumbing errors. If the sight is being given with a plumb bob, sight high on the plumb bob string to minimize the effect of plumb bob oscillations. Using the lower tangent screw, position the vertical cross hair on the target, once again making the last adjustment motion against the tangent screw spring.

3. **Turn the angle.** Loosen the upper clamp and turn clockwise to the final point (E). When the sight is close to E, tighten the upper clamp. Using the upper tangent screw, set the vertical cross hair precisely on the target using techniques already described. Read the angle by using (in this case) the left-side vernier, and book the value in the appropriate column in the field notes (see Figure 4.8).

4. **Repeat the angle.** After the initial angle has been booked, *transit (plunge) the telescope*; loosen the lower motion; and sight at the initial target, station B. The simple act of transiting the telescope between two sightings can eliminate nearly all the potential instrument errors associated with the transit, whether you are turning angles or producing a straight line.

The first three steps are now repeated. The only difference is that the telescope is now inverted and the initial horizontal angle setting remains at $101°24'$ instead of $0°00'$. The angle that is read as a result of this repeat of steps 1 to 3 should be approximately double the initial angle. This double angle is booked and then divided by two to find the mean value, which is also booked.

If this procedure has been executed properly, the mean value should be the same as the direct reading or half the least count ($30''$) different. In practice, a discrepancy equal to the least count ($01'$) is normally permitted. Although doubling the angle is sufficient for most engineering projects, precision can be increased by repeating the angle several times. Due to personal errors of sighting and scale reading, this procedure has practical constraints for improvement in precision. It is generally agreed that repetitions beyond six or perhaps eight times will not improve the precision.

When multiple repetitions are being used, only the first angle and the final value are recorded. The number of repetitions divides the final value to arrive at the mean value. It may be necessary to augment the final reading by 360° or multiples of 360° prior to determining the mean. The proper value can be determined roughly by multiplying the first angle recorded by the number of repetitions.

G.4 Stadia

G.4.1 Principles

Stadia surveying is an outmoded form of distance measurement that relies on a fixed-angle intercept to measure the distance optically along the sight path. With a limited accuracy of 1:400, stadia was ideally suited for the location of natural features.

The theodolite and some levels have a cross-hair reticle that has, in addition to the normal cross hairs, two additional horizontal hairs [see Figure G.13(a)], one above and the other equally below the main horizontal hair. These stadia hairs are positioned in the reticule so that, if a rod were held 100 ft (m) away from the theodolite (with the telescope level), the difference between the upper and lower stadia hair readings (rod interval) would be exactly 1.00 ft (m). You can see in Figure G.13(b) that horizontal distances can be determined by (1) sighting a rod with the telescope level, (2) determining the rod interval, and (3) multiplying the rod interval by 100 to compute the horizontal distance:

$$D = 100S \qquad \text{(G.1)}$$

Elevations can be determined using stadia in the manner illustrated in Figure G.13(c). The elevation of the instrument station (A) is usually determined using a level and rod. When the theodolite is set up on the station in preparation for a stadia survey, the height of the optical center of the instrument above the top of the stake (hi) is measured with a tape and noted in the field book. A rod reading can then be taken on the rod with the telescope level. The elevation of the point (B) where the rod is being held is computed as follows:

$$\text{Elevation of station A } (\bar{\wedge}) + \text{hi} - \text{RR} = \text{elevation of point B (rod)} \qquad \text{(G.2)}$$

Note that hi (height of instrument) is the vertical distance from the station to the optical center of the theodolite in stadia work, as it is in total station work, whereas in leveling work, HI (height of instrument) is the elevation of the line of sight through the level.

The optical center of the theodolite is at the center of the telescope on the horizontal axis. The exact point is marked with a cross, colored dot, or screw. Measuring the hi (height of instrument) with a tape is not exact because the tape must be bent over the circle assembly when measuring; however, this error does not significantly affect the stadia results.

Figure G.13(d) illustrates that the location of point B (rod) can be tied in by angle to a reference baseline XAY. The plan location of point B can now be plotted using the angle from the reference line and the distance, as determined by Equation G.1. The elevation of point B can also be plotted, either as a spot elevation or as a component of a series of contours. You

Cross Hair Reticle

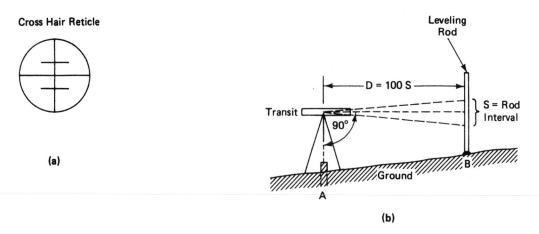

(a)

(b)

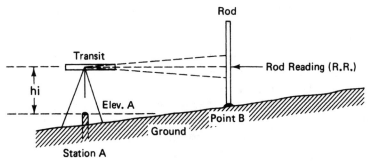

Elevation A + hi − R.R. = Elevation B.

(c)

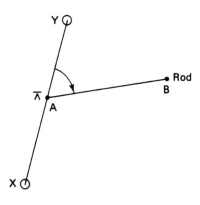

(d)

FIGURE G.13 Stadia principles. (a) Stadia hairs. (b) Distance determination. (c) Elevation determination. (d) Angle determination.

can see that the stadia method (like the total station method) permits the surveyor to determine the three-dimensional position of any point with just one set of observations.

G.4.2 Inclined Stadia Measurements

The discussion thus far has assumed that the stadia observations were taken with the telescope level; however, the stadia method is particularly well suited for the inclined measurements required by rolling topography. When the telescope is inclined up or down, the computations must be modified to account for the effects of the sloped sighting. Inclined sights require consideration in two areas: (1) the distance from the instrument to the rod must be reduced from slope to horizontal, and (2) the rod interval of a sloped sighting must be reduced to what the interval would have been if the line of sight had been perpendicular to the rod.

Figure G.14 llustrates these two considerations. The value of hi and the rod reading (RR) have been made equal to clarify the sketch. The geometric relationships are as follows: (1) S is the rod interval when the line of sight is horizontal, and (2) S' is the rod interval when the line of sight is inclined by angle θ. Equation G.1 was illustrated in Figure G.13(b):

$$D = 100S$$

Figure G.14 illustrates the relationship between S and S':

$$S = S' \cos \theta \qquad (G.3)$$

The following equation can be derived from Equations G.3 and G.1:

$$D = 100S' \cos \theta \qquad (G.4)$$

Figure G.14 also illustrates the following relationship:

$$H = D \cos \theta \qquad (G.5)$$

The following equation can be derived from Equations G.4 and G.2:

$$H = 100S' \cos^2 \theta \qquad (G.6)$$

The following relationship is also illustrated in Figure G.14:

$$V = D \sin \theta \qquad (G.7)$$

The following equation can be derived from Equations G.7 and G.4:

$$V = 100S' \cos \theta \sin \theta \qquad (G.8)$$

Some theodolites read only zenith angles ($90° - \theta$), which makes it necessary to modify Equations G.6 and G.8 as follows:

$$H = 100S' \sin^2 (90° - \theta) \qquad (G.9)$$
$$V = 100S' \sin (90° - \theta) \cos (90° - \theta) \qquad (G.10)$$

Equation G.9 is a modification of Equation G.6; Equation G.10 is a modification of Equation G.8. Equations G.6 and G.8 can be used for computing the horizontal distance and difference in elevation for any inclined stadia measurement.

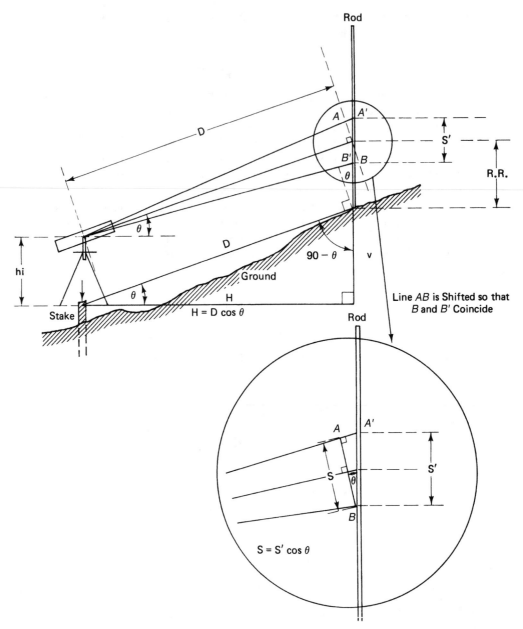

FIGURE G.14 Geometry of an inclined stadia measurement.

In the past, special slide rules and/or tables were used to compute H and V. With the universal use of handheld calculators and computers, however, slide rules and stadia tables have become almost obsolete. The computations can be accomplished just as quickly working with Equations G.6 and G.8. Stadia reductions are given in Table G.1. The table is entered at the value of the vertical circle reading (VCR), with the horizontal distance factor

Table G.1 STADIA TABLES*

Minutes	0° Hor. dist.	0° Diff. elev.	1° Hor. dist.	1° Diff. elev.	2° Hor. dist.	2° Diff. elev.	3° Hor. dist.	3° Diff. elev.	Minutes	4° Hor. dist.	4° Diff. elev.	5° Hor. dist.	5° Diff. elev.	6° Hor. dist.	6° Diff. elev.	7° Hor. dist.	7° Diff. elev.
0	100.00	0.00	99.97	1.74	99.88	3.49	99.73	5.23	0	99.51	6.96	99.24	8.68	98.91	10.40	98.51	12.10
2	100.00	0.06	99.97	1.80	99.87	3.55	99.72	5.28	2	99.51	7.02	99.23	8.74	98.90	10.45	98.50	12.15
4	100.00	0.12	99.97	1.86	99.87	3.60	99.71	5.34	4	99.50	7.07	99.22	8.80	98.88	10.51	98.49	12.21
6	100.00	0.17	99.96	1.92	99.86	3.66	99.71	5.40	6	99.49	7.13	99.21	8.85	98.87	10.57	98.47	12.27
8	100.00	0.23	99.96	1.98	99.86	3.72	99.70	5.46	8	99.48	7.19	99.20	8.91	98.86	10.62	98.46	12.32
10	100.00	0.29	99.96	2.04	99.86	3.78	99.69	5.52	10	99.47	7.25	99.19	8.97	98.85	10.68	98.44	12.38
12	100.00	0.35	99.96	2.09	99.85	3.84	99.69	5.57	12	99.46	7.30	99.18	9.03	98.83	10.74	98.43	12.43
14	100.00	0.41	99.95	2.15	99.85	3.89	99.68	5.63	14	99.46	7.36	99.17	9.08	98.82	10.79	98.41	12.49
16	100.00	0.47	99.95	2.21	99.84	3.95	99.68	5.69	16	99.45	7.42	99.16	9.14	98.81	10.85	98.40	12.55
18	100.00	0.52	99.95	2.27	99.84	4.01	99.67	5.75	18	99.44	7.48	99.15	9.20	98.80	10.91	98.39	12.60
20	100.00	0.58	99.95	2.33	99.83	4.07	99.66	5.80	20	99.43	7.53	99.14	9.25	98.78	10.96	98.37	12.66
22	100.00	0.64	99.94	2.38	99.83	4.13	99.66	5.86	22	99.42	7.59	99.13	9.31	98.77	11.02	98.36	12.72
24	100.00	0.70	99.94	2.44	99.82	4.18	99.65	5.92	24	99.41	7.65	99.11	9.37	98.76	11.08	98.34	12.77
26	99.99	0.76	99.94	2.50	99.82	4.24	99.64	5.98	26	99.40	7.71	99.10	9.43	98.74	11.13	98.33	12.83
28	99.99	0.81	99.93	2.56	99.81	4.30	99.63	6.04	28	99.39	7.76	99.09	9.48	98.73	11.19	98.31	12.88
30	99.99	0.87	99.93	2.62	99.81	4.36	99.63	6.09	30	99.38	7.82	99.08	9.54	98.72	11.25	98.30	12.94
32	99.99	0.93	99.93	2.67	99.80	4.42	99.62	6.15	32	99.38	7.88	99.07	9.60	98.71	11.30	98.28	13.00
34	99.99	0.99	99.93	2.73	99.80	4.47	99.61	6.21	34	99.37	7.94	99.06	9.65	98.69	11.36	98.27	13.05
36	99.99	1.05	99.92	2.79	99.79	4.53	99.61	6.27	36	99.36	7.99	99.05	9.71	98.68	11.42	98.25	13.11
38	99.99	1.11	99.92	2.85	99.79	4.59	99.60	6.32	38	99.35	8.05	99.04	9.77	98.67	11.47	98.24	13.17
40	99.99	1.16	99.92	2.91	99.78	4.65	99.59	6.38	40	99.34	8.11	99.03	9.83	98.65	11.53	98.22	13.22
42	99.99	1.22	99.91	2.97	99.78	4.71	99.58	6.44	42	99.33	8.17	99.01	9.88	98.64	11.59	98.20	13.28
44	99.98	1.28	99.91	3.02	99.77	4.76	99.58	6.50	44	99.32	8.22	99.00	9.94	98.63	11.64	98.19	13.33
46	99.98	1.34	99.91	3.08	99.77	4.82	99.57	6.56	46	99.31	8.28	98.99	10.00	98.61	11.70	98.17	13.39
48	99.98	1.40	99.90	-3.14	99.76	4.88	99.56	6.61	48	99.30	8.34	98.98	10.05	98.60	11.76	98.16	13.45
50	99.98	1.45	99.90	3.20	99.76	4.94	99.55	6.67	50	99.29	8.40	98.97	10.11	98.58	11.81	98.14	13.50
52	99.98	1.51	99.90	3.26	99.75	4.99	99.55	6.73	52	99.28	8.45	98.96	10.17	98.57	11.87	98.13	13.56
54	99.98	1.57	99.89	3.31	99.74	5.05	99.54	6.79	54	99.27	8.51	98.94	10.22	98.56	11.93	98.11	13.61
56	99.97	1.63	99.89	3.37	99.74	5.11	99.53	6.84	56	99.26	8.57	98.93	10.28	98.54	11.98	98.10	13.67
58	99.97	1.69	99.88	3.43	99.73	5.17	99.52	6.90	58	99.25	8.63	98.92	10.34	98.53	12.04	98.08	13.73
60	99.97	1.74	99.88	3.49	99.73	5.23	99.51	6.96	60	99.24	8.68	98.91	10.40	98.51	12.10	98.06	13.78

*Example, VCR, −3°21′; rod interval, 0.123 m; from tables: $V = 5.83 \times 0.123 = 0.72$ m; $H = 99.66 \times 0.123 = 12.3$ m.

(continued)

Table G.1 *(continued)*

Minutes	8° Hor. dist.	8° Diff. elev.	9° Hor. dist.	9° Diff. elev.	10° Hor. dist.	10° Diff. elev.	11° Hor. dist.	11° Diff. elev.	Minutes	12° Hor. dist.	12° Diff. elev.	13° Hor. dist.	13° Diff. elev.	14° Hor. dist.	14° Diff. elev.	15° Hor. dist.	15° Diff. elev.
0	98.06	13.78	97.55	15.45	96.98	17.10	96.36	18.73	0	95.68	20.34	94.94	21.92	94.15	23.47	93.30	25.00
2	98.05	13.84	97.53	15.51	96.96	17.16	96.34	18.78	2	95.65	20.39	94.91	21.97	94.12	23.52	93.27	25.05
4	98.03	13.89	97.52	15.56	96.94	17.21	96.32	18.84	4	95.63	20.44	94.89	22.02	94.09	23.58	93.24	25.10
6	98.01	13.95	97.50	15.62	96.92	17.26	96.29	18.89	6	95.61	20.50	94.86	22.08	94.07	23.63	93.21	25.15
8	98.00	14.01	97.48	15.67	96.90	17.32	96.27	18.95	8	95.58	20.55	94.84	22.13	94.04	23.68	93.18	25.20
10	97.98	14.06	97.46	15.73	96.88	17.37	96.25	19.00	10	95.56	20.60	94.81	22.18	94.01	23.73	93.16	25.25
12	97.97	14.12	97.44	15.78	96.86	17.43	96.23	19.05	12	95.53	20.66	94.79	22.23	93.98	23.78	93.13	25.30
14	97.95	14.17	97.43	15.84	96.84	17.48	96.21	19.11	14	95.51	20.71	94.76	22.28	93.95	23.83	93.10	25.35
16	97.93	14.23	97.41	15.89	96.82	17.54	96.18	19.16	16	95.49	20.76	94.73	22.34	93.93	23.88	93.07	25.40
18	97.92	14.28	97.39	15.95	96.80	17.59	96.16	19.21	18	95.46	20.81	94.71	22.39	93.90	23.93	93.04	25.45
20	97.90	14.34	97.37	16.00	96.78	17.65	96.14	19.27	20	95.44	20.87	94.68	22.44	93.87	23.99	93.01	25.50
22	97.88	14.40	97.35	16.06	96.76	17.70	96.12	19.32	22	95.41	20.92	94.66	22.49	93.84	24.04	92.98	25.55
24	97.87	14.45	97.33	16.11	96.74	17.76	96.09	19.38	24	95.39	20.97	94.63	22.54	93.82	24.09	92.95	25.60
26	97.85	14.51	97.31	16.17	96.72	17.81	96.07	19.43	26	95.36	21.03	94.60	22.60	93.79	24.14	92.92	25.65
28	97.83	14.56	97.29	16.22	96.70	17.86	96.05	19.48	28	95.34	21.08	94.58	22.65	93.76	24.19	92.89	25.70
30	97.82	14.62	97.28	16.28	96.68	17.92	96.03	19.54	30	95.32	21.13	94.55	22.70	93.73	24.24	92.86	25.75
32	97.80	14.67	97.26	16.33	96.66	17.97	96.00	19.59	32	95.29	21.18	94.52	22.75	93.70	24.29	92.83	25.80
34	97.78	14.73	97.24	16.39	96.64	18.03	95.98	19.64	34	95.27	21.24	94.50	22.80	93.67	24.34	92.80	25.85
36	97.76	14.79	97.22	16.44	96.62	18.08	95.96	19.70	36	95.24	21.29	94.47	22.85	93.65	24.39	92.77	25.90
38	97.75	14.84	97.20	16.50	96.60	18.14	95.93	19.75	38	95.22	21.34	94.44	22.91	93.62	24.44	92.74	25.95
40	97.73	14.90	97.18	16.55	96.57	18.19	95.91	19.80	40	95.19	21.39	94.42	22.96	93.59	24.49	92.71	26.00
42	97.71	14.95	97.16	16.61	96.55	18.24	95.89	19.86	42	95.17	21.45	94.39	23.01	93.56	24.55	92.68	26.05
44	97.69	15.01	97.14	16.66	96.53	18.30	95.86	19.91	44	95.14	21.50	94.36	23.06	93.53	24.60	92.65	26.10
46	97.68	15.06	97.12	16.72	96.51	18.35	95.84	19.96	46	95.12	21.55	94.34	23.11	93.50	24.65	92.62	26.15
48	97.66	15.12	97.10	16.77	96.49	18.41	95.82	20.02	48	95.09	21.60	94.31	23.16	93.47	24.70	92.59	26.20
50	97.64	15.17	97.08	16.83	96.47	18.46	95.79	20.07	50	95.07	21.66	94.28	23.22	93.45	24.75	92.56	26.25
52	97.62	15.23	97.06	16.88	96.45	18.51	95.77	20.12	52	95.04	21.71	94.26	23.27	93.42	24.80	92.53	26.30
54	97.61	15.28	97.04	16.94	96.42	18.57	95.75	20.18	54	95.02	21.76	94.23	23.32	93.39	24.85	92.49	26.35
56	97.59	15.34	97.02	16.99	96.40	18.62	95.72	20.23	56	94.99	21.81	94.20	23.37	93.36	24.90	92.46	26.40
58	97.57	15.40	97.00	17.05	96.38	18.68	95.70	20.28	58	94.97	21.87	94.17	23.42	93.33	24.95	92.43	26.45
60	97.55	15.45	96.98	17.10	96.36	18.73	95.68	20.34	60	94.94	21.92	94.15	23.47	93.30	25.00	92.40	26.50

and the difference in elevation factor both being multiplied by the rod interval to give H and V (see Examples G.1 and G.2).

Figure G.15 shows the general case of an inclined stadia measurement, where the elevation relationships between the theodolite station and rod station can be stated as follows:

$$\text{Elevation at } (\barwedge) \text{ station K} + \text{hi} \pm \text{V} - \text{RR} = \text{elevation at (rod) station M} \qquad \text{(G.11)}$$

The relationship is valid for every stadia measurement. If the hi and RR are equal, Equation G.11 becomes:

$$\text{Elevation at } (\barwedge) \text{ station K} \pm \text{V} = \text{elevation at (rod) point M} \qquad \text{(G.12)}$$

because the hi and RR cancel each other. In practice, the surveyor reads the rod at the value of the hi unless that value is obscured (e.g., by a tree branch, rise of land, etc.). If the hi value cannot be sighted, usually a value that is an even foot (decimeter) above or below is sighted, which allows for a mental correction to the calculation. Of course, if an even foot (decimeter) above or below the desired value cannot be read, any value can be read and Equation G.11 is used.

G.4.3   Examples of Stadia Measurements

There are three basic variations to a standard stadia measurement:

1. The rod reading is the same as the hi.
2. The rod reading is not the same as the hi.
3. The telescope is horizontal.

Examples G.1 through G.3 illustrate each of these variations.

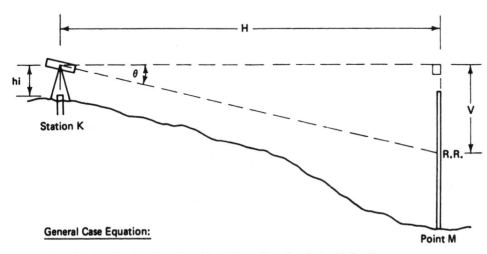

General Case Equation:

Elevation Station K ($\barwedge$) + hi ± V − RR = Elevation Point M (Rod)

FIGURE G.15   General case of an inclined stadia measurement.

■ EXAMPLE G.1

This example, where the rod reading has been made to coincide with the value of the hi, is typical of 90 percent of all stadia measurements (see Figures G.16 and G.19). Here, with the theodolite set up at 0 + 40 and sighting at 0 + 00, the VCR is + 1°36″ and the rod interval is 0.401. Both the hi and the rod reading are 1.72 m. Find the horizontal distance and the elevation of 0 + 00.

Solution

From Equation G.6:

$$H = 100 \, S' \cos^2 \theta$$
$$= 100 \times 0.401 \times \cos^2 1°36'$$
$$= 40.1 \text{ m}$$

From Equation G.8:

$$V = 100S' \cos \theta \sin \theta$$
$$= 100 \times 0.401 \times \cos 1°36' \times \sin 1°36'$$
$$= +1.12 \text{ m}$$

From Equation G.12:

$$\text{Elev.} (\barwedge) \pm V = \text{elev. (rod)}$$
$$185.16 + 1.12 = 186.28$$

See station 0 + 00 on Figure G.19.

Or, from Table G.1:

$$\text{VCR} = + 1°36'$$
$$\text{Rod interval} = 0.401$$
$$H = 99.92 \times 0.401 = 40.1 \text{ m}$$
$$V = 2.79 \times 0.401 = +1.12 \text{ m}$$

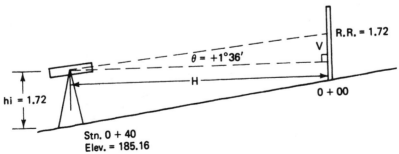

FIGURE G.16 Example G.1: RR = hi.

■ **EXAMPLE G.2**

This example illustrates the case where the value of the hi cannot be seen on the rod due to some obstruction (see Figures G.17 and G.19). In this case a rod reading of 2.72 with a vertical angle of $-6°37'$ was booked, along with the hi of 1.72 and a rod interval of 0.241. The theodolite is set up at $0 + 40$, and the rod sighting is at station 3. Find the horizontal distance and the elevation at station 3.

Solution
From Equation G.16:

$$H = 100 \, S' \cos^2 \theta$$
$$= 100 \times 0.241 \times \cos^2 6°37'$$
$$= 23.8 \text{ m}$$

From Equation G.8:

$$V = 100 \, S' \cos \theta \sin \theta$$
$$= 100 \times 0.241 \times \cos 6°37' \times \sin 6°37'$$
$$= -2.76 \text{ m (the algebraic sign is given by the VCR)}$$

From Equation G.11:

$$\text{Elev.} (\barwedge) + \text{hi} \pm V - RR = \text{elev. (rod)}$$
$$185.16 + 1.72 - 2.76 - 2.72 = 181.40 \text{ (elevation of station 3)}$$

See station 3 in Figure G.19.
 Or from Table G.1:

$$\text{VCR} = -6°37'$$
$$\text{Rod interval} = 0.241$$
$$H = 98.675 \text{ (interpolated)} \times 0.241 = 23.8 \text{ m}$$
$$V = 11.445 \text{ (interpolated)} \times 0.241 = -2.76 \text{ m}$$

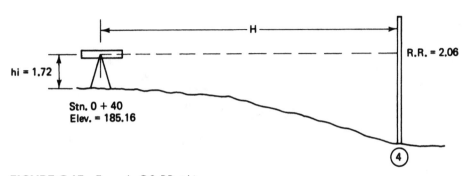

FIGURE G.17 Example G.2: RR ≠ hi.

■ EXAMPLE G.3

This example illustrates the situation where the ground is level enough to permit horizontal rod sightings (see Figures G.18 and G.19). The computations for this observation are quite simple. The horizontal distance is simply 100 times the rod interval (D = 100S, Equation G.1). Because there is no vertical angle, there is no triangle to solve (i.e., V = 0). The difference in elevation is simply + hi − RR.

Surveying a level area where many observations can be taken with the telescope level speeds up the survey computations and shortens the field time. However, if the survey is in typical rolling topography, the surveyor does not normally spend the time necessary to see if a single horizontal observation can be made. Instead, the surveyor continues sighting the rod at the hi value to maintain the momentum of the survey (a good instrument operator can keep two rod holders busy).

Solution
In this example, a rod interval of 0.208 was booked, together with hi of 1.72 and a rod reading of 2.06. From Equation G.1:

$$D = 100\ S$$
$$= 100 \times 0.208$$
$$= 20.8 \text{ m (horizontal distance)}$$

From Equation G.11:

$$\text{Elev.}\ (\overline{\wedge}) + \text{hi} + V - RR = \text{elev. (rod)}$$
$$185.16 + 1.72 + 0 - 2.06 = 184.82$$

See station 4 on Figure G.19.

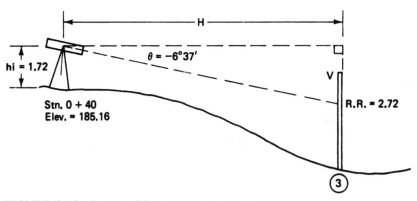

FIGURE G.18 Example G.3: telescope horizontal.

G.4.4 Stadia Field Practice

In stadia work, the theodolite is set on a point for which the horizontal location and elevation have been determined. If necessary, the elevation of the theodolite station can be determined after setup by sighting on a point of known elevation and working backward through Equation G.11.

The horizontal circle is zeroed and a sight is taken on another control point (1 + 40 in Figure G.19). All stadia sightings are accomplished by working on the circle and using the upper clamp and upper tangent screw (on two-clamp instruments). It is a good idea to check the zero setting periodically by sighting back on the original backsight; this practice ensures that the setting has not been invalidated by inadvertent use of the lower clamp or lower tangent screw. At minimum, a zero check is made on the backsight station just before the instrument is moved from the theodolite station. If the check proves that the setting has been inadvertently moved from zero, all the work must be repeated. Before any observations are made, the hi is measured with a steel tape (sometimes with the leveling rod) and the value is booked, as shown in Figure G.19.

The actual observation proceeds as follows. With the upper clamp loosened, the rod is sighted. A precise setting can be accomplished using the upper tangent screw after the upper clamp has been locked. The main cross hair is sighted approximately to the value of

STADIA SURVEY OF SENECA DRIVE   CLEAR 17°

Date .... Oct. 6, 2006 .... Page 18

STA.	H.C.R.	ROD INTERVAL	V.C.R.	HORIZ. DISTANCE	ELEV. DIFFERENCE	ELEV.
		⊼ @ 0 + 40; hi = 1.72				185.16
1+40	0° 00'	1.002	−1° 24'	100.1	−2.45	182.71
0+00	180° 01'	0.401	+1° 36'	40.1	+1.12	186.28
①	5° 48'	0.911	−2° 18'	91.0	−3.65	181.51
②	18° 18'	0.562	−3° 38'	56.0	−3.55	181.61
③	21° 58'	0.241	−6° 37' on 2.72	23.8	V = −2.76 / −1.00 / −3.76	181.40
④	156° 02'	0.208	0° 00' on 2.06	20.8	−2.06 / +1.72 / −0.34	184.82
	P. VASSALLO−NOTES					
	R. CURTIS − ⊼					
	J. BODMAN−ROD					

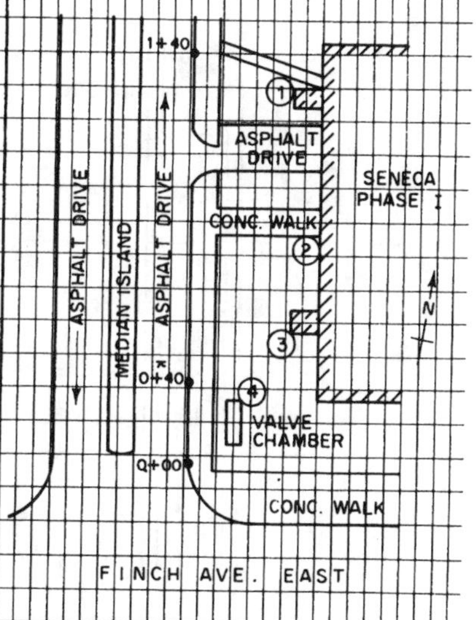

FIGURE G.19 Stadia field notes.

the hi, and then the telescope is revolved up or down until the lower stadia hair is on the closest even foot (or decimeter) mark (see Figure G.20). The upper stadia hair is then read, and the rod interval is determined simply by subtracting the lower hair reading from the upper hair reading. After the rod interval is booked (see Figure G.19), the main cross hair is then moved to read the value of the hi on the rod. The rod holder is then waved off, and while the rod holder is walking to the next point, the instrument operator reads and books the VCR and the HCR (see Figure G.19). The calculations for horizontal distance and elevation are usually performed after field hours.

The technique of temporarily moving the lower stadia hair to an even value to facilitate the determination of the rod interval introduces errors in the readings, but these errors are not large enough to affect the results significantly. The alternative to this technique is to lock the main hair initially on the value of the hi and then read and book the upper and lower hair readings. When the lower hair is subtracted from the upper hair, the result can then be booked as the rod interval. This alternative technique is more precise, but it is far too cumbersome and time consuming for general use.

If the value of the hi cannot be seen on the rod, another value (e.g., the even foot or decimeter above or below the hi) can be sighted, and that value is booked along with the vertical angle in the VCR column (see Figure G.19, station 3). If the telescope is level, the rod reading is booked in the VCR column alone or together with 0°00′ (see Figure G.19, station 4). Sometimes the entire rod interval cannot be seen on the rod (e.g., because of tree branches, intervening ground, extralong shots). When this happens, half the rod interval can be read. That value is doubled and then entered in the rod interval column. Reading only half the rod interval reduces the precision considerably, so extra care should be taken when determining the half-interval. Generally, if the relative accuracy ratio of 1/300 to 1/400 is

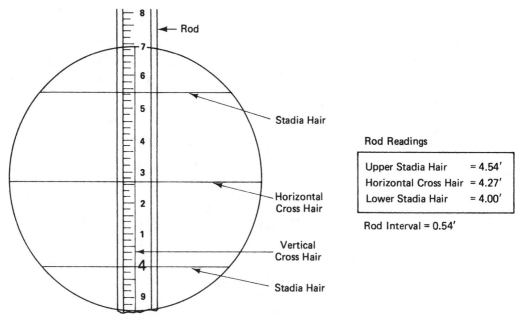

FIGURE G.20 Stadia readings. (Courtesy of the Ministry of Transportation, Ontario)

to be maintained on long sights and on steeply inclined sights, extra care is required (particularly in reading the rod, plumbing the rod, etc.).

G.4.5 Summary of Stadia Field Procedure

1. Set the theodolite over the station.
2. Measure the hi with a steel tape or rod.
3. Set the horizontal circle to zero.
4. Sight the reference station at 0°00′.
5. Sight the rod being held on the first stadia point.
6. Sight the main horizontal cross hair roughly on the value of hi. Then move the lower hair to the closest even foot (or decimeter) mark.
7. Read the upper hair. Determine the rod interval and enter that value in the notes.
8. Sight the main horizontal hair precisely on the hi value.
9. Wave off the rod holder.
10. Read and book the horizontal angle (HCR) and the vertical (VCR) angles.
11. Check the zero setting for the horizontal angle on the backsight before moving the instrument.
12. Reduce the notes (compute horizontal distances and elevations) after field hours. Check the reductions.

Problems

G.1. The data shown below are stadia rod intervals and vertical angles taken to locate points on a topographic survey. The elevation of the theodolite station is 371.21 ft, and all vertical angles were read with the cross hair on the rod at the value of the height of instrument (hi).

Point	Rod Interval (ft)	Vertical Angle
1	3.48	+0°58′
2	0.38	−3°38′
3	1.40	−1°30′
4	2.49	+0°20′
5	1.11	+2°41′

Compute the horizontal distances and the elevations using Table G.1.

G.2. The data shown below are stadia rod intervals and vertical angles taken to locate points on a watercourse survey. Where the hi value could not be sighted on the rod, the sighted rod reading is booked along with the vertical angle. The hi is 1.83 and the elevation of the transit station is 207.96 m.

Point	Rod Interval (m)	Vertical Angle
1	.041	+2°19′
2	.072	+1°57′ on 1.43
3	.555	0°00′ on 2.71
4	1.313	−2°13′
5	1.111	−4°55′ on 1.93
6	0.316	+0°30′

Compute the horizontal distances and the elevations using Equations G.6 and G.7.

G.3. A transit is set up on station K with hi 5.36 ft. Reading are taken at station L as follows: rod interval = 1.3 ft, cross hair at 4.27 with a vertical angle of +3°12′. The elevation of station L = 293.44 ft.

(a) What is the elevation of station K?

(b) What is the distance *KL*?

G.4. Reduce the following set of stadia notes using either Table G.1 or Equations G.6 and G.8.

Station Point	Rod Interval (ft)	Horizontal Angle	Vertical Angle	Horizontal Distance	Elevation Difference	Elevation
⊼ @ Mon. 36; hi = 5.14						373.58
Mon.37	3.22	0°00′	+3°46′			
1	2.71	2°37′	+2°52′			
2	0.82	8°02′	+1°37′			
3	1.41	27°53′	+2°18′ on 4.06			
4	1.10	46°20′	+0°18′			
5	1.79	81°32′	0°00′ on 8.11			
6	2.61	101°17′	−1°38′			
Mon. 38	3.60	120°20′	−3°41′			

G.5. Reduce the following cross-section stadia noted.

Station Point	Rod Interval (m)	Horizontal Angle	Vertical Angle	Horizontal Distance	Elevation Difference	Elevation
⊼ @ Ctrl. Point K; hi = 1.82 m						167.78
Ctrl. Pt.L.		0°00′				
0+ 00 ₡	0.899	34°15′	−19°08′			
South ditch	0.851	33°31′	−21°58′			
North ditch	0.950	37°08′	−20°42′			
0 + 50 ₡	0.622	68°17′	−16°58′			
South ditch	0.503	64°10′	−20°26′			
North ditch	0.687	70°48′	−19°40′			
1 + 00 ₡	0.607	113°07′	−13°50′			
South ditch	0.511	109°52′	−16°48′			

North ditch	0.710	116°14′	14°46′
1 + 50 ₵	0.852	139°55′	−10°04′
South ditch	0.800	135°11′	−11°22′
North ditch	0.932	144°16′	−10°32′
2 + 00 ₵	1.228	152°18′	−6°40′
South ditch	1.148	155°43′	−8°00′
North ditch	1.263	147°00′	−7°14′

Appendix H

Illustrations of Machine Control and of Various Data-Capture Techniques

Appendix H appears at the end of this textbook in a full-color insert. Figures H.1, H.2, and H.3 illustrate the most common construction machine guidance and control techniques. Figure H.1 illustrates the use of guide wires to provide horizontal and vertical guidance to the machine operator. Figure H.2 illustrates the use of a computer-controlled robotic total station in a local positioning system (LPS) providing guidance or control to a reflecting prism-equipped motor grader. Figure H.3 illustrates the use of the global positioning system (GPS) in providing horizontal and vertical guidance and control. The GPS receiver-equipped bulldozer's in-cab computer compares the surface coordinates received from the satellites to the proposed design coordinates at each particular point. The software program provides the operator with the necessary up/down and left/right guidance movements, or the software can control machine valves directly to achieve the design layout.

Figures H.4 through H.8 illustrate various data capture techniques. Figure H.4 is an excerpt of a hydrographic map where the data was captured using manual field techniques. Figures H.5 and H.6 are aerial photographs of the Niagara River—at the falls and at the mouth of the river at Lake Ontario. Figure H.7 is a lidar image of Niagara Falls, and Figure H.8 is a Landsat satellite image showing western New York and the Niagara region of Ontario.

Note that these data capture techniques result in images of the earth with varying ground resolution. You are invited to view these figures and to consider a range of applications for which these images can be employed.

Sonic detection machine guidance. This conventional machine guidance system uses a high-accuracy sonic transmitter to locate a string line set to the proper grading specification.
(Courtesy of Topcon Positioning Systems, Inc., Pleasanton, Calif.)

FIGURE H.1

Total station guidance and control. A hybrid Topcon robotic total station tracks a motor grader equipped with a Topcon 3D-LPS automatic control system. This local position system uploads plan information directly from the attached field computer and relays the elevation and slope data via a laser beam emitted from the total station. This system permits high-speed, automatic grading to an accuracy of a few millimeters.
(Courtesy of Topcon Positioning Systems, Inc., Pleasanton, Calif.)

FIGURE H.2

GPS machine guidance and control. **(A)** In-cab display. The Topcon System 5 3D in-cab display/control panel features touch-screen operation and the ability to monitor the position of the controlled equipment over the entire job site in multiple views. **(B)** GPS-controlled bulldozer. Rough grading is performed by a bulldozer equipped with a Topcon 3D-GPS+ control system without any need for grade stakes.
(Courtesy of Topcon Positioning Systems, Inc., Pleasanton, Calif.)

FIGURE H.3

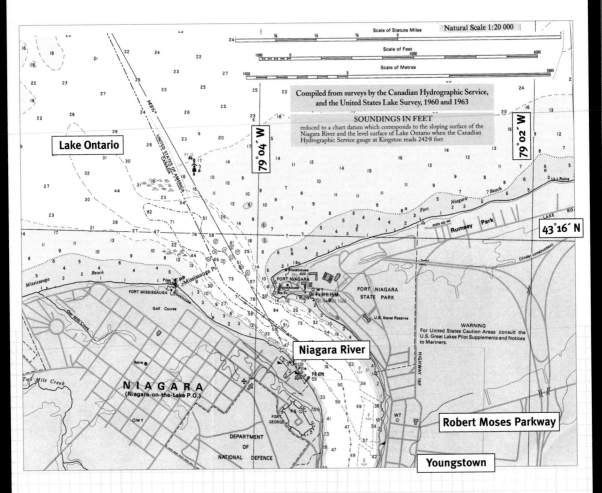

Hydrographic map of the lower Niagara River. This map is adapted from one produced by the Canadian Hydrographic Service and the United States Lake Survey, 1960 and 1963, respectively. Also, refer to Figures H.5 through H.8 for additional Niagara River coverage.

FIGURE H.4

Aerial photograph, at 20,000 ft, showing the mouth of the Niagara River.
(Courtesy of U.S. Geological Survey, Sioux Falls, S.Dak.)

FIGURE H.5

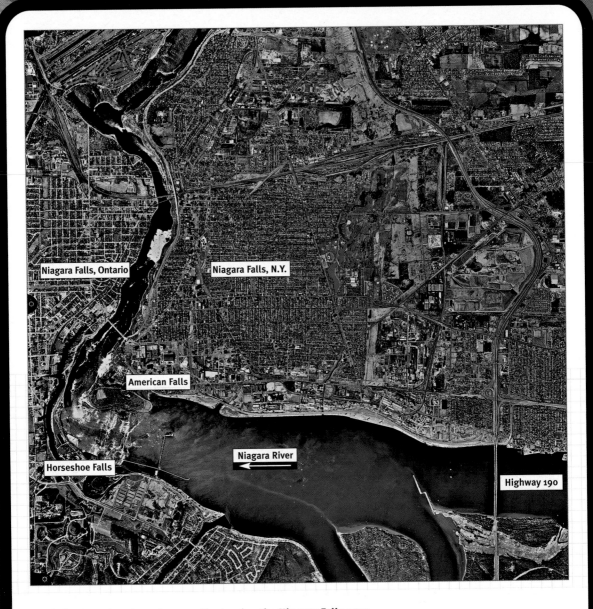

Aerial photograph, taken at 20,000 ft, showing the Niagara Falls area.
(Courtesy of U.S. Geological Survey, Sioux Falls, S.Dak.)

FIGURE H.6

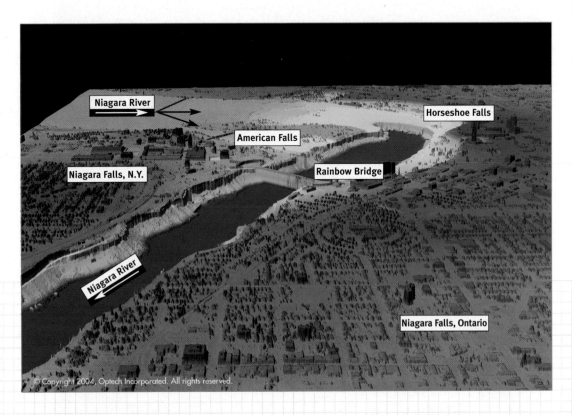

Color-coded elevation data of Niagara Falls, surveyed using lidar techniques by ALTM 3100.
(Courtesy of Optech Incorporated, Toronto)

FIGURE H.7

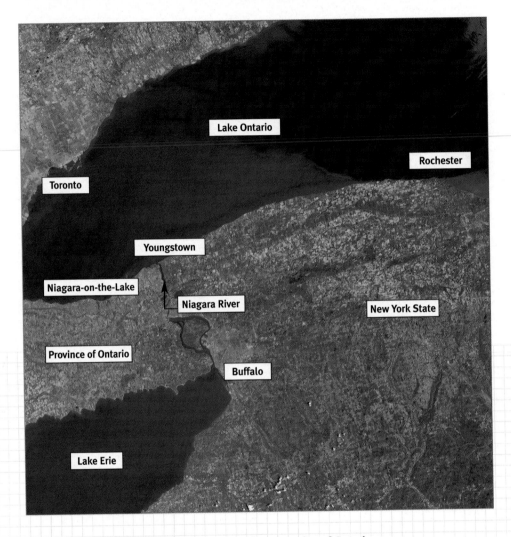

Landsat image showing western New York and the Niagara region of Ontario.
(Courtesy of U.S. Geological Survey, Sioux Falls, S.Dak.)

FIGURE H.8

Index

AASHTO Survey Data Management System (SDMS), 143
Abney hand level (clinometer), 24
Absorbed energy, 238
Accuracy, 14,
  construction surveys, 366
  control surveys, 58, 59
  EDM, 44, 159
  positional (ISO 4463), 366
  real-time positioning, 377
Accuracy ratio, 14
Adjacency, 261
Aerial cameras, 222
Aerial surveys, 222
Airborne and satellite imagery, 233
Alidade, 108, 647
Altitude, 227
American Congress of Surveying and Mapping (ACSM), 324
American Society of Civil Engineers, 3
AM/FM, 250
Anchor bolts, 543
Angle
  adjustment of, 167
  deflection, 99
  exterior, 99
  horizontal, 91
  interior, 99
  laying off, 117
  measurement, 9, 109
  nadir, 97
  vertical, 98
  zenith, 97
Antenna reference height (ARH), 296
Arc of circular curve, 385
Area, 554
  by coordinates, 191
  by cross sections, 569
  by grid squares, 561
  by planimeter, 561, 566
  by Simpson's one-third rule, 558
  by total station, 137
  by trapezoidal technique, 554
  by trigonometric formulae, 561
Arithmetic check (page check), 75
Arterial road, 448, 468
As-built (final) drawings, 575
Augmentation Services, 279
Automatic level, 62
Automatic target recognition (ATR), 154
Average end area, 569
Azimuth, 134, 172

Backsight, 71, 76
Barcode, 65
Batter boards, 462, 498
Bearing, 169
Benchmark, 71
Bore holes, 519
Borrow pit leveling survey, 80, 83
Breaking tape, 29, 459
Break line, 212
Bridge layout, 519
Bucking-in, 120, 550
Building layout, 534, 624
Building measurements, 610
Building sections, 535
Bulldozer, 515

CAD, 209, 249
Capstan screws, 113, 645
Carpenter's level, 461
Catchbasin (CB), 494, 504
CD-ROM, 198
Central meridian, 168
Chain, Gunter's, 19
Chainage, 12
Chaining pins, 26
Charge-coupled device, 65, 222
Chord, of circular curve, 384, 393
Chorobate, 642
Circular bubble, 117
Circular curve, 384
    arc (L), 384
    arc definition, 386
    beginning of curve (BC), 384
    chord (c), 384, 393
    chord definition, 386
    compound, 403
    curve deflection angle, 385
    deflection angles for, 391
    degree of curve (D), 386
    end of curve (EC), 384
    external distance (E), 384
    field procedure, 394
    geometry, 385
    metric considerations, 394
    mid-ordinate distance (M), 384
    moving up on the curve, 395

offsets, 396
    passing through a fixed point, 443
    Point of Intersection (PI), 384
    project, 625
    reverse, 405
    set-ups on, 395
    stations on radial lines, 397
    tangent (T), 384
Classification and feature extraction
    (images), 243
Classification of ALTA-ACSM Title
    Surveys, 325
Classification of roads and streets, 467
Clearing, 445, 454
Clinometer, 24
Close the horizon, 362
Coding for field-generated graphics, 146
COGO, 160, 179
Collector road, 448, 468
Collimation, 62
Collimation correction, 85, 115
Columns, building, 540
Combined factor, 345
Compass rule, 183
Compound curve, 443
Connectivity, 261
Construction projects, list of, 12
Construction lasers, 541
Construction surveys, 4, 6, 149, 370, 381
Construction volumes, 553, 567
    cross section end areas, 569
    from contours, 568
    geometric formulas, 575
    mass diagram, 568
    prismoidal formula, 573
    shrinkage, 567
    swell, 567
Containment, 261
Continuously Operating Reference Station
    (CORS), 288
Contours, 215
    break lines, 212, 218
    characteristics of, 218
    ridge lines, 218
    valley lines, 218

Contract drawings, 519
Control surveys, 4
    Father Point, 57
    Positioning accuracy standards, 325
Convergence of meridians, 349
Coordinates, 187
    area by, 191
    geometry of, 585
    grid, 251
    traverse stations, 187
CORS, 288
Cross hairs, 114
Cross section
    highway, 449
    road, 469, 477
Cross sections for leveling, 78
    field notes for, 81, 82
    for footing excavations, 533
    for topographic surveys, 205, 447, 466
Crown
    pipeline, 491, 494
    road, 476
Culvert layout, 516
Curvature error in leveling, 60
Cut and Fill, 444, 459, 479
Cycle slip, 285, 291

Database, 247
    layers, 247
Database management, 252
Data coding, 141
Data collection, 197
Data collector (electronic field book), 6, 15, 125
    data collector menu structure, 304
Data file construction, 380
Data plotting
    conventional methods, 209
    digital plotters, 209
Deflection-angle, 99, 165
Degree of curve (D), 386
Departure of a traverse course, 176
    adjustment of, 183
Differential leveling, 58
    field project, 614
Digital imagery, 233
Digital level, 65

Digital plotting, 209
Digital number (DN), 241
Digital terrain model (DTM), 212, 374
DIN (Deutsches Institut fur Normung), 105
Distance measurement, 6
Double centering, 118
Drafting, 205
    computer assisted (CAD), 209, 212
    contours, 215
    map symbols, 213
    plan symbols, 214
    plotting - manual, 210
    scales, 210
    standard drawing sizes, 211
    title block, 212
Dumpy level, 644
Dual-axis compensation, 107, 127, 132, 138

Electromagnetic spectrum, 236
    visible spectrum, 237
Electronic angle measurement, 5, 40
Electronic distance measurement (EDM),
        6, 18, 40
    accuracy, 46
    atmospheric corrections, 43
    geometry of, 50
    principles of, 42
    range of measurements, 40
    without reflecting prism, 53
Electronic distance measurement instrument
        (EDM)
    characteristics of, 44
    instrument/prism constant, 47
    operation, 47
    prism, 45
Electronic field book (data collector), 6
Electronic leveling, 127, 542
Elevation, 57
    of building, 536
Elevation factor, 341
Engineering surveying, 3
Eratosthenes, 641
Error
    random, 13
    systematic, 13
Evolution of surveying, 639

Existing ground (EG), 454
Excavation equipment, 515
External distance (E), 385

4-foot mark, 543
Father point, 57
Fiducial marks, 228
Field notes, 15, 609
    for angles by direction, 112
    for building layout, 624
    for closed traverse, 167
    for cross sections (highway format), 82
    for cross sections (municipal format), 81
    for fencing measurements, 588
    for GPS field log, 297
    for leveling, 75
    for open traverse, 166
    for removal of sewer pipe, 559
    for pile driving, 560
    for prepared coordinates for layout, 363
    for profile, 80
    repeated angles, 105
    for sodding measurements, 557
    for stadia surveys, 665
    for station visibility diagram, 295
    for taping, 38, 39
    for three-wire leveling, 87
    for topography, single baseline, 203
    for topography, split baseline, 203, 204
    for topography, 203
Field Projects, 610
Final surveys, 372, 553, 575
Flight lines, 229
Flying heights and altitude, 226
Focal length, 226
Forced centering, 45, 108
Foresight, 72
Free station, (resectioning), 133, 551
Freeway, 448
Front-end loader, 515
Functional planning (highways), 383, 444

Galileo, 268, 276
Geocenter, 320
Geocoding, 252
Geodetic Reference System, 1980
    (GRS 80), 317

Geodetic surveying, 3, 317
Geodimeter, 643
Geographic information system (GIS), 140,
    197, 246
Geographic meridian, 168
Geoid, 310
    Geoid03, 312
    Geoid99, 76, 312
    CGG2000, 312
Geoid modeling, 312
Geoid undulation, (N), 310
Geomatics, 2, 197
Geomatics data model, 198
Geometric Dilution of Precision
    (GDOP), 287
Geometry, 640
    (coordinate) review, 585
Georeferencing, 251
Geospatial data, 207
Global navigation satellite system
    (GNSS), 277
Global Positioning System (GPS), 6, 199,
    268, 270
    Active Control Station (ACS), 290
    applications, 302
    carrier phase, 283
    Coarse Acquisitions (CA) code, 278
    code phase, 282
    Continuously operating reference station
    (CORS), 288
    cycle slip, 285
    DGPS, 280
    differencing, 285
    differential positioning, 278
    Epoch, 285
    errors, 286
    field procedures, 140, 296, 297
    geoid, 310
    Geometric Dilution of Precision
    (GDOP), 287
    initial ambiguity resolution, 285
    kinematic techniques, 293
    L-band frequencies, 278
    planning, 290
    Precision code (P), 278
    pseudorange, 383

rapid static technique, 298
real time kinematic technique (RTK), 270,
   280, 299
receivers, 5, 272
relative positioning, 278, 285
satellite signals, 282
Selective Availability (SA), 288
static techniques, 290
station visibility, 295
status reports, 281
vertical positioning, 308
visibility diagram, 295
wide area augmentation system
   (WAAS), 282
GLONASS, 268, 275
Glossary, 597
Gon, 12, 130
GPS Glossary, 313
GPS field log, 297
GPS/total station instrument, 160
Ground control for airborne imaging, 232
Grade, definitions, 371
Grade rod
   sewer contruction, 497
   slope staking, 457
Grade sheet
   roads, 478, 480
   sewer, 496, 498, 500, 503
Grade stake, 495
Grade transfer, 460
Gradient, 32
Gravity, 57
Grid meridian, 97, 168
Gridzone designation (GZD), 339
Ground floor plan, 535
Ground-truthing (accuracy assessment), 243
GRS80 ellipsoid, 308
Grubbing, 445, 454
Gunter's chain, 19
Gyrotheodolite, 509

Handheld total station, 161
Hand level, 23
HARN, 318
Harpedonapata, 639
Hauler, 461

Height of instrument (HI), 72
Height of instrument (hi), 132
Hertz, 236
High Accuracy Reference Network
   (HARN), 318
Highway curves, 384
Highway design, 448
Highway layout, 448
Horizon, closing the, 362
Horizontal
   angle, 98
   control, 357
   distance, 6
   line, 57, 98
Hydraulic excavator, 515

Image analysis, 240
Incident energy, 237
Inertial measurement unit (IMU), 235, 270
Initial ambiguity (N), 285
Interlining, 120
International Organization for Standardization
   (ISO), 78
International Terrestrial Reference Frame
   2000 (ITRF00), 289, 308, 321
International system of units (SI), 11
Internet Web Sites, 595
Intersection computation
   straight line and curved line, 592
   two straight lines, 120, 589
Intersection (street) construction, 480
Intersection tie-in, 5
Invar tape, 22
Invert, pipe, 494
ISO (International Organization of
   Standards), 78
ITRF 2000, 289, 308, 321

Lambert projection, 332
Landsat 7, 245
Laser alignment, 462, 502, 541
Laser plummet, 106
Latitude of a course, 176
   adjustment of, 183
Layout surveys, 4, 136, 149, 305, 459, 495,
   519, 534
Layout survey accuracies, 366

Level, 6
    automatic, 62
    compensator, 62
    digital, 65
    dumpy, 644
    loop, 88
    peg test, 84
    precise, 324
    tilting, 66
Leveling, 57
    adjustment of closure error, 88
    backsight in, 71
    benchmark, 76
    borrow pit, 80, 83
    cross sections, 78
    definitions, 71
    differential, 58
    foresight, 72
    height of instrument (HI), 72
    instrument operator, 90
    intermediate sight, 72
    mistakes in, 91
    loop adjustments, 88
    peg test, 84
    profile, 78
    reciprocal, 83
    rod, 6
    rod holder, 89
    specifications, 58, 59
    techniques, 72
    three-wire, 86, 132
    trigonometric, 87
    turning point, 71
    waving the rod, 74
Leveling head
    four screw, 646
    three screw, 64
Leveling rod, 67
    graduations, 69
Level line, 57
Level tube (vial),
    coincidence type (split bubble), 67
    sensitivity of, 68
Lidar, 6, 235, 270
Line and grade, 6, 459
Linear error of closure, 179

Local positioning system (LPS), 373
Local road, 448, 468
Location drawing, 535
Location ties, 4
Low point on vertical curve, 408
Lump sum bids, 553

Machine guidance and control, 371
    GPS RTK, 373, 376
    total station, 375
Magnetic meridian, 97, 169
Map scales, 210
Mean sea level (MSL), 57
Measurement units, 10
Meridian, 97, 168
Metadata, 253
Meter, 11
Mid-ordinate distance (M), 384
Missing course computation, 186
Mistakes, 15, 91
Motorized Total Station, 153
    automatic target recognition, 152, 154
    remotely controlled surveying, 154
Multispectral scanners, 6, 235
Multistory building construction, 545
Municipal street construction, 466

NAD83, 310
Nadir, 97
Nadir angle, 98
NASA, 245
National Geodetic Survey (NGS), 111
    NGS Tools, 289, 329, 341
National Geodetic Vertical Datum (NGVD
    1929), 57
National Spatial Reference System
    (NSRS), 289
Navstar satellite system, 6
Normal tension, 34, 612
North American Datum (NAD '83), 317
North American Vertical Datum (NAVD '88),
    57, 319

Odometer, 18
Offset layout, 475
Omitted measurement, 186
On-line positioning service (OPUS), 289

Optical plummet, 106, 109, 117, 118
  precise, 509, 551
Original ground (OG), 454
Orthometric height (H), 3, 309

Pacing, 18
Parabolic (vertical) curve, 407
Parallax, 74
Peg test, 84
Photogrammetry, 222
Photograph overlap, 229
Photographic scale, 224
Pipejacking, 491
Pipeline layout, 491, 626
Plan and profile, 468, 470, 487, 489,
  513, 514
Plan, profile, cross section—relationships
  between, 78
Plane coordinate grids, 325
  State plan coordinate grids, 326
Plan surveying, 2
Planimeter, 561
Plat, 472
Plumb bob, 23
Plumb bob target, 27
Polar layout, 448
Polar tie-in, 5, 177
Position measurement, 9
Positioning, 284, 366
  Accuracies (ISO 4463), 363
  tie-ins, 5
Precision, 14
  techniques for "ordinary" taping, 36
Precision ratio, 179
Preliminary surveys (preengineering surveys),
  4, 444, 466
Profile leveling, 78, 79, 205
  field notes for, 80
Progress payments, 553
Project control, 358
Prolonging a straight line (double
  centering), 118
  past an obstacle, 121, 123
Pseudorange, 383
Pseudo random noise (PRN), 282

Quantity surveys, 553

Radar, 6
Radiation, 237
  radiation sensors, 240
Radius of circular curve, 385
Random error, 13, 35
Range pole, 26
Raster model, 255
Reciprocal leveling, 83
Rectangular tie-in, 4, 177, 201
Reference Azimuth Point (RAP), 360
Reflected energy, 238
Refraction error in leveling, 60
Relief displacement, 228
Remote object elevation, 135
Remote sensing, 6, 223
Remote sensing satellites, 245
Resection, 133, 151, 323
Right of way (ROW), 383, 466
Road allowance, 468
Road cross sections, typical, 469, 477
Rod, 6, 67
Rod level, 68, 70
Route surveys, 383
RTK (Real Time Kinematic), 270, 280, 370

Sag of tap, 34, 634
Satellite, 268
  constellations, 274
Satellite imagery, 233
Satellite signals, 282
Satellite web sites, 262
Scale factor, 343
Scraper, 459
Sewer layout, 493
Shrinkage, 567
SI system, 11
Side friction (f), 432
Sidewalk construction, 482
Simpson's one-third rule, 558
Site grading, 403
Skew angle
  bridges, 525
  culverts, 517
Skew number, 517
Slope corrections, 32
Slope distance, 6

Slope gradient, 32
Slope stake, 455
Sonic detector, 372, 474
Spatial entities or features, 254
Spiral curves, 414
    approximate formulas, 430
    computations, 416
    layout procedure summary, 423
    tables, 417-421
Springline, 494
Stadia, 655
    computations, 657
    field notes for, 665
    principles of, 655
    tables, 659
State Plane Coordinate Grid System,
    SPCS27, 326
    SPCS83, 326
Stationing, 12
Steel tape, 5, 20
    measurement project, 613
Street intersection, 480
String line level, 461
Stripping, 454
Subtense bar, 19
Superelevation (e), 432
    design, 432
    tangent run-out, 433
Surveying, defined, 2
Survey drafting, 207
Surveying instruments, 5
Swell, 567
System International d' Unites, (SI), 11
Systematic error, 13

Tangent (T), 384
Tape, 5, 20
    add, 22
    breaking, 29, 459
    clamp handle for, 26
    cut, 22
    drag, 21
    fiberglass, 20
    invar, 22, 633
    steel, 20
    tension handle, 26

Taping (chaining)
    alignment, 35
    breaking tape, 29
    errors, random, 31, 35
    errors, systematic, 31, 631
    field notes, 38, 39
    marking and plumbing errors, 36
    Measuring procedure, 28
    mistakes, 37
    normal tension, 34
    plumbing, 28
    procedure, 28
    standard conditions for, 31
    techniques for "ordinary precision", 36
Taping corrections
    correction formulas, 632, 634
    erroneous tape length, 631
    sag, 34, 634
    slope, 32
    temperature, 632
    tension, 34, 634
Target
    plumb bob, 27
    tribrach mounted, 108
Telescope, 114, 645, 651
Temperature
    effect on taping, 632
Tendering units, 554
Tension on tape, 26
Theodolite, 5, 99, 642
    adjustments of, 112
    angle measurement, 109, 110, 117
    direction, 109
    electronic, 100, 102
    geometry of, 111, 113
    optical, 100, 107, 117
    plate bubbles, 113
    setting up, 106
    telescope, 114
    typical specifications of, 105
Tie-ins, 447
TIGER, 197
Tilting level, 66
Topographic surveys, 200, 302
    field projects, 617, 622
    precision requirements, 200

Topology, 261
Total station, 6, 99, 105, 125
  accuracies, 159
  area computation, 137
  automatic target recognition, 154
  characteristics of, 137, 158
  coding for survey station descriptors, 141
  construction layout using, 149
  data transfer and editing, 145
  data collectors (controllers), 125
  distance offset measurements, 135
  field generated graphics, 146
  field procedures, 106, 131, 139
  guide lights, 153
  handheld total station, 161, 305, 306
  instrument errors, 131
  layout, 136
  motorized, 153
  point location, 132
  remote controlled surveying, 154
  remote object elevation, 135
  resection, 133
  survey station descriptors, 140
Transit, 5, 99, 643, 647
  adjustment of, 652
  measuring angles by repetition, 654
  set-up, 653
Transmitted energy, 238
Transverse Mercator projection, 328, 332
Trapezoidal technique of area
      computation, 554
Traverse, 164
  accuracy, 182
  area, 191
  closed, 99, 166
  computations, 178, 181
  coordinates, 191
  error of closure, 179
  field notes, 167
  open, 100, 164
  precision, 181, 182
  specifications, 182, 322

Triangulation, 121, 321
Triangulated irregular network (TIN), 218
Tribrach, 45, 108
Trigonometric leveling, 87, 132
Trigonometry review, 581
Trilateration, 321
Tunnel layout, 505
Turning point, 71

Units of measurement, 10
Universal Transverse Mercator Grid
      System (UTM), 333
U.S. Geological Survey (USGS), 250
United States National Grid
      (USNG), 337

Vector model, 255
Vernier readings, 651
Vernier transit, 5, 100, 648
Vertical angle, 98
Vertical control
  specifications, 58, 59
Vertical cross hair, 114
Vertical curves, 406
Vertical distance, 6
Vertical line, 57, 98
Virtual reference station (VRS), 377

Waving the rod, 74
Web sites,
  civil design, 382
  GPS, 316
  remote sensing, 262
  surveying and mapping, 595
WGS84, 310
Wide area augmentation system, 282
World geodetic system (WGS), 317
Wild, Heinrich, 643

Young, William, 643

Zenith, 97
Zenith angle, 97
Zero velocity update, 263

SYMBOLS

ℬ	baseline
℄	centerline
ℒ	street line
Δ N	change in northing
Δ E	change in easting
Δ λ″	change in longitude (seconds)
Δ hi	difference in height between transit and EDM
Δ R	difference in height between reflector and target
φ, λ	latitude, longitude
⊼	instrument
P	occupied station (instrument)
P	reference sighting station
P	point of intersection
=	is equal to
≠	is not equal to
>	is greater than
<	is less than
≈	is approximately equal to
Σ	the sum of

THE GREEK ALPHABET

Name	Uppercase	Lowercase
alpha	A	α
beta	B	β
gamma	Γ	γ
delta	Δ	δ
epsilon	E	ε
zeta	Z	ζ
eta	H	η
theta	Θ	θ
iota	I	ι
kappa	K	κ
lambda	Λ	λ
mu	M	μ
nu	N	ν
xi	Ξ	ξ
omnicron	O	o
pi	Π	π
rho	P	ρ
sigma	Σ	σ
tau	T	τ
upsilon	Y	υ
phi	Φ	φ
chi	X	χ
psi	Ψ	ψ
omega	Ω	ω